Designing with Alternative Building Materials and Review of Building Materials

This book explores alternative building materials and their test results. It includes a review of common building materials, earth engineering and construction, suitable soil types, soil mechanics and production, masonry construction, rammed earth construction, the use of ferro-cement units in projects, and the application of fibre-reinforced concrete. Other topics such as project management, computer-aided design, education and training for engineers, construction work, quality control, safety aspects, communication, applications, and case studies, accompanied by construction photographs, are covered in this book.

Features:

- Focuses on alternative building materials and their applications in projects.
- Provides a useful overview of earth and concrete alternative construction materials.
- Comprehensively covers different types of fibre-reinforced concrete.
- Explains relevant sustainability and environmental benefits.
- Includes a case study on carbon fibre composite material used for strengthening a column.

This book is aimed at professionals in civil and construction engineering.

Designing with Alternative Building Materials and Review of Building Materials

Kalyan Kumar Ganguly

CRC Press
Taylor & Francis Group
Boca Raton London New York

CRC Press is an imprint of the
Taylor & Francis Group, an **informa** business

Designed cover image: Shutterstock

First edition published 2025
by CRC Press
2385 NW Executive Center Drive, Suite 320, Boca Raton FL 33431

and by CRC Press
4 Park Square, Milton Park, Abingdon, Oxon, OX14 4RN

CRC Press is an imprint of Taylor & Francis Group, LLC

Library of Congress Cataloging-in-Publication Data
Names: Ganguly, Kalyan Kumar, author.
Title: Designing with alternative building materials and review of building materials / Kalyan Kumar Ganguly.
Description: First edition. | Boca Raton : CRC Press, 2025. | Includes bibliographical references and index. |
Identifiers: LCCN 2024022032 (print) | LCCN 2024022033 (ebook) |
ISBN 9781032333182 (hbk) | ISBN 9781032333199 (pbk) | ISBN 9781003319108 (ebk)
Subjects: LCSH: Building materials. | Green products. | Substitute products. | Sustainable construction.
Classification: LCC TA403 .G36 2025 (print) | LCC TA403 (ebook) |
DDC 691—dc23/eng/20241031
LC record available at https://lccn.loc.gov/2024022032
LC ebook record available at https://lccn.loc.gov/2024022033

ISBN: 9781032333182 (hbk)
ISBN: 9781032333199 (pbk)
ISBN: 9781003319108 (ebk)

DOI: 10.1201/9781003319108

Typeset in Times
by codeMantra

Contents

Preface

It was realised that designing and utilising alternative building materials would be necessary for projects. Efforts should be made to identify suitable alternative building materials, particularly concerning earth engineering and construction, which are vital for the development of knowledge. Engineers and architects have applied various alternative building materials in their projects. Conducting a review to compare alternative building materials with traditional ones, such as concrete and structural steel, is essential for understanding their advantages and limitations..

Such materials could be used for at least a few projects if they meet the necessary suitability and requirements. Furthermore, efforts should be made to utilise local labour and materials.

In this context, the author was fortunate to be selected by the late Chairman of DCPL to attend a course, along with two colleagues, at Auroville, India. This residential course focused on earth engineering and architecture, requiring the author to stay in Auroville for a couple of days. During this time, the author observed various applications of compressed earth blocks and ferro-cement units successfully implemented by engineers and architects for different purposes. Earth engineering was truly enlightening for the author, fostering a strong interest in employing suitable alternative building materials for diverse applications. This experience instilled a desire to attempt earth engineering in future projects when possible.

With this background, the author had the opportunity to use compressed earth blocks (CEB) and ferro-cement elements for projects, primarily in West Bengal, following the guidance and instructions of both the late Chairman and the current Chairperson of DCPL.

The applications were associated with a light industrial unit and a research centre focused on the study and improvements of roofing units, various types of doors, wall units, water tanks, etc., using compressed earth blocks and ferro-cement. The compressed earth blocks were used in construction of walls, domes, tiles, and other decorative items for the projects. The strengths of the compressed earth blocks were tested in-house and at a local university, specifically for compressive strengths and for comparisons with local bricks and records.

Rammed earth construction had been used for a project located in the northern part of the country.

In all these applications, local labourers were trained and employed for the construction work. This also enhanced local employment and encouraged the community to be more serious and efficient in their work. The responses from the local people were very encouraging, and they felt proud to become an essential part of the construction activities. Local materials were extensively used for earth engineering in all the projects. The applications will also alleviate the pressure on the use of conventional construction and its constituent materials, resulting in savings from their use. Transportation costs for traditional materials were eliminated.

The use of suitable alternative materials is relevant and seems to be gaining importance currently.

There is a strong belief that efforts should be made to prepare and publish materials related to alternative building materials for students and young practising engineers. These resources could be beneficial after studying, collecting information, summarising, and extending the knowledge

The author would be greatly satisfied if the concept of "Designing with Alternative Building Materials" in the publication is accepted and understood by one and all.

Acknowledgements

It is very challenging to adequately acknowledge all the help and guidance received from various publications in preparing this text.

I appreciate the support and opportunities given by Development Consultants Private Limited (DCPL) for my work with alternative building materials on different projects, specifically those involving compressed earth blocks, rammed earth constructions, and ferro-cement.

The author would also like to thank DCPL and various other clients for providing the opportunity to work on different projects over an extended period.

The knowledge and training programme arranged by Auroville, as well as the book *Building with Earth*, is invaluable. Without this assistance, this publication would not have been possible.

I acknowledge with gratitude the help received from other important publications, mainly the following publications/books that have been consulted:

- **Concrete Society, UK –** Various publications are mentioned with the respective author's name, publication year, and month in the list of references.
- **Institution of Structural Engineers, UK –** Various publications are mentioned with the respective author's name, publication year, and month in the list of references.
- **The Institution of Civil Engineers, UK –** Various publications are mentioned with the respective author's name, publication year, and month in the list of references.
- The contributions of IS codes are well recognised. Guidance from IS codes is referenced in different chapters and included in the list of references. Assistance has also been obtained from ACI and BS codes for relevant sections.
- Wikipedia publications.
- Constructor publications.
- Other significant publications from various sources and included in the list of references.
- Information collected from the internet.

The following books have been mainly referred to:

- Doat P, Hays A, Houben H, Matuk S, and Vitoux F (CRATerre, France), *Building with Earth*, translated by Asha Puri.
- A Training Manual for Technicians and Entrepreneurs, Auroville Building Centre.
- Neville A M, *Properties of Concrete*, ELBS.
- Perkins P H, *Repair, Protection and Waterproofing of Concrete Structures*, Elsevier Applied Science Publishers, London & New York.
- Allen R T L, *The Repair of Concrete Structures*, Blackie, Glasgow, London.
- Mehta P K and Monteiro P J M, *Concrete: Microstructure, Properties and Materials*, Cement – Ambuja.
- Terzaghi K and Peck R B, *Soil Mechanics in Engineering Practice*, Asia Publishing House.
- Teng W C, *Foundation Design*, Prentice-Hall of India (Pvt.) Ltd.
- Hollaway L, *Polymer Composites for Civil and Structural Engineering*, Blackie.
- Swammay R N, *New Reinforced Concrete*, Surrey University Press.
- Faber J and Mead F H, *Reinforced Concrete*, ELBS.
- Murashev V I, Sigalov E, and Baikov V N, *Design of Reinforced Concrete Structures*, MIR Publishers.
- Bresler B and Lin T Y, *Design of Steel Structures*, John Wiley & Sons Inc.
- Leech L V, *Structural Steelwork for Students*, Butterworths, London.
- Subramanian N, *Design of Steel Structures*, Oxford University Press.

- *Steel Designers' Manual*, Crosby Lockwood and Son Ltd.
  - Mero Information Document, Applications of Steel in Buildings and Industrial Projects.
  - Other published papers from universities, journals, and books related to the subject.

All these have been properly mentioned in the list of references.

I would like to apologise in advance if some references have been unintentionally omitted from the list.

I am grateful to my wife and family members for the support, tolerance, and encouragement extended to me in completing this book.

I must extend my appreciation and recognise the help and cooperation received from the staff of the publisher (Engineering), Editorial Team, CRC Press, Taylor & Francis Group.

SPECIAL COMMENTS

I am grateful to those who encouraged and cared for me throughout. Their support is invaluable, to say the least. I remember and respect this with humble gratitude. I dedicate this publication to these respected individuals.

Author Biography

Dr. Kalyan Kumar Ganguly (Gangopadhyay) received his BE in civil engineering from Jadavpur University in 1964. He is a qualified Chartered Engineer (London) and obtained his PhD from Teesside, UK, in 1972.

He has more than 57 years of experience in India and the UK, covering design, site supervision, project management, research, and teaching. The spectrum of projects includes bridges, power projects, fertiliser projects, chemical plants, heavy civil engineering projects, university complexes, hotels and resorts, research centres, housing, and commercial complexes. Some of the special projects involving alternative building materials will include ferro-cement, compressed earth blocks, rammed earth construction, etc. He has also handled repair and rehabilitation projects on buildings, bridges, and heritage structures.

He is a corporate member of the Institute of Structural Engineers (London), the Institution of Civil Engineers (London), and the Concrete Society (London), as well as a Fellow of the Institution of Engineers (India), a Member of the Indian Concrete Institute, and a Fellow of the Indian Association of Structural Engineers.

As an external faculty member, he was associated with Jadavpur University for approximately 15 years, teaching Design of Industrial Steel Structures and Repair of Structures.

He also taught special engineering courses in the UK, preparing students for the Part-III examination of the Institution of Structural Engineers, London.

He retired from Development Consultants Private Limited (DCPL) in October 2020.

He has published 81 technical papers in national and international journals, including those of technical institutions. He has also published one book on reinforced concrete design, repair/restoration, strengthening of reinforced concrete structures, and concrete technology.

SYNOPSIS

As mentioned earlier, there is a strong belief to study and collect information on possible alternative building materials, with an emphasis on earth engineering and construction, which may be considered along with test results for our learning and for students and young engineers.

The list of alternative materials outlined below has been focused on, studied, and considered in this publication:

- Review of common building materials
- Earth engineering and construction: different methods
- Suitable soil, soil mechanics, and CEB production, uses, and associated items
- Masonry construction – adobe, clay bricks, fly ash bricks, etc., and their applications
- Rammed earth construction
- Use of ferro-cement units for projects
- Fibre-reinforced concrete for projects and applications

There are limitations for each of these alternative materials with respect to different applications. These will be discussed to a limited extent in this publication.

We have observed that the applications of CEB and ferro-cement have been implemented by engineers and architects at Auroville, India, to meet various requirements quite satisfactorily. The use of such materials is economical and environmentally friendly and can involve local people and materials as needed. The use of fibre-reinforced concrete for projects is currently quite limited in our country. However, fibre composite wrapping materials have been used for strengthening and retrofitting work of structures. The applications of fibre-reinforced concrete are more prevalent in advanced and European countries. It is expected that these applications will increase in the future for industrial floor construction, particularly with fibre composites for strengthening and retrofitting work.

Using a building material for design and construction is not easy. Various considerations and checks are required by designers to address different issues and ensure the structures are durable and safe. Environmental and economical applications are very important, to say the least. Provisions in codes are necessary for the general acceptance of materials and for guiding designers. As examples, two well-known conventional materials we typically use – namely concrete and steel – and their composite actions have been selected and briefly mentioned, along with some photographs for comparison and assessment. The common design requirements that must be addressed for both materials are included. The respective codes for the design work with these materials are also referenced.

Other topics such as project management, computer-aided work, education and training for engineers, construction work, quality control, safety aspects, communication, applications, and case studies have also been covered to a limited extent in this publication.

Construction photographs have been included from some projects that faced difficulties due to lack of proper preservation. Finally, the photographs are arranged and summarised, covering mainly concrete, steel structures, their combinations, and alternative materials like CEB, rammed earth, and ferro-cement from projects primarily carried out in India.

The emphasis is on the use of alternative building materials, with particular reference to earth.

Earth engineering and architecture has been used by many countries for a long time. There are various earth engineering techniques to meet specific requirements. One must assess what will be suitable for a particular situation, considering both advantages and limitations. The design and construction with earth can be regarded as reliable and practical in most cases.

It may be noted that the combination of rammed earth/compressed earth blocks with ferro-cement can produce sustainable and elegant building construction. The photographs from projects will demonstrate this.

The important criterion would be to assess the extent to which it is possible to use selected alternative building materials to replace at least part of the consumption of traditional materials.

The study has been conducted, and some information has been collected. These are summarised as clearly as possible for convenience and for sharing with young engineers. Improvements based on research and test results will be required in many cases, as indicated, and this needs to be highlighted. An attempt has been made to do this to a limited extent. With this in mind, the documents for publication have been prepared.

References are included after each section for further study and guidance.

This publication will be useful to students and young practising engineers who have developed an interest and are seeking information on suitable alternative building materials for design and applications in projects.

In that case, the efforts put into the publication will be worthwhile.

1 General Items and Brief Comments on Building Materials and Earth Technology

1.1 ALTERNATIVE BUILDING MATERIALS

1.1.1 INTRODUCTION

With developments taking place in different countries and the assessment of shortages of basic raw materials for construction, it has become important to consider alternative building materials for future projects. Some of these materials may have already been used in the past. The materials can be further reviewed and utilised with improvements wherever possible. Furthermore, knowledge of these materials will be relevant for engineers, as the applications may not be fully known to many people. Initially, a review will be conducted to understand how humans began using building materials in stages to meet their diverse needs in the past. It is needless to say that tremendous efforts were made by our predecessors to improve the performance of materials for various applications. It is appropriate to start with earth engineering for structural applications.

The author would like to make the following comments:

- Searches have been conducted to collect information on alternative materials from various publications, including older ones, to understand different perspectives and the work done by many individuals wherever possible.
- Most of the information has been collected from technical journals of different institutions, universities, books, and internet publications.
- The emphasis is on the use of alternative materials, particularly in reference to earth. Attempts have been made to suggest the use of alternative building materials to replace at least a portion of traditional materials in their current form, covering constituent materials including reinforcing bars. Recycling materials in the process will be an added advantage.
- In this context, one needs to understand the details of alternative materials, including their specific advantages and costs.
- Studies have been conducted to compile a list of alternative materials that have been used in the past and are still in use for projects. A brief overview of these materials has been included. Hopefully, this may be helpful.
- It would be interesting to discover what authors and researchers have communicated on different or similar topics over time. Hence, some comments may appear to be repetitive in places. Even so, the author believes it will be an important exercise to understand the progress and development work carried out on the item.
- Studies have been conducted to learn and gather information on a topic, compiling and organising it for sharing with others.
- Certain portions may seem redundant and repetitive, but this has been done intentionally so that the reader does not have to refer to earlier chapters to quickly look up a topic.
- Emphasis is placed on earth engineering and construction.
- Depending on the author's interests, more importance may have been given to some particular items compared to others, but this does not necessarily reflect the actual importance of the items.

DOI: 10.1201/9781003319108-1

- New ideas need to be explored, especially considering the environment, durability of structural construction, life cycle cost, and making steady progress, like other developing countries. We have been doing this in our country for a long time. Further actions are still needed for quick development to catch up with others; otherwise, we will fall behind.
- A lot of references have been included from developed countries, particularly the Concrete Society of the UK. We need to improve our knowledge for progress, and our codes must be updated in line with others for applications and confidence. Adequate studies supported by research with sufficient funding will be required for this. Of course, we must do this in stages.
- Selecting materials will require coordination with the concerned architects and engineers from other disciplines, which is important.
- The author understands that many items are not included in the publication, primarily due to time limitations. There is hope to include more items in the future.

Some photographs have been included, although they have not been preserved properly for publication. However, despite difficulties, some of them have been retrieved. These are presented to demonstrate different types of structures and projects the author has had the opportunity to work on, as well as the requirements for suitable materials to meet design needs. It is hoped that this will be interesting to students and young engineers trying to learn and looking for opportunities to use alternative materials wherever possible in projects.

The publication is expected to be useful to students and practising engineers, particularly in our country, who will be involved with design, construction, detailing, monitoring, diagnosis, assessment, and rehabilitation work, especially with alternative materials for structures.

A list of references has been provided for further reading and information with each chapter.

1.2 ITEMS TO BE INCLUDED FOR THE PUBLICATION

The items that will be considered broadly for study and discussion are as follows:

- Review of common building materials
- Earth engineering and construction
- Use of compressed earth blocks (CEBs) for construction
- Masonry construction – adobe, clay bricks, etc.
- Rammed earth construction
- Ferro-cement construction
- Fibre-reinforced concrete
- Traditional building materials, i.e. concrete and steel – Some design items with the materials
- Computer-aided work, education and training, communications
- Construction work, site supervision, safety, project management
- Hazards, risks, structural safety
- Repair work to structures
- Photographs from projects
- Summary of comments

1.3 FOCUSING ON SOME ALTERNATIVE BUILDING MATERIALS

The alternative building materials, with particular reference to earth engineering/construction, considered in this publication include the following items:

- Earth engineering and construction
- Use of CEBs for construction
- Masonry construction – adobe, clay bricks, etc.

- Rammed earth construction
- Ferro-cement construction
- Fibre-reinforced concrete

These items have been discussed in detail in various chapters.

Other building materials have been reviewed and mentioned briefly. However, discussions/comments on wood, bamboo, straw, and bio-composite materials have also been included to some limited extent.

1.4 COMMENTS ON OTHER BUILDING MATERIALS

This review has been conducted to identify items that may be considered suitable alternative materials for future use. An attempt has been made to compile an approximate list of materials that have already been tried.

The applications of these materials have also been explored in recent times. These materials have been identified as alternatives that may be used instead of traditional materials in some cases. With the development of further knowledge about these materials and their proper applications, the confidence of engineers, architects, and clients will definitely increase, allowing for execution at competitive prices and reduced time in some instances.

While selecting materials, emphasis has been placed on the availability of local resources, the use of not highly skilled labour, easy application after brief training for workers, and finally, the cost/energy consumption of execution to reduce reliance on traditional materials that are mostly used at present. Depending on the situation, the use of these materials may be restricted to specific projects if necessary. Even then, this will be a significant step in the right direction and should be appreciated; sincere efforts should be made to continue using these materials in the future as applicable.

The above items have been covered separately, with photographs from projects to demonstrate the versatility and flexibility of the materials, showcasing impressive elevations. From the photographs, it can be observed that clever combinations of ferro-cement and earth blocks can produce very attractive structures. The units can be arranged in various ways to create pleasing aesthetics and elevations.

1.5 ADDITIONAL COMMENTS ON BUILDING MATERIALS

1.5.1 Introduction [1]

Over time, it has become clear that a serious review of building materials is necessary. This should aim to explore and identify the list of building materials that have been used/tried by humans over many years.

It is understood that this will require tremendous effort, given the vastness of the subject, and a realistic presentation will not be easy. However, a moderate selection of building materials may be attempted based on the author's limited knowledge and experience in this area. Hence, the review has been conducted using a selective approach, reflecting the author's experience and preferences. It should be noted that this will not be exhaustive and will not cover all items by any standard.

However, this publication places greater emphasis on earth engineering and construction, particularly the applications of rammed earth and compressed earth blocks. The details will be highlighted with brief descriptions and photographs as applicable.

Furthermore, efforts have been made to use ferro-cement items for various applications in different projects. The details of a few other materials are also included.

We have also observed that various types of building materials have been used for different development work in many countries. These may include earth, stone, masonry brickwork, timber, concrete, steel [1], ceramic, polymer, and their combinations, as per the requirements. These materials have been extensively used by engineers and architects successfully in recent years throughout the world. Each material has its advantages, disadvantages, and limitations.

The basic design aspects of two traditional materials, i.e., concrete and steel, have been included to understand and briefly discuss the flexibility and versatility of the building materials necessary to meet design issues.

The term durability for a building material is closely related to its satisfactory performance without deterioration over time. Lack of durability may be due to the environment to which the structure is exposed or internal weaknesses within the material itself. Like others, the structure will be subject to different conditions during its service life, which must be assessed carefully and accurately during design and construction. It further requires that the quality of the material and workmanship be verified and checked at different stages, including inspection and supervision, to ensure safety, serviceability, and durability.

Structural engineers are familiar with using both structural steel and reinforced concrete in design and construction related to various projects. In most cases, composite action using structural steel and concrete has been considered in the design to meet specific requirements and for economic reasons.

With the development and constraints of available space, particularly in urban areas, the requirements for innovative design and construction are becoming increasingly complicated and challenging for engineers. An attempt has been made to provide some brief comments on the building materials that have been tried so far, including some photographs.

The projects mentioned in this presentation relate to infrastructural development, specifically building projects, covering educational institutions, schools, hostels, hotels, resorts, hospitals, auditoriums, science centres, research centres, handicraft workshops, etc. Concrete structures will require proactive inspection, maintenance, and monitoring to extend their service life. Damage and repair work, whenever required, should be carried out promptly for the structures [2].

Source:

- Ganguly [1,2]

1.6 FURTHER REVIEW OF COMMON BUILDING MATERIALS

The comments will focus on some building materials, starting with earth construction as stated.

1.6.1 History of Earth Construction [3]

Earth engineering was revived by the well-known Egyptian architect Hassan Fathy [3] in the 1940s. He promoted adobe after the Second World War. Earth has been used and promoted for social building programs by governmental organisations, particularly in Germany.

CEBs were developed in 1957 by the Colombian team of Raoul Ramirez [3]. They developed the first CEB press known as the Cinvaram.

In India, earth construction is mainly recognised through the use of compressed earth blocks, often known as mud blocks [3]. There are other methods to construct buildings with earth. Most of these types of construction may be summarised as follows:

- Compressed earth blocks (bricks)
- Masonry bricks, adobe
- Rammed earth
- Cob
- Wattle-and-daub
- Other types

The contributions of earth blocks have transformed the use of earth from traditional methods to modern technology and applications. They can significantly contribute to earth engineering and architecture by taking into account studies and research conducted in laboratories and in the field in many countries.

1.6.2 Scientific Research [3]

Research has been ongoing in many countries since the fifties, with Raoul Ramirez [3] focusing on the following areas:

- Stabiliser and stabilisation methods
- Suitable soil
- Improved production of CEB
- Equipment for production and improvement
- Industrialisation and mechanisation
- Building techniques
- Standardisation

Source:

- Auroville [3]

1.7 IMPORTANT INFORMATION ON EARTH ENGINEERING [3]

A few items include

- 30% of the world's population lives in earth homes.
- A majority of the rural population and at least 20% of urban areas may be associated.
- 50% of the population in developing countries.
- In India, about 72.2% of buildings are made with earth.
- In Africa, most rural and even some urban areas are constructed with Banco (West Africa) and Thobe (Egypt and North Africa).
- In China, more than 10 million people live in dug-out dwellings.
- In the Middle East, the oldest dwellings discovered date back to 8000 BC.
- In France, 15% of houses are made with mud.
- In Egypt, the Ramasseum has stood for about 4,000 years.
- In the USA, especially in California and New Mexico, adobe construction is increasing by about 30% per year.

Source:

- Auroville [3]

1.8 ADDITIONAL INFORMATION ON EARTH ENGINEERING [3,4]

The comments are as follows:

- Earth is undoubtedly one of the most ancient construction materials known to humanity.
- For 10,000 years, earth has been used as a building material.
- 40% of the world population lives in earthen dwellings [4].

- One-third of the world's population lives in earth buildings.
- 25% of the world population does not have access to decent housing [4].
- As a building material, it continues to dominate in Africa, China, the Middle East, and Latin America. Mud buildings are also found in some European countries.
- 17% of the 'world cultural heritage sites' are built with earth [4].
- 25% of the 'world heritage sites in danger' are built with earth [4].
- 14% of the 'hundred most endangered world heritage sites' are constructed with earth [4].

1.8.1 Common Types of Applications

- Compressed earth
- CEBs
- Rammed earth construction

These types of constructions are discussed separately.

Source:

- Auroville [3]
- Auroville Earth Institute [4]

1.9 EARTH CONSTRUCTION, METHODS, AND FLEXIBILITIES

The following construction methods are typically involved [3]:

Techniques	Hydrous States
• Dug-out	Solid or dry state
• Covered	
• Filled in	
• Cut	
• Compressed: rammed earth/CEBs	Humid state
• Shaped	Plastic state
• Stacked	
• Moulded	
• Extruded	
• Daubed	Liquid state
• Poured	
• Formed	

Source: Auroville Building Centre [3]

1.10 MORE ON EARTH ENGINEERING/CONSTRUCTION

Some common building materials that have been used will be briefly mentioned below:

1.10.1 Earth Structures [1,3]

Earth structure is made of locally available soil that may be sieved and stabilised by the addition of lime or cement and mixed with straw to gain specific advantages. These types of construction are economical and environmentally friendly.

The following types are well-known:

- **Stabilised Earth:**
  A higher sand content in the soil is required for good results, and the soil may be stabilised by adding cement, lime, or pozzolana for constructing load-bearing walls.
- **Rammed Earth:**
  Natural damp soil with stabilisers may be placed using temporary formwork. The soil is rammed, and construction is done in layers, resulting in a monolithic structure. The wall constructed will be able to carry loads. Further discussion on rammed earth is provided in Chapter 5.
  A photograph of rammed earth construction from a project is shown in Figure 1.1.
- **Compressed Earth Blocks:**
  The blocks are prepared using natural and locally available soil with stabilisers and are compressed in a hand-operated or powered machine. After curing, the CEBs are used for constructing building walls to carry loads. Further discussion on this is provided in Chapter 3. CEB applications and photographs from a project are shown in Figures 1.2 and 1.3.
- **Earth Bags**
  In this method, the soil is packed into woven polypropylene or burlap sacks. The bags are laid in courses, with barbed wire between each course to prevent slipping.
  More comments on this have been made in Chapter 2.
- **Adobe, Fired Clay Bricks, etc. –**
  The use of adobe and fired clay bricks is quite old, and the units are laid with mortar prepared with cement/lime and sand for constructing buildings.

Further comments on the above items have been covered separately.
Source:

- Ganguly [1]
- Auroville Building Centre [3]

FIGURE 1.1 Rammed earth and timber roof.

FIGURE 1.2 Inside of a CEB dome.

FIGURE 1.3 CEB arches.

1.11 SUMMARY AND REVIEW ON EARTH ENGINEERING/ARCHITECTURE AND APPLICATIONS [4,5]

Earth is one of the most ancient materials known to man, as mentioned already.

Examples of applications of earth can be found in the Persian, Egyptian, Assyrian, and Babylonian civilisations. People have constructed various structures with earth that are well known.

Earth architecture is not necessarily meant for archaeologists only. It seems that more than 50% of the world's population lives in earth buildings, according to recent assessments. As a building material, it is popular and dominates countries like those in Africa, the Middle East, and Latin America, and may be considered a particular form of vernacular dwelling in China. Applications of earth have also been made in many European countries, including Sweden, Denmark, Germany, Eastern European countries, Spain, and Great Britain.

People have used earth and lived in harmony with nature. The earth was used as a raw material with sensitivity and respect. Modern development has spoiled and abused Mother Earth. It will be important to re-establish the harmonious link between man, nature, and spirit [4].

The important tasks and challenges will include [4]

- How to use and build with earth on a large scale while respecting nature.
- How to realise engineering/architecture that includes simplicity, ample light, suppleness, imagination, attractive expressions, and beauty with heavy and formless mud.
- To maintain sustainable development.

The earth had been used in different forms by people depending on the condition of the material at the time of its use. The uses may cover solid, plastic, liquid, or dry states.

The possible applications of earth according to the state of the material are indicated in Table 1.1 [5]:

Some of the **applications** will be discussed further separately.

1.12 DIFFERENT TYPES OF APPLICATION METHODS WITH EARTH

The types may be mentioned as [6] follows:

Cut Blocks, Poured Earth, Super-adobe: The technique uses long or short sandbags that are filled with moistened earth and arranged in layers or coils, adobe, Cob, compressed stabilised earth blocks, rammed earth, wattle-and-daub, straw clay, shaped earth, etc.

TABLE 1.1
Possible Applications [5]

	Earth Type	State of Earth	Type of Earth	Applications
1	Un-processed earth	Solid state	Excavated earth	Trogloditic dwelling
			Cut turf	Sod
			Cut earth	Blocks cut-out of the ground
2	Processed earth	Liquid state (pouring)	Earth poured into form work	Poured earth in shuttering
			Monolithic	Adobe
			Earth poured in a mould bricks	
		Plastic state (shaping)	Moulded earth bricks	Adobe
			Shaped earth bricks	Adobe
			Earth shaped in position	Direct moulding
			Monolithic wall	Cob
			Earth built up in position	Lath and plaster
			Monolithic wall	
			Earth shaped on frame work composite wall	
		Dry state (compaction)	Earth compacted within formwork Monolithic wall	Rammed earth
				Tamped blocks
			Earth compacted in mould blocks	Compressed blocks (bricks)
			Compressed earth bricks	

Source: Auroville Earth Institute [4]; Building with Earth [5]

The above items have been discussed in different chapters that will follow, i.e., between Chapters 2–5.

Source:

- Biju [6]

1.13 ADVANTAGES OF EARTHEN BUILDINGS

The advantages of earthen buildings have been covered in different chapters that will follow. Some basic items, in this connection, will be [7,8] as follows:

- Ecological advantage
- Ergonomic advantages, control of interior temperature
- Economic advantage, low-cost material
- Aesthetics/beauty – aesthetic quality will be attractive
- Flexibility and strength
- Performance – can survive for a long period of time

Source:

- Williamson [7]
- Build your home with earth [8]

1.14 EARTHEN ENGINEERING/ARCHITECTURE FOR THE FUTURE [8]

Some creative architects/engineers and builders have been attempting to construct earthen buildings following ancient methods with some improvements wherever possible.

It may be possible to construct efficient, inexpensive, and attractive earth buildings, as stated. A mechanised system of construction will also be possible as per the requirements.

We have to assess carefully what may be taken from the past to move forward without abandoning everything that we are presently practising. With proper judgment, we must understand and accept the items/knowledge based on our past activities/experience that may be kept and used. Others may have to be improved to meet modern requirements, and some items may have to be set aside and not used. Further comments on this have been made in Chapter 16. It may be possible to see more applications of earthen buildings in the future.

Source:

- Building your home with earth [8]

1.15 OTHER CONSTRUCTION MATERIALS WITH SOME PHOTOGRAPHS

A few of other construction materials are as follows:

1.15.1 Stone Construction [1,9]

This is an old method of construction, and stone has been recognised as a building material of excellent durability and good artistic quality. The use of stone has decreased in recent years. However, it maintains its status as a material for aristocratic applications. The different types of stones may be mentioned as follows:

- **Igneous –** Non-porous and hard rock developed from slow and rapid cooling of molten magma. Granite is an example.

- **Sedimentary** – Soft and reasonably porous rock developed from eroded pre-existing rock, which was deposited mostly from eroded pre-existing rock and settled in layers, mainly on sea beds, and got compacted with time. Examples of this type of rock include sand stone and lime stone.
- **Metamorphic** – These are non-porous and hard type of rock developed from pre-existing rock, which has been changed by intense heat or pressure. Examples of this type include slate and marble.

There may be a lot of variations within each type of the above rock types. Any type of stone may be used as a building material for construction and can be suitable for different purposes.

It should be noted that granite, sandstone, and limestone can be used for walls in a building, while slate may be suitable for roofs and floors.

1.15.2 Dry Stone Stacking [9]

This is the earliest method of stone construction, known as dry stone or dry stacking.

Examples of applications of this type of construction include buildings, field walls, and bridges. Irregular stone pieces are selected and placed carefully to construct walls, with pieces of stone fitting together properly without slipping. The structural walls are wider at the base and taper as they increase in height. To support the structure, the weight of the stone pieces pushes inwards. Furthermore, the stone pieces in the structure will lock together following the loads, making the wall stronger. Dry stone construction is very durable, and the structure can be easily repaired if required. There is unlikely to be any damage to the stone while draining water through them. The construction will not need any special tools. The important requirements are the skill of the worker/craftsman, who must be able to select the stone pieces properly and place them according to the requirements.

Source:

- Gray [9]

1.15.3 Dry Stone Applications [10]

A few uses of dry stone include

- Dry stone buildings
- Dry stone monuments
- Dry stone bridges
- Dry stone markings
- Ancient walls
- Gallery

Source:

- Wikipedia [10]

1.15.4 Stone Masonry [9]

This is a traditional method where stone blocks are used in dry stone stacking. The stone blocks are typically laid in rows of even courses or uneven uncoursed heights and put in place with the use of mortar. A cement or lime-based mortar is placed between the stone blocks.

The stones for building may be extracted from surface quarrying, drilled, and split with diamond saws or iron wedges. The stones may later be polished and shaped according to specifications. Common hand tools used for shaping the stones include chisels, mallets, and a metal straight edge.

However, modern methods may include the use of power tools like angle grinders and compressed air chisels, which can save time and money. The stones may be shaped to form blocks known as ashlar masonry or may be kept rough and cut irregularly, which is known as rubble masonry. Dry stone construction is generally more durable than mortared stone structures. Water may get trapped between the stone pieces and push them apart.

The traditional method of stone construction is not commonly used in recent years. It can be expensive to quarry, cut, and transport the stone pieces to the construction site. Furthermore, building construction and operation require skilled labour and are labour-intensive. In modern applications, stonework is often used as a veneer (i.e., thin and flat pieces) glued to concrete block surfaces. This is referred to as veneered stone or simply stone cladding.

Slip-form stone structure construction is associated with a cross between veneered and traditional masonry. This will involve the use of short forms, which may need to be placed on either side of the wall to act as guides for the structural construction. The stones are to be placed inside the forms with the flat face outward. The concrete may be poured within the form and then placed behind the stone blocks, compacted to hold them together. Stone buildings can be constructed easily using this method and can be completed quickly.

1.15.5 Sustainable Stone [9]

Stone may be considered a low-maintenance building material that is very durable and has high thermal mass. It is versatile and available in various shapes, sizes, textures, and colours. Applications may include floors, arches, walls, and roofs. Stones typically suit the natural landscape well, and the material can be recycled for other buildings later if required.

The stone industry is under threat due to the increasing influx of cheap, imported stone and synthetic imitations. For sustainable use, necessary checks are required to ensure the proper use of stone reclaimed from demolished buildings or sourced from a local stone quarry, if possible, to protect the stone industry and retain a proper and sustainable building material.

Source:

- Gray [9]

1.15.5.1 Photographs

The photographs from projects with different applications are shown below (Figures 1.4–1.6):

1.15.6 Masonry Construction [11]

This type of construction may be executed using individual units that are laid and bonded together with mortar.

Common materials for masonry construction include brick, building stone such as marble, granite, limestone, concrete blocks, cast stone, glass block, and cob. The quality of mortar, workmanship, and pattern of laying the units properly are important for durable construction.

Masonry construction requires skilled labour for execution. Masonry walls can provide good compressive strength and are suitable for structures with nominal lateral loads when the cores are not filled. By filling the cores with concrete or with concrete and reinforcing bars, the same will provide better tensile and lateral strength to the structure.

1.15.6.1 Advantages [11]

The following items may be mentioned:

- Masonry structures may not require painting for treatment, which can help reduce life cycle costs.

FIGURE 1.4 Stone wall construction. (Source: Ganguly [1]).

FIGURE 1.5 Stone wall construction. (Source: Ganguly [1]).

- Items like bricks and stones can improve the thermal mass of the building and are suitable for protecting the building against fire.
- Masonry is a non-combustible material.
- Masonry structures can be built primarily to take compressive loads with lime mortar and may provide a service life of more than 500 years [8]. This is likely more than modern construction materials used for structures with reinforced concrete or steel.

FIGURE 1.6 Stone wall construction in progress. (Source: Ganguly [1]).

- Masonry structures are more resistant to projectiles from debris during hurricanes or tornadoes.
- Fire resistance for bricks is considered good.

1.15.6.2 Disadvantages [11]

The following items are relevant:

- A strong foundation system will be required since masonry construction can be heavy.
- Degradation of masonry can occur due to expansion and contraction forces from freeze–thaw cycles and other causes.
- Loose components are associated with masonry construction. Hence, it will have less tolerance to oscillations from vibrations compared to other materials.
- Less mechanisation is involved, and it will require more skilled labour than stick framing.

1.15.6.3 Limitations [11]

The masonry construction will have low tensile strength unless reinforced and has good compressive strength for resisting vertical loads. The tensile capacity of masonry walls can be improved by increasing the thickness of the wall or by introducing vertical columns or ribs with appropriate spacing. Steel reinforcing bars may also be used wherever possible, as mentioned. Masonry construction is covered further in Chapter 4.

Source:

- Wikipedia [11]

1.15.7 Timber

The use of timber in constructing buildings and bridges has been practised for a long time. Figures 1.7–1.9 may be referred to for timber roofs in buildings.

FIGURE 1.7 Stone wall construction in progress. (Source: Ganguly [1]).

FIGURE 1.8 Stone wall and timber roof. (Source: Ganguly [1]).

1.15.8 Steel

Steel has been successfully used for various structural applications. Tall buildings and long-span bridges are good examples. Further discussion on steel will be made later. Some applications of steel are presented below (Figures 1.10–1.13):

FIGURE 1.9 Timber roof.

FIGURE 1.10 Space structures, ball joints. (Source: Ganguly [1]).

FIGURE 1.11 Space structure with welded joints. (Source: Ganguly [1]).

FIGURE 1.12 Gate complex area steel space structure with circular concrete columns. (Source: Ganguly [1]).

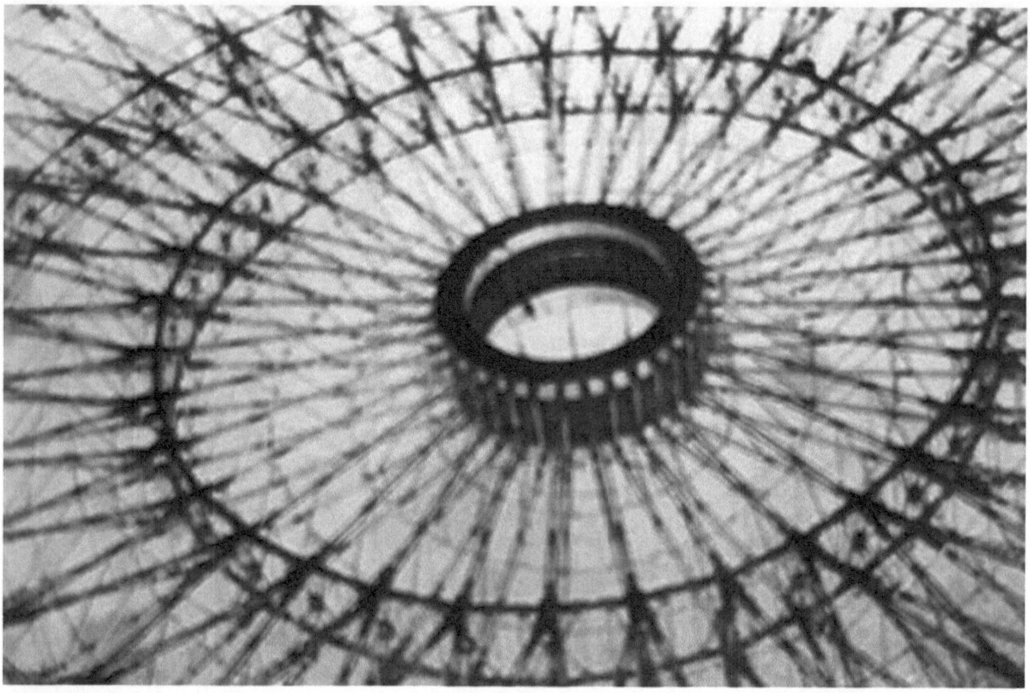

FIGURE 1.13 Steel space structure. (Source: Ganguly [1]).

1.15.9 Ceramic [12]

This is a very old material, and the composition of clay can be divided into two groups: Primary clay and Secondary clay. Primary clay is essentially earthen clay from the ground, treated by firing in a kiln at high temperatures after air drying. It may absorb more water compared to secondary clay. The secondary type of clay is generally transported by wind, water, or both from their areas of formation and is normally easier to form and fire.

The main advantages of ceramic include durability, relatively low cost, environmental friendliness, and resistance to fading.

Disadvantages may include that ceramics can be cold when used for flooring, especially in cold climates. Furthermore, ceramics may be quite heavy, and different mortars, grouts, and tools may be necessary for execution, leading to potentially high labour costs.

Uses of ceramics include bowls and plates, terracotta, stoneware, fireclay, earthenware, roofing, and porcelain. Additionally, ceramics may cover the preparation of different types of floor tiles, flower pots, thermal/electrical insulation, wastewater pipes, bricks, and surface beautification in building construction. They may be available in natural or brownish-orange colours.

Source:

- Ceramic presentation of building materials [12]

1.15.10 Polymers [13,14]

Brief comments will be presented.

Polymer materials include plastics, rubber, thermoplastic elastomers, adhesives, foams [10], sealants, and paints and may be used for wall panels, roofing materials, piping systems, and coatings.

Polymer composites were developed during the 1940s for military and aerospace applications [11].

Polymer materials have been used as construction materials in various projects.

The application areas of polymers include flooring, cladding, rainwater pipes, sealants, membranes, glazing, signage, and insulation. The advantages of using polymers compared to traditional materials include lightness, resilience to corrosion, and simple processing techniques [13]. Polymers may be combined with fibres to improve material properties. Such composite materials have been successfully applied for repair and retrofitting work. Polymer composites have been used for load-bearing and infill panels. Glass-reinforced plastics have been used for pressure pipes, tank liners, and roofs [11].

Polymer composites have been applied for bridge repair, bridge design, structural strengthening, mooring cables, etc. [13,14].

There may be some concerns related to the long-term performance of the polymer material, including aging and pollution, considering that the material is relatively new compared to traditional materials that have been used for a long time [14].

Source:

- Sandberg [13]
- Polymer composites as construction materials [14]

1.15.11 Concrete

This section will cover the discussion of using reinforced concrete as a building material, which has already been successfully applied in the design and construction of various projects. A brief discussion of this material has been provided in Chapter 9. A few selected items are mentioned below:

1.15.11.1 Concrete Blocks [11]

Ordinary concrete blocks, cinder concrete blocks, or hollow tiles are often known as concrete masonry units (CMUs). They are generally lighter than ordinary bricks, and walls can be constructed quickly with these blocks. Cinder and concrete blocks have lower water absorption rates than conventional bricks. They can be used for constructing the walls of factories, garages, and other industrial buildings where the appearance of the blocks is aesthetically acceptable. Extra strength can be provided to the blocks by using synthetic fibres as reinforcement with surface bonding cement.

Surface bonding cement may be coloured, or the treatment can be stained, painted, and finished to achieve a stucco-like surface as per the requirements.

An important advantage of concrete blocks is that CMU walls can be reinforced by filling the voids with concrete and steel reinforcing bars as needed. This reinforcement is especially required at corners, wall ends, and openings, where filling and the use of reinforcing bars are necessary. Other voids in the blocks may not require treatment. Such reinforcement will enhance the stability of the wall and increase its strength. Generally, CMUs will have the top course of the blocks filled with concrete, which will tie with the reinforcing bars to form a bond beam. The need for a bond beam and tying may be required as per the code. Ladder-type reinforcing bars may be used and placed in horizontal mortar joints of the concrete block walls for tying. This treatment will provide higher lateral and tensile strength to CMU walls compared to unreinforced walls.

Various appearances may be possible during the manufacture of CMUs. They can be produced as coloured, stained, or painted after installation. The blocks may be given a rough face to resemble natural stone, such as brownstone. CMUs can be scored, sandblasted, polished, ribbed, raked, or brushed during the manufacturing process, including the addition of decorative appearances.

Glazed CMUs may be manufactured using polyester resins, other chemicals, silica sand, etc., to create a smooth and impervious surface.

Glass bricks or glass blocks are units made from glass, providing a translucent to clear vision through the block.

Source:

- Wikipedia [11]

1.15.11.2 Ferro-Cement [15]

The discussion on ferro-cement has been covered separately in Chapter 6. Some well-known ferro-cement applications are shown below (Figures 1.14–1.17):

1.15.12 Reinforced Concrete

In reinforced concrete members, steel reinforcing bars are placed in the concrete elements to act compositely, particularly to handle tensile, shear, and sometimes compressive stresses. Further discussion on this topic can be found in Chapter 9.

FIGURE 1.14 Pre-cast ferro-cement units used as waterproofing and decorative cover to the main structure, by Ove Arup & Partners. (Source: Walkus and Kowalski [15]).

FIGURE 1.15 Pre-cast ferro-cement elements used as shuttering for cast-in-situ ribs and topping, by Prof Nervi. (Source: Walkus and Kowalski [15]).

FIGURE 1.16 Ferro-cement fishing vessel, Hong Kong. (Source: Walkus and Kowalski [15]).

FIGURE 1.17 Hull reinforcement. (Source: Walkus and Kowalski [15]; Ganguly [1]).

1.15.12.1 Applications of Various Structures in Reinforced Concrete from Projects

See Figures 1.18–28.

FIGURE 1.18 Entry to a project site.

FIGURE 1.19 Tall reinforced concrete building.

FIGURE 1.20 A group of buildings.

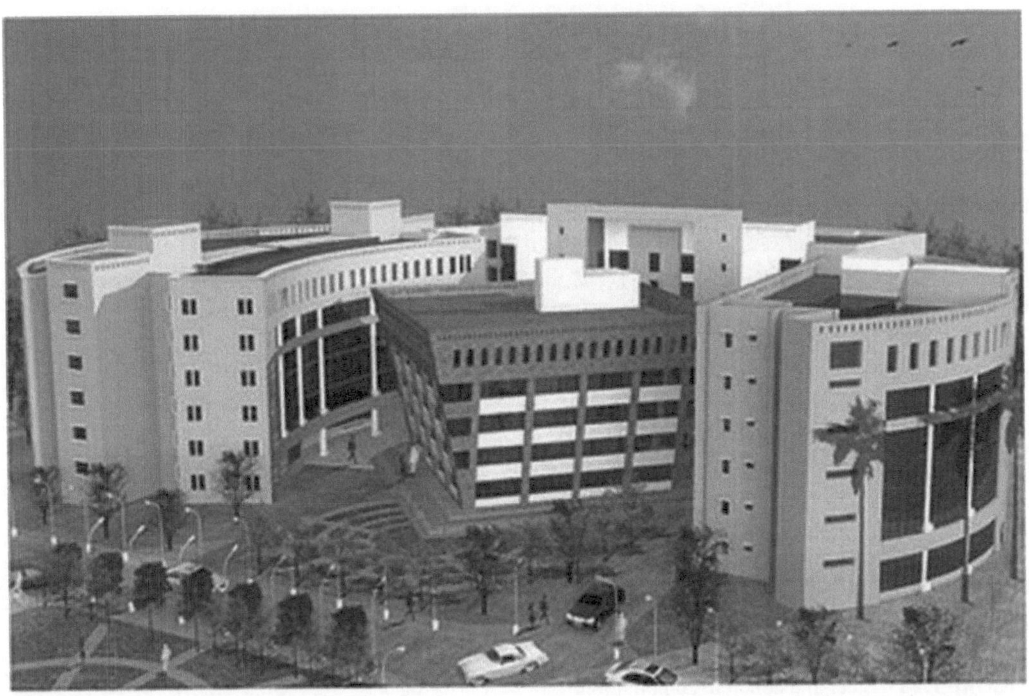

FIGURE 1.21 Tall RC buildings.

FIGURE 1.22 Tall reinforced concrete building.

FIGURE 1.23 Steel and concrete combination.

FIGURE 1.24 Tall reinforced concrete buildings.

FIGURE 1.25 External staircase.

FIGURE 1.26 RC buildings.

FIGURE 1.27 Auxiliary structure.

FIGURE 1.28 Secondary structures.

1.15.13 Use of Concrete and Steel

The combination of both concrete and steel materials has been used in various projects (Figures 1.29 and 1.30). This is quite common in industrial projects, and a few applications are shown below:

1.15.14 Pre-cast Concrete

Pre-cast concrete structural members are manufactured in factory conditions and then transported to the construction site for placement/erection. Pre-casting increases the durability of members and decreases construction time and cost (Figure 1.31).

The photographs of a few applications are as follows:

1.15.15 Pre-stressed Concrete

Pre-stressing involves pre-tensioning or post-tensioning of concrete members, allowing them to carry greater loads or span larger distances than ordinary reinforced concrete. In pre-tensioning, steel wires or cables are placed in empty moulds and stretched. The concrete is then placed after the stretching. In post-tensioning, the stretching of cables is done after the concrete has hardened (Figures 1.32 and 1.33).

1.15.16 Fibre-reinforced Concrete [16–18]

Concrete containing discontinuous and discrete fibres is known as fibre reinforced concrete. The concept of adding fibres to the concrete mix is not new. The use of straw and asbestos fibres for strengthening units has been successfully implemented for many years [16].

FIGURE 1.29 Industrial project.

FIGURE 1.30 Industrial project.

FIGURE 1.31 Bridge, pre-cast and pre-stressed concrete.

FIGURE 1.32 Bridge, pre-cast and pre-stressed concrete.

FIGURE 1.33 Bridge, pre-cast and pre-stressed concrete.

The idea of improving the physical properties of concrete by adding fibres originated in advanced countries.

Various shapes and sizes of fibres made from steel, aramid, plastic, carbon [16,17], glass, and natural materials have been tested with concrete. Studies have shown that the ultimate tensile strength of concrete with fibres does not increase significantly compared to normal concrete. Furthermore, fibre reinforced concrete is much tougher and more resistant to impact than normal concrete.

It is noteworthy that, based on the load–deflection curve, plain cement concrete may fail suddenly when the deflection corresponding to ultimate strength is exceeded. In contrast, fibre reinforced concrete continues to support loads at deflections well beyond the fractured deflection of plain concrete [16].

The interface region is crucial for the performance of the composite material [17].

Fibre reinforced concrete has been applied in various building projects to gain benefits related to crack resistance, improved concrete quality, better installation, and enhanced impact and abrasion resistance. Engineers have utilised this material in ground floor slab construction with the intention of eliminating the commonly used steel reinforcement. The design can also accommodate slabs supported on piles using steel fibre reinforcement.

This publication attempts to study several aspects of fibre reinforced concrete, including the applications of fibre reinforced polymer composite wrapping materials, which have been increasingly applied in recent years for repair and retrofitting applications. Further comments and discussions on fibre reinforced concrete are covered in Chapter 7.

Source:

- Mehta and Monterio [16]
- Hollaway [17]
- Ganguly [18]

1.16 FURTHER INFORMATION ON SOME OTHER BUILDING MATERIALS

A few other building materials will be briefly mentioned [19]:

1.16.1 Shotcrete

This is a method of spraying concrete to the required location. For shotcreting, a mixture of aggregate, cement, and admixture is combined and conveyed by compressed air to the spraying gun, where water is added for application. The wet mix is then sprayed to the desired area and can be carved or trowelled quickly. There are various structural applications for shotcreting.

1.16.2 Plaster

Plaster can be prepared from lime or gypsum, water, and sand, and is used for coating walls, ceilings, and partition walls. Proper curing is recommended. The plaster hardens after drying. This is one of the ancient materials used by people. Primitive individuals also used plaster for reed or sapling shelters in the past. Some fibres may be incorporated into plastering work.

1.16.3 Tile

Tiles can be thin, flat slabs or blocks and have been used for structural work or for decorating buildings. Tiles may be made from glazed or unglazed fired clay. Modern tiles can be manufactured using plastic, asbestos, glass, cement, or asphalt and are available in different colours to meet various requirements. Fibreboard or similar types of tiles may be used for acoustic purposes.

1.16.4 Aggregates

Aggregates are construction materials that can be combined with cement, bitumen, gypsum, lime, and other adhesives to create concrete or mortar. They provide stability, volume, resistance to wear or erosion, and the required physical properties to the finished product. Aggregates come in various sizes and are essential components in concrete construction.

1.16.5 Slabs

Slabs can be made of stone or concrete and are used for floor or roof construction. Different types of concrete slabs reinforced with steel are designed by engineers in conjunction with beams for various projects to meet design requirements.

Further comments on concrete will be addressed separately.

1.16.6 Wattle-and-Daub

This method of constructing walls involves vertical wooden stakes (wattle) woven with horizontal twigs and branches, followed by daubing with mud or clay. It is considered one of the oldest methods of construction for creating weatherproof structures. Many applications of this technique have been carried out in the past. More comments on wattle-and-daub are made in Chapter 2.

1.16.7 Shingle

These are thin pieces of building material with the butt end thicker than the other. Shingles may be used for roof covering on residential buildings and also for siding. The stock sizes may be of various materials and may include items such as wood, slate, and asphalt. The pieces are connected in the overlapping areas.

1.16.8 Lath

Lath is associated with materials that may be fastened to the structural member to prepare the base for plaster. Lath may be made of wood, gypsum, metal, or insulated board. Narrow wood strips were used for older buildings in the past. Gypsum lath is considered the most common item.

1.16.9 Brick and Tile

These are structural clay products produced as standard units and used for building construction. The items have different uses in building projects. The selection of suitable clay is important for production. Further comments on these are made separately.

1.16.10 Plaster of Paris

It is a quick-setting gypsum plaster made from a fine white powder (calcium sulphate hemihydrate) that is moistened before use and hardens when the material is allowed to dry. The material has been successfully applied for a long time. The name comes from its use of material from the abundant gypsum site located near Paris.

1.16.11 Gypsum Plaster

This is a white cementing material prepared by partial or total dehydration of mineral gypsum, normally with the use of special retarders or hardeners. It is applied in a plastic condition using water. It sets and hardens through the chemical action of gypsum with water.

1.16.12 Mortar

This is required in building construction to bond brick, tile, stone, or concrete blocks to the structure. Mortar is prepared from inert siliceous sandy material mixed with cement, admixtures, and water in the required proportions. The material is initially plastic and provides a proper bond with brick, tile, stone, or concrete blocks after curing.

1.16.13 Fired Brick

This is a refractory material made from non-metallic minerals and can be produced in different shapes for use at high temperatures. It is required in structures for metallurgical applications and the manufacturing of glass. The main raw materials for the bricks include fire clays, such as hydrated aluminium.

1.16.14 Adobe

These are clay blocks that may be used as bricks after drying in the sun. The term is Spanish and refers to bricks. The blocks are made using a mixture of clay, sand, and silt with plastic qualities that dry to form a uniform hard mass. These constructions are suitable for arid or semi-arid environments. Further discussion on adobe is provided in Chapter 4.

1.16.15 Clapboards

Clapboards may be used to treat the external surface of a framed building. They may be bevelled toward one edge. Clapboards can be placed horizontally, with each one overlapping the next one down. These may be 150–200 mm in width and 16 mm (approx.) thick at the lower edge, with a finer dimension at the upper edge.

1.16.16 Siding

This material may be used to treat the external surface of a building, especially for protecting building elements to prevent heat loss and to provide a uniform look for the façade. The siding may include wood units and imitation wood products used for house building. There are various types of siding that can be considered for treatments.

1.16.17 Asphalt Tile

Asphalt tiles are intended for smooth floor surfacing. It is prepared from a mixture of asphalts or synthetic resins, pigments, asbestos fibres, and mineral fillers. Normally, it may be about 3 or 4.8 mm thick. It is non-porous, non-flammable, easy to maintain, and low in cost.

Source:

- Britannica [19]

1.17 MORE ON BUILDING MATERIALS [20]

The earth material will need treatment or processing as per the requirements for applications. Stabilising materials like lime or cement may be added to make the earth hard when compacted, increasing its strength. Further comments on earth construction have been made in Chapters 3–5.

A few more items for building materials may be mentioned below, covering different types of construction.

1.17.1 Mud

Mud construction techniques may be considered the oldest form of construction for building walls. The base of the wall may be formed with moist mud prepared by hand and allowed to dry. Additional mud is placed in stages for the courses after each course has dried, and the method continues until the wall is complete. For the puddle mud method of construction, a hand-made mud form may be used and filled with wetter mud, which will then be allowed to dry. Puddled mud walls are also known as chine construction in Iran. Each course may be approximately 460–610 mm high. This method was primarily used for garden walls and not for house building due to concerns about the strength of the wall construction. The disadvantages of this method include the time required for each course of the wall to dry. This method may be used for infill walls as insulation for a wooden-framed house. A similar technique was used in England with cob. Mud wall construction has also been practised in India, especially in rural areas.

1.17.2 Sod or Turf

Cut sod bricks, also known as terrones in Spanish, may be used for durable and tough wall construction. These are prepared by cutting from soil that has a heavy mat of grass roots. The material may be found in river bottom lands. It is placed on edge for drying before being used for construction.

Based on experience from the North American Prairies, sod is not likely to deteriorate due to freezing or rain from dried sloughs.

Turf was extensively used for wall construction in houses in European countries. If maintained properly, a turf house may last more than 50 years or longer. The best quality of turf would be Strengur (top 50 mm grass turf), according to Icelanders.

1.17.3 Cob

Cob is also known as "monolithic adobe" and is prepared using soil that includes clay, sand or small stones, and straw as organic material. Cob walls may be constructed using courses with no mortar joints and require about 30% or more clay in the soil. Cob walls may also be applied in framed buildings and can be used as load-bearing walls up to two stories. The width-to-height ratio of the wall should not be more than one to ten, with a minimum wall thickness of 410 mm. The walls can be plastered on both the inside and outside using a mixture of lime, soil, and sand. Cob walls are fireproof, and the thermal mass helps control and stabilise indoor temperatures. Such construction has some resistance against seismic loads but may not be considered suitable material according to building codes. This item has been discussed further in Chapter 2.

1.17.4 Stabilised Earth

The clay becomes soft when it absorbs water and can be hard and strong in dry conditions. The dry clay walls may stand and will help to keep the structure together. However, when the clay wall is exposed to rain or water leaking from the roof, it may become saturated, soft, and can finally collapse. To avoid this, the earth must be stabilised to make it resistant to weather. The method of adding burnt lime to soil for stabilisation has been used for a long time in the past. However, stabilised earth will not be as strong as fired clay or concrete. To improve strength, Portland cement or bitumen may be included in the earth. Mixtures of lime and cement or pozzolana and lime may be applied for stabilisation.

It is preferable for the sand content of the soil to be about 65%–75%. Suitable stabilised earth should have low clay content, with no more than 15% non-expansive clay. The clay percentage in the soil may be modified by adding sand appropriately. If the clay content exceeds 15%, it may be necessary to use more than 10% cement to stabilise the soil.

Low-density stabilised earth may be porous and weak. In that case, the earth may need to be compacted by a machine, similar to preparing blocks for use, or rammed within the wall, as in rammed earth construction. More comments will be available on soil stabilisation in Chapter 3.

1.17.5 Concrete

Brief comments on concrete have already been made.

The Romans were responsible for introducing durable concrete for supporting load-bearing walls. Roman concrete was prepared with rubble from broken bricks and rocks that were set in mortar. The mortar contained pozzolana, lime, and volcanic materials to achieve the required strength.

The discussion on concrete is addressed separately in Chapter 9.

1.17.6 Rammed Earth

A few comments have been made before.

The rammed earth method of construction requires natural materials like earth, chalk, lime, or gravel. Rammed earth walls are constructed by placing wet soil in temporary formwork. The soil is sufficiently compacted, either manually or mechanically, to build the walls. The formwork is removed after the compaction process is complete. Rammed earth structures can last for a long time.

This topic is discussed further in Chapter 5.

1.17.7 Building Units

1.17.7.1 Mud Bricks or Adobe Bricks

Mud bricks, or adobe bricks, are modular masonry units that are preformed and dried in the sun. Sun-dried mud units have been experimented with in different areas, covering various parts of the world. Construction with preformed bricks avoids delays compared to puddle mud construction and the drying time for the courses. It seems that adobe production and construction advanced in Egypt by 2500 BC. Adobe bricks are normally prepared from sand and clay with water to achieve a plastic consistency. Straw and grass may be used as binders for the material. Wooden forms are used for the mud, which is tamped and levelled in the form and removed later to dry for several days. The bricks are air-cured by being kept on end for a period of a month or more.

The use of adobe has been carried out in the USA and Mexico, with thick walls for structures that are generally no more than two stories high. Adobe mission churches typically do not exceed 11 m in height. Coatings may be required to protect the surface and should be checked and renewed periodically. The coatings may include mud plaster, lime plaster, stucco, or whitewash. Adobe walls are normally constructed by laying the bricks with mud mortar. Modern adobe may be treated with cement and may be built with cement mortars. The cement mortar may cause unstabilised adobe bricks to deteriorate due to variations in rates of thermal expansion and contraction.

The discussion on adobe has been presented in Chapter 4.

1.17.7.2 Earth Bags

Some comments have already been made.

This is an old construction method utilised as a military construction technique for bunkers. An adobe mix for the earth may be preferred, but any composition of local soil may also be used. The soil is moistened and placed into woven polypropylene or burlap sacks for compaction and proper construction.

Structures built this way will be very strong and can be compared to rammed earth construction. Earth bags may be used for the construction of retaining walls and domed-shaped buildings. More comments can be found in Chapter 2.

1.17.7.3 Compressed Earth Block

Brief comments on this have been made before.

CEBs are prepared by using a stick to ram soil into a wooden mould. Recently, hand-operated or power-operated machines are used instead to compress sub-soil material to form the blocks. The manual machines may be a cost-effective solution for making uniform building units.

Further discussion on CEBs will be addressed separately in Chapter 3.

1.17.7.4 Fired Clay Bricks

A few comments on bricks have been made before.

The use of fired clay bricks in a kiln has been ongoing for a long time. Modern fired clay bricks are prepared from clays or shales and are fired in a kiln for a period of 8–12 hours at a temperature of 900°C–1,150°C. The porosity of the brick will depend on the quality of the material used, the firing temperature, and the duration. The colour of bricks may vary and depends on the amount of iron and calcium carbonate present in the material used, as well as on the amount of oxygen present in the kiln. The bricks may be damaged due to crystallisation of salts on the brick and in its pores from the action of frost and acidic gases.

The bricks are laid in courses with mortar prepared from cement/lime and sand. There are different brickwork bonds or patterns of stretches and headers available, including English, Dutch, and Flemish bonds.

Further discussion has been conducted separately in Chapter 4.

Source:

- Wikipedia [20]

1.18 COMMON CONSTRUCTION AND BUILDING MATERIALS USED IN INDIA

The common construction materials that may be identified in this connection are as follows [21]:

1.18.1 Wood

This material has been used in the construction industry for a long time. Applications have been made in different areas, covering rural and urban regions. The advantages include high tensile strength, heat and electrical resistance, sound absorption, good acoustical properties, and aesthetic appeal. Different types of materials are easily available for use. The disadvantages include deforestation, prone to damage from natural elements, and potentially high maintenance cost.

Further information on wood is given in Chapter 2.

1.18.2 Glass

Glass is a very important material for facade design of buildings and complexes in urban areas. Various treatments can be developed to meet aesthetic requirements. Different reflective surfaces may be used as needed. The advantages include pleasant and attractive looks from the outside, maximum use of daylight, energy-saving materials, privacy, and UV resistance. It is a costly material that must be handled carefully to meet specific requirements. The material will require regular cleaning and maintenance, and the costs may be high, especially for large glass walls. Security may need special design considerations and attention for safety.

1.18.3 Plastics

These may be used as polymers for construction activities. They can be moulded into any shape or form, are hard, heat tolerant, resilient, not damaged by water, and cost-effective. However, they are not biodegradable and will pollute the environment; they are flammable and can be damaged by fire.

More comments on recycled plastics are provided in Chapter 2.

1.18.4 Metals

Both steel alloys and aluminium have been extensively used in various projects associated with developmental work. The material is durable and strong, with the ability to withstand adverse environmental conditions. Special design for fire protection will be necessary. The discussion on steel as a building material is covered in Chapter 10.

1.18.5 Cement

Cement has been used as a binder for concrete in various construction activities. This topic is discussed further in a separate section.

1.18.6 Concrete

Concrete has been used as a building material for a long time, as mentioned. The material is durable and can be used for construction across different projects. More comments are included later in Chapter 9.

1.18.7 Bricks and Blocks

Clay or mud bricks have been extensively used for construction activities, as stated. Alternative building blocks and the use of fly ash have also been explored. Soil erosion from excavation can damage the environment. Further comments on soil are provided separately. More information on bricks and blocks for masonry work is included in Chapter 4.

Source:

- Gupta [21]

1.19 NATURAL AND ANCIENT MATERIALS STILL IN USE

The items that may be mentioned are [22] as follows:

1.19.1 Bamboo

Bamboo is a sustainable natural building material that has been used for a long time. It is relatively easy to plant and grows quickly. However, treatment is required against rot and pests. It is a durable material that has better tensile strength than many other building materials. It may have applications for both indoor and outdoor architectural treatments, such as facades, trusses, and plumbing.

More comments are provided in Chapter 2.

1.19.2 Cob

Cob is made from moist earth and fibrous materials like straw. As mentioned, cob is placed on top of a stone or concrete foundation to create a monolithic wall construction. Comments on cob are provided separately in Chapter 2.

1.19.3 Adobe

A few comments have been made previously.

Adobe bricks are prepared from a mixture of clay, sand, straw, and asphalt. Mud mortars are typically used for constructing thick walls. Further comments are included separately in Chapter 4.

1.19.4 Thatch

This is a construction technique where straw or grasses are used as building materials. Thatched roofs may be preferred for their good insulating properties, durability, and attractive appearance. This natural building method is time-consuming, which may increase construction costs.

1.19.5 Straw Clay

Straw clay is prepared with loose straw coated with clay slips. It may be used as an infill item for a timber structure. Wood chips may be used in clay to make bricks. Straw clay buildings may resemble conventional structures and may be preferred for construction as a natural building method.

Further comments are provided in Chapter 2.

1.19.6 Hemp and Other Fibres

Fibre-producing plants like hemp may be considered building products. Hemp can be used to make roof tiles, wall boards, and fibre boards. It can also be utilised to insulate homes, exhibiting good resistance to heat flow and the ability to absorb moisture. Hemp and other fibres are good renewable natural building materials, growing in the soil with minimal chemical treatment.

Additional comments will be available in Chapter 2.

1.19.7 Earth Bags

Brief comments on earth bags have been made before.

Typically, earth bags may be used for flood control or military bunkers; they are fabric sacks and tubes filled with soil, as mentioned. They can be plastered for the construction of walls, domes, etc. They may also be useful for foundations, especially in flood-prone areas. Further comments are provided separately in Chapter 2.

1.19.8 Wood

Some comments on wood have been made previously.

It is a popular material for projects, as indicated. Cross-laminated timber is prepared by placing layers of lumber in alternate directions and gluing them together. It is more stable and much stronger than regular wood, and it is lighter than concrete and steel.

More comments on wood are provided in Chapter 2.

1.19.9 Baled Straw

Baled straw may be used as an infill material or as a load-bearing arrangement for roofs. This material is renewable and may be preferred as a natural building material for various applications. It is lightweight and provides good insulation and sound absorption, and it is easy to work with. Furthermore, the work will require simple types of tools.

This item has been discussed further in Chapter 2.

Source:

- Robinson [22]

1.20 WHAT IS NEXT FOR TRADITIONAL BUILDING MATERIALS? [23]

The list of materials given below may be relevant:

Earth Building Materials, Concrete, Wood as a Building Material, Cement, Bricks, Plastic.

These materials may play an important role in decreasing the cost of construction materials while maintaining quality as per requirements.

Source:

- Lets Build [23]

1.21 APPLICATIONS OF EARTH STRUCTURES [20]

The applications of earth structures will cover various items, including Earth Shelters, Old American Earth Lodges, Mud Brick Buildings, Pyramids and Mounds, and Toulous.

A few well-known applications of earth structures are mentioned below with brief details [20]:

1.21.1 Earth Sheltering

The use of earth for sheltering has been employed for many years, resulting in energy-efficient dwellings. There may be various types of designs. Earth-sheltered dwellings may be underground, with an open courtyard to provide light and air. Earth houses may be built on slopes with windows and door openings on the sides. The buildings can also be constructed on the ground with earth mounded against the walls and finished with an earth roof.

Pit houses were built by Hohokam farmers between 100 and 900 AD in the Southwest of the USA. In Matmata, Tunisia, different ancient homes can be found, constructed with surrounding courtyards approximately 12 m (39 ft) square and about 12 m (39 ft) below the ground. The homes can be accessed through tunnels. Cliff-based dwellings for hot and cold weather may be found in Turkey, North China, the Himalayas, and the southwest USA. Similar applications using earth and other materials were made for Buddhist monasteries built into cliff sides or caves in the Himalayan regions covering Tibet, Bhutan, Nepal, and Northern India. These structures may not have been well maintained and can be risky.

By constructing earth houses into the ground, the houses will be cooler in warm temperatures and warmer in winter.

1.21.2 Wattle-and-Daub

This is an old method of construction where vines or smaller sticks are interwoven between upright poles and followed by plastering with mud mixed with straw and grass to create the walls. This method of construction has been used in different parts of the world, from the Nile Delta to Japan. Bamboo can also be used for making the wattle. It seems that wattle-and-daub houses were constructed 0.3–0.91 m (1–3 ft) below the ground. A different technique of construction known as bajareque is used in Colombia. In Britain, simple circular wattle-and-daub shelters were constructed in the past when suitable clay was available. In timber-framed buildings, wattle and daub can be seen as panels. The walls are not normally intended as structural elements, and the interior design was modified later using lath, plaster, and gypsum wallboards.

1.21.3 Native American Earth Lodge

This type is an old earth lodge and is a circular building constructed by native North Americans. They are made using a wooden post and beam construction method and are dome-shaped. The structural system consists of four or more central posts embedded in the ground, with cross beams connected at the top. At the centre, there is a smoke hole that is kept open. The structural arrangement may include larger ring beams supported on shorter posts in the central portion, connected by cross beams. The rafters radiate from the central cross beams to the outside cross beams. The structure is made with sticks, brush, or grass and is covered with a thick layer of earth or sod. The whole structure may be plastered with mud to give the appearance of a shell structure after drying.

1.21.4 Prairie Sod House

These were constructed by farmers in the prairies of North America by making a dugout cave in the side of a hill or ravine with a covering over the entrance. They built sod houses when time allowed. They used a plough to cut the sod into bricks measuring 0.3 m×0.61 m (1×2 ft), which were arranged to build the walls. The sod strips were placed grass side down and staggered like brickwork. The sod walls were constructed around door and window frames. The corners were secured by rods placed vertically through them. The roofs were constructed with poles or brush covered with prairie grass and finally sealed with a layer of sod. Sod houses can be strong and may last many years. Unless internal walls are plastered, they may appear damp and dirty. The roofs may leak and could collapse in a rainstorm.

1.21.5 Mud Brick Buildings

There are various applications of mud brick or abode buildings in different parts of the world. Shibam in Yemen is known for its ten-storied un-reinforced mud brick buildings and was designated a World Heritage Site in 1982.

Some examples are mentioned below:

- **Djinguereber Mosque of Timbuktu, Mali –** Built first at the beginning of the 14th century AD using round mud bricks and a stone–mud mixture, it has been reconstructed several times since.
- **Great Mosque of Djenne –** Located in southern Mali, this mosque is a prime example of Sahel mud brick architecture and was built in 1907. It is similar in design to an earlier Great Mosque that was first constructed on the site in 1280. Mud brick structures require maintenance work from time to time.
- **Casa Grande Ruins –** Now a national monument in Arizona, this site is protected by a modern roof system. It is a massive four-story adobe structure constructed by the Hohokam

people between 1200 and 1450 AD. The first European Jesuit priest, Father Eusebio Kino, visited the site in 1694. After that, it was abandoned for a long period. A temporary roof was provided in 1903. The adobe building remained empty and unmaintained for many years.

- **Huaca de la Luna –** This large adobe temple, located in Peru, was constructed by the Moche people. It reaches a height of about 32 m (105 ft) and features three main platforms, four plazas, and many small rooms and enclosures. The walls were covered and treated with attractive multi-coloured murals and friezes.

1.21.6 Mounds and Pyramids

Ziggurats are elevated temples constructed by the Sumerians between the end of the 4th millennium BC and the 2nd millennium BC, featuring a series of terraces leading to a temple with a height of about 61 m (200 ft) above the ground. The Ziggurat of Ur was also a significant project.

The largest ziggurat was located in Babylon, which was destroyed by Alexander the Great, leaving only the foundation of the structure. Originally, it stood 91 m (300 ft) high on a base of about 200 m (660 ft). Sun-dried bricks were used for the interior, while kiln-fired bricks were used for the facing. Clay or bitumen materials were employed to bind the bricks together.

The Pyramid of the Sun in Teotihuacan, in Mexico, which reaches a height of 63 m (207 ft), is also noteworthy. Construction began in 100 AD, and it was a stone-faced rammed earth structure.

1.21.7 Toulous

The Fujian Toulous is a type of rural dwelling for the Hakka people in the mountainous regions of southeastern Fujian in China, constructed between the 13th and 20th centuries. A toulous is a large, enclosed, and fortified rammed earth load-bearing wall building, typically rectangular or circular, and between three and five stories high. The walls are very thick, and a toulous can accommodate 80 families. Smaller interior buildings may be enclosed by large peripheral walls containing halls, storerooms, wells, and living areas. The structure may resemble a small fortified city. The walls are built by compacting earth mixed with stone, bamboo, wood, and other materials, and can be 1.8 m (6 ft) thick. These buildings were designed to be well-lit, ventilated, wind- and earthquake-proof, and to remain warm in winter and cool in summer.

1.21.8 Forts and Trenches

Earth constructions have been used to build fortifications for many years, creating strongholds and walls protected by ditches. Traces of earth fortifications can be observed from the Roman era through medieval times. Offa's Dyke, which stretches along the disputed border between England and Wales, is an example of significant earthwork.

Further Comments

- Trenches were useful for besieging forces to approach a fortification and provided protection from missiles.
- Sappers would construct saps or trenches.
- Excavated dirt was used to make a protective wall or gabion, which was common.
- The total trench depth and gabion height might be 2.4–3.0 m (8–10 ft).
- Saps or trenches were used for protection, as mentioned. Saps could also be used as tunnels and constructed a few metres below the surface.
- Sappers were highly skilled and well paid for the difficult and risky work.
- In the American Civil War (1861–1865), the use of trenches played an important role.
- An extensive network of trenches was constructed and adopted, stretching from Switzerland to the North Sea and used by the end of 1914.

- The soldiers had to live in trenches under difficult conditions of cold, dampness, and filth.
- The Germans used concrete blocks to create secured deep underground shelters with electrical lighting and heating.

1.21.9 Earthworks

The innovative engineering works were done with earth. The method of such activities had been carried out using earth generated through moving or processing quantities of soil or unformed rock. The material may be transported to a place to meet specific needs. A levee, flood bank, or stop bank may be used as a natural ridge or an artificially constructed dirt-filled wall for regulating and controlling water levels.

Examples include

- It may be an earthen wall that runs parallel to the course of a river in the floodplain or along low-lying coastlines.
- Mechanically stabilised earth (MSE) retaining walls have been used for embankments. Different types of designs for the wall have been tried.
- MSE walls may consist of concrete levelling pads, wall-facing panels, soil reinforcement, coping, and selected backfill material.
- The reinforcement for the earth may be tensioned polymer or galvanised metal strips or grids and is connected to the facing panels. The method of laying reinforcement over earth may be repeated with successive layers of panels, earth, and reinforcement. The panels may be tied into the earth embankment to provide a stable structure for maintaining balanced stresses.
- MSE has been used for a long time, and further development in its present form took place in the 1960s. The reinforcement types can vary and may include materials like steel and geo-synthetics.
- MSE construction will be labour-intensive but can be fast and inexpensive, not requiring high levels of skill. It may be suitable for both developed and developing countries.

1.21.10 Embankment Dams

An embankment dam may be constructed to create a massive artificial water barrier. Generally, it is developed by placing and compacting a complex semi-plastic mound of different proportions of soil, sand, clay, and/or rock.

It may have a semi-permanent waterproof natural covering on the surface and a dense waterproof core. With such treatment, the dams can be made impervious to surface or seepage erosion. The waterproofing issue is very important for the underlying foundation of the dam at the interface between the dam and the stream bed. The dams may be made of fragmented independent material particles. The friction and interaction of particles can bind the particles together into a single, stable mass instead of using a cementing substance.

The Syncrude Mildred Lake Tailings Dyke in Alberta, Canada, is an embankment dam approximately 18 km long and 40–88 m high. It is considered to be the largest earth structure in the world.

Source:

- Wikipedia [20]

1.21.11 Historical Monuments/Structures

There are many historical monuments/structures constructed with earth in different parts of the world.

Some of these include [24] -

- Temple of Ramses II at Gourna, Egypt
- The Citadel of Bam, Iran
- Sun Pyramid in Teotihuacan, Mexico
- Potala Palace in Tibet
- Fortified village of Ait Ben-Haddou, Ouarzazate, Morocco

The remains of rammed earth foundations, dating back to 5000 BC, were located in Assyria.

For these constructions, locally available excavated materials like clay, sand, and gravel were mixed efficiently and used.

Source:

- Vyncke, Kupers, and Denies [24]

1.21.12 Tarbela Dam [25]

Tarbela Dam was constructed over the River Indus near the small town of Tarbela in the Haripur District of Pakistan.

It is the second largest earth and rock-fill dam in the world, with a reservoir capacity of 14.3 billion m^3.

The project was managed by the Water and Power Development Authority on behalf of the Government of Pakistan. The construction work was executed under a single civil works contract, and the project began in 1968. The initial operations of the dam commenced in 1978. The project cost was $1.49 billion and was fully completed in 1984. The World Bank was involved in the 1960s, providing the main financial contributions for the Tarbela Dam project.

The project aimed to provide a significant reserve of water resources for the Indus basin, primarily for irrigation purposes.

A European consortium was involved from 1968 to construct Tarbela's system of dams, an open-air powerhouse, and the tunnels necessary for the main plant.

Dam – The dam is an earth and rock-fill structure constructed on the River Indus, equipped with a power plant. The maximum height of the main embankment from the lowest foundation level is 143.5 m (470 ft), and its length is 2,744 m (9,000 ft) at the crest.

The World Bank participated in the Tarbela Dam Project as part of the Settlement (Replacement) plan under the Indus Water Treaty of 1965. The construction of the dam was carried out in three stages to meet the diversion requirements of the river.

Furthermore, to fulfil the main purpose of the dam, which is water supply for irrigation, Tarbela Power Station has generated 438.978 billion kWh of affordable hydel energy since commissioning. During the years 1998–1999, a record annual generation of 16.463 billion kWh was achieved. The annual generation for the year 2013–2014 was 15.181 billion kWh, and the plant shared a peak load of 3,605 MW during that year.

Tarbela Dam was a joint venture project consisting of European contractors responsible for the construction. The project was led by the Italian contractor Impregilo.

Source:

- CE & CR [25]

1.22 BRIEF REVIEW OF CLAY AND EARTH MATERIALS [26]

The composition of clay and earth materials can vary significantly depending on local sub-soil conditions. The sub-soil composition may indicate different ratios of clays and aggregates, such as sands, gravels, and silts. Clay-based materials are likely to be plastic in nature when moist and will be free from organic materials unless added intentionally. Turf materials from the topsoil will contain organic matter.

Additives in the form of chopped straw are often used with the soil, as previously mentioned. Other plant types may also be used. For internal plaster, animal hair has been used with clay.

1.22.1 Clay-Rich Materials [26]

The addition of vegetable fibres or hair may be made to provide tensile strength clay, which will reduce the problems of shrinkage during drying. Other additives, including dung, have been used to change the properties of earth materials for different applications. Similar materials may be used for repair work when required.

Local variations in clay and earth materials used for construction may be present.

Proper analyses and tests will be required to assess the suitability of the material, especially for repair work.

It will be possible to analyse samples of earth and clay materials to assess the type of materials used initially and any other materials that have been added to the mix. The list may indicate the use of aggregates and fibrous re-enforcement, such as straw or hair, in clay/earth to provide better binding. It is likely that the original earth building materials used for construction would have come from a source close to the construction site. There may also be suppliers who stock similar earth-based building materials for the purpose of supplying replacement materials.

Test panels may be constructed and studied for confirmation.

1.22.2 Lime Harling and Plaster [26]

The external surfaces of clay and earth masonry structures may be treated and finished with thin lime harl or lime wash. If present, the surface should be treated with a lime finish after completing the repair work on the earth masonry surface. Adequate time should be allowed for drying the earth material before the application of lime harling. Cement harling should normally be avoided on earth surfaces, as this treatment is likely to be impermeable and inflexible, leading to cracking and potentially allowing moisture to penetrate behind the render, which may prevent the evaporation of moisture. This may have long-term effects, such as loss of cohesion in earth-based materials and can damage the earth structure.

Source:

- Earth and Clay Construction [26]

1.23 MAINTAINING EARTH STRUCTURES [26]

Care must be taken to prevent water from entering the fabric of clay or earth structures, which may cause damage later. It is to keep the roof of such a structure in good condition, ensuring that rainwater outlets are well-maintained and that no water collects on the roof after rainfall. Earth structures are normally constructed on a plinth of stonework to control dampness rising from the ground. When this detail is not present in the original structure, it is better to avoid providing a damp proof course during the repair operation, as it may act as a barrier and collect or concentrate moisture at the base of the wall. Existing details and treatments must be maintained properly during the repair work if the same treatment was used before with lime harling and/or lime wash.

To control excess moisture around the wall, the use of a French drain may be considered, as this could be more effective.

Source:

- Earth and Clay Construction [26]

1.24 REPAIRING EARTH BUILDINGS [26]

The following steps will be relevant:

- A detailed survey of the building and the surrounding areas must be conducted.
- Note any changes that may have been made since the original construction and whether the structure is less protected against rain due to these changes.
- The problems are to be identified carefully, such as ground drainage, defects in the roof structure including roof drainage, the condition and composition of external coatings/treatments, and the form and material used for construction.
- The deterioration in the base course will require special attention. The repair work should be done to ensure that the wall does not deform. The use of lime or clay mortar for re-bedding masonry may be the proper choice when repairing the damaged base course.
- The external ground level and the plinth course should be checked. The external ground levels should not be higher than the masonry. Water cannot be allowed to enter by passing above the original stone base course. If necessary, the ground levels should be lowered below the level of the original base course. The base course may be repaired or re-pointed as required.
- Care should be taken to repair earth walling using materials and methods that match the original construction.
- Patch repairs to earth structures with cement-based renders may not be effective and can cause additional problems related to moisture movement in the adjacent areas of the building, potentially leading to damage.

Source:

- Earth and Clay Construction [26]

1.25 PHOTOGRAPH

A photograph of earth buildings is presented in Figure 1.34.

FIGURE 1.34 Earth buildings.

REFERENCES

1. Ganguly K K, Durable concrete will be more suitable for different applications in projects, *Indian Concrete Journal (ICJ)*, July 2017, Vol. 91, pp. 80–96.
2. Ganguly K K, Reinforced concrete design, repairs/restoration, strengthening of RC structures and concrete technology, *Scientific International*, January 2020, pp. 1, 208–233, 313–365.
3. Auroville Building Center, A Training Manual for Technicians and Entrepreneurs, Production and use of Compressed Earth Blocks, Auroville. July 1992.
4. Auroville Earth Institute. Visitors Center, finished in 1992 in Auroville.
5. Doat P, Hays A, Houben H, Matuk S and Vitoux F, *Building with Earth*, Translated by Asha Puri, The Mud Village Society, New Delhi, India, 2021.
6. Biju A, 10 Construction Techniques Used in Earth Architecture, RTF, 2021, pp. 1–29.
7. Williamson J A, *Earth Construction: Poured Earth*, University of Colorado, 2006, pp. 1–26.
8. Alexander Sumerall, Earth Building Techniques - Building Your Home with Earth, 2019, pp. 1–22.
9. Gray J, Stone Construction, Home, Types of Construction, September 2015, pp. 1–2.
10. Dry Stone - Wikipedia, the free encyclopaedia, May 2016, pp. 1–8.
11. Masonry - Wikipedia, the free encyclopaedia, 2015, pp. 1–7.
12. Lee Jian Ru, Darren Loong Chi Yoong, Chong Kai Xiang, Kiing Kiu Chun et al., Ceramic Presentation of Building Materials, Ceramic-Group Presentation of Building Materials, June 2015, pp. 1–7.
13. Marion Ingle, Polymers in Construction, SANDBERG, pp. 1–2.
14. NGCC Steering Group, Polymer Composites as Construction Materials, DTI, pp. 1/1.
15. Walkus R and Kowalski T G, Ferro cement: a survey, *Concrete*, February 1971, Vol. 5, pp. 48–52.
16. Mehta P K and Monteiro P J M, *Concrete, Microstructure, Properties, and Materials*, June 1997, Indian Concrete Institute, supported by Cement - Ambuja.
17. Hollaway L, *Polymer Composites for Civil and Structural Engineering*, Blackie Academic & Professional. First edition 1993.
18. Ganguly K K, Fibre Reinforced Concrete, ACE-2011, KL University, Vijayawada, AP, October 21–23, 2011.
19. Building Materials, Britanica, Britannica.com, 2019, pp. 1–5.
20. Earth Structure - Wikipedia, the free encyclopaedia, March 2016, pp. 1–17.
21. Gupta M, Seven (7) Most Commonly Used Construction and Building Materials in India, November 2017, pp. 1–7.
22. Robinson J, 9 Ancient Natural Building Materials We Still Use Today - Super draft, 2019, pp. 1–7.
23. Lets Build, Ten (10) Innovative Construction Materials That Could Revolutionise the Industry - Infographic, May 2019, pp. 1–19.
24. Vyncke J, Kupers L and Denies N, Earth as building material - an overview of RILEM activities and recent innovations in geotechnics, *MATEC Web of Conferences,* 2018, Vol. 149, pp. 1–12.
25. Tarbela Dam, Civil Engineering and Construction Review, Infrastructure & Civil Engineering - CE & CR, 2020, pp. 1–9.
26. Earth and Clay Construction, *INFORM, Information for Historic Building Owners*, Historic Scotland, ALBA AOSMHOR, June 2015.

2 Earth/Clay, Alternative Building Materials, Cob, Wattle-and-Daub, and Some Other Methods

2.1 EARTH ENGINEERING FOR STRUCTURES [1]

The structures constructed in this category were made mostly with soil. The soil has been used for construction for a long time by architects and engineers. Soil is a variable material and can be improved by adding other materials or by compressing or baking it to achieve better properties, including strength, as highlighted separately. Soil is widely available in all countries and is an economical material, with low environmental impact during and after construction for different applications.

The materials for earth structures may include mud or mud mixed with straw to create cob. Strong residential units may be constructed with sod or turf. Stabilisation of soil may be achieved by including lime or cement and compacting it to produce rammed earth construction. The construction process may be made quicker by adopting pre-formed mud bricks or adobe, compressed earth blocks, earth bags, or fired clay bricks and other blocks as mentioned. Some of these items will be discussed further in subsequent chapters.

Source:

- Wikipedia [1]

2.2 IMPROVEMENTS NECESSARY FOR EARTH ARCHITECTURE, DESIGN AND CONSTRUCTION [2]

The items mentioned below will be relevant:

- To improve traditional methods – Studies including tests and experiments will be necessary to develop guidelines for better design and construction with earth.
- To study soil characteristics and to determine the suitability of the materials for building construction.
- To develop and adopt proper equipment for efficient use and for the production of earth made construction materials.
- The study areas and research information must be transferred to the people who will be involved with the use of the materials for construction at different sites.
- To communicate and educate decision-makers with knowledge from the study/research so that earth may be accepted as a building material.

Source:

- A Training Manual, Auroville [2]

DOI: 10.1201/9781003319108-2

2.2.1 Comments on Soil [1]

Soil is developed from rock that has been chemically or physically weathered, transported, deposited, and precipitated in different places. The soil particles generally consist of sand, clay, and silt. The sand particles are larger, i.e., 2–0.05 mm in diameter, and the smallest particles will be clay with a diameter of less than 0.002 mm.

Sand and silt are normally inert rock particles and will also include quartz, feldspar, calcite, and mica.

The clay typically consists of phyllosilicate minerals with a sheet-like structure. The small clay particles will interact with each other both physically and chemically. A small proportion of clay may affect the physical properties of soil. Clays such as kaolinite will not expand or contract under wet or dry conditions and may be useful for manufacturing bricks. Other types, such as smectites, may expand or contract significantly when wet or dry and may not be suitable for buildings.

Further comments on soil have been made separately in Chapter 3.

Source:

- Wikipedia [1]

2.3 IDENTIFYING CLAY AND EARTH CONSTRUCTION [3]

With rendered surfaces, it may not be easy to identify earth construction. This may be noticeable, in some cases, when the rendered surface is damaged or detached from the building. Other indications may include a rubble stone base course at the bottom of the wall or a high slope provided to the wall. Earth-based building materials need to be properly identified to understand the construction type and composition of the wall, and this information will be important for planning and executing repair work, including the selection of repair materials. Analyses, tests, and consultations with experts may be necessary for making correct and reasonable decisions.

Source:

- INFORM, Historic Scotland [3]

2.4 REQUIREMENTS OF BUILDING MATERIALS INCLUDING ALTERNATIVE ONES

It is expected that the ideal building materials, including similar alternative ones for construction, will satisfy the following issues to the maximum extent possible for efficient and successful applications:

- The material, including the ingredients, should have low embodied energy covering delivery to the construction site.
- Environmentally friendly to meet green building requirements.
- Should be a low-cost material, and comparisons of costs with other materials may be relevant/important.
- Easily available
- Long-lasting and sustainable
- Eco-friendly
- User-friendly
- It would be beneficial if the material is biodegradable
- Renewable
- Can be recycled and reused
- The material should be lightweight

- The construction should be carried out without major problems. The material may be used with minimal training for workers, if required.
- Lower impact on the environment.
- Special or heavy equipment should be avoided to a large extent for construction.
- Local labour may be utilised for execution.
- Execution time will be important for efficient operations; it should not be too long.
- Hazards and safety during execution should be addressed.
- Durability under different environmental and load conditions should be satisfactory. The structure should satisfy different limit states as per design assumptions.
- The mechanical behaviour of the material, including the ingredients, should be suitable for meeting design requirements. The material should have adequate strength to provide reasonable structural dimensions against design loads, comparable to those of other alternative materials.
- No health hazards should be present in using the material.
- The availability of different test results on the performance of the material will be an advantage and may be required or referred to.
- Records/experience related to the performance of the material over a long period will be important.
- Guidance on design and construction from codes will be relevant/required.
- Repair and maintenance work will be nominal and can be done easily.
- Permeability must be low against moisture and harmful chemicals/gases.
- Fire resistance should be adequate.
- Resistance against insects, termites, moulds, etc.
- Good insulation properties for heat and sound are required.
- Good performance against impact and abrasion is essential.
- The life cycle cost will be within reasonable limits compared to other materials.
- Finally, the structure should be aesthetically pleasing after construction without excessive spending on finishing items.

It may not be possible for all building materials, including alternative materials, to satisfy all the above requirements when selected for a project.

In this connection, the role of building materials covering sustainable and green building design/construction requirements, CO_2 emissions, embodied energy, reduced waste, efficient use of materials, low cost, health protection, etc., has been reviewed to some limited extent as described below:

2.5 SUSTAINABLE BUILDING DESIGN

2.5.1 Some Basic Issues of Sustainable Building Design

The relevant items are listed below [4,5]:

2.5.1.1 Proper Use of the Site Potential

Proper site selection will be important, covering the reuse and rehabilitation of existing buildings as applicable. Factors such as location, orientation, and landscaping of a building may influence the local ecosystem, energy use, and transportation system. Efforts will be required to control and reduce rainwater runoff.

2.5.1.2 Water Resources to Be Protected

The management of waste and drinking water must be done efficiently to minimise effects on the fresh water supply [4–7]. Stormwater and drainage in the surrounding area should be monitored to ensure proper water utilisation without waste.

The water usage of the building must also be examined.

2.5.1.3 Efficient Use of Energy

The items are as follows:

- To reduce energy use at all stages of the building's life cycle [7] and to aim for operating at zero energy [4]. This will reduce the reliance on fossil fuels [4].
- Use renewable and low-carbon technology for energy supply, utilising built-in and natural efficiencies [7].

2.5.1.4 Indoor Environmental Quality to Be Improved [4,5]

Indoor environmental quality must be maintained in good condition for the well-being of the users. Important factors will include adequate daylight, good acoustic performance, proper ventilation and moisture control, comfortable temperature, and avoiding materials with high emissions of volatile organic compounds (VOCs).

2.5.1.5 Material and Building Space Are to Be Properly Utilised [4,5]

For sustainable building design, proper use and reuse of materials will be important throughout the life span of the structure. The selection of sustainable materials may mitigate impacts on the environment and human health. Building space must be utilised effectively in the design to meet requirements.

2.5.1.6 Optimise Operational and Maintenance Activities [4,5]

Key considerations in building design would include safety, security, cost-effectiveness, and a productive environment.

The engineers, architects, and material manufacturers may recommend materials for sustainable design with reduced maintenance activities for the building.

Following their guidance, it will be possible to reduce water and energy requirements while avoiding the use of toxic chemicals in the design and construction of the building, thereby creating a safe environment.

Furthermore, the whole procedure is expected to be cost-effective and will reduce life cycle costs.

Source:

- 6 Fundamental Principles [4]
- WBDG [5]
- The benefits of green buildings [6]
- World Green Building Council [7]

2.5.2 Further on Sustainable Architectural Design Approach [8]

It may be relevant to comment that the design approach must avoid any harmful effects on the environment from building design and construction. Architects and engineers must take into account various factors such as planning, energy management, landscaping design, stormwater management, and the use of appropriate building materials to achieve and contribute to sustainable design. Some related items will be briefly mentioned.

2.5.3 Passive Sustainable Design

For window placement, sun orientation and climate conditions will need to be considered, along with daylighting and natural ventilation. In some cases, thermal mass may be utilised, where thick walls are provided to absorb heat from the sun during the day and release it into the building at night as needed.

2.5.3.1 Green Building Materials and Finishes

Proper use of green building materials must be ensured. Building material manufacturers may be consulted for guidance.

2.5.3.2 Active Sustainable Design

Small environmental footprints may be achieved by providing efficient electrical, HVAC, and plumbing designs.

2.5.3.3 Renewable Energy Systems

A good option for the building would be to use solar and wind energy wherever possible.

2.5.3.4 Native Landscaping

Different and efficient uses of trees, plants, and grasses may be included under this heading. The use of trees for shading roofs and windows should be considered during hot weather.

2.5.3.5 Stormwater Management

When water from the surface area does not evaporate, it will be absorbed or will go to the ground to replenish the water table. By efficiently designing the stormwater system, runoff water may be retained in ponds. The collected water may be released to the ground slowly or may be used for other purposes.

Source:

- Sustainable Architecture Strategies [8]

2.5.4 Review for Possible Scope for Sustainable Architectural Building Design [9]

The items may be summarised as follows:

2.5.4.1 Use of Open Rain Screen

An open rain screen will allow air to vent through the siding and help pull out stagnant air.

2.5.4.2 Community Outdoor Space

Locally available wood for the deck spaces and raised benches may be used.

2.5.4.3 Natural Siding Materials

Wood cladding, along with tree canopies, may be considered for protecting the environment.

2.5.4.4 Renewable Hydraulic Energy Production

This may be considered along with other types where possible.

2.5.4.5 Rooftop Decks and Terraces

Flat spaces should be utilised. They may be designed as gathering spaces and areas for green roofs, treated with planting trees. This will also reduce the direct impact of sunlight and help keep the building cool.

2.5.4.6 Low-Maintenance Exterior Siding

Long-lasting and low-maintenance siding materials may be selected to ensure the long life of the structure.

2.5.4.7 Use of Eco-friendly Design

An eco-friendly environment must be developed to bring communities together, and it should be approachable.

2.5.4.8 Small Spaces, Big Ideas

The carbon footprint of the home should be lowered without sacrificing design and functionality.

2.5.4.9 Efficient, LED Lighting

The quality of indoor air and light will be important for the design. An LED lighting system may be considered.

2.5.4.10 Using Recycled Materials

Recycled plastics have been used to construct boats for water bodies. Other recycled materials may be used in the design.

2.5.4.11 Sustainable Wood Decking

Wood decking may be applied to pedestrian bridges and other suitable areas as applicable.

2.5.4.12 Environment Not to Be Disturbed

The site should be kept green, open, and natural. The footprint of man-made structures should be minimised to maintain a healthy environment.

2.5.4.13 Natural Light

A sufficient amount of natural light should be introduced into the interior to reduce the need for artificial lighting.

2.5.4.14 Traditional Materials to Be Used in Innovative Ways

Different traditional materials may be used efficiently, in the design, with proper detailing for aesthetics.

2.5.4.15 Natural Ventilation

To reduce energy consumption, attempts may be made to lessen reliance on active heating and cooling systems. Natural cool air (to the extent possible) should be allowed to flow freely through the interior using openings/windows.

2.5.4.16 Wood Pool Decking

The sustainable design may include the use of a wood deck surrounding the pool for a natural look.

2.5.4.17 Passive Heating and Cooling

The use of sustainable wood siding with strategic solar shading may be considered in the design.

The shape of the structure will also need careful planning, considering buildable space.

Source:

- Sustainable Architectural Design [9]

2.6 GREEN BUILDINGS AND DESIGN [4,10]

The design and construction of green buildings will be very relevant for engineers/architects, in the current situation, to protect the environment and the community in the future. In this connection, a few items will be reviewed/presented including the advantages of green building design.

2.6.1 What is Green Building and Why It is Important [4,10]

Green building, along with sustainable design, will be connected with the use of suitable building materials in the design, which will have a nominal impact on the environment and on human health throughout the total service life of the building [4]. This will be very important for future consequences

Furthermore, it will include other items like construction, operation, maintenance, repair/renovation, and eventual demolition [4]. It will also efficiently utilise land, water, energy, and natural resources while protecting/restoring the surrounding habitat to sustain life [5].

2.6.2 More on Green Building Design [5,6]

The following items will be relevant:

2.6.2.1 Maximum Reuse and Minimum Waste

The items [4,7] include

- Less and more durable materials are to be used, generating less waste.
- Demolition waste recovery and reuse should be practised.
- Building users are to be trained to recycle and reuse
- Avoid using materials with high emissions of VOCs.

2.6.2.2 Making Our Environment Green [7]

- Creating more green spaces in urban areas
- More landscaping for the surrounding areas of the buildings on polluted land.

2.6.2.3 Safeguarding Health and Well-being

The items are [7] as follows:

- Fresh air must be brought inside. Toxic emissions to the outside must be avoided.
- Maximise natural light while considering user comfort to reduce lighting energy requirements.
- Acoustics and sound insulation will be important for education and health.
- Comfort with the right indoor temperature must be ensured through passive design, building management, and monitoring methods.

2.6.2.4 Designing for Resilient and Flexible Structures [7]

The items are as follows:

- Resilient structures must be designed for flooding, earthquakes, and fire to ensure the safety of building users.
- Spaces should be designed for flexibility, considering changes that may take place in the future. Demolition or re-building of the structure should be avoided to the extent possible.

2.6.2.5 Considering All Stages of a Building Life Cycle [7]

- The entire life cycle stages of the building, i.e., design, construction, operation, maintenance, and finally demolition, must be considered.
- It is essential to check and ensure that the embodied energy or water used to make the materials and transport them to the building site is controlled and reduced to achieve a low impact on the environment.

2.6.2.6 Connecting Communities and People [7]

- Create environments that connect and engage the community in economic and social activities.

- The transport system and the distance to amenities may be important. Environmentally friendly options like walking or cycling should be encouraged.
- Smart information communication will be relevant. The requirements for smart electricity grids and towers for internet facilities should be considered.

Source:

- GREEN BUILT [10]
- 6 Fundamental Principles [4]
- World Green Building Council [7]
- The benefits of green buildings [6]

2.6.3 Review of Advantages, Green Buildings

There are many benefits that can be achieved with green building design and construction. A few of them will be addressed here.

2.6.3.1 Environmental Benefits [10]

These are as follows:

- Increases and protects biodiversity and ecosystems
- Reduces waste with proper design
- Improves air and water quality
- Saves/restores natural resources

2.6.3.2 Improved Quality of Life [11]

The surrounding environment significantly impacts our health. The improvement of health may be expected for building users from green building design, such as the application of improved lighting sources, ergonomic features, thermal conditions, and upgraded air quality. Experience has shown that health stress levels and overall quality of life have improved [11].

2.6.3.3 Saving of Water [6,11,12]

Water waste may be reduced [11] through efficient design of plumbing fixtures and minimising strain on shared water resources. Water recycling using purifiers and alternative water sources like rainwater will be relevant. Recycled rainwater and greywater from kitchens and bathrooms may be used for toilet flushing and landscape requirements [12].

Recycling and reuse of water in buildings may have a large impact on lowering water consumption [6]. Water may be treated in a ground tank using a filter to reduce the harmful effects of wastewater on the surrounding areas [6].

2.6.3.4 Energy Efficient [11]

The energy-efficient design will be the main objective of green building design.

The use of energy from natural sources like the sun, wind, and water will be very helpful for the environment and for protecting the ecosystem from the pollution associated with the use of non-renewable types, i.e., energy from coal and oil. Furthermore, non-renewable energy sources may be more costly and toxic compared to energy from natural sources.

2.6.3.5 Economic Benefits [10]

These are as follows:

- Will reduce operating costs
- Will improve occupant productivity

- Increase the value of assets and profits
- Life cycle economic performance may be optimised

2.6.4 Further on the Use of Natural Energies [6]

The use of renewable energy sources like solar heaters, wind energy, solar photovoltaic panels, bio-energy, and other types of clean energy will help keep the environment clean. The careful selection of windows, insulating glass, and insulating materials for other areas such as walls and ceilings will be very important.

2.6.4.1 Social Benefits [11]

It may be mentioned that

- Will improve indoor air quality as mentioned previously
- Health and comfort will be increased
- Strain on local utility infrastructure will be reduced
- The quality of life will be improved.

2.6.4.2 Extended and Better Health [6,11]

People living in buildings made with eco-friendly materials can enjoy better health. Many pollutants from the air inside the building will be eliminated through green building design.

2.6.4.3 Less Waste [6]

Green architecture will mean less waste by using renewable and sustainable materials like sustainable wood and plant materials. Green construction will include and cover proper growth while considering safety and human health.

2.6.4.4 To Maintain Cleanliness, Protecting Our Ecosystem [11]

Sustainable and green architecture can reduce our dependence on non-renewable energy resources (i.e., oil and coal) and can promote the maintenance of a clean environment.

2.6.4.5 Preserving Our Infrastructure [12]

By properly using energy and water in buildings, the capacity of local infrastructure will be preserved.

2.6.4.6 Cost/Money Saving [6,12]

The cost of green building will be comparable and may be high for some items. It may be prudent to consider the lifecycle cost of the building when assessing the savings.

2.6.4.7 Efficient Use of Material – Minimum Waste [11]

By recycling and reusing resources and tapping into existing resources, we can reduce carbon footprint and protect natural resources. By following this approach, it may be possible for engineers and architects to achieve sustainable design.

Preservation of natural resources like water and wood and protecting air quality by controlling energy consumption efficiently and in a limited manner will be important for the design of buildings.

2.6.4.8 Carbon Foot Print Reduction [11]

It will be a positive action to attempt to increase sustainability in design for businesses and society as a whole through green initiatives.

2.6.4.9 Improved Productivity [6]

Studies have shown that people living in green buildings with healthy conditions are more productive.

2.6.4.10 Sound Design and Noise Avoidance [6]

The sound effects on human beings may be of two types:

- **Good Effects –** produced by pleasant sounds
- **Harmful Effects –** loud sounds and noise from sources such as television, washing machines, and electrical appliances.

Noise from outside can enter the building through open windows, doors, and small openings.

2.6.4.11 Method of Cutting Out Noise Inside the Building [6]

This may be done by

- Increasing the distance between the noise source and the construction
- Planting trees near the source of noise
- Using materials that are not affected by noise in the vicinity of the noise source
- Using noise barriers to absorb sound

2.6.4.12 Popularity of Sustainable Architecture [11]

Considering the environment, the popularity of green architecture is growing. Efforts are being made to protect our ecosystem and reduce carbon footprints through green architecture and construction. This is essential for our survival and for the protection of future generations.

2.6.4.13 Aesthetics

Elegant, visually appealing structures can be constructed with green building design. This approach may serve as a fundamental solution for housing issues in many areas.

2.6.5 Disadvantages of Green Buildings [12]

2.6.5.1 Location [12]

The design of the building will primarily depend on sunlight for energy. The location of such buildings may interfere with neighbouring houses.

2.6.5.2 Availability [12]

The materials for construction may not be readily available in areas where environmental protection is not a priority. The transportation of these materials can also be costly.

2.6.5.3 Long Time to Construct [12]

Green building design and construction may take a longer time. Surrounding environmental conditions must be studied first.

2.6.5.4 No Air Cooling Features [12]

Heat will be required to generate power for the buildings. Air conditioners may be necessary in hot environments, which could affect eco-friendliness.

2.6.5.5 Cost of Construction [12]

The construction of green buildings may be more expensive compared to regular buildings, as previously mentioned. Natural materials used in the design may not be available in all locations at all times.

Source:

- 10 Benefits of Green Building [11]
- GREEN BUILT [10]

- Blog Weetas [12]
- The benefits of green buildings [6]

2.7 ALTERNATIVE BUILDING MATERIALS FOR GREEN CONSTRUCTION

A few materials, as indicated in this context, may be considered alternative types that have a lower impact on the environment when used for construction purposes. Some of these materials are listed below:

2.7.1 Straw Bale [13]

Using straw bales can provide better insulation against climate change. They can be used in their raw state without further processing, making them an affordable option. Buildings constructed with straw bales will have good insulation for climate variations. This is a low-impact and low-carbon building material. However, straw bale construction may not be suitable for areas with extreme humidity and high rainfall.

2.7.1.1 Further on Straw Bale [14]

The method of construction will require bales of straw, such as wheat, rice, rye, and oats, to be used as structural elements as well as for building insulation. This represents a sustainable method of building construction.

The advantages may include the renewable nature of straw, its easy availability, low cost, and its properties as a natural fire retardant with good insulation value. The disadvantages may include susceptibility to rot and high space requirements for the straw itself.

Straw houses have been built in the African plains in the past. Straw bales were used in Germany 400 years ago. Thatched straw roofs have been used in many countries.

2.7.1.2 Method of Construction [14]

The method involves stacking rows of bales on a raised footing or foundation with a moisture barrier between the bales and the supporting platform.

The bales may be bound together with two or three strings, with three-string bales being larger in size.

The bale walls may be tied together with wooden or bamboo pins or with surface wire meshes. The surface may be stuccoed or plastered using either lime-based material or earth/clay rendering. The bales may provide the structural support for the buildings. The plastered bale system may be designed to withstand lateral loads from wind and seismic effects.

Alternatively, the bale house may have a structural framework using other materials like timber or lumber, with the bales acting as a plastered surface providing insulation. For wet weather conditions, a proper vapour-permeable treatment will be required, and a cement-based finish will not be necessary.

Furthermore, a skeleton framing arrangement of wood or metal may be considered for the construction of the roof prior to raising the bales, protecting the bale walls during construction.

A combination of load-bearing systems and framing may be implemented, referred to as hybrid straw bale construction.

The inclusion of mass and insulation may be suitable for passive solar building design for winter and summer.

The compressed and plastered straw bale walls are fire resistant.

2.7.1.3 Disadvantages [14]

The main issues in this context are as follows:

- Moisture
- Mould

The walls must be protected against rain as mentioned. When exposed to water, compressed straw may expand due to moisture absorption, potentially causing cracking later and allowing moisture go inside.

Furthermore, the walls may be damaged by mould, which releases toxic spores into the wall cavities and the air.

Rats and mice can enter straw bale homes during construction.

There may be breathing difficulties from straw dust for individuals with allergies to straw or hay.

Source:

- Mahendriyani [13]
- Wikipedia [14]

2.7.2 Hempcrete

This material may be prepared from the woody, balsa-like interior of the *Cannabis sativa* plant ([13]) mixed with lime and water. Hempcrete can be mould-free and pest-resistant, and is almost fireproof. The material provides natural airtight, breathable, and flexible insulation. This is a sustainable building material, as hemp can be grown and replenished quickly.

2.7.2.1 Review on Hempcrete [15]

Hempcrete is a biocomposite material made from a mixture of hemp hurds (shives), sand, lime, and pozzolanas, which can be used for construction and insulation work.

It may be easier to work with hempcrete than with lime mixes. It is not brittle like concrete, and expansion joints may be avoided in construction.

The lime-based binder can be natural hydraulic lime or hydrated lime.

Hydraulic binders may be combined with regular hydrated lime to reduce the setting time.

Hempcrete might take a few weeks to a month to gain sufficient strength. A small amount of cement or pozzolanic binder may be used to reduce the setting time. The material, as processed, will be solid, light, and durable.

For block making, the mixture of lime, hemp, and water is prepared, and blocks may be produced using a machine. The blocks are cured and packed with polyethylene packaging film after proper curing and made ready for dispatch to sites.

There may be two types of uses for hempcrete construction:

- By using forms, where hempcrete may be cast or sprayed directly into place at the site.
- The other type involves using pre-fabricated blocks similar to masonry construction.

The material may be applied between timber frame construction, drywall, or plastered to increase durability and aesthetics.

The typical compressive strength of the material may be about 1 MPa. It is a low-density material, resistant to cracking when subjected to movement, and may be suitable for seismic zones.

Hempcrete walls must be used along with the frames of another material to support the vertical loads of the building. Hempcrete walls will offer good acoustic performance, be fireproof, resist mould, and transmit humidity. Hempcrete material may be used as a filling item for infill walls, as mentioned. The hempcrete blocks may need to be treated with a finishing plaster using the same hempcrete mixture with different mix proportions. Hempcrete walls should be constructed with a joint between the wall and the supporting ground to avoid capillary rise of water. Hempcrete blocks should be used above ground level. The external surfaces of the walls may be protected against rain with lime and sand plaster. The interior surfaces of the walls need not be treated similarly.

Hemp will absorb CO_2 from the atmosphere while growing. Hence, the material may be considered as carbon-negative material.

Source:

- Mahendriyani [13]
- Wikipedia [15]

2.7.3 Grasscrete

This will be a green alternative treatment to concrete surfaces, i.e., for parking areas, driveway, and access roads; and benefit drainage; and improve stormwater absorption. It will be a natural bio-filter for eliminating pollutants. This can be referred to as sustainable urban drainage method [13].

2.7.3.1 More on Grasscrete [16]

This is a reinforced concrete system with voids created in the middle of the block.

The advantages of this type over the pre-cast concrete and plastic system include

- Control differential settlements.
- May a take load of up to 40 tonnes gross weight of the vehicle.
- Can reduce sub-base depths.
- Avoid kerb edges.
- No reliance on grass for stability.
- The drainage capabilities of the area will be maximum.
- Lightweight types of grasscrete moulds can be transported and used suitably in other areas.

2.7.3.2 Applications [16]

Some of the applications include

- Emergency access
- Maintenance access
- Vehicle parking areas for daily or overspill applications
- Highway verges and pull-ins

The self-venting paving system with the capacity of taking the traffic loads may be considered very suitable for water engineering covering

- Reservoirs
- Storm/drainage channels
- Spillways
- River and sea defences
- Swales

The applications for the parking areas are shown in Figures 2.1–2.6:

2.7.4 BAMBOO as a Building Material [13]

Bamboo is very useful for various projects and has been used for a long time. It is an eco-friendly material for green construction. Bamboo produces more oxygen and absorbs more carbon dioxide, which is necessary for controlling global climate change. Bamboo can be easily grown and is a cost-effective construction material. Further comments on this have been made separately.

FIGURE 2.1 Car parking. (Source: Mahendriyani [13]; Grasscrete Paving [16]).

FIGURE 2.2 Car parking. (Source: Mahendriyani [13]; Grasscrete Paving [16]).

FIGURE 2.3 Car parking. (Source: Mahendriyani [13]; Grasscrete Paving [16]).

FIGURE 2.4 Car parking. (Source: Mahendriyani [13]; Grasscrete Paving [16]).

FIGURE 2.5 Car parking. (Source: Mahendriyani [13]; Grasscrete Paving [16]).

FIGURE 2.6 Car parking. (Source: Mahendriyani [13]; Grasscrete Paving [16]).

TABLE 2.1
Basic Information

Serial No.	Item	MPa	MPa	kg/m^3
1	Bamboo	Tensile strength = 120	Compressive strength = 55	Density 700
2	Concrete M-30 Grade	Tensile strength = 3.8	Compressive strength = 38	2,400
3	Mild steel	Ultimate strength = 410	Yield strength = 250	7,850

2.7.4.1 Additional Comments on Bamboo [17]

Some important information includes

- Production of 1 tonne of cement emits more than 1 tonne of CO_2 into the atmosphere.
- Production of 1 tonne of steel emits more than 2 tonnes of CO_2 into the atmosphere.

2.7.4.2 Advantages of Bamboo

- Production of 1 tonne of bamboo consumes more than 1 tonne of CO_2 from the atmosphere.

A few basic pieces of information are provided in Table 2.1.

Bamboo offers a competitive strength-to-mass ratio.

Disadvantages – The material may be attacked by termites. This problem can be addressed by providing suitable protective treatment.

2.7.5 Design Method

The following points should be noted:

The application such as for a shed or industrial structure:

- Different loads and combinations need to be considered in the design
- The working stress method of design may be employed
- Linear/elastic behaviour is assumed
- Factor of safety = 4
- Allowable stresses
  Tension = 30 MPa
  Compression = 13 MPa (l/r = 80)

2.7.5.1 Summary of Comments

- Analyses can be conducted using a conventional approach for the frame.
- Different loads must be considered as stated.
- The working stress method of design should be followed.

2.7.6 Simple Uses

The simple uses may include the design of bamboo structures suitable for cottage industries, cattle sheds, warehouses, and poultry sheds, etc.

Source:

- Bhalla [17]

2.7.6.1 Other Applications

It should be noted that delicate and elegant structures have been successfully constructed with bamboo.

Two-storey buildings with split bamboo material used as reinforcement for the concrete columns have been attempted. Steel bars have been used for the stirrups in the columns. Brick walls have been placed between the columns at the periphery.

2.7.7 Guidance from IS Code

The following IS codes can be referenced for designing with bamboo:

2.7.7.1 IS 15912: 2012 – Structural Design Using Bamboo – Code of Practice

This code covers the design of structural bamboo in buildings, focusing on mechanical resistance and durability of the structures. It addresses minimum strength data, grading requirements, dimensional stability, and traditional bamboo for the structures.

The following items are not covered in the standard:

- Bamboo foundations.
- Scientific design of bamboo joints and their fastenings.
- Limit state design and performance of structures.
- Bamboo-reinforced cement concrete/mud structures.

Further comments on bamboo will be made separately.

2.7.7.2 IS 6874: 2008 – Method of Tests for Bamboo (Amendment–June 2012)

This code addresses the standard tests for bamboo related to the physical and mechanical properties of round bamboo.

These tests include

2.7.7.2.1 Physical Properties

- Moisture content (MC)
- Shrinkage
- Density or basic mass per volume

2.7.7.2.2 Mechanical Properties

- Static bending strength
- Tensile strength parallel to grain
- Compressive strength parallel to grain
- Shear strength parallel to grain

2.7.7.3 IS 8242: 1976 – Method of Tests for Split Bamboos

The items are as follows:

2.7.7.3.1 Physical Properties

- MC
- Specific gravity

2.7.7.3.2 Mechanical Properties

- Static bending
- Compression parallel to grain
- Shear parallel to grain

As reinforcement, split bamboo are being used for construction.

2.7.7.4 IS 9096:2006 – Preservation of Bamboo for Structural Purposes – Code of Practice

The preservatives and treatments for bamboo are recommended for structural

The recommendations for preservatives are as follows: Coal tar creosote

- Acid–curprick–chromate composition
- Copper–chrome–arsenic composition
- Boric–acid–borax
- Copper–chrome–boron composition
- Copper–zinc–naphthalene/abietates

2.7.7.5 Choice of Treatment

The concentration and absorption of recommended preservatives related to the choice and method of treatment of bamboo for structural purposes, covering dry and green bamboo, are given in the above IS code. The preservation of bamboo for non-structural purposes is included in IS 1902.

2.7.8 Further Review on Structural Use of Bamboo [18]

It is a low-cost sustainable building material used for structures for many years. The cost will be lower for the material compared to many conventional items like timber. Many pleasant and impressive structures have been constructed with bamboo.

Bamboo may have strength similar to high-grade hardwood. Tests for determining strength may be required before the final and detailed design. Some tests may be more relevant for the design.

For simple structures, a conservative design approach/values may be used, avoiding testing.

Bamboo appears promising for structural applications in the future.

When using bamboo in design, and for safe and durable construction, the items mentioned below will be relevant:

- General information, availability, type for use
- Durability and preservation
- Design values
- Method of design, consideration of limit state method
- Elements of design equations/approach
- Connections
- Tests

Source:

- Kaminski, Lawrence, Trujillo, and Feltham [18]

2.7.9 International Code

Referring to International code only a few specific comments as given below may be noted:

2.7.9.1 Design Concept [19]

The bamboo design concept shall be based on calculations. The limit state conditions or permissible stresses shall not be exceeded.

2.7.9.2 Comments on Limit States

Both ultimate and serviceability limit state requirements have to be satisfied.

2.7.9.3 Joints

The joints shall be designed to ensure continuity between the elements covering force transmission, deflection, etc.

The items are as follows:

- Force transmission shall be as per the design assumptions.
- The deflection of the elements shall be within the specified limits.
- The joint design of bamboo shall be based on calculations.

2.7.9.4 Tests

The full-scale joints or components shall be tested as per the relevant code.

Source:

- Appendix 1, Bamboo Structural Design Code [19]

2.7.10 Wood as a Building Material [13]

This is a durable and green building material, aesthetically pleasing, and may last for many years. By using this material for construction, the applications will be energy efficient, and the finished products will have lower embodied energy. Large environmental advantages may be achieved by using this material.

2.7.10.1 More on Use of Wood in Construction [20]

2.7.10.1.1 General Comments

Wood may be considered a suitable material, considering a wide range of properties depending on the type. It may be readily available, economical to use, competitive, and very strong with respect to its weight.

2.7.10.1.2 Types and Uses

Wood may be classified as hardwood and softwood. Hardwood is generally heavier and denser than softwood and is typically utilised for the construction of walls, floors, ceilings, etc. High-grade hardwood may contain 83% usable material, making it suitable for high-quality furnishings, solid wood mouldings, and interior joinery.

Generally, when the wood is smaller than 125 mm wide by 125 mm thick, it may be referred to as lumber. The items may be specially prepared for different requirements and used in residential construction.

Pieces that are more than 200 mm wide and 200 mm thick are referred to as beams. Timber pieces with larger dimensions may be used for the frames of structures in buildings and bridges, as well as for railroad ties, mine shaft support, and cross beams on utility poles.

2.7.10.1.3 Benefits of Wood

These benefits include

- Thermal insulation
- Adaptability
- Acoustic insulation
- Structural stability
- Easy to manufacture
- Strengthening rural development and industries
- Health benefits from comfort
- Easy to repair
- Better fire resistance due to low thermal conductivity

Further comments on the above items will follow.

Source:

- Designing Building Wiki [20]

2.7.11 Further on the Benefits of Wood as a Building Material [21]

Wood as a building material has been used for a long time, as already mentioned. The advantages of wood as a building material will outweigh those of other materials when considering environmental impact and performance.

2.7.11.1 Characteristics of Wood

The items mentioned below are relevant:

2.7.11.1.1 Tensile Strength

Wood is a lightweight material that performs better than steel in terms of breaking length (i.e., self-supporting length). It can cover larger spaces with fewer supports.

2.7.11.1.2 Sound Absorption

The properties of wood are suitable for reducing echo in living and office areas, as it absorbs sound.

2.7.11.1.3 Electrical and Heat Resistance

Wood has natural resistance to electrical conduction when dried to standard MC limits, which may be 7%–12% for most types of wood.

2.7.11.1.4 Aesthetics/Beauty

Wood can provide a large range of aesthetic options with different mechanical, acoustic, and thermal properties as per the requirements.

2.7.11.1.5 Green Benefits of Wood

There may be arguments against using wood considering the issues of deforestation and greenhouse gas emissions. However, wood materials have various advantages that contribute positively to building construction and the environment.

2.7.11.1.6 Wood Will Be Renewable

Wood may be grown and re-grown through natural processes, including re-plantation, unlike concrete or steel. With the development of forests, they can also eliminate CO_2 from the environment.

2.7.11.1.7 Wood Emits Lower Volatile Organic Compounds

Being a natural building material, wood emits fewer VOCs and carbon dioxide off-gases compared to steel, concrete, aluminium, and plastics.

2.7.11.1.8 Wood is More Readily Processed for Applications

The manufacturing process is less complicated and cost-effective.

2.7.11.1.9 Wood Enhances Energy Efficiency

With its natural cellular structure, wood has a higher insulation rating than steel or plastics. This results in reduced energy requirements in buildings for maintaining heating and cooling.

2.7.11.1.10 Wood is Biodegradable

Under natural climatic conditions, wood degrades/breaks down at a faster rate and replenishes the soil in the process. Wood is a biodegradable material.

2.7.11.1.11 Better Use of Wood with MC Measurement

Wood's interaction with moisture is important and may affect the performance of the material. Moisture-related deterioration in wood may be prevented by controlling the MC.

The correct MC level is particularly relevant for different applications like wood floors and other wood-based building materials. This is also important for wood's performance.

2.7.11.1.12 The Focus on MC

Wood is a hygroscopic material and has a close relationship with the internal MC and the ambient conditions surrounding it. Wood needs a proper balance between MC and the surrounding ambient conditions.

When this balance is achieved, it is referred to as equilibrium MC. The proper equipment for measuring MC is a moisture meter.

When this requirement is not met, wood may experience problems such as warping, splitting, twisting, or large gaps and buckling.

Wood has built-in advantages when selected as a material to meet environmental and aesthetic requirements, provided that MC is properly controlled. Wood can retain and exhibit its properties over a long period of time.

Source:

- Loffer [21]

2.7.12 Review and Summary of Comments on Wood as a Building Material [22]

Wood is an organic, anisotropic, and hygroscopic material.

It is suitable for building construction considering its thermal, mechanical, electrical, and aesthetic properties, among others. It may be possible to construct a suitable house using only wood products. However, wood also has some disadvantages.

2.7.12.1 Advantages

2.7.12.1.1 Thermal Properties

Materials are likely to change in size and volume with changes in temperature. This includes both linear and volumetric expansion. The expansion may cause a decrease in the strength of the material.

Wood does not expand significantly when exposed to heat. On the other hand, the effect of heat will dry out wood, leading to an increase in strength. It can expand slightly when the humidity level is below 0%, which may be scientifically significant. In practice, the humidity level of wood does not drop below 5%, even in the driest conditions.

The coefficient of thermal conductivity of wood is low, while its specific heat is high.

2.7.12.1.2 Acoustic Properties

Sound isolation is related to the mass of the surface. Wood is a lightweight material and may not be very effective for sound isolation. However, it is good for sound absorption. Wood may prevent echoes and noise by absorbing sound. The material may be selected for concert halls.

2.7.12.1.3 Aesthetic Properties

For aesthetic requirements, wood may be a decorative material. Wooden materials will be suitable to match according to colour and design preference. It may be painted in darker colours and varnished, and may be provided with bright and matte touches.

2.7.12.1.4 Mechanical Properties

Wood is a light material, as mentioned, but its strength is reasonably high. Considering these properties, wood and laminated wood have been applied in construction, such as health centres and sports halls.

2.7.12.1.5 Electrical Properties

The resistance of dry wood to electrical current is satisfactory. Oven-dry wood is a very good electrical insulator, and air-dried wood will have similar performance. However, the electrical resistance will decrease with an increase in MC. In general, wood may be considered and selected as a healthy material.

2.7.12.1.6 Working Properties

It is relatively easy to repair and maintain wooden materials. Old wooden items can be renewed by special methods or treatments.

2.7.12.1.7 Oxidation Properties

Wood performs better against oxidation compared to metals. The use of wood is preferred to avoid rust when necessary.

2.7.12.1.8 Variations of the Material

There are various types of wood available in different countries. The material's specific gravity, macroscopic, and microscopic structures may vary. Hence, their physical, electrical, thermal, acoustic, and mechanical properties may also differ. It may be possible to find the right kind of wood to meet specific requirements. For heat insulation and sound absorption, lightweight wood materials may be preferred. For construction purposes, heavier types are used. Checked 27/7.

2.7.12.2 Disadvantages of Wood

2.7.12.2.1 Shrinkage and Swelling

The material is hygroscopic, as stated. It absorbs vapours from the surrounding areas and loses moisture to the air below the fibre saturation point (FSP).

2.7.12.2.2 Deterioration of Wood

The deterioration and destruction of wood may be due to

- Biotic (i.e., biological) reasons
- Abiotic (i.e., non-biological) effects

Biotic factors include decay and mould fungi, bacteria, and insects. Abiotic factors encompass sunlight, water, wind, certain chemicals, and fire.

2.7.12.2.3 Biotic Deterioration of Wood

Wood, being an organic material, is a nutritional source for some plants and animals. Certain fungi and insects can digest wood, leading to partial or total decay.

Biological damage to wood due to attacks by decay fungi, wood-boring insects, and marine borers during processing and its service life can have technical and economic significance.

2.7.12.2.4 Fungi

It is important to address problems related to fungi.

The physiological requirements of wood-damaging and wood-inhabiting fungi include a favourable temperature. The temperature must be within 25°C–30°C for optimal growth of wood-rotting fungi. Furthermore, some can tolerate temperatures between 0°C and 45°C. Oxygen is necessary for fungal growth; without it, fungi cannot grow. Keeping wood submerged in water may provide good protection against fungal attacks.

2.7.12.2.5 Nutrients

Wood is an organic compound with 50% carbon.

This will make wood a suitable nutrient for fungi because fungi get their energy from the oxidation of organic compounds.

2.7.12.2.6 Moisture

Normally, wood may not be attacked by common fungi when the MCs is below the FSP. The FSP for different woods varies and may lie between 20% and 35%. However, 30% may generally be considered an acceptable value.

2.7.12.2.7 Insects

Insects are the second to decay wood, which may be responsible for damage to lumber and wood during its service life. The insects may be grouped into powder-post beetles, termites, carpenter ants, and marine borers.

2.7.12.2.8 Powder-Post Beetles

Powder-post beetles can attack both soft and hardwood. The problem may arise with seasoned wood, un-dried wood, and freshly harvested wood.

2.7.12.2.9 Termites

There are two types of termites.

- **Subterranean Termites –** these can damage untreated wood, especially when moist and in contact with soil, standing water, and other sources of moisture.
- **Dry Wood Termites –** problems may begin when the moisture level is between 5%–10%. Deterioration caused by drywood termites may be less severe than that caused by subterranean termites.

2.7.12.2.10 Carpenter Bees

Carpenter bees mainly damage unpainted wood and make large tunnels in the material to lay eggs.

2.7.12.2.11 Carpenter Ants

Carpenter ants can tunnel through wood to create shelters. The attack may occur on wood in contact with the ground or wood that is wetted from time to time.

2.7.12.2.12 *Marine Borers*

Marine borers can attack wood in salt and brackish water.

2.7.13 Reducing and Controlling the Problems of Wood

The suitable techniques for protecting wood include drying, impregnation, or coating.

2.7.13.1 Careful Selection of Wood

Careful selection of wood is necessary. Some types are decay-resistant hardwoods, such as sweet chestnut, oak, and juniper.

2.7.13.2 Coatings

Coatings can protect wood used both indoors and outdoors. They may control the rapid uptake and loss of moisture and can slow down shrinkage and swelling that lead to surface cracking and other problems. Coatings may not fully control changes in MC but can reduce it.

Coatings containing solid colour or pigmented stains can protect wood against ultraviolet rays.

The inclusion of fungicides in coatings can provide protection against decay and mould fungi.

2.7.13.3 Drying

Normally, wood will not be attacked by common fungi when the moisture level is below the FSP. The FSP for different types of wood may vary between 20% and 35%. However, 30% is generally considered an acceptable value.

2.7.13.4 Treating with Wood Preservatives

It may be possible to prevent decay in wood by treating it with wood preservatives.

Some wood preservatives may not be suitable for humans and other creatures. Hence, careful selection of the material is important.

Wood preservatives can be divided into two groups:

- Water-borne
- Oil-based or oil-borne chemicals

Most of the woods are often treated with waterborne salts.

Oil based or oil-borne preservatives are often used outdoors in industrial applications like piling, ties, and poles.

2.7.13.5 Remedial Treatment

During service life, wood items must be checked regularly and treated by brushing or with other methods.

The treatment of wood in contact with the ground may be carried out by applying paste and wrapping it with impregnated preservative bandages.

2.7.14 Abiotic Deterioration of Wood

2.7.14.1 Fire

Wood may catch fire easily. Wood is made of organic compounds composed primarily of carbon and hydrogen, which can combine with oxygen and burn. Wood is often referred to as combustible material.

2.7.14.1.1 *Fire Retardants*

It may not be possible to make wood totally non-combustible.

Fire retardants are available for applications and can be divided into two groups:

- Coatings and
- Chemicals

These are water-soluble salts, and the wood structure is saturated/impregnated with them.

Coatings may be used to control the development of volatile, flammable gases by promoting the quick decomposition of the wood surface to charcoal and water.

Fire retardants can only slow down the flammability of wood or eliminate progressive combustion.

They may not prevent burning when an external source is present. When the external source of fire is removed, the wood will stop burning.

Source:

- Ozen [22]

2.7.15 Recycled Plastic [13]

These are energy-efficient materials, and their lifespan may cover 30–50 years when used for building and construction.

2.7.15.1 Additional Comments on Plastic Waste for Construction Materials [23]

Efforts have been made to utilise plastic waste in eco-friendly construction materials.

The items include

- Tiles manufactured using plastic waste. Tests are required for the suitability of the product, and some tests have been carried out to check against water absorption, mechanical strength, and flammability.
- Other applications related to converting plastic waste into building materials have been considered.

  This includes the selection of plastic waste, grinding it into flakes, putting it into a mould, and melting it in a plasma state. Pressure may be applied to make the raw material suitable for construction purposes. Building houses with this material may be cheaper than using concrete.
- A building material has been developed in one advanced country using cigarette butts, plastic waste, bitumen, and paraffin wax. This material may be used for construction and laying roads as a suitable lightweight construction material. It can support heavy traffic loads and reduce thermal conductivity.
- **Plastic Waste in Road Construction**

  In India, plastic waste has been used in various areas for road construction.

  Plastic road surfacing may make the roads more heat resistant and sturdier. The melting point of plastic roads may be about 66°C compared to 50°C C for conventional roads. Furthermore, it is expected to be more economical than conventional roads.
- **Paper from Plastic Waste**

2.7.15.2 Further Comments

Our efforts should focus on using plastic waste efficiently. The large-scale utilisation of plastic waste, which is a non-biodegradable material.

Source:

- Mahendriyani [13]
- Khanna [23]

2.7.16 Mycelium

This material is concrete with the ability of self-healing [13] and will reduce the requirements for costly repair work. It is an eco-friendly insulation material and may also be considered a biomass green building material.

2.7.16.1 Further on Mycelium Products

2.7.16.1.1 Mycelium Bricks [24]

This is a new item and looks encouraging for meeting sustainability requirements in the future.

Mycelium brick is an organic brick that may be produced from organic waste and the mycelium of fungus. Mycelium materials are the thin, root-like fibres from fungi that grow underground. After drying, it is strong and can be used as a durable, water-, mould-, and fire-resistant construction material. It can be grown into specific forms without any processing needs.

A few comments are given below:

- The material will be sturdy inside the building but may not be suitable for outside exposure.
- Mycelium has already been used in other countries for packaging items as an alternative to polystyrene and for furniture items indoors. The material can be recycled after use.
- Other uses may include composite board (Myco-board).
- By adding mycelium to crop waste, the material can be put into a mould, grown, and dried to become a hard material.
- It may take about 5 days for the bricks to grow and become suitable for use.
- The compressive strength of mycelium bricks may be about 0.207 N/mm^2 (30 Psi) and 1 m^3 of mycelium bricks will weigh around 430 N (43 kg).
- By using mycelium, there will be a significant reduction in the use of fossil fuels related to embodied energy for fabrication, erection, and a large reduction in building waste.
- Mycelium items will be fully biodegradable and can be used as soil.
- Mycelium products will be termite-proof.

Mycelium products may play an important role in reducing the embodied energy of products and the net waste of materials after the end of their service life, making them very useful in the construction of buildings.

Source:

- Mahendriyani [13]
- Bonnefin [24]

2.7.17 Rammed Earth

It is a suitable and eco-friendly building material for green construction [13]. This item has been covered separately, in detail, in Chapter 5.

2.7.18 Timbercrete

The material is a green building material [13] made from a mixture of timber waste from different sources and concrete. It is lighter than solid concrete, with better strength and insulating properties. Timbercrete is also bushfire-proof, with minimum heat transfer and radiation. The applications of the material will be user-friendly, as it can be screwed, nailed, and sawn easily.

2.7.18.1 More on Timbercrete [25]

The material is a viable concrete alternative.

Timbercrete may be prepared in blocks, panels, and paving, and may be used for construction purposes.

2.7.18.1.1 Timbercrete Blocks

This can be produced by

- Three parts of sawdust, i.e., wood low in tannin, gums, and oils for the best results.
- Two parts of sand
- One part of cement

2.7.18.1.2 Preparation of Timbercrete [25]

For smaller panels, four parts of sawdust may be adequate.

- Mixing of the sawdust and sand properly must be done.
- Cement may be added and blended.
- The mix should have a consistent colour.
- Water may be added, and mixing should be done again. Water should be added slowly and uniformly.

2.7.18.1.3 Applications of Timbercrete [25]

This may be used for residential and commercial projects and for paving related to landscaping and other areas. Like concrete, it can be easily moulded into different shapes along with various textures and colours, offering flexibility.

2.7.18.1.4 Comparison of Timbercrete and Concrete [25]

The following may be mentioned:

- Timbercrete will have a higher insulation value as compared to traditional concrete or bricks.
- Timbercrete is strong, and its load-bearing capacity will exceed that of concrete.
- It traps carbon dioxide.
- It is lighter than concrete or clay.
- It can be nailed or screwed into place as per the requirements.
- It may be easier to work with than concrete.
- It will be more fire-resistant.

2.7.18.1.5 Disadvantages of Timbercrete [25]

- It may be costly.
- The investment may be justified in the long run, considering the insulation value of the product.

2.7.18.1.6 Timbercrete Construction [25]

- It will consume less energy to prepare compared to clay-fired bricks
- Clay bricks are fired at high temperatures, which is not required for timbercrete.
- Timbercrete material has been used for prefab homes and roof tiles.

The potential of timbercrete is promising, and we may see more uses of the material in the future.

Source:

- Mahendriyani [13]
- What is Timbercrete [25]

2.7.19 Ferrock

It is a carbon-negative cement alternative [13] and may significantly reduce carbon emissions from manufacturing. Eco-friendly advantages accompany this material when used as an alternative to cement. Structures constructed with Ferrock cement will last long and will not require repair or replacement.

2.7.19.1 Further on Ferrock [26]

Ferrock is a carbon-negative, environmentally friendly material that may be considered a suitable replacement for concrete. It is primarily made from recycled materials and may be cost-effective to produce. Ferrock is manufactured by mixing steel dust waste and silica. Following a chemical reaction, mixing with water and exposure to carbon dioxide allows it to harden into a solid similar to concrete. Ferrock may be five times stronger than Portland cement concrete. The material is flexible and will not crumble under compressive loads or seismic movements. Ferrock absorbs CO_2 while hardening and can reduce greenhouse gases.

2.7.19.1.1 Advantages

- Five times the strength of concrete.
- **Chemically inactive**– the material will not deteriorate when exposed to chemicals and gases.
- **Flexible** – it can withstand more pressure and movements.
- The material will be carbon neutral.
- **Comments on Material –** eco-friendly alternative to concrete. Ninety-five percent of Ferrock's ingredients are recycled materials.
- **Costs –** it may be more expensive initially but could become cheaper over time.

2.7.19.1.2 Problems

- More uses supported by test results will be required
- **Untested Item –** more tests may be needed, as long-term performance like concrete is not yet known.

Source:

- Mahendriyani [13]
- Builder Space [26]

2.7.20 Construction Boards with Carbonate [27]

The construction boards may have various applications. The board is prepared by mixing magnesium oxide, magnesium chloride, perlite, a bonding agent (i.e., wood shavings), recycled board scraps, and water to form a paste. The paste is then placed into a mould to create construction boards. The mould is allowed to run through a series of rollers to spread the paste uniformly across the mould to achieve the required thickness. The technique may also include fibreglass are polyester paper sheets for boards. The applications of construction boards may include interior wall board, soffit board, structural sheathing, barrier sheathing, fascia board, tile backer board, structural wrap, window wrap, stucco wrap, ceiling tile, etc.

The construction boards will have different advantages, such as water resistance, fire resistance, and may be more durable than normal gypsum wallboard and other similar wallboard materials.

No silicate will be present in the making of construction boards. Hence, harmful effects from silicate-based products may be avoided in construction boards.

Source:

- Construction Boards with Carbonate [27]

2.7.21 Magnesium Oxide Wall Boards [28]

Magnesium oxide is a very useful material when used as part of the cement mixture and may be cast into thin cement panels. This item may be used for residential and commercial buildings. The material will be suitable for fire resistance, mould, mildew, and sound control. Magnesium boards will have high strength and resistance due to the development of a strong bond. Magnesium boards may replace gypsum drywall, ceiling covering material, and sheathing. Furthermore, they may be used for shaft liners, area separation, tile backing, and substrates for coatings and insulating arrangements, finishing systems, etc. Magnesium cement boards may be available in different sizes and thicknesses. Normally, the boards may be available in light grey, white, or beige colour. Different grades may be available covering smooth face, utility, rough texture, etc., with varying densities and strengths to meet various requirements. The magnesium boards may be used mainly as wallboard instead of gypsum-based drywall and plywood as mentioned. The material may be scored and snapped, drilled, screwed, and may be fixed to steel or wooden frames. The material may be found useful for satisfying the requirements of safety and durability.

2.7.21.1 Review of Applications

These may be mentioned as

- Interior and ceiling board
- Soffits
- Exterior wall and fencing board
- Fascias
- Materials for trimming
- Shaft-liner and separation wall boards
- Exterior sheathing
- Tile backer boards
- Substrates for coatings, insulating, and finishing

2.7.21.2 Review of Advantages

These may be

- Fire-resistant
- Energy efficient
- Water-resistant
- Mould/fungus/bug-free
- May be used instead of cement board or dry wall. No special tools will be necessary for applications.

- Silica/asbestos-free
- Impact-resistant
- Hard and non-absorbent surface.
- May be available in different colours.
- Suitable for biologically friendly construction technique.
- The cost is compatible with cement boards.

Etc.

2.7.21.3 Review of Disadvantages

These may be

- Good quality magnesium boards may be more expensive than paper-faced gypsum dry wall material.
- Strict quality control will be required for raw materials, mixing, curing, and handling of fresh and semi-fresh items.
- By using magnesium oxy-chloride, in the mix, high early strength may be obtained. However, such products may be susceptible to water damage and can give inconsistent results.
- Different products may be available with variations in production and cost. This may influence the mix design and curing method.
- The boards may have special installation requirements. Manufacturer's recommendations will be necessary to avoid problems.
- The boards may be produced with Sorel cement (i.e., magnesium oxy-chloride) as mentioned. The use of such boards may cause problems in humid climates.
- The chloride in Sorel cement may create corrosion environment for steel studs and fasteners.

Source:

- Wikipedia [28]

2.7.22 Bio-Composites [29]

Bio-composites are composite materials that may be manufactured using resin and reinforced with natural fibres. The comments/discussions on natural fibres are covered in Chapter 7.

2.7.22.1 Characteristics

These materials are biodegradable and do not pollute the environment. They come from renewable sources, are inexpensive, and some types may be completely recycled.

2.7.22.1.1 Classification

Bio-composites may be divided into

- Non-wood fibres and
- Wood fibres

Non-wood fibres (natural fibres) may be suitable for industries due to their mechanical and physical properties. These fibres are comparatively long and have high tensile strength. Natural fibres can also react with water molecules, causing them to swell.

Wood fibres may consist of 60% of the mass of root elements. They can be softwood (i.e., long and flexible) or hardwood fibres (i.e., shorter and stiffer). Straw can serve as a low-cost reinforcement for bio-composites. Wood fibres may be recycled or non-recycled. For wood composites, many polymers like polyethylene, polypropylene (PP), and polyvinyl chloride may be used for industrial applications.

2.7.22.1.2 Flax Application

Different applications have been explored.

2.7.22.2 Green Composites

These are bio-composites that contain natural fibres with biodegradable resins. They are sustainable items and can be disposed of without affecting the environment.

2.7.22.3 Hybrid Composites

Hybrid composites are made with different types of fibres in a single matrix. The fibres may be natural or synthetic and can be used randomly to create hybrid composites.

The properties of this type of bio-composite depend on various factors such as fibre content, types/arrangement, length, and bonding to the matrix. The strength of the hybrid composite relies on the failure strain of the individual fibres.

2.7.22.4 Hemp Composite

Hemp fibre composites are suitable where weight reduction and increased stiffness are required. Hemp fibre composites have been developed to replace plastics. Items such as suitcases, mobile phone cases, and cosmetic packaging may be made with hemp fibre composites.

2.7.22.5 Processing

The items listed for the manufacture of plastics or composite materials include

- Machine press
- Pultrusion
- Filament winding
- Injection moulding
- Extrusion
- Resin transfer moulding
- Compression moulding
- Sheet moulding compound

Source:

- Wikipedia [29]

2.8 ALTERNATIVE LOW-COST BUILDING MATERIALS

A few items will be discussed below under this heading [30]:

2.8.1 Natural Materials

Some of the items in this connection have already been covered to some extent. However, the list of low-cost building materials will include

2.8.1.1 Further on Bamboo [30]

This material is readily available in countries like China, Japan, and India, and may be used as a structural material. It is considered a structural item due to its affordability, good durability, and ease of assembly. Bamboo has a circular profile and may not be very easy to connect to other structural members. It is also hollow in the middle; hence, the connection details have to be organised carefully and accordingly for transferring loads. As a bamboo product, corrugated bamboo roofing sheets may be referred to. The details on the product are

2.8.1.1.1 *Corrugated Bamboo Roofing Sheet*

These roofing sheets will be eco-friendly, strong, lightweight, durable, and have less fire hazard compared to thatch and other roofing materials. The sheets may be used for roofing, door and window shutters, walling, etc., in building components related to construction. Some additional comments have been made on bamboo in item no. [7].

2.8.1.2 Random-Straw or Coconut Fibre Stabilised Soil [30]

Soil–straw mix is an old concept of construction material and method that has been used in many countries, as mentioned. By applying this method and using modern geotechnical methods, both the strength and durability of the material may be improved. It seems that the tensile strength of the material may be improved three times compared to soil without straw by adding 1% by weight of straw. The ductility behaviour may also be improved by using soil–straw mix material. The maximum increase in tensile strength may be possible with 1.6% by weight of straw. From wet–dry tests, it has been observed that coconut fibres (CFs) may provide better durability performance. The water resistance of the material may be improved by applying sulphur coating to compacted soil–straw mix.

Source:

- Vidya/Radha [30]

2.8.2 Coir Material

2.8.2.1 General Comments [31]

The term coir has been derived from KAYAR, which means cord or rope in Tamil and Malayalam. Coir or CF is a natural fibre made from the outer husk of coconut and may be used for floor mats, mattresses, door mats, brushes, and construction materials. It is a fibrous material obtained from between the hard internal shell and the outer coat of coconuts. There are two types of coir: brown and white. The brown coir is prepared from ripe coconuts and may be used for sacking, upholstery padding, and horticulture. The white coir comes from unripe coconuts and could be used for string, finer brushes, and fishing nets.

2.8.2.2 Additional Comments on Coir

2.8.2.2.1 *Structure*

The coir fibre is obtained from between the hard internal shell and the outer layer of the coconut, as mentioned. When immature, they are pale and become hardened and yellowed later. The brown coir is thick, strong, and has high abrasion resistance. The coir fibre is comparatively waterproof and resistant to damage by saltwater. Freshwater is used to process the brown coir. Seawater and freshwater may be used for the production of white coir. Coir is also known as coprah in some countries.

2.8.2.2.2 *Bristle Coir*

This is the longest type of coir fibre. It is made from retted coconut husks using a process known as de-fibreing. The fibres are cleaned with a steel comb to remove the short fibres. Bristle coir fibre may be used as bristles in brushes and utilised for domestic and industrial applications.

2.8.2.2.3 *Applications*

These may be beneficial in

- Cordage, bedding, packaging, flooring etc.
- Agriculture and horticulture
- Oil and fluid absorption
- Animal bedding

2.8.2.2.4 Construction Materials

Natural fibres are eco-friendly types. Furthermore, it is a tough, resistant to thermal conductivity, inexpensive, durable, and renewable material. It has been observed that by using CF and replacing 2% of cement in the mix, the compressive strength of concrete increased.

2.8.2.2.5 Bio-control

Bio-security risks

Coco fibre may harbour organisms that could pose biosecurity risks to the importing country. Trichoderma, a naturally occurring fungus, may develop in coco peat. It works in symbiosis with plant roots for protection from pathogenic fungi like Pythium. It may not be present in sterilised coco peat.

India and Sri Lanka are the main producers of coir fibres.

Source:

- Wikipedia [31]

2.8.3 COIR – Traditionally Used Fibres for New Construction Applications

2.8.3.1 Coir Material [32]

Coir fibres are by-products of copra production.

2.8.3.1.1 As Insulating Material

The fibres when applied, are capable of protecting the building from heat from intensive solar radiations. Coir had been used successfully for the purpose of insulating the buildings for many years. In European countries, coir has been used as insulating material for a long time. The cavity may be 30 μm in the middle of the matured fibres.

The special advantages include heat and sound protection. Furthermore, it has a slow decomposition rate, making it suitable for the protection of geo-textiles.

2.8.3.1.2 Other Applications

This may include clay building boards reinforced with coir fibres, which may be useful. Coir may also be used for plaster base boards.

2.8.4 Review of Construction Materials from Coir

2.8.4.1 Coir Ply

Walls and roofs of buildings have been constructed using coir ply, which can be protected from UV rays with weather-resistant paints and tiles. The boards have good insulation properties. Coir may be a good substitute for plywood and can be prepared using coir bonded with a phenol-based resin. The material is termite, water, and insect resistant.

2.8.4.2 Corrugated Coir Ply Roofing Material

Corrugated coir ply roofing material can be made using phenol resin-bonded coir. This serves as a low-cost alternative to corrugated galvanised iron and asbestos cement sheeting. The material can be prepared using short, randomly distributed coir fibres in a cement matrix.

2.8.4.3 Particle Boards

All parts of the coconut tree, including its fruits, can be utilised for the manufacture of particle boards. Lignin can be used as a natural binding agent to develop a simple and efficient high-density material from whole coconut husks without the inclusion of chemical binders. It may be possible to combine coir with other types of biomass. These boards can be very efficient with good properties and can be compared to superior wood-based panels. For coconut boards, common woodworking equipment can be used for sawing, drilling, planing, and polishing. Prior drilling will be required for screwing. The board material will be very hard for nailing.

2.8.4.4 Insulation Materials

Coir has been successfully used for insulating items for a long time, as stated. The fibres are durable and work well in tropical conditions with high humidity, providing protection from fungi and insects.

2.8.5 Other Construction Materials

Clay building boards reinforced by coir fibres may be used. Coir may also be applied for plaster baseboards as mentioned.

2.8.5.1 Ecological Issues

It is the by-product of copra, as mentioned. There will not be any need for additional space for production compared to materials such as hemp, jute, sisal, etc. The coir raw material is available from Asian countries, as stated. A lot of this material is burnt or used as mulch.

2.8.5.2 Further Comments

Due to the development of other homegrown renewable materials like hemp and wood fibres, the perception and versatility of coir have not been properly appreciated. Considering the labour along with the need for income, especially in tropical countries, and with the ecological and social advantages of coir, the utilisation of this resource is expected to increase in the future.

Source:

- Kursten [32]

2.8.6 Straw Material [30]

Straw is an agricultural by-product associated with plant stalks (i.e., cereals) after the elimination of grain and chaff. Rice straw has a high silica content and may be considered the toughest when compared with other cereal straws. Burning of straw may create environmental problems associated with breathing difficulties. The production of straw and straw bales is high in India, with good scope for proper utilisation. Straw is considered to have high potential and may contribute significantly to alternative building materials. It does not support combustion and has fire-resistant properties. It is thermally insulating and has sound and moisture insulation properties, and it is not toxic in nature.

2.8.6.1 Other Alternatives with Straw and Straw Bales [30]

2.8.6.1.1 Improved Thatch Roofing

Development work has been carried out by CBRI to improve the fire hazard risk of thatch roofs and to create a treatment to make them water-repellent. It is necessary that the thatch layers are treated with special plasters of mud to make them fire-resistant and durable.

2.8.6.1.2 Life Extended Thatch Roofing

It may be locally available as an alternative to corrugated sheets, which are also environmentally friendly. The life may be extended by controlling the effects of biodegradability by treating the material with copper sulphate solution. For additional requirements, the roof surface may be treated with phosphorylated spray or CNSL oil to provide performance for waterproofing, termite proofing, fire resistance, and weathering resistance. The work has been developed by CBRI/RRL-TVM.

Source:

- Vidya/ Radha [30]

2.8.6.2 Further Review on Straw [33]

Straw is a by-product of the food-growing industry. For many years, straw has been used for house construction, as fuel, as feed for animals, and as preparation material for mattresses/upholstery, etc. It is an organic material that is environmentally friendly, sustainable, and sequesters carbon. It is carbon neutral, like wood and other biomass. It is a cheap product with different applications, as briefly mentioned below:

2.8.6.2.1 Applications

These may be mentioned as below:

- **Bio-fuels**
  It is a carbon-neutral energy source.
- **Composite Wood**
  Wheat straw may be utilised as a filler with polymer to make composite wood.
  Enviro-board may be prepared with straw. Enviro-board as construction panels using agricultural wastes has been used for many years. Pressed leaves and straw may be considered for structural and insulating items.
  Straw blocks will be straw bales that may be recompressed to achieve the density of wood blocks.
  This item will be solid with adequate strength for use as load-bearing walls, and it will be easy to transport.
- **Construction Material**
  In many countries, straw has been used to bind clay and concrete. A mixture of clay and straw (known as Cob) has been used for building construction for many years. The discussion on Cob has been done separately in this chapter.
  Simple houses may be constructed with straw bales as stated. Comments on straw bales and their applications with advantages have already been made.
  Thatching with reed, straw, or similar materials has been used to make roofs lightweight, waterproof, and well-insulated.
- **Other applications**
- May include animal feed, rope, horticulture, clothing, bedding, basketry, packaging, and paper.

Source:

- Sustainable Straw [33]

2.8.7 Earth as a Building Material

This is the oldest building material used by people over a long period of time [30]. There are problems associated with damage to earth construction from water penetration, erosion of walls due to water splashing, and attacks by termites and pests, and earth structures may need regular maintenance. Different items have been developed over time for improvements. Development activities have covered the use of compressed earth blocks and non-erodible mud plaster, among others. Some comments will be made on the items below:

2.8.7.1 Non-erodible Mud Plaster [30]

CBRI, India, has developed non-erodible, economical mud plaster for protecting mud walls. The plaster is prepared by mixing bitumen cutback (i.e., a mixture of bitumen and kerosene oil) with a specified mud paste. Walls with non-erodible plaster will be resistant to water erosion. The development also includes the application of potter-prepared tile lining for the protection of mud walls against rain and moisture.

2.8.7.2 Compressed Earth Block (CEB) [30]

The development has taken place from moulded earth blocks, often referred to as adobe blocks. The use of earth blocks is economical and environmentally friendly. Earth blocks may be produced by compacting soil mixed with a small amount of cement or lime and by pressing the soil mix manually or using a machine. Further details on this have been covered separately in Chapter 3.

2.8.8 Further Review of Different Types of Building Construction with Earth [34]

The development of different types of construction includes

2.8.8.1 Rammed Earth

The Rammed Earth method of construction involves mixing wet mud, sand, and gravel, placing it into formwork, and compacting it layer by layer. By using different coloured soil, various effects may be displayed externally. This method has been covered further in Chapter 5.

2.8.8.2 Cob Construction

In this method, soil mixing is done, with layers placed one on top of the other. The mixture may be prepared with soil, water, sand, and straw. This method has been discussed further later in this chapter.

2.8.8.3 Earth Blocks

2.8.8.3.1 Abode Blocks

These are sun-dried bricks. The soil may contain cut straw, and formwork is used for the bricks. This method has been discussed in detail in Chapter 4.

2.8.8.3.2 Cut Blocks

When the soil is adequately cohesive, the blocks may be cut directly from the soil. This technique is referred to as cut block. The blocks will be hard when exposed to air due to dehydration. Further comments on this will follow.

2.8.8.3.3 CEBs

The blocks may be prepared using a press (which may be operated manually). To increase the strength of the blocks, cement or lime may be added to the soil mixture as mentioned.

2.8.8.3.4 Mix Design and Production of CEB

By using more cement content, the compressive strength of the blocks may be increased.

Discussion and information on compressed earth blocks have been included in Chapter 3, as stated.

2.8.9 Improvements in Earth Construction

Mix design and production related to improvements in strength and durability will be important for earth construction. The application of alternative binders may need to be considered. Emphasis may be placed on using eco-friendly binders. Importance is given to specific admixtures and surface treatments. A few items in this connection will be mentioned.

The mix design for earth blocks will be important for achieving workability, shrinkage, and strength.

2.8.9.1 Alternative Binders

The common binders that are used will be lime or cement, as stated. Stabilisation with lime will be more effective if the soil has a high clay content. For sandy soil, stabilisation with cement may be recommended.

An alternative may be to use the alkaline activation method. The alkaline-activated materials are called geo-polymers. This may be preferred instead of cement due to lower CO2 emissions. If the soil contains clay, it may act as a binder as well.

The use of laterite as a potential alkaline-activated material has been attempted. The curing temperature will be important for strength development.

2.8.9.2 Admixtures

Proper selection of admixtures will be critical to controlling water penetration. The application of a low dosage of hydrophobic silicone-based admixture could control water absorption without influencing vapour permeability.

2.8.9.3 Surface Treatments

The proper use of surface treatments would be important to improve the surface quality of earth construction. Different types of surface treatment materials may be available for applications, including water repellents. By using these products, the construction may become resistant to water penetration and dust, as well as provide sealing renderings and coatings.

The development of nano-acrylic emulsions is also relevant. Further attempts have been made to use fluorinated silicones or fluorinated acrylic polymer sealers for areas that will need regular cleaning.

2.8.9.4 Deep Soil Mixing

The method of deep soil mixing may be used for buildings and structures, and this method will not require excavation of earth. In this technique, the earth is mechanically mixed in place and in depth using an auger or a cutter machine, and the binder (i.e., normally cement) is injected into the mix.

Depending on the application and the nature of soil, 100–600 kg of cement/m^3 of mix may be required.

2.8.9.5 Further Comments

For improvements, attempts have been made to study/research for admixtures and surface treatments as mentioned. Further, the important item will be to use eco-friendly binders.

Some brief comments are made below:

Different ground improvements have been made, covering cylindrical columns, continuous barriers, rectangular panels, etc. for global mass stabilisation.

The deep soil mixing technique is known as a ground improvement method.

This method may be suitable for global stabilisation related to soft cohesive soil like silt, clay, and peat to obtain adequate bearing capacity for the foundation of structures and to increase the general stability of the soil.

The soil mix walls may be used as temporary or permanent geo-structural elements and may satisfy the requirements for earth-water retaining, bearing, and cut-off, etc.

Source:

- Vidya/Radha [30]
- Vyncke, Kupers, and Denies [34]

2.8.10 Additional Information

2.8.10.1 Advantages of Earth Construction [35]

These may be mentioned as follows:

- Will meet modern design requirements
- It will be possible to have sustainable construction and contemporary design

- Clay as a binder uses low levels of primary energy, with no supply issues
- No harmful substances will be released into the environment
- It will be a non-allergenic and reusable material
- Waste building materials may be recycled

2.8.10.2 Modern Earth Buildings

The contemporary design would be associated with rational working techniques, innovative approaches, and processing methods.

There are many pre-fabricated earth building materials/items available.

These may include

- Earth panel boards, ready-to-use mix bagged materials, earth bricks, etc.

Earth may be considered an innovative and competitive construction material with flexibility.

Clay surfaces and plaster work can provide good living surfaces after interaction with light, colour, and other materials.

2.8.10.3 Vernacular Earth Buildings

- Earth building skills were very common in the past.
- Observing/studying the old and existing buildings, it can be noted that builders in the past used this material very efficiently, covering both design and construction.
- The aesthetics and performance of the buildings have influenced local culture and the environment.
- The process is largely based on local people's participation in site activities
- In the process, people would be trained to take on similar assignments confidently.
- All these activities are associated with vernacular architecture and sustainability in the process.

Our knowledge, experience, and education on earth design and construction have been diluted over the years in many countries. Hopefully, we will be able to assess the versatility of the material and see more structural applications with earth in the future.

Source:

- ECVET Earth building [35]

2.8.11 Jute Material

2.8.11.1 General Comments [36]

Jute is a natural fibre and an affordable item for use. Jute material is soft, long, shiny, and belongs to the bast fibre group, which can be spun into coarse and strong threads.

2.8.11.2 Types

These are

- White jute (*Corchorus capsularis*)
- Tossa jute (*Corchorus olitorius*)

2.8.11.2.1 White Jute

There are different uses of white jute, such as types for carrying grains or agricultural items. Jute products are used in India for ropes and twines for household and other purposes.

2.8.11.2.2 Tossa Jute

Tossa jute fibre is softer, silkier, and stronger than white jute. Tossa jute is also known as Paat in Bengal/Bangladesh. Bangladesh is the largest producer of Tossa jute.

2.8.11.3 Production

Jute fibres are obtained from the stem and ribbon (outer skin) of the jute plant. The fibres are extracted first by retting. The method involves bundling the jute stems together and immersing them in slow-running water. There are two types of retting: stem and ribbon. After retting, the stripping operation begins. In this process, non-fibrous material is scraped off.

Jute is a rain-fed crop with minimal requirements for fertiliser and pesticides.

2.8.11.4 Uses

Rope, twine, and matting can be prepared from jute. It is used for making gunny sacks for storing rice, grains, wheat, etc.

Other applications include curtains, carpets, hessian cloth, and chair coverings.

The fibre matting may be used as a preventive measure against flood erosion. The demand for jute is high due to its cost, length, softness, lustre, and uniformity of the fibre.

It is a natural and biodegradable fibre.

2.8.11.5 Special Comments

- Jute fibre is 100% biodegradable, recyclable, and environmentally friendly.
- It is a natural fibre with a golden and silky shine, known as golden fibre.
- It is the cheapest vegetable fibre typically available from the bast or skin of the plant's stem.
- The requirements for pesticides and fertilisers are low for jute.
- It is the second most important vegetable fibre after cotton. With the fibres, it is possible to make top-quality industrial yarn, nets, fabric, and sacks. It can be used in packaging, construction, textiles, and agriculture.
- It has high tensile strength, low extensibility, and provides better breathability for fabrics.
- The advantages of jute include good insulation and anti-static properties, low thermal conductivity, and moderate moisture regain qualities.
- Jute can be blended with other fibres, both synthetic and natural, and can accept dyes, including pigments.

2.8.11.6 Disadvantages

These include

- Poor drapability and crease resistance, brittleness, fibre shedding, and yellowing in sunlight. Treatment with castor oil lubricants may reduce yellowing and fabric weight loss while increasing dye brilliance.
- Jute can lose its strength when wet and may be subject to microbial attacks in humid conditions.
- With enzyme treatment, jute may lose its brittleness and stiffness.

Source:

- Wikipedia [36]

2.8.11.7 Study for Jute Fibre for Earthquake Resistant Earthen House Construction [37]

Earthen buildings must perform well under different seismic forces. A large number of people live in earthen houses, as mentioned. Earthen buildings have many advantages, such as low cost, readily available materials, ease of construction, low energy requirements, and environmentally friendly properties. They are a suitable alternative building material for the comparatively low-income group.

It seems, according to comments, that earthen houses will continue to exist in the future for a long time in developing countries, including Bangladesh. At the same time, earthen houses are being damaged due to high winds and seismic effects. From a survey of damaged structures, it was observed that the weaknesses of such buildings are due to their heavy weight, brittle behaviour, low tensile strength of the material, poor strength of the mortar, and inadequate bonding between the blocks and the mortar.

One of the main causes of failures in earthquake areas is poor bond or mortar strength.

With the formation of shrinkage cracks, the mortar loses its tensile strength rapidly. The shrinkage cracks also cause separation/de-bonding of mortar from the block surface. The formation of shrinkage cracks depends on clay mineralogy, clay content, and the gradation of the clay–silt–sand fraction of soil that may be used for the mortar. Natural fibres have been used for a long time to improve the performance of earthen structures.

In this study, the performance of jute natural fibre was used to reinforce soil for the earthquake resistance of earthen houses. The effects of soil composition on the strength and shrinkage properties were observed in relation to earthen construction using small-scale earthen blocks.

Three types of soil were used for this study. Earthen blocks were made with different compositions, with or without jute reinforcement. After drying, the blocks were bonded with different soil mixtures, with or without reinforcement. Following the natural drying of the block mortar, the development of cracks was studied under loads. Unconfined compression tests were also conducted on cylinder samples with or without jute reinforcement. Different strength parameters were included and compared for the study.

2.8.11.7.1 Material Used

Three types of soil were used:

- Rupgonj soil
- Kaligonj soil
- Soil–sand mix

The soils from these areas have been used by people for earth construction.

2.8.11.7.2 Jute Reinforcement

Jute was cut into short pieces, 30 mm (average).

2.8.11.7.3 Shrinkage Tests

The linear shrinkage test or Alcock's test was carried out.

2.8.11.7.4 Comments from the Study

The development of reinforcement for the protection of earthen structures against seismic effects is very important for saving lives and structures. A few comments from this study are as follows:

- The composition of soil is relevant for the development of cracks in blocks and mortar. With proper composition, the strength properties of the material may be improved.
- Some clayey soil (montmorillonite) showed high shrinkage cracks and may not be suitable when the soil comes in contact with water.

- The soil must be strengthened and stabilised. The composition of soil will have minor effects on bond strength.
- Jute may be effective in increasing bond and mortar strength. Furthermore, it can improve the ductility and toughness of the earthen blocks.
- Considering availability, cost, and ease of construction, jute fibres may be useful as reinforcing material for earthen structures against lateral forces. Furthermore, it will improve mortar bond.

Additional work supported by tests will be required for jute fibre-reinforced earthen structures.

Source:

- Rahman and Islam [37]

2.8.12 Fibre Cement Composites [30]

Research and applications have been ongoing to use natural fibres in various construction activities related to housing for a long time. This may include bagasse, corn stalk, cereal straw, cotton stalk, rice husk/rice straw, kenaf, etc., as composite building materials for civil engineering construction.

A few products may be mentioned as follows

- Bagasse–PVC boards
- Bagasse–cement Boards and Panels
- Coir products
  - Coir-CNSL Thermal Insulation Board
  - Coir-CNSL Board
  - Jute–coir composites
  - Coconut and wooden chips roofing sheets

2.8.13 Man-Made Materials [30]

2.8.13.1 Structural Materials

By using by-product materials from industries like fly ash, rice husk, etc., it may be possible to produce alternative building materials for various uses. This will also solve the problems of disposing of by-product materials. These materials have good pozzolanic properties.

Some of the alternative materials are as follows:

2.8.13.1.1 Pozzolana Materials as Blending Materials with Cement (i.e., Fly Ash/Slag/Calcined Clay)

Suitable materials like fly ash have been used with about 35% replacement of cement as blending materials. This satisfies the design requirements. The quality and properties of concrete durability can be significantly improved by adding fly ash to the mix. The advantages of using blended cement in concrete are known to engineers. These include ease of pumping, better microstructure and reduced permeability, better performance in aggressive environments (i.e., sulphates and chlorides), less heat of hydration, greater long-term strength, reduced risk of alkali–silica reaction, and better electrical resistance with fewer risks of reinforcement corrosion. Portland pozzolana cement may save energy by 20%, and for lime pozzolana mixtures, it may achieve savings of up to 70% energy.

2.8.13.1.2 Polymer-Bamboo-reinforced Concrete

Reinforced concrete with bamboo may present issues such as high volume change associated with expansion and shrinkage due to water content, low bond strength between bamboo and concrete, and cracks at service loads in the tensile zone of the concrete elements due to low modulus of elasticity and decay. Different techniques have been attempted to improve the situation of low-cost material by using bitumen, paint, cement, etc.

Impregnation methods to increase the bond strength with sulphur are being tried for the material. While this material may not be suitable for the main structure, it may be used for secondary structures, especially when steel is not available. Other low-cost materials like sulphur-treated bricks or masonry blocks are also under consideration for applications.

2.8.13.1.3 Recycled Steel Reinforcement [30]

Steel reinforcement may be recycled and used for construction purposes. For reuse, materials may be sourced from automobiles, steel-reinforced structures such as buildings, concrete pavements, bridges, defence scrap, scrap generated from shipbreaking, discarded structures, tested billets from mini steel plants, and main producers, etc. Obviously, the quality of scrap materials must be checked, which is important before deciding to reuse them.

2.8.13.1.4 Ferro-Cement and Pre-cast Components

Pre-cast items may be recycled and generated with low carbon dioxide production and are energy efficient.

- **Pre-cast RCC/Ferro-Cement Frames –** Pre-cast RCC door frames may be constructed with welded reinforcement. There may be different types of applications.

Ferro-cement has been discussed separately in Chapter 6 with photographs.

2.8.13.2 Bricks and Blocks

The applications of bricks and blocks will be required in the design and may be considered an important issue for building construction. Attempts have been made to use alternative items for the same purpose in order to protect our natural resources. Brief comments will be made on other items as alternative materials for bricks and blocks.

2.8.13.2.1 Use of Fly Ash

Bricks may be prepared using Class ‘C’ or Class ‘F’ fly ash and water. The raw materials include fly ash, sand or stone dust, ordinary Portland cement, etc., which are compressed. There may be a high concentration of calcium oxide in Class ‘C’ fly ash, which will offer self-cementing ability. The bricks will be energy efficient, mercury pollution resistant, lightweight, less permeable to water, cost-effective, and possess thermal insulation properties. Fly ash may contain unhealthy elements like silica, arsenic, iron oxides, cadmium, aluminium, and mercury. Due to this, the mechanical bonding may be weak, resulting in a poor appearance. Fly ash brick has been discussed separately in Chapter 4.

2.8.13.2.2 Bricks/Blocks from Mine Waste and Industrial Waste

Bricks and blocks may be made using rejects, mine waste, and industrial waste, which may replace fine aggregates and partially replace cement in concrete. The utilisation of waste for brick production may be energy efficient and cost-effective and reduces the negative effects of waste on the environment. Large quantities of solid waste can also come from mining industries, which may be used for brick production.

2.8.13.2.3 Bricks from Coal Washery Rejects

The bricks may be produced from coal washery rejects. The waste from coal washery plants is hazardous and should be handled carefully to eliminate harmful effects on the surrounding areas. These bricks may be eco-friendly and will help in waste utilisation. They can control or reduce air, land, and water pollution, and are energy efficient and cost-effective.

2.8.13.2.4 C′ Brick

The bricks are manufactured using a C-brick machine developed by CBRI. The materials required for the production of bricks are fly ash–sand–lime, fly ash–sand–cement, and cement–sand aggregates.

2.8.13.2.5 Aerocon Panels

These may be prepared from two fibre-reinforced cement sheets with a lightweight core of Portland cement, binders, and aggregates from a mixture of silicaceous and micaceous types of aggregates. The material thus produced will be environmentally friendly by using fly ash as a replacement for timber-based products. The advantages of these applications will include quick construction, no wet plastering and on-site curing, eco-friendliness, lightweight, fire resistance, good thermal insulation, and resistance to water, termites, and weather. They can also be used in seismic and cyclone-prone zones. These are thin walls that can be relocated easily with minimal ground preparation for foundation work and can be executed easily.

2.8.13.2.6 Cement Concrete Hollow Blocks

These are produced as masonry units of concrete. The work is carried out based on the densification of a lean concrete mix and is used to produce regular and uniformly shaped high-performance masonry units. The blocks are cost-effective and may be considered a good alternative to burnt clay bricks for fire resistance, better durability, thermal insulation, partial resistance to sound, less dead load, and faster construction speed at sites.

2.8.13.2.7 Ferro-Cement

This item has been discussed separately in Chapter 6.

2.8.13.2.8 Rice Husk

Two types may be considered in this connection and may be mentioned as follows:

- **Low-Cost Sand–Crete Block –** Rice husk is produced using charcoal from rice husk, and it is pozzolanic in nature, which may be used for block making.
- **In Concrete –** Proper quality rice husk, which is highly siliceous, may be used as an admixture in concrete.

2.8.13.3 Plaster

2.8.13.3.1 Fibre-reinforced Clay Plaster

The sticking ability of clay plaster may be improved by using fibres. The fibres may be natural or artificial PP. Cracking due to drying shrinkage and thermal movement may be controlled by using fibres.

The dried plaster will be less brittle and will have a greater ability to withstand small movements of the substrate compared to conventional plasters.

2.8.13.3.2 Calcium Silicate Plaster

Calcium silicate plaster may be used, which will be economical, eco-friendly, with less wastage, lower energy consumption, and will provide a smart finish on the surface without emitting toxic fumes.

This may be safe for handling, require no skilled manpower, be durable, fast drying, and be carried out with reduced water consumption.

2.8.13.4 Roofing

2.8.13.4.1 Micro Concrete Roofing Tiles

Micro concrete roofing tiles are prepared using a controlled mix of cement, sand, fine stone aggregates, and water, and may be a cost-effective alternative. The tiles will be durable, aesthetically acceptable, and may be a good alternative for sloping roofs.

2.8.13.4.2 Bamboo Matt Corrugated Roofing Sheets

The roof cladding materials may include burnt clay, Mangalore tiles, corrugated sheets of galvanised iron, asbestos cement, aluminium, thatch, etc. The Building Materials and Technology Promotion Council (BMTPC) and the Indian Plywood Industries Research and Training Institute (IPIRTI) have jointly developed the material for manufacturing bamboo mat corrugated sheets.

Source:

- Vidya/Radha [30]

2.9 SHALLOW CORRUGATED ASBESTOS CEMENT SHEETS [38]

For this material, reference may be made to IS 13008:1990. Pozzolanic materials like pigments and fillers compatible with asbestos cement may be included. Colouring materials may be used in the mix for coloured products.

2.9.1 Dimensions and Tolerances

The sheets may be manufactured with

- Overall width, B = 1,015 mm
- Depth of corrugation, D = 20 mm
- Pitch of corrugation = 75 mm
- Nominal thickness, T = 4.2 mm
- Length of sheet, A = 1,500 – 2,250 mm

For the tolerances related to production, guidance will be available in the IS code.

2.9.2 Load Bearing Capacity

The load bearing capacity should not be less than 1.8 N/mm width of the specimen if tested as per IS code.

2.9.3 Finish

The finished products should have clean and straight edges. Furthermore, the products shall have rectangular shapes, good appearances, smooth surfaces on the weathering side, and no visible defects. The corrugations should be regular and correct.

Inspection, Manufacturer' Certificate, Testing, Marking, etc. – IS code may be referred.

Source:

- IS 13008 [38]

2.10 LOW-COST MATERIALS SUITABLE FOR HOUSING IN INDIA [39]

With the increase of urbanisation, the requirements for affordable low-cost housing are rising.

It is important to consider the energy consumption and environmental impact of the materials used for constructing houses. Low-cost building materials will meet the housing needs of middle and low-income groups and may contribute to sustainability, particularly when the materials are available locally.

There are different types that may be considered for low-cost construction. This will be briefly mentioned below:

2.10.1 Bamboo

Bamboo is a low-cost, lightweight, and durable material. The material is produced in India and must be treated with chemicals to make it termite resistant. Houses in earthquake-prone areas have been constructed with bamboo. Further comments on bamboo have been made separately.

2.10.2 Pre-fabricated Houses

The components of the house may be made in the factory and assembled on-site according to the design. The items may include steel frames, wooden panels, cement, and gypsum for walls, floors, doors, windows, etc., which may reduce construction costs. Some work related to plastering, plumbing, and wiring may be done during the casting stage.

2.10.3 Concrete Blocks

The blocks may be either hollow or solid and are normally lightweight, durable, and fire-resistant. They can be used for basement walls, foundations, and partition walls. Steel rods may be placed in the cores to improve strength. They are termite resistant, soundproof, and provide natural insulation against cold and heat. Additional comments have been made separately.

2.10.4 Interlocking Bricks

This type may be used with a projection on one side and a depression on the other side, allowing similar types of bricks to be laid like a jigsaw puzzle for constructing walls.

2.10.5 Compressed Earth Bricks

The discussion on compressed earth blocks has been covered in Chapter 3

2.10.6 Mud Bricks Reinforced by Natural Fibres from Straw and Coconut

This is a sustainable building material often used in rural areas. With these bricks, it is possible to construct durable, waterproof homes with adequate strength.

Bricks reinforced with natural fibres like straw and coconut can increase the durability and strength of the walls. Furthermore, a coating of sulphur can improve the water resistance of the walls. Additionally, other types of fibres for the bricks may include bagasse, cereal straw, corn straw, and rice husk, along with cement, to make the bricks strong and corrosion resistant.

2.10.7 Magnesium Oxide Cement

This is also known as eco-cement and requires only 20%–40% of the energy needed for the production of Portland cement. Different waste materials may be used for this and it can be considered a sustainable building material.

2.10.8 Shipping Container Houses

Homes made of steel shipping containers can be considered recycled building materials. It is a cost-effective pre-fabricated home that must be assembled on a prepared foundation.

2.10.9 Fly Ash Hollow Bricks

This type can be used for load-bearing walls associated with low-cost housing. These are prepared using fly ash, slag, stone powder or sand, and cement/gypsum for bonding. They may be energy efficient, water resistant, and provide natural thermal insulation related to low-cost housing. Further comments on fly ash bricks have been made in Chapter 4.

2.10.10 Autoclave Concrete or AAC Blocks

It is made of gypsum, quartz sand, lime, water, and aluminium powder. The blocks are manufactured using heat and pressure within an autoclave as per the specifications.

The blocks may be used for exterior and interior wall applications and will be lightweight and heat resistant. More comments have been made on this separately in Chapter 4.

Source:

- Deo [39]

2.11 SUSTAINABLE LOW-CARBON MATERIALS FOR BUILDINGS [40]

For sustainable low-carbon alternative building materials, the items mentioned below will be relevant:

- Energy conservation
- The use of local resources/materials to cut down transportation costs
- Minimising the use of high-energy materials
- Maximum use of local skills
- The use of industrial and mine wastes for the manufacture of building materials
- Recycling building waste/materials
- The use of renewable energy sources

2.11.1 Types of Low-Carbon Building Materials

The suitable materials for design and construction must satisfy the requirements of low carbon emission, minimal carbon footprint, and have the potential for recycling and reuse. The use of natural materials such as stones, soil, and timber/biomass may be considered.

To reduce carbon emissions, it will be important to produce building materials with minimal energy expenditure.

Some of these techniques are mentioned below:

2.11.1.1 Blended Cement

These are cements containing one or more complementary cementing materials like fly ash, ground granulated slag, silica fume, and reactive rice husk ash. The emission of CO2 may be reduced by using complementary cementing materials.

2.11.1.2 Stabilised Mud Blocks

Comments on stabilised mud blocks have been made in Chapter 3.

2.11.1.3 Compacted Fly Ash Blocks

Discussions on compacted fly ash blocks have been made in Chapter 4.

2.11.1.4 Rammed Earth Construction

The review of rammed earth construction has been done in Chapter 5.

2.11.1.5 Low-Energy Intensity Floor and Roofing System

The items may include

- The use of composite masonry jack-arch roof and floor systems
- RC filler slab
- Unreinforced masonry vaults for the roof, etc.

Source:

- Venkatarama Reddy [40]

2.12 ENERGY IN BUILDING MATERIALS/SYSTEM [40]

The energy requirements for building materials and the functioning of the building will be a critical issue for environmental impact and green construction.

The energy requirements for building materials may be divided into

- **Embodied Energy (Primary Energy)** – that is required for the production of the material and the assembly of the building.
- **The energy** required for the maintenance/servicing of the building to meet different requirements.

The first one will be a one-time investment and can vary depending on the type of material and the techniques used.

The second one would depend on the environment and its variations in the particular region and uses of the building.

It may be mentioned that rammed earth walls will have lower embodied energy requirements as compared to burnt clay masonry bricks.

Source:

- Venkatarama Reddy [40]

2.13 FURTHER COMMENTS ON EMBODIED ENERGY [41]

The embodied energy is the energy required for all the processes connected with the making of the building, covering mining and processing of natural resources, manufacturing, transporting, and finally the delivery of the material/product to the construction site. The embodied energy does not include the operation and disposal of the building material.

The total embodied energy of a building may be obtained from the combination of many processed materials.

The critical factor to reduce the impact of embodied energy is to design for long life, durability, and adaptability in buildings.

Alteration, renovation, and maintenance work will also be included in the embodied energy over building's life time.

It may not be easy to assess the embodied energy component within the structure/building.

The embodied energy is not dependent on the occupant; it is the energy built into the materials. The operational energy requirements will depend on the occupants and the types of uses of the building.

The embodied energy occurs once (i.e., not considering repair/maintenance), while the operational energy will accumulate over a period of time covering the total life of the building.

Source:

- Embodied energy, Your home [41]

2.14 REVIEW OF LOW-ENERGY BUILDING MATERIALS [42]

The following items may be identified as energy-efficient options in this context.

2.14.1 Recycled Steel

Recycled steel may be used instead of wood to satisfy design requirements. The use of steel is most suitable for designing structures that withstand seismic activity and high wind effects.

2.14.2 Plant-based Polyurethane Foam

Plant-based polyurethane foam may be prepared from natural materials like bamboo, hemp, and kelp, and may be used as an insulating material. It would have good resistance to moisture and heat and will provide protection against mould and pests.

2.14.3 Insulating Concrete Forms

This material may be used for freestanding walls and building blocks. It can be used by pouring concrete between insulating layers and allowing it to harden. The applications will contribute to energy savings.

2.14.4 Cool Roof

Using this material will improve heat dissipation and lower the temperature in the house during summer. It reduces heat from the atmosphere and is safe for the environment. It is capable of reflecting sunlight.

2.14.5 Straw Bales

Comments on straw bales have been made previously under green construction.

2.14.6 Plastic Composite Lumber

This material may be prepared from waste plastic and wood fibres. It will be durable and less toxic compared to treated wood. It will be resistant to mould and rot, being rigid in cold and pliable in heat compared to purely plastic building materials. The product may be suitable for bathrooms and outdoor decks.

2.14.7 Vacuum Insulation Panels

These are made of a textured silver rectangle that covers a core panel in an airtight envelope. Heat loss will be reduced, making it possible to achieve greener homes.

2.14.8 Low-E Windows

These are also referred to as high-performance windows and can replace normal glass, reducing heat in summer and controlling infrared radiation.

2.14.9 Structural Insulated Panels

These panels are prepared using foam insulation sandwiched between plywood or cement panels. They are fire-resistant and may be used for floors, load-bearing walls, basements, and foundations. There may be different types. Energy consumption will be reduced by using the material.

Earth – The comments have been made separately.

Energy-efficient materials will be good options for building design and construction.

Source:

- Dengarden [42]

2.15 ADDITIONAL COMMENTS AND REVIEW ON TYPES OF EARTH AND CLAY CONSTRUCTION

There may be different types of applications and used under different names as observed in other places. These are as follows [3]:

2.15.1 Mud Wall

This type of wall is constructed with earth mixed with straw. The method of construction will be in lifts or courses between 150 and 550mm, dependent on the material properties used. Mud wall is also known as Cob in Britain. Further comments on Cob will follow.

2.15.2 Clay Wall

This method involves the insertion of stones into the mud–wall mixture, which would be placed between the formwork. Good quality stones are normally placed on the outside face of the wall to provide the appearance of clay-mortared masonry finish.

2.15.3 Clay and Bool

In this method of construction, rounded stones of a type are placed in courses between the earth material. The rounded stones may not be suitable in other forms of building construction.

2.15.4 Shuttered Clay

This is formed within the shuttering faces of wood. Similar raw materials for mud wall, clay, or earth mixed with straw may be used for construction. Bricks may also be used to create a permanent face and to hold the clay or earth fill.

2.15.5 Turf

This may also be referred to as sod. Some comments on sod have already been made. Turf construction involves using earth blocks to build the wall. The earth blocks may be used alone or alternating with stone pieces. To provide additional strength, the turf is normally constructed in a herring-bone pattern.

2.15.6 Other Types

Timber framing along with clay has been used to form wattle-and-daub and other types of thin walls. Some comments on wattle-and-daub have already been made. Clay has been used for plasters mixed with straw, and the material may be used for deafening the timber floors.

Source:

- Earth and clay construction [3]

2.16 REVIEW AND SUMMARY OF ALTERNATIVE BUILDING MATERIALS THAT ARE ALREADY USED

The list of materials that are already used may be summarised as follows [43]:

2.16.1 General

- **Aluminium –** Windows, doors, facades, wall cladding, door handles, staircases, ladders, catches, roofs, support structure for solar systems, etc.
- **Soil Conditioning Agents –** Bentonite slurry, polymer liquids, etc.
- **Bitumen Materials –** Paving system, asphalt paving, tunnelling, pipe jacking, etc.
- **Crumb Rubber –** Noise reduction, reduces skid.
- **Tempered Glass –** Automobile windows, bulletproof glazing.
- **Ferro-Cement Construction –** Different applications with photographs have been included in Chapter 6 for ferro-cement.
- **Fibre-Reinforced Polymer –** Further discussion has been covered for fibres in Chapter 7.
- **Glass Fibre-Reinforced Plastics –**
- **Polyester Fibres –** Comments have been made in Chapter 7.
- **Bamboo Reinforced Plastics –**
- **Stabilised, Compressed Earth Blocks –** These are mostly covered in Chapters 3 and 4 with photographs.
- **Stabilised Adobe**
- **Fly ash/Sand–Lime Bricks**
- **Clay–Fly Ash Burnt Bricks**
- **Fly Ash–Lime–Gypsum Stabilised Mud Blocks –** Developed by the INSWAREB Building Centre
- **Fly Ash – Lime–Gypsum Items –** Developed by the INSWAREB Building Centre
- **Lato Blocks –** Developed by CBRI
- **Clay Red Mud Burnt Bricks –** Developed by CBRI
- **Pre-cast Hollow Concrete Blocks –** Developed by CBRI
- **Pre-cast Stone Blocks –** Developed by CBRI
- **Pre-cast Concrete Blocks –** Comments on pre-cast have been presented separately.
- **Pre-cast Aerated/Cellular Concrete Walling Blocks and Roofing Slabs –** Developed by CBRI
- **Fly Ash-Based Lightweight Aerated Concrete Walling & Roofing Block –** Developed by CBRI/CFRI
- **Rat-Trap Bond –** Developed by Laurie Baker
- **Bamboo/Timber Mat-Based Walls (EKRA Walling)**
- **Composite Ferro-Cement System –** Details on Ferro-cement have been discussed in Chapter 6.

2.16.2 Roofs and Other Applications

- **Life Extended Thatch Roofing –** Developed by CBRI RRL-TVM
- **Coconut Fibre and Wooden Chips Roofing Sheets –** Developed by RRL-TVM
- **Guna Tile Roof**
- **Corrugated Bamboo Roofing sheets**
- **Country Tile Roofing**
- **Micro Concrete Roofing Tile –** Developed by ITDG/Tara Nirman Kendra

- **Pyramidal Brick Roof –** Developed by CBRI
- **Cement-Bonded Fibre Roofing Sheets –** Developed by CBRI
- **Extruded Structural Clay Unit Floor/Roof –** Developed by CBRI/SERC
- **Stone Patti Roofing**
- **Pre-cast Brick Panels –** Developed by CBRI
- **Pre-cast Brick Arch Panel System –** Developed by CBRI
- **Pre-cast L-Panel Elements –** Developed by CBRI
- **Ferro-Cement Channel Shell Units –** Developed by the SERC/Auroville Building Centre
- **Pre-cast Plank and Joist Flooring/Roofing –** Developed by CBRI
- **Funicular Shells over Edge Beams –** Developed by SERC
- **Pre-cast Cored Units –** Developed by CBRI
- **Pre-cast Waffle Units –** Developed by CBRI
- **Filler Slabs –** Used by Laurie Baker
- **Pre-cast Channel Units –** Developed by CBRI
- **Pre-cast In-Situ Thin Ribbed Slab –** Developed by CBRI

Discussions on ferro-cement and fibre-reinforced concrete have been presented in Chapters 6 and 7, respectively, with photographs.

Source:

- Constructor [43]

2.17 DIRECT MOULDING AND COB [44]

In this method, the construction is carried out directly with earth, in layers, and without the use of moulds or shuttering, utilising the plasticity of the soil when it is wet to provide a monolithic construction. The prepared earth is applied directly to provide the shape as used in pottery.

Some well-known and unique applications have been made in Africa and Yemen in the past using this technique. The method is also well known in other parts of the world, especially related to the developments that took place in European countries. Cob cottages have been found in England and Scotland. In France, this method of construction is referred to as bauge.

2.17.1 Materials Used

The materials may be listed as follows:

2.17.1.1 Earth

Reasonably heavy sandy soil will be suitable for Cob buildings. Earth with high clay content will be difficult to work with and would take a longer time to dry out. The particle size analyses of the earth samples collected from a house in England indicated that the proportions that may have been used were sand –33%, clay –21%, and straw –1.6%. The proportions may be similar to those of rammed earth construction.

2.17.1.2 Fibres

Vegetable fibres have been used with soils. The fibres may include straw, wheat chaff, heather, etc. and the inclusion of these fibres will serve different functions. The fibres in the earth would improve the tensile strength and flexibility of the material. Reinforced soil with fibres may be subjected to deformations and can perform without cracking, especially during the drying period. Due to the presence of fibres, the density will be reduced, and the earth with fibres would have improved thermal insulation properties.

The proportion of fibres may be in the range of 25 kg/m^3 of earth (i.e., 1.6%), which is about 250 L of fibres for every 1,000 L of soil. The straw, before mixing, will have to be chopped into pieces with dimensions of 150–400 mm.

General Comments – A few salient items are mentioned below:

- The foundation work may generally consist of heavy footing made of stone or brick, covered with material that has resistance against humidity.
- The cob construction will take a longer time to dry in rainy weather. In some cases, it might take 3 weeks between each course.
- Completing the construction of a house will be a lengthy process.
- A team of four men may construct 15 m^2 of wall, 600 mm thick, in a day.
- The use of artificial drying methods is uncommon, other than lighting a good fire inside.
- The critical points where cracks might occur are at the corners of windows and doors. Wooden sticks coated with mud may be used in the width of the wall for control.
- The houses are generally single storey. The walls may be 450 mm thick and may be increased to 800 mm for two- or three-storey houses.
- Special care will be required to maintain the line, verticality, levels, and working at the corners as per good engineering practice.
- Excellent protection against humidity will be required for cob houses. Cob buildings may last for a long time.
- This type of construction may be used in cold and windy winters.
- Cob buildings may last for a long time.
- Additional treatment may be required to prevent leakages from the roof (i.e., use of tiles, roofing felt etc.)
- Careful detailing will be required (i.e., use of projecting ledges) to prevent water from running down the wall.
- Rendering may be done for finish, protection, and aesthetics.

Further Comments – These will cover both advantages and disadvantages of the method as mentioned below:

2.17.1.3 Advantages

These may include:

- Different types of shapes are possible for a building, as the method does not need formwork or moulds.
- The method requires only easily available tools.
- The technique can provide an efficient solution when the soil is too wet for rammed-earth construction.
- The method will require minimal labour, making the construction inexpensive.

2.17.1.4 Disadvantages

The following may be relevant:

- Drying time will be longer in cold and humid climates, as stated.
- The mechanical performance of the material and the construction against bad weather has been found to be rather poor.

Further study/research and applications will be necessary to improve the performance of cob construction. Modern stabilisation and mechanisation methods may be helpful and should be properly considered to control/eliminate the disadvantages associated with the method.

Source:

- Building with Earth [44]

2.17.2 Further Review of Advantages and Disadvantages of Cob [45]

2.17.2.1 Advantages

- It is a simple construction technique, easy to learn, and may require a little or no experience.
- It is a low-cost material application.
- The structure will be strong and sturdy.
- This type of construction will last a long time.
- With this technique, very attractive structures can be constructed.
- This will provide high thermal mass.

2.17.2.2 Disadvantages

- It will be a labour-intensive activity.
- It may be subject to more cracking compared to other building methods.
- It will be a slow construction technique.
- The flexibility and convenience of modular bricks or forms will be missing.
- The construction may have poor insulation value.

Source:

- Building your Homes with Earth [45]

2.18 WATTLE-AND-DAUB [45]

It is a simple earth building method that has been used for a long time in the past. This type of construction can be found in South Asia and Africa and has also been used in Europe.

It may be used as an infill item for traditional timber-framed buildings.

The walls of wattle-and-daub will be thin and primarily made of lattice material with mud packed from both sides. Considering the thickness, it will have less thermal mass.

2.18.1 Construction Method

As stated, it is the method of pressing mud into a lattice of sticks, boards, bamboo, or reeds. The mud may contain additional sand or fibres depending on the requirements.

It will require a few workers for the operation and has been used for small buildings and privacy walls.

2.18.1.1 Advantages

- The construction can be quick.
- It is a low-cost material.
- It will be suitable for small structures.
- It is lightweight.
- Structural flexibility may be advantageous for earthquake resistance.

2.18.1.2 Disadvantages

- The construction will have low thermal mass.
- Insulation performance will not be good.
- It is more suitable for tropical environments (i.e., warm to hot).
- It is not a monolithic construction like rammed earth or cob.
- Rotting may occur in the lattice if kept exposed to water.
- For heavy roofs, it will function as a non-load-bearing element.
- It is not convenient for extensions or second stories.

Source:

- Building your Homes with Earth [45]

2.19 POURED EARTH METHOD [46–49]

The poured earth method of construction will be similar to ordinary concrete [46], using formwork and cement. In this technique, sand or gravel aggregates in concrete will be replaced by the appropriate type of ordinary soil. The poured earth method will require a smaller amount of Portland cement in the mix.

Standard types of formwork may be used for the construction [46].

Poured earth may be referred to as moderate-strength concrete. Tests may be carried out for compressive strength and shrinkage control.

Magnesium oxide may be used to reduce the amount of Portland cement in the mix [46].

The formwork may be removed after the material has set.

About 10% of cement may be used, in the mix, to control shrinkage and to achieve durability [47].

It is mentioned that with the addition of natural or synthetic fly ash and lime, the cement content in the mix may be reduced by approximately 50% [46].

The New Zealand standard deals with the tests and quality control covering poured earth, adobe, rammed earth, and pressed blocks [47,48].

The poured earth method provides the opportunity to use on-site materials to construct walls, interior plaster, floors, and some roofing systems [49].

Poured earth can reduce the use of high-energy-intensive Portland cement by replacing it with low-energy materials like fly ash [49].

The earth construction methods are generally labour-intensive. Poured earth will have the highest potential to be a mechanised system to reduce labour [48].

The soil mixing may be done using concrete construction equipment, improving the uniformity of earth construction. Very nominal maintenance work may be required for poured earth walls, which will have high resistance against sun and water [46].

2.19.1 Review of the Method [50]

In this method, the earth is poured into frameworks and allowed to dry as mentioned. Using frameworks, the gaps between them will be filled with mud. This type of construction is also referred to as Encajonado, which in Spanish means the method of stuffing earth/mud into a lath framework.

This method of construction has been carried out in the UK, where a mixture of chalk, soil, and straw has been poured into the framework. The framework may act as primary structural support for the building. After the material has dried, the formwork may be removed to expose the walls.

Source:

- Building Today for Tomorrow, Poured Earth [46]
- Poured Earth – Earth Building Association of New Zealand [47]

- Williamson [48]
- Living Systems, Poured Earth Sustainable Architecture [49]
- Niroumand, Zain and Jamil [50]

2.20 CUT BLOCKS METHOD [51]

In this construction method, cohesive soils containing concretions of carbonates are cut into blocks and may be used as bricks or stones. These blocks may be located in tropical areas where lateritic soils serve as suitable material for construction. Lateritic soils may be found in two conditions: soft soil, which hardens when exposed to air over time, and in the form of hard crust, which has already hardened over the years. In some places, the soil may not be cohesive enough; in such situations, a mixture of topsoil and grass may be used to make the blocks. The blocks may be stacked one on top of another.

Source:

- Biju [51]

2.21 SOIL CEMENT METHOD [46]

This method is similar to ordinary concrete, where sand/gravel aggregates are replaced by soil. The soil-cement method has been used for constructing walls, paved roads, stabilising river banks, making floors, etc.

Soil may be properly mixed with Portland cement and water. The cement content in the mix may vary depending upon the type of soil, typically ranging from 6% to 16% of cement by volume based on the density of the soil. It may be tamped after placement and covered with plastic for proper curing.

Source:

- Building Today for Tomorrow, Poured Earth [46]

2.22 STRAW CLAY TECHNIQUE [51]

In this method, clayey soil is mixed with water and then placed on straw. The earth helps to bind the straw together. Straw clay is suitable for fabricating wall panels, large bricks, and flooring. This construction method may be considered a substitute for drywalls. Furthermore, it will be a lightweight construction with insulation and good thermal properties.

Source:

- Biju [51]

2.23 SHAPED EARTH [51]

The method of construction is similar to a potter's technique for making pots. In this method, plastic clay is placed directly into position without using a mould or formwork. The important factors will be the quality of soil, water consistency, and soil preparation. This type of construction may be executed with minimal use of tools as well as the labour force. Shaped earth technique will provide flexibility in architectural design to a large extent. A few items related to this method will need careful understanding, including soil quality and controlling shrinkage once the structure dries.

Source:

- Biju [51]

2.24 FURTHER ON SUPER-ADOBE (EARTH BAG) [45]

In this method of construction, PP bags are used, filled with soil and compacted. With this technique, arches, vaults, and domes may be constructed.

2.24.1 Method of Construction

The method of construction involves the use of tube or PP bags filled with a mixture of soil, sand, and often cement. The bags are tamped to provide compression.

Barbed wire is placed between the layers of filled bags to obtain tensile strength.

2.24.1.1 Advantages

- Will give high thermal mass.
- It is a low-cost material.
- It will be strong and sturdy after construction.
- Suitable for domes, vaults, and arches.

2.24.1.2 Disadvantages

- The construction will not be monolithic.
- It is a labour intensive activity.
- It will have poor insulation.

Source:

- Build Your Homes with Earth [45]

2.25 HAND-FORMED LAYERED CONSTRUCTION [52]

The guidance from the IS code is briefly mentioned below:

This is an old form of earthen wall construction and may be considered the weakest of all earthen walls. Fissures may develop in horizontal and vertical directions, as enough moisture for full dispersion of clay may not be possible. Straw may be used in the mix to improve strength and eliminate fissures.

The quality of construction will be better if the clay–water–straw mixture is kept for 7 days (minimum 3 days) before being used for walls. This will help disperse moisture into the clay and facilitate the decomposition of straw into fibres.

The lower layer of the walls should be moistened properly before adding the new layer to prevent developing horizontal fissures at the joints.

Source:

- IS-13827 [52]

2.26 SEISMIC STRENGTHENING OF BEARING WALL BUILDINGS

The comments on seismic strengthening of bearing walls [52] have been covered in Chapter 5.

Source:

- IS-13827 [52]

2.27 SUMMARY OF COMMENTS ON EARTH TECHNOLOGY [44]

The main comments may be summarised as follows:

- Earth is an ancient and common construction material used by people in many countries.
- Our forefathers used this material without hesitation for various constructions with success.
- It has dominated as a building material in many countries.
- The applications of earth technology largely cover the use of cob, adobe, compressed earth blocks, rammed earth, wattle-and-daub, etc., or combinations with other materials to achieve composite actions that satisfy design requirements. These are based on traditional methods with technological simplicity.
- A wide range of possible applications may be observed, and the buildings may differ significantly based on their intended purposes.
- It is possible to have a natural house using earth technology.
- The desire for a more human environment may be achieved with earth engineering.
- The ecological advantages of earth construction are easily appreciated.
- Earth architecture/engineering need not be considered an archaeological curiosity.
- Considering the rise in energy costs, earth construction may be preferred over cement and fired bricks.
- Will require less capital investment for applications and may be suitable and adaptable for use.
- Different types of production may be organised, i.e., industrial or workshop production, owner–builder cooperatives, rural self-help builders, etc.
- Modern technology may be becoming increasingly remote from ordinary people.
- Earth construction and technology are well known and are currently positioned to compete with traditional materials in many areas.
- The use of earth in some projects will also help control the depletion of the constituent items of traditional materials, to a limited extent, depending on the applications.
- Depending on the state of the material at the time of application, earth may be used in solid, plastic, dry, or liquid forms, as mentioned.
- The use of locally available materials will be possible for the projects. The utilisation of local people with minimal training may be employed for construction. This will educate the local population and may help address employment issues to some extent. Furthermore, this will also aid in self-development. The workers do not have to travel long distances for jobs.

Source:

- Building with Earth [44]

REFERENCES

1. Earth Structure, Wikipedia, the free encyclopaedia, pp. 1–17.
2. Sylvain Arnoux, Alexandre Douline and Serge Maini, A Training Manual for Technicians and Entrepreneurs, Auroville Building Center, Auroville, July 1992.
3. Earth and Clay Construction, *INFORM, Information for Historic Building Owners*, Historic Scotland, Alba Aosmhor, June 2015.
4. Energy Engineering Company, 6 Fundamental Principles of Sustainable Building Design, May 2021.
5. WBDG, Whole Building Design Guide, August 2018, pp. 1–8.
6. IEREK News, The benefits of green buildings, in *International Conference on Resilient and Responsible Architecture and Urbanism (RRAU)*, Groningen, Netherlands, April 2018, pp. 1–7.
7. How can we make our buildings green? World Green Building Council, 2012, pp. 1–4.
8. HMC Architects, The Top 6 Sustainable Architecture Strategies for Public Building Deign, October 2018, 2021, pp. 1–10.

9. 17 Sustainable Architectural Design Ideas, Kebony, USA, pp. 1–34.
10. Importance of Green Building, Green Built Alliance, Green BUILT, pp. 1–4.
11. 10 Benefits of Green Building, New School of Architecture and Design, September 2021, pp. 1–5.
12. Green Buildings: Advantage and Disadvantages, Blog Weetas. September 2021,
13. Mahendriyani D, Alternative Building Materials for Green Construction, Asia Green Building, July 2016, pp. 1–3.
14. Wikipedia, Straw-bale construction, pp. 1–9.
15. Wikipedia, Hempcrete, pp. 1–5.
16. Grasscrete Paving, Grass Concrete Limited, 2021, pp. 1–6.
17. Bhalla S, Scientific Design of Bamboo Structures, Department of Civil Engineering, IIT Delhi.
18. Kaminski S, Lawrence A, Trujillo D J A, Feltham I, Structural use of bamboo, part 3: Design values, *The Structural Engineer,* December 2016, Vol. 94, pp. 1–14.
19. Appendix 1, Bamboo Structural Design Code, INBAR, 1998.
20. Heritage Builders Ltd., Designing Building Wiki, ICE, October 2020.
21. Loffer L, *The Advantages of Wood as a Building Material*, Wagner Meters, 2021, pp. 1–12.
22. Ozen R, Kultur Ve Turizm T C, Bakanligi, Wood as a Building Material; It's Benefits and Disadvantages, Zonguldak Karaelmas University, 2021, pp. 1–5.
23. Khanna S K, *Turning Plastic Waste into Eco-friendly Building Materials & Products*, NBM & CW, August 2019, pp. 1–16.
24. Bonnefin L, EME Mycelium Brick, Emerging Materials, February 2017.
25. What is Timbercrete? Just Wood Furniture, August 2020, pp. 1–5.
26. BuilderSpace, How is Ferrock Made? Builder Space, pp. 1–13.
27. Feigin M E, Wisenbaker, Jr, J W, Wambaugh J A, Construction boards with carbonate. Google patents, US7867597B2, January 2011, pp. 1–12.
28. Wikipedia, Magnesium oxide wallboard, pp. 1–4.
29. Wikipedia, Bio-composites, pp. 1–4.
30. Vidya/Radha, Alternative Low-Cost Building Materials, School of Architecture, DSATM, Bangalore, India.
31. Wikipedia, Coir, pp. 1–10.
32. Kursten E, Coir - Traditionally used fibres for new construction applications, in *First International Conference on Bio-based Building Materials*, France, June 2015, pp. 1–9.
33. Sustainable Straw, Uses of Straw, for People and Planet, 2021, pp. 1–6.
34. Vyncke J, Kupers L and Denies N, Earth as building material - an overview of RILEM activities and recent innovations in geotechnics, *MATEC Web of Conferences*, 2018, Vol. 149, pp. 1–12.
35. ECVET Earth Building, Earth as a Construction Material, 2021, pp. 1–4.
36. Wikipedia, Jute, pp. 1–11.
37. Rahman M M and Islam M S, Use of Jute Fibre for Earthquake Resistant Earthen House Construction, BUET, Dhaka, Bangladesh, 2021, pp. 1–11.
38. IS 13008: 1990, Shallow Corrugated Asbestos Cement Sheets - Specification.
39. Deo R, What are the best materials for Low cost housing in India? November 2019, pp. 1–19.
40. Venkatarama Reddy B V, Sustainable materials for low carbon buildings, *International Journal of Low-Carbon Technologies*, September 2009, Vol. 4, pp. 1–21.
41. Your Home, Australian Government, Australia's Guide, Embodied Energy, 2021, pp. 1–15.
42. 10 Energy Efficient Building Materials, Dengarden, June 2021, pp. 1–5.
43. Alternative Building Materials Used in Construction Industry, The Constructor, Civil Engineering Home, 2019, pp. 1–15.
44. Doat P, Hays A, Houben H, Matuk S and Vitoux F, *Building with Earth,* Translated by Asha Puri, The Mud Village Society, New Delhi, India.
45. Earth Building Techniques - Build your Homes with Earth - This Cob House, 2021, pp. 1–22.
46. Building Today for Tomorrow, Poured Earth, Greenhomebuilding.com, Sustainable Architecture
47. Poured Earth - Earth Building Association of New Zealand, Natural Building Technique.
48. Williamson J A, Earth Construction: Poured Earth, University of Colorado, 2006, pp. 1–26.
49. Michael Frerking, Living Systems, Poured earth Sustainable Architecture, 2021, pp. 1–8.
50. Niroumand H, Zain M F M and Jamil M, Assessing of critical parametrs on earth architecture and earth buildings as a vernacular and sustainable architecture in various countries, in *2nd Cyprus International Conference on Educational Research,* National University of Malaysia (UKM), Malaysia, Various Types of Earth Buildings, Science Direct, Elsevier.
51. Biju A, 10 Construction Techniques Used in Earth Architecture - RTF, Rethinking The Future, pp. 1–29.
52. IS-13827: 1993, Improving Earthquake Resistance of Earthen Buildings - Guidelines.

3 Compressed Earth Blocks (CEB), Soil/Earth, Stabilisers, Fibres, and Applications

3.1 INTRODUCTION

An important aspect of earth engineering is the type and suitability of the soil. Adequate knowledge and experience in using the proper types of soil are critical for the production of compressed earth blocks (CEB). A brief discussion on soil follows.

3.2 MAIN TYPES OF SOIL

Broadly, soil can be divided into two groups [1]:

- Formation due to chemical and physical rock weathering, including residual soil or transported soil.
- Organic origin. When rock weathering materials are found at the location where they originated, they form residual soil; otherwise, the soil may be classified as transported, regardless of the method of transportation.

The depth of residual soil can vary and primarily depends on climatic conditions and time of exposure. In some cases, the depth may exceed 30–50 m. In moderate/temperate zones, residual soils are typically stable and stiff. Transported soil, on the other hand, may be loose and soft in places and can extend to significant depths. Such transported soils may pose challenges for foundation design and other construction activities.

Organic origin soils can be formed by the growth and decay of plants, such as peat mosses, or by the accumulation of fragments from inorganic skeletons or shells of organisms. Soils can be of organic or inorganic origin. Generally, organic soils refer to transported soil made from rock weathering, with visible admixtures of decayed vegetable matter.

Soil conditions at the construction site must be assessed before designing the foundation of the structure. Explorations should be conducted using test borings or test shafts, supported by other laboratory tests in accordance with IS Codes. Construction activities will significantly influence the type of soil at the site where the structure is built.

Source:

- Terzaghi and Peck [1]

3.3 FURTHER ON TYPES OF SOILS

The following types may be mentioned [1]:

Sand and Gravel – These can be described as cohesionless aggregates consisting of rounded, sub-angular, or angular fragments of unaltered rocks or minerals. Particles up to 3 mm (1/8 in) are referred to as sand, while sizes from 3 mm to 150 or 200 mm (6 in or 8 in) are classified as gravel. The pieces with a diameter greater than 200 mm (8 in) are referred to as boulders.

DOI: 10.1201/9781003319108-3

Inorganic Silt – This type may be defined as fine-grained soil with hardly any plasticity. The less plastic type is made of more or less equidimensional grains of quartz and may be referred to as rock flour. The most plastic varieties may contain a sufficient amount of flake-shaped particles and are known as plastic silt. Due to its smooth texture, inorganic silt may be mistakenly assumed to be clay. However, this can be easily checked; when shaken in the palm of the hand, a portion of the saturated inorganic silt will expel water, making the surface look glossy. When the pat is bent between the fingers, the surface becomes dull again. This method is known as the shaking test.

Organic Silt – Organic silt consists of fine-grained plastic soil (more or less) mixed with finely divided particles of organic matter. Noticeable fragments of partly decayed vegetable matter and shells may be present. The colour of the material may range from light to dark grey and may contain a large quantity of H2S, CO2, and other gaseous products of decayed organic matter, which are responsible for providing the characteristic odour. The permeability of organic silt will be very low, and compressibility may be very high.

Hard Pan – This type of soil offers high resistance to the penetration of drilling items. Generally, hardpan types are very dense, well-graded, and made of cohesive aggregates of mineral particles.

Clay – Some comments on this have been made before. Clay is composed of a combination of microscopic and sub-microscopic particles developed from the chemical decomposition of rock constituents. The material may be plastic within a moderate to large range, depending on the amount of water content. The dry samples of clay may be very hard, and no powder can be removed by rubbing the surface with fingers. The permeability of clay will be very low. At higher water contents, the material can be sticky.

Organic Clay – This type of clay material may depend on the presence of finely divided organic matter for some of its relevant physical properties. When saturated, organic clay may be very compressible, but when it is dry, its strength will be high. It is normally dark grey or black in colour and may have a specific odour.

Till – This is an unstratified glacial deposit of clay, sand, silt, gravel, and boulders. It is associated with rock surface regions that were glaciated during the ice age.

Peat – This is a combination of fibrous macroscopic and microscopic elements of decayed vegetable material. The colour may vary from light brown to black. Peat will be compressible, and hence the material may not be suitable under foundations and supporting earth embankments.

Loess – This is a wind-blown sediment with uniform cohesive properties. The size of the particles may be between 0.01 and 0.05 mm, and the cohesive performance is due to the presence of a calcareous binder. The common colour will be light brown. It may be recognised by observing vertical root holes, and it may stand nearly vertical slopes. With more water, the bond between the particles may dissolve, and the surface of the deposit may settle.

Tuff – This may be defined as a fine-grained water- or wind-laid aggregate having very small mineral or rock fragments coming from volcanoes during explosions.

Modified Loess – The typical characteristics of the material, i.e., loess may have been lost due to secondary processes such as covering erosion, temporary immersion, and further deposition. The chemical changes may have occurred with the destruction of the bond between the particles or through chemical decomposition that may have taken place due to perishable items like feldspar. Thorough chemical decomposition may provide greater plasticity than other types of modified loess.

Lake Marl or Boglime – This is associated with the white fine-grained powdery calcareous deposits precipitated by plants in ponds. It is generally connected with beds of peat.

Diatomaceous Earth – This refers to a deposit of fine (Kieselguhr), normally white siliceous powder containing mainly the remains of diatoms. The term diatom includes a group of microscopic unicellular marine or freshwater algae containing silicified cell walls.

Marl – This is an alternative term used for different types of fairly stiff or very stiff marine calcareous clays with a greenish colour.

Adobe – This is associated with different types of light-coloured soils ranging from sandy silts to very plastic clays. Further details on adobe have been covered in Chapter 4.

Caliche – This type may refer to layers of soil where soil grains are connected together by lime as carbonates. The layers may form at a depth of a few metres below the ground surface. The thickness of the layers may vary from a few millimetres to a few metres. It seems that a semiarid climate may be suitable for their formation.

Bentonite – It is a clay product containing a large amount of montmorillonite. Most bentonites are produced by the chemical changes of volcanic ash. The dried bentonite will swell in contact with water more than other dried clays. However, saturated bentonite will shrink more upon drying. Bentonite has many applications in pile construction work.

Varved Clay – This is a combination of alternating layers of medium grey inorganic silt and darker silty clay. The thickness of the layers may not normally exceed 40 mm (1½ in); however, thicker varves may also be observed. They may have been transported to freshwater lakes by the melting of water at the close of the ice age. They may be affected by the properties of silts and soft clays.

Source:

- Terzaghi and Peck [1]

3.4 SIZE AND SHAPE OF SOIL PARTICLES [1]

The size of soil particles may vary from boulders to small particles.

It should be noted that grains larger than 0.06 mm may be seen by the naked eye or with the help of a hand lens.

Grains from 0.06 mm to 2 μm (i.e., 1 μm = 0.001 mm) may be examined only under a microscope. They are referred to as the fine fraction.

Grains smaller than 2 μm will be classified as very fine fraction. Grains sized between 2 μm and approximately 0.1 μm may be studied under a microscope, but their shape cannot be distinguished. The shape of grains smaller than 1 μm can be ascertained using an electron microscope. The molecular structure of these may be studied or investigated by means of X-ray analysis.

Source:

- Terzaghi and Peck [1]

3.5 TEXTURE, STRUCTURE AND CONSISTENCY [1]

Texture – The term texture is associated with the degree of fineness and uniformity of the soil. Various terms are also used to describe it, such as floury, smooth, gritty, or sharp. This can be determined from the experience of rubbing the soil between the fingers.

Structure – The structure deals with the pattern of combining the soil particles in the aggregate. This may be a single-grained structure where the grains touch several of their neighbours. A single-grain structure may be loose or dense.

Consistency – The term is connected with the degree of adhesion between soil particles and the resistance provided by the soil particles against the forces that may tend to deform or rupture the soil formation/collection. Consistency may be referred to as hard, stiff, brittle, friable, sticky, plastic, or soft.

3.5.1 Other Items

These may be mentioned as

- **Porosity** – It is the ratio of the volume of voids to the total volume of the soil aggregate. The volume of voids refers to the portion of the volume of the soil not occupied by mineral particles.
- **Void Ratio** – The void ratio is the ratio of the volume of voids to the volume of the solid substance.
- **Water Content** – This is the ratio of the weight of water to the dry weight of the aggregate. It is normally expressed as a percentage.
- **Unit Weight of Soil** – This is defined as the weight of soil material (i.e., soil plus water) per unit volume.
- **Liquid Limit** [2] – It is defined as the water content at the boundary between the liquid and plastic states of soils.
- **Plastic Limit** [2] – It is the water content at the boundary between the plastic and semi-solid states of soils.

Source:

- Terzaghi and Peck [1]
- Teng [2]

3.5.2 Further on Clays

The additional items may include

Consistency – The consistency of clays and other cohesive soils may be expressed as soft, medium, stiff, or hard. The measurement of consistency will be the load per unit area at which unconfined prismatic or cylindrical samples fail in a simple compression test. The value is termed the unconfined compressive strength of soil.

The values for consistency and unconfined compressive strength are shown in Table 3.1.

TABLE 3.1
Consistency of Clay, Unconfined Compressive Strengths

Sl. No.	Consistency	Unconfined Compressive Strength (kg/cm^2)
1	Very soft	<0.25
2	Soft	0.25–0.5
3	Medium	0.5–1.0
4	Stiff	1.0–2.0
5	Very stiff	2.0–4.0
6	Extremely stiff	>4.0

When it is extremely stiff clay, it is brittle and may be referred to as hard type.

Sensitivity – This indicates the effect of remoulding on the consistency of clay, irrespective of the physical nature responsible for the changes.

Source:

- Terzaghi and Peck [1]

3.6 ENGINEERING PROPERTIES OF SOILS [1,2]

The properties of soils are complex and may vary.

The important engineering properties are

- **Basic Properties** – related to unit weight, void ratio, water content, and grain size distribution
- **Seepage** – permeability
- **Strength** – shear strength
- **Compressibility** – consolidation
- **Compaction Characteristics** – may be required for improving the bearing capacity of soil by compaction

3.6.1 Engineering Properties of Cohesive Soils [1,2]

The following items are relevant:

- It may often have low shear strength.
- It can lose part of its shear strength on wetting.
- It may become compressible and plastic.
- It can lose part of its shear strength upon disturbance.
- It is an impervious material.
- It can deform plastically (creep) under constant loads.
- It can shrink on drying and expand upon wetting. Seasonal changes may occur and should be duly considered in the design.
- It is a poor material for backfill and can generate large lateral pressure.
- It may be considered a poor material for embankments due to low shear strength, and it may be difficult to compact.
- Clay slopes may have landslide problems.

3.6.2 Engineering Properties of Granular Soils [1,2]

The following may be mentioned:

- It is a good material for foundation work and for supporting structures and roads. Other than loose sand, the bearing capacity will be high, and the settlement will be less. The settlement will occur after the application of loads.
- It is a good backfill material for retaining walls and basement walls, etc. It will exert less lateral pressure, is easy to compact, and easy to drain.
- Due to its high permeability, the material cannot be used alone in earth embankments for dikes and reservoirs, etc. Dewatering may be required when working below ground level.
- It is a good embankment material because of its high shear strength, ease of compaction, and resistance to frost action.
- It may be subjected to settlement under vibration from loads.

3.6.3 Engineering Properties of Organic Soils [1,2]

When the soil contains a large amount of organic matter, it may be called organic soil, and the engineering properties may be affected.

The following will be relevant:

- Low shear strength.
- Spongy structure that may deteriorate quickly.
- High compressibility.
- Acidity and harmful effects on construction materials.

3.6.4 Engineering Properties of Silt and Loess [1,2]

The following items may be noted:

- Low shear strength just after the application of loads.
- Low permeability.
- High capillarity and frost susceptibility.
- Low relative density, which may make compaction difficult.

3.6.5 Engineering Properties of Rocks [1,2]

The following items will be important:

- Texture and structure.
- Mineral constituents.
- Weathering conditions.
- Joints, bedding planes, and foliations.
- Faults.
- Cementation.

Source:

- Teng [2]
- Terzaghi & Peck [1]

3.7 REVIEW AND MORE ON SOILS

3.7.1 Soil Components [3]

These may be described as follows:

- **Gaseous Components** – air, carbon dioxide, etc.
- **Liquid Components** – water, acids, alcohol, etc.
- **Solid Components** –
    - **Organic Particles** – humus and roots, typically found in the top layer. These items should be eliminated.
    - **Clay Particles** – considered as a cementing agent for the earth. The clay particles may swell and shrink as mentioned.
    - **Inert Mineral Particles** – pebbles, gravel, sand, and silt. These are considered to be the skeleton of the earth.

TABLE 3.2
Types of Soil

1	Gravely soil	—	May have all the components, i.e., gravel, sand, silt, and clay in different proportions
2	Sandy soil	—	May have all the components, i.e., gravel, sand, silt, and clay in different proportions
3	Silty soil	—	May have all the components, i.e., gravel, sand, silt, and clay in different proportions
4	Clayey soil	—	May have all the components, i.e., gravel, sand, silt, and clay in different proportions

3.7.2 Sizes of the Components [3]

These include

- **Pebbles** – 200–20 mm
- **Gravel** – 20–2 mm
- **Sand** – 2–0.02 mm
- **Silt** – 0.02–0.002 mm
- **Clay** – 0.002–0.000 mm

3.7.3 Composition of Soil [3]

The composition of good soil may be considered as a combination of the following:

Good Soil	=	Gravel	+	Sand	+	Silt	+	Clay
(100%)		(15%)		(50%)		(15%)		(20%)

Source:

- Auroville Building Center [3]

3.8 FURTHER ON SOILS [3]

The soil types may be divided into four groups.
These are mentioned in Table 3.2.

3.8.1 Soil Behaviour [3]

The four typical soils with the same quantity of water in the soil sample may be described as follows:

a.	Gravely Soil and Water	–	When moulded, it collapses into pieces
b.	Sandy Soil and Water	–	When moulded, it keeps the shape but very crumbly
c.	Silty Soil and Water	–	When moulded, it keeps the shape but crumbly
d.	Clayey Soil and Water	–	When moulded, it keeps the shape

3.8.2 Review and Further Comments on the Basic Properties of Soil

The following items are related to

• Texture	–	Grain size distribution (i.e., percent content of different grain sizes) of soil will be relevant. The items are pebbles, gravels, coarse sand, fine sand, silts, fine silts, and clays from sedimentation
• Plasticity	–	Related to the properties of soil to accommodate deformations without elastic failures, i.e., cracking or disintegration
• Compactibility	–	Compacting ability of the soil to the maximum value against the optimal moisture content With the increase of density, the porosity of the soil will be reduced. The compactibility of soil is measured by Proctor compaction test
• Cohesion	–	The capacity of the grains of the soil to stay together when the tensile stress is applied on the material. It depends on the adhesive or cementation properties of its coarse mortar. The types of coarse mortars may be • Sandy mortar • Average mortar • Lean mortar • Fat mortar • Clays

Source: Auroville Building Center [3]

3.9 SOIL IDENTIFICATION [4]

The following items will be relevant for the identification of soil:

- **Collecting Samples** – Samples should be collected in sufficient quantities, and each sample must be placed in sealed bags.
- **Laboratory Testing** – Various tests may be conducted for the assessment.
- **Site Tests** – Different tests may be involved.

Source:

- Building with Earth [4]

3.10 THE COMMENTS ON TESTS [3,4]

The identification of soil may be done through different tests. These may be listed as follows:

Visual Examination	–	The soil may be examined visually with the naked eye to assess the relative proportions of the sandy and the fine particles to some extent. The particle size up to 0.08 mm may be seen with the naked eye. The individual grains of clay and silt may not be visible. Large stones, gravel and coarse sand may be removed to help the assessment.
Nibble/ Chewing Tests		This is related to chewing/crushing lightly a pinch of soil between the teeth. Different types of soil will indicate different sensations The comments on the types are • **Sandy Soil –** Hard sand particles will provide gritty feeling between the teeth even when the sand is fine • **Silty Soil –** Similar feelings but silt will be less gritty than sand as the particles are smaller than sand • **Clayey Soil –** The clay particles will not be gritty. Clay will provide smooth and floury feeling between the teeth and can stick to the tongue and may be difficult to work free
	–	The care will be required to ensure that such samples will be safe to put into the mouth

(Continued)

(Continued)

Smell Tests	–	This may be done immediately after the removal of the soil. When the organic matters are present, it will smell musty. The smell may become stronger when the soil is wetted or heated. Organic soils should be avoided for construction.
Washing Test	–	When the soil is sandy, the hands may be easily rinsed and cleaned. With silty soil, the feeling will be powdery and the hands can be rinsed and cleaned without much problems. Wet clayey soils will have a creamy, soapy feeling and it will be difficult to rinse them off from the hands. The clay soils will give clayey sensation and the hands may be cleaned with difficulty.
Touch Test	–	The soil may be crumbled by rubbing the same between the fingers and palm of the hand. The rough sensation may be felt when the soil is sandy and no cohesion when it is moist. With slightly rough sensation along with moderate cohesion feeling, when the soil is moistened, it will indicate silty soil. When it is a clayey soil, it becomes plastic and sticky when moistened.
Lustre Test/ Observing the Shine	–	A ball may be prepared with the damp soil. The ball may be cut with a knife to make it in to two halves. The cut surface may be examined and when the cut surface is shiny, it will mean that the earth contains clay in plastic state. If the surface is dull, it is likely to be mostly silty.

Source: Auroville Building Center [3]; Building with Earth [4]

A few other tests may cover [4] –

- **Simplified Sedimentation Tests** [4]
  The test may be done using a glass beaker filled with earth (1/4th height) and then with water. The beaker has to be thoroughly shaken and kept on a flat surface for the particles to settle. Shaking the beaker may be repeated before leaving it to settle. Sand will settle at the bottom, while the silt and clay will remain in suspension at the top. After about 8 hours, different layers may be measured. The information will indicate the proportions of each constituent of the soil.
- **Tests on the Finer Soil Components** – Samples have to be isolated previously by sieving or decanting. The tests may be carried out by –
  Decanting, Tapping (or seepage) test, String test, Ribbon test, Dry strength tests.
- **Quick Clay Identification Test** [4]
  (Emerson test)
  The test may be carried out using a piece of earth the size of a pea. This should be put in a glass receptacle filled with water. The piece of earth should not be handled unnecessarily before putting it into the water. Rainwater or distilled water may be used for the test. After keeping the piece of earth in water for ten (10) minutes, it may be examined. It may still be intact or disintegrated. The test may provide very preliminary information on the nature of the clay. The test should not be considered a substitute for a thorough mineral analysis.

Source:

- Building with Earth [4]

3.11 FURTHER ON FIELD TESTS

Few tests are given as follows [3]:

• The Linear Shrinkage	–	Plastic soil may be placed in a box. The soil will shrink in the box after drying. If the shrinkage is more than 5%, the soil may be classed as too much silty or clayey
• The Cigar		A cigar may be rolled with the plastic soil in the hand or on a table. By pushing the cigar to the end of the table, a piece will fall down. The length of the piece which had fallen may be measured. If the length is less than 40 mm, the soil is too sandy. If the length is more than 150 mm, the soil may be classified as too clayey.
	–	The test may be done quickly during the production of CEB provided the quality of soil does not change.

Source: Auroville Building Center [3]

3.12 LABORATORY TESTS [3]

The items below may be noted:

• **Sieving**	–	This will involve passing the soil through a series of standard set of sieves placed one on the top of another with the finest sieve at the bottom. The method will determine the distribution of grain sizes from the amount retained by each sieve
• **Further on Sedimentation**	–	The method uses the suspension technique of the particles, suspended in water and the particles will fall at different rates. Largest particles will settle first and the finest will settle last as mentioned
• **Atterberg's Limit**	–	Five limits may be measured, the important items are liquid limit and plastic limit
• **Proctor**	–	The QC of the material may be determined by Proctor compacting test. This information will be necessary for effective soils compaction
• **Cohesion '8' Tests**	–	Eight tests may be carried out to determine wet tensile resistance

Source: Auroville Building Center [3]

3.13 TESTS WITH A PRESS [3]

For the tests, three CEBs may be made with four typical soils (approximately 25 per soil). The relevant information will be:

- Preparation of the soil
- Moist state of the block
- Compression
- Dry state of the block (compared to the block sample)

The typical soils available may not always be suitable for CEBs. The properties of sandy soil may be improved to meet specific requirements. Sandy soil can be enhanced by including clayey soils to increase cohesion and a portion of gravelly soil to improve the resistance of the skeleton. For example, sandy soil may be improved by using cement as a stabiliser, which is considered a suitable stabiliser.

Source:

- Auroville Building Center [3]

3.14 THE CHARACTERISTICS OF EARTH AS A BUILDING MATERIAL

The following values may serve as guidance for design and recommended performance [4]:

- Compressive strength = 20 N/cm^2 (2 kg/cm^2)
- Shearing strength = 3 N/cm^2 (0.3 kg /cm^2)
- Tensile strength = 0

The values may be used for single-storey construction.

3.14.1 Values for Design Guidance

Compressive Strength

- Cement stabilised: 50–100 kg/cm^2
- Lime stabilised: 30–80 kg/cm^2
- Fibre stabilised: 5–20 kg/cm^2
- Bitumen stabilised: 15–60 kg /cm^2
- Stabilised with chemical products: 20–40 kg/cm^2
- Stabilised with powerful chemical products: 150–400 kg/cm^2

Wet Compressive Strength – may be considered approximately half of the dry compressive strength.

Young's Modulus – 7,000–70,000 kg/cm^2

Tensile strength: 1/5th of the compressive strength (Brazilian test)

Linear Shrinkage on drying of tamped stabilised earth: 2 mm/m

Permeability – 1×10^{-6} cm/ s

Thermal expansion: 0.012 mm/m /°C

Linear horizontal shrinkage due to mortar of a wall built with bricks 30×30×30 cm = 1.07–2 mm per 5 m.

Thermal characteristics

The coefficient of conductivity: 0.44–0.57 Kcal/h m°C

Specific heat C = 0.2 Kcal/kg

Overall coefficient of transmission K for walls in stabilised earth may be assumed as:

- 200 mm wall:
  - 1.6 Kcal/h m^2°C – 1.3 W/m^2°C
- 300 mm wall:
  - 1.2 Kcal/h m^2°C – 1.0 W/m^2°C
- 400 mm wall:
  - 1.0 Kcal/h m^2°C – 0.86 W/m^2°C
- 500 mm wall:
  - 0.8 Kcal/h m^2°C – 0.7 W/m^2°C

Damping for a 400 mm wall

m = 10%

Time lag for a 400 mm wall: 8–12 hours.

Acoustic Characteristics

400 mm wall: attenuation for the frequency of 500 Hz = 56 dB.

Note – 1 kg ≈ 10 N

Source:

- Building with Earth [4]

3.15 REVIEW OF STANDARD TESTS [4]

The following will be relevant:

3.15.1 Compression Test

The tests may be conducted using cylinder test samples that are 50 mm in diameter and 50 mm high, treated in a similar way, i.e., wet moulded for adobe, compacted for rammed earth, and compressed blocks. The tests are done on five samples, and the average value is considered. The pressure from the testing machine can vary from 500 kg for the weakest samples to 1,500 kg for the strongest samples.

3.15.2 Bending Strength (Modulus of Rupture)

Large samples/complete bricks may be tested. The samples are placed on two parallel tube supports. A concentrated load is applied in the middle of the test sample. The force necessary to break the brick is recorded.

The modulus of rupture may be calculated from

$$R = 3/2 \times 25 \times P/(w\,h^2)$$

where

R = modulus of rupture in kg/cm^2
25 = distance between two supports in cm
w = width of the brick
h = height of the brick

For modulus of rupture computation, an average of five test results may be considered.

3.15.3 Water Absorption

Cubic samples with 100 mm sides may be tested. The samples are to be dried first in the oven until the weights do not change. Water losses (i.e., water content) and water gains are to be measured for the units after placing the samples in water. The increase in weight of samples (i.e., water gains) is expressed as a percentage of their dry weight. Five tests may be considered. The IS code may be referred to for the tests.

Erosion – The earth sample of suitable size may be kept on end and placed 170 mm from the spray of water using a horizontal jet at a pressure of 1.6 kg/m for two hours. The weight loss of the sample/brick and the holes can be measured for indicative information.

Wetting and Drying Cycles – May be done as per ASTM or IS codes. After completing 12 wetting cycles, the maximum weight loss should not be more than 10%.

Source:

- Building with Earth [4]

3.16 HOW TO IMPROVE SOIL [3]

It may be possible to improve the soil type with excessive large or small content of gravel, sand or fines by adding or reducing the required particles accordingly.

The comments are -

Excessive Gravel Content – By sieving the coarsest of particles may be removed from the soil. The largest stones may be removed by hand also.

Excessive Fine Content – The initial treatment may be carried out by washing and removing the fine materials from the soil. The care has to be taken to ensure that all the fines are not washed out from the soil. It may be prudent to wash a part of the soil first and then mix with the original soil after drying. A few trials may be required to get the expected results. Another option may be to mix a soil with more coarse grains with the original soil and checks may be made for obtaining a suitable distribution of the soil particles.

Source:

- Auroville Building Center [3]

3.17 STABILISATION OBJECTIVES [3]

The items mentioned below are relevant:

• **Fundamental Problems**	–	It may be connected with the selection from the following: • The site's soil may be used by changing the design as per the soil condition • To use a different soil that will be more suitable for the project (i.e., the material to be brought from a place to the site) • The local soil may be modified to suit project requirements Normally a third option is preferred, which may mean soil stabilisation methods
• **Objectives**	–	The items are • To reduce the porosity, i.e., by reducing the volume of voids from the soil • To reduce permeability, i.e., by filling the voids which cannot be removed • To increase the mechanical strength, i.e., by increasing the bond between the soil grains
• **Definition**	–	The items that may be involved are • To plan the improvements • The soil properties to be stabilised • Economy for the project to be considered • Construction methods • Performance and maintenance of the buildings
• **Procedure**	–	• **Mechanical –** By compacting the soil to reduce the porosity and to improve compressibility • **Physical –** By working on the texture, i.e., mixing different grain fractions • **Chemical –** By adding chemicals/other materials to improve the properties of the soil

Source: Auroville Building Center [3]

3.18 REVIEW AND MORE ON SOIL STABILISATION [4]

The construction with earth will be associated with the following:

- The soil that is available on site should be used. The design of the project may be carried out accordingly.

- To obtain a better quality of soil to meet design needs if possible.
- To improve the characteristics of the local soil as per the design requirements.

3.18.1 More on Objectives [4]

The stabilisation of soil will be connected with the improvement of the characteristics of the soil.

For the stabilisation problem and its solution, it will be necessary to know -

- Earth properties that are to be improved/treated.
- The materials/products that may be considered for use.
- The required improvements.
- The application methods that may be used.
- Restraints like time available and cost.
- Performance and the requirements of maintenance once the building is open and used.

The stabilisation techniques that have been used in the past have met with limited success. Attempts were made to address issues like

- To reduce the volume of spaces between the solid particles, i.e., porosity.
- To improve the existing binding between particles, i.e., mechanical strength.
- To fill the spaces between particles which will be difficult to eliminate, i.e., permeability.

The improvements planned and carried out must function as a permanent solution.

Source:

- Building with Earth [4]

3.19 STABILISATION METHODS [4]

The stabilisation techniques may be applied in the ground or on the extracted soil.

The methods are given below.

3.19.1 Stabilisation of Soils in the Ground [4]

The items mentioned below will be relevant:

- **Drainage** – This will mean the elimination of water from between soil particles and preventing its return. The removal of water will take longer when the soil has high proportions of fines, silt, and especially clay. The withdrawal of water will affect the shrinkage of the soil and improve the attachments between the particles.
- **Injection Method** – In this method, a product is injected into the soil. The injected material will replace the water present between the soil particles and will remain there to set and prevent any water from coming back later. The products that may be considered for injection are clays and various types of lime and cement, which may fill the voids and improve the links between the particles. The products may be in powder form; hence, the materials must be mixed with water to create an injection-quality fluid.

3.19.2 Stabilisation of Extracted Earth [4]

In this method of application, the earth is extracted from one place and used at another site. The stabilisation operations may be easier. This type of stabilisation may be required for the construction of dams, roads, buildings, etc.

3.19.3 Stabilisation without Using a Stabiliser

This will mean reducing the porosity of the material.

This may be achieved by the following methods:

Compaction – The method may be carried out by adopting a mechanical process such as

- **Static Method** – Use of cylindrical rollers, ridged rollers, wheeled rollers, etc.
- **Dynamic Approach** – Use of vibrating rollers, vibrating pick hammers, tampers, and rammers.
- **Complex Technique** – projection and crushing.

The quality of compaction will depend on the particle distribution of the material and the type of compaction applied.

Desiccation – This method may be used for improving the cohesive strength of the material by drying the fine particles to avoid compacting the material. The results will be comparable with the method done with slight compaction, and the undesirable shrinkage cracking will be mostly avoided by using small units. Good compaction will reduce the risk of shrinkage cracks.

3.19.4 Stabilisation by Using a Stabiliser [4]

The types that may be included are as follows:

3.19.4.1 Chemically Inert Stabilisers

The materials that may be considered for inclusion with the earth will control shrinkage effects.

These may include

- **Sand and Gravel** – The use of these materials may provide good results and are used in rammed-earth construction.
- **Vegetable or Animal Fibres** – These applications may provide effective performance and are cheap and suitable for brick making.
- **Reinforcement** – Straps prepared from polymer or aluminium alloys may be used. This method is also known as reinforced earth construction.

These may be applied for the construction of reinforced earth embankments. Apart from controlling shrinkage, such uses will meet other requirements.

3.19.4.2 Physico-Chemical Stabilisers

Good compaction is important as a requirement. The inclusion of physico-chemical stabilisers can be considered favourably for this purpose.

The relevant items for the products to work will be

- Low concentrations may be effective (less than 10%) for execution.
- Water-soluble materials may be preferred.
- Simple equipment may be used for the operation.
- The method may be applied to different types of soils.
- It can perform in various environmental conditions.
- The price will be compatible with the benefits and use of soil as the material.
- Water content in the soil may not be critical at the time of treatment.
- Setting time will be reasonable and will not affect the construction work schedule.

3.19.5 Main Physico-Chemical Stabilisers [4]

The important requirement of physico-chemical **stabilisation** will be to make clayey soils water-resistant. Good compaction, in most cases, may be adequate for this purpose, along with sufficient mechanical properties of the soil for applications. Due to the presence of clay, it will help in achieving water resistance performance, as clay will be less water absorbent.

The products are listed below.

3.19.5.1 Hydrophobic Agents

Amine Derivatives – These may be from ammonia and are ammonium salts or amines in a solution of an organic solvent and should be diluted in water. The performance will be good in low proportions, i.e., 1% or less.

Bitumen and Hydrocarbons – Fluidised bitumen (cut back) or bitumen emulsions may be used for stabilisation.

Resins – Due to polymerisation of resin and the formation of an efficient membrane that takes place, in the soil, it will increase the resistance of earth against water.

The items mentioned below may be considered:

- **Abietic Resins** – May be available as a waste product of wood processing for the manufacturing of paper pulp.
- **Furfural Aniline and Resorcinol–Formaldehyde Resins** – The effect will be cementation.

They work at low concentrations and may be produced from vegetable products.

3.19.5.2 Binding Agents

These products may be used for clayey soils when it is necessary to have special mechanical characteristics.

Non-hydraulic Lime – The following items may be relevant:

- **Quicklime and Hydrated (Slaked) Lime** – This may be considered to stabilise fine clayey soils. Lime in the proportion of 4%–10%, in the soil may significantly improve the mechanical properties of the material. The compressive strength can be increased significantly, up to a factor of 4 or 5.
- **Sodium Silicate** – The use of sodium silicate along with a reagent will form a gel and may provide additional cohesion in some clayey soils.
- **Hydraulic Limes and Cement** – These materials, when used in the proportion of 4%–10%, have shown good results in sandy soils. The inclusion of secondary materials like sodium, sodium meta-silicate, and sodium sulphate can improve the cement treatment in some soils.
- **Lignosulphates** – These are by-products from the paper pulp industry, as mentioned before. They may be cheaper than cement and can work as binding agents, being effective at 1%–2% concentrations. The mixed material may lose strength when exposed to water, but this situation can be improved by adding chromo-lignin salts.
- **Resins** – The resins may be of vegetable origin and, being water-repellent, can act as binding agents. Synthetic resins may also serve as effective binding agents, but they are quite expensive, with content levels comparable to that of cement used.
- **Bitumen** – Generally used in fluidised form, it will be water-repellent for clayey soils and can act as a binding agent for sandy soils as well.

Treatment for clayey soils may be very effective by using physico-chemical stabilisers to provide water resistance properties. Binding agents may be required for sandy materials to improve mechanical properties.

Source:

- Building with Earth [4]

3.20 UNCONVENTIONAL STABILISERS [4]

These may be reviewed under the following headings:

- Natural products: vegetable or animal
- Industrial products
- Commercial stabilisers

3.20.1 Natural Products

The applications may also be done in traditional construction, with items obtained from natural materials without much processing.

The types for stabilising effects may include:

Tannic acid, natural rubber, humic acid, latex, casein, wood ash, vegetable oils, copal de manille, gum Arabic, straw – May be suitable for both plastic and liquid earth, bulls' blood, palmo-copal, sisal, termite mounds – May be suitable for unfavourable weather conditions.

Industrial Products – These materials are still under study for further information and effective performances and the products may not be used largely.

These may include

- **Acids –** Not easy to handle in concentrated form
- **Salts**
  - Chlorides
    - **Sodium** – flocculant
    - **Calcium** – may be less effective than sodium chloride
    - Ferric
    - Dialkyl dimethyl ammonia chloride
  - Calcium sulphate (or gypsum), precaution and care will be required for the use.
  - Aluminium Salts – The types will be electrolytes and will provide electrochemical stabilisation.
- **Resins –** Different types of synthetic resins are available.
- **Silicates** – Potassium, calcium, and sodium may be referred to. The industrial waste products may be used for stabilisation. The materials may not be economical.

Other items may include

- **Sump Oil** – The performance may not be reliable and may be washed out by rain.
- **Lignin** – The cost for the application should be checked.
- **Molasses** – Wood molasses, sugar, or lime sucrose.
- **Blast Furnace Slag** – May be effective depending on the composition.

Source:

- Building with Earth [4]

3.21 COMMERCIAL STABILISERS [4]

Many commercially available stabilisers may be used for applications. The performances and effectiveness of the materials have to be properly checked prior to use. Their prices are expected to be comparable with conventional stabilisers.

Source:

- Building with Earth [4]

3.22 REVIEW AND FURTHER COMMENTS ON STABILISERS [3]

The review and comments may be listed on the items mentioned below:

Stabilisers may include

- **Cement** – Cement is a good stabilising item. The amount may be 6% of cement in the mix. This quantity may be necessary for obtaining satisfactory results. Good results may be expected with sandy soil as well.
- **Lime** – Lime may be used for stabilisation. The addition of 2%–3% of quick lime in soil will reduce the plasticity of the soil and break up lumps. Satisfactory results may be achieved with silty soils.
- **Fibres** – The use of straw is very common for stabilisation as stated. The straw may be considered a reinforcing item. The fibres will control cracking during drying, accelerate drying, increase tensile strength, and lighten the material.
- **Resins** – They may include processed natural products and furfural-based resins, formaldehyde, polyvinyl, urea, etc. The products may be expensive but are effective.
- **Bitumen** – Bitumen for stabilisation may be in the form of cut-back or emulsion. Often 2%–3% of bitumen may be used but the amount may increase to 8%. Good performance may be achieved with sandy or silty soil.
- **Synthetic Products** – These stabilisers may be synthetic or industrial products and may include industrial wastes. Items such as acids, sodas, silicates, salts, paraffins, adhesives, oils, etc. may be mentioned. Economically, the products may not be attractive, and effective performance may be doubtful.
- **Natural Products** – Mineral products may be sand, pozzolanic sand, animal products, blood, fur, and vegetable products like ashes, fats, oils, sap, tannins, etc. The stabilisation materials may be available locally but may not yield effective results in the long run.
- **When to Stabilise?** –Stabilisers may not always be required when part of the structure is not exposed to water and the design is carried out accordingly. In this respect, maintenance has to be done properly and effectively.
- **Stabilisers May Be Required –** to cover poor design, severe climatic conditions, use of high compressive strength in the design, and for applications with poor maintenance.

Source:

- Auroville Building Center [3]

3.23 TYPES OF EQUIPMENT FOR PRODUCTION

The different types may be mentioned as follows [3]:

- **Manual Press** – The operations for production covering compression and turning out will be done by the machine, and the operations are carried out manually.

- **Mobile Production Units** – The compression and turning out, including the preparation of the materials and evacuation of finished products, are motorised and automated.
- **Motorised Press** – The operations, including compression and turning out by the machine, are motorised.
- **Fixed Production Units** – The production units are not meant for transport. The operations covering compression, turning out, preparation of materials, evacuation of finished products, etc., are motorised and automated.

Source:

- Auroville Building Center [3]

3.24 CHOICE OF EQUIPMENT

The following items are relevant [3]-

- **Specific Requirements** – Daily output, quality of production, etc.
- **Level of Project** – Individual size and limitations.
- Financial condition.
- **Local Conditions** – Availability of energy, skilled labour, appropriate materials, technical level, geological knowledge, climatic conditions, etc.
- Availability and quality of equipment.
- Other associated items.

It may be mentioned that priorities should be given to investment, daily output, and the availability and quality of labourers.

Source:

- Auroville Building Center [3]

3.25 STEPS FOR THE PROCEDURE RELATED TO PRODUCTION

The procedure/steps may be as mentioned below [3]:

Extraction, transportation, screening, measuring, dry mixing, moist mixing, moulding, curing, quality control checks, storing, and finally despatch/use.

Source:

- Auroville Building Center [3]

3.26 REVIEW OF BUILDING DESIGN WITH CEBs

3.26.1 Introduction [5]

The CEBs are also known as pressed earth blocks [5] or compressed mud/soil blocks and may be used as a building material as stated. They are manufactured using damp soil that is compressed to produce blocks/bricks.

CEB may be produced by a mechanical press using the proper mixture of fairly dry inorganic subsoil, non-expansive clay, and aggregates [5]. The blocks may be stabilised with a chemical binder like Portland cement/lime and may be referred to as compressed stabilised earth blocks or stabilised earth blocks (SEB).

CEBs are not the same as mud bricks, as mud bricks are not compressed and become solid through chemical changes that occur during drying. The compressive strength of well-made CEB can be similar to or exceed the values typical of cement or mud bricks.

CEBs are assembled for wall construction following standard bricklaying and masonry building methods. The mortar may be a simple slurry prepared from the same soil/clay mix without aggregate, spread or brushed very thinly between the blocks to develop the bond, or cement mortar may be applied to achieve high strength. This may be important for freeze-thaw cycle conditions and to maintain stability.

Interlocking blocks may also be produced as per requirements [5].

3.26.2 Developments [5]

The development occurred primarily to meet the requirements of low-cost construction and as an alternative to adobe, with some advantages related to building construction. The method of production has been appreciated due to the eco-friendly approach of the process and the development of mechanical presses for manufacturing.

CEB technology has been used in different countries including the advanced countries of the world.

Different types of CEB production machines are available. This will range from manual to semi-automatic and fully automatic, with increased costs to improve production rates and decrease the utilisation of labour. The automatic machines are more popular and are used in developed countries, while manual machines may be used in developing countries. Further comments have been made on the production of manually operated machines separately.

CEB may be considered as a suitable and practical building material [5,6] for the construction of load-bearing walls. Domes and arches [6] may be constructed with CEB, as shown in the photographs.

Rendering may be considered for exposed walls to achieve the required appearance.

Source:

- Wikipedia [5]
- Ganguly [6]

3.27 FURTHER COMMENTS ON CEBs

Further comments are given below:

3.27.1 Advantages of CEB [5]

The following may be mentioned:

- Traditional techniques may be improved through the use of CEB.
- Economical in building techniques and design.
- The drying time for CEB will be shorter.
- The cost of CEB will depend on local conditions, i.e., the cost of labourers, materials, stabilisers etc.
- Uniform sizes of blocks can be produced.
- The blocks will be strong, stable, and will last for a longer period.
- The shorter distance to the supply source of CEB will be an advantage.
- The waiting time for the material may be nominal.
- It is a low cost construction material.

- CEB can be manufactured to the required size with true flat sides and 90° angle edges.
- The requirements of mortar for be reduced due to uniform block sizes.
- The materials are natural, non-toxic and do not release any gases.
- It will be an energy-efficient material.
- The construction with CEB will be sound-resistant.
- Fire resistant – earth walls will not burn.
- Insect-resistant items.
- The units will be mould-resistant.
- In India, cement stabilisation techniques appear to be popular.

Source:

- Wikipedia [5]

3.28 MORE ON ADVANTAGES [4]

The additional comments are listed below:

It may be mentioned that compared to rammed earth construction, CEBs (i.e., bricks) will have similar advantages to adobe bricks.

This may be reviewed as follows:

- Phasing of production over a long period of time may be done as per requirements.
- Greater flexibility will exist in architectural design and applications.
- The chances of cracks in the wall will be reduced, as shrinkage may occur in individual bricks during drying. This can be checked before application.

3.28.1 Disadvantage

- A greater number of operations may be required with the material, which may affect production.

3.28.2 Advantages over Adobe Bricks

The items may be summarised as follows:

- Immediate storage may be possible.
- The bricks are more regular and uniform.
- The area required for production and drying will be less and may be controlled/covered.
- One side of the block may be stabilised.
- Different types of bricks may be produced, i.e., hollow, interlocking blocks, drainage tiles, or roof tiles, etc.
- CEB will have a better finish.
- Greater compressive strength may be possible.

The production process may be longer and will depend on the type of machine, which may be costly.

The discussion on adobe has been done in Chapter 4.

Source:

- Building with Earth [4]

3.29 COMMENTS ON FURTHER ITEMS [5]

The comments are made on the following issues:

3.29.1 Strength

After pressing and curing, the blocks are to be submerged in water for a period of 4 hours. They are then removed from the water and tested for compression. The minimum strength must be 2 MPa, which will be higher than that of adobe, which typically has a strength of at least 2 MPa.

3.29.2 Finishing

The walls will require a reinforced bond beam or a ring beam on the top or between the floors when the blocks are not stabilised. A plaster finish may be provided, typically using stucco wire, stucco cement, or lime plaster. The stabilised walls may be left exposed with no exterior plaster finish. In tropical conditions, polycarbonate varnish may be used to provide an extra layer of wet-weather protection.

3.29.3 Foundations

The normal design of foundations for brick walls may be applicable. A CEB wall will be heavy. The footings must be at least 250 mm thick and must satisfy the width requirement, which should be at least 33% greater than the width of the wall it will support. When a stem wall is used, it must extend to an elevation of no less than 200 mm above the exterior finish level/grades. A rubble-filled foundation trench type with a reinforced concrete grade beam above may be used to support CEB wall construction.

Source:

- Wikipedia [5]

3.30 QUALITY CONTROL, CEB [3]

Checks on quality and daily control must be conducted properly, as this will be very important.

The following items will be relevant:

- Everyone should be aware of the quality requirements. Necessary interactions will be required for the local workers.
- Simple tests may be conducted daily during production.
- Checks on production at every stage.
- Daily production records.
- Weekly/monthly production checks with field tests.
- Records/output data should be kept properly with comments if applicable.
- Laboratory tests on samples and checks will be required regularly.
- Interactions with workers on production improvements if required.

Source:

- Auroville Building Center [3]

3.31 CHECKS AFTER PRODUCTION

The following items for checking may be useful after production [4]:

- Number of workers - soil preparation, machine operation, stacking, etc.
- Number of hours worked by the workers.

- Number of units produced.
- Type of blocks produced.
- Number of blocks per mix and per bag of cement.
- Number of mixes used.

To check the exact percentage of cement consumed according to the number of mixes and blocks produced.

Source:

- Auroville Building Center [3]

3.32 TYPES OF PRESSING MACHINES [3]

The types of press may include [4]

- Manual
- Mechanical hydraulic
- Pneumatic

3.32.1 Method of Compression [4]

- Static pressure
- Dynamic pressure

3.32.2 Guidance on Using the Press

It should be done strictly as per the manufacturer's recommendations.

The following types, as mentioned in Table 3.3, may be considered for an initially operated press:

A hand-operated machine is shown in Figure 3.1

Checks on the following items will be relevant:

- Manufacturer's references.
- Description of the press (compression ratio, compression pressure, output, moulds, etc.)
- Weight and encumbrance
- Dimensions of the press block
- Energy sources and degree of automation
- Maintenance guidance
- Factory price
- Training contract

TABLE 3.3
Types of Press

Sl. No.	Types	Light/Heavy Manual
1	Astram/Auram	
2	Cinvaram/Ceraman	
3	Balram/Ellson b.m.	
4	Unata/Terstaram	
5	Brepak	
6	Geo 50	

FIGURE 3.1 Hand-operated machine. (Source: Auroville Building Center [3]; Building with Earth [4]).

- Servicing practices/formalities
- Availability of basic machine parts when required

3.33 GUIDANCE FROM IS CODE

The guidance may be available for stabilised soil blocks related to general building construction from IS-1725: 2013 [7].

3.33.1 Information

The following will be relevant:

Fresh Block Density – This is the ratio of the mass of the fresh block (i.e., immediately after taking the block out of the mould) to the net volume of the block.

Dry Density of the Block – It is the ratio of the mass of the dry block divided by the net volume of the block.

Wet Compressive Strength – The compressive strength tested in the saturated condition of the block.

Quality – The blocks must be prepared from a proper soil-sand mixture using stabilisers like ordinary Portland cement or a combination of lime and ordinary Portland cement. The homogeneous mixture of soil-sand-stabiliser must be compacted to produce a high-density block at optimum moisture content.

All blocks must be sound and free from cracks, nodules of free lime, or defects that may interfere with placing the units and construction.

3.33.2 Materials

Materials are listed below:

- **Cement** – as per IS-269 or IS-1489 (part-1) or IS-8112 or IS-12269.
- **Lime** – as per IS-712.
- **Sand** – as per IS-383.
- **Water** – quality should not affect durability.
- **Soil** – the soil must be proper and suitable for the manufacture of stabilised soil blocks.

3.33.3 Dimensions and Tolerances

3.33.3.1 Dimensions

For Modular Bricks

- **Length** – from 190 to 290 mm
- **Width** – from 90 to 240 mm
- **Height** – from 40 to 90 mm

- For Non-modular Bricks
  - **Length** – from 230 to 305 mm
  - **Width** – from 105 to 190 mm
  - **Height** – from 75 to 100 mm

Tolerances – The maximum variation in dimensions of the block shall not be more than = ± 2 mm.
Frog – IS-1725 may be referred.

3.33.4 Physical Characteristics

- **Dry Density of Blocks** – Should not be less than 1,750 kg/m^3
- **Compressive Strength** – The minimum average compressive strength tested as per IS code shall be 3.5 MPa.
- The compressive strength of any individual block shall not be less than the minimum average compressive strength by more than 15%.
- **Water Absorption** – This shall not be more than 18% by weight, in 24 hours, after immersion in cold water.

Linear Expansion on Saturation – Following the test as per code, it should not be more than 0.10%
Weathering – The maximum loss of weight shall not be more than 3%.
Sampling for Conformity – IS-1725 may be referred.
Marking – IS-1725 may be referred.
Annexures – A–E of IS-1725 may be consulted for further information.
Source:

- IS-1725: 2013 [7]

3.34 DESIGN GUIDELINES

The following items will be important for the design [3]-

Foundation Work	–	Good foundation/basement and good overhangs
Compressive Strength	–	Relevant for the design of the walls, i.e., thickness and stability. A factor of safety of five may be used on the wet strength
Shear Strength	–	Important near windows and verandas. Portions for the change of stresses may be checked
Shrinkage Stress	–	• Special details of using tie beams. Proper use of tie beams at salient places and to control shrinkage of mortar • Length of plain walls without doors or windows may be reduced to 5–6 m
Module of Blocks	–	Proper use of module dimensions of blocks should be done in the design
Water Absorption and Erosion	–	Water should not be allowed to accumulate any part of the building or to travel along the walls
Module of Blocks	–	Proper use of module of blocks should be used in the design

3.34.1 Comments for Treatment at Important Places

- Tie beam on stabilised rammed earth foundation
- Window or door frame treatment
- Window sill treatment
- Arches
- Lintels

Requirements/ Checks for Mortar	–	• May be stabilised 1.5 times more than the CEB strength as the blocks • Coarse sand may be added (0.2–2 mm) to control shrinkage while drying • The mix should not be too dry or too wet
Tests for Mortar	–	• Percentage of cement to be finalised (say 1.5 × % of blocks) • May be started with 50% of soil and 50% of sand • May be applied on a block and necessary checks after curing • When the mortar develops fine cracks, sand may be added and the tests may be redone • When the mortar is too crumbly, soil may be added to the mix and the tests may be repeated

Source: Auroville Building Center [3]

3.35 STUDY ON COCONUT FIBRE-REINFORCED CEMENT STABILISED RAMMED EARTH BLOCKS

Coconut fibres may be used along with cement in the soil mix to increase strength and durability of the material. The study included the use of ordinary Portland cement with different proportions of fibres in the mix. Coconut fibres of 15 mm length were also used with varying proportions by weight of the soil as reinforcement.

Rammed earth blocks were prepared by ramming the mass of soil in the mould.

Tests of the samples were conducted.

TABLE 3.4
Properties of Soil

Grain Size Distribution	Values
Sand	78.1%
Clay	9.1%
Silt	12.8%
Density	2,610 kg/m^3
Optimum moisture content	9.5–11.0%

3.35.1 Review of the Materials

Soil – The soil was collected from a depth of 0.5–1 m from the National Institute of Technology Agartala campus, India.

Cement – Ordinary Portland Cement, 43 grade, was used as per IS-8112.

Coconut Fibres – Coconut fibres were collected from the local Agartala market. The average diameter of fibres was about 0.2 mm and they were made into pieces of 15 mm in length.

3.35.1.1 Properties of Soil

These are provided in Table 3.4:

3.35.2 Preparation of Test Samples

The test samples were made using steel moulds with dimensions of 100×100×100 mm and with water content of 1%–2% more than the optimum moisture content level.

The fibre content in the mix was varied and included 0%, 0.2%, 0.4%, 0.6%, 0.8%, and 1% by weight of dry soil. The cement content of the mix was also changed such as 0%, 2.5%, 5.5%, 7.5%, and 10% by weight of dry soil. The soil was properly mixed.

The blocks were removed from the mould after 24 hours. The items prepared with cement were cured for 28 days using dampened gunny bags. The other samples were kept in the shade for drying at room temperature.

3.35.3 Tests and Comments

The following tests were conducted:

3.35.3.1 Unconfined Compressive Tests

The blocks demonstrated 1.87, 1.93, 4.59, 6.69, and 8.61 times the compressive strength of raw soil blocks for 0%, 2.5%, 5%, 7.5%, and 10% cement content, respectively.

3.35.3.2 Split Tensile Test

The tests were carried out as per the relevant IS code.

3.35.3.3 Failure Pattern

The failure pattern was important for cement-stabilised blocks with or without fibre reinforcement.

Brittle types of failures were observed for both types. The units with fibre reinforcement remained attached firmly after the failure.

Abrupt and rapid failures were observed for the raw blocks.

Furthermore, it was noted that for blocks with fibre reinforcement, the fibres were pulled out during the failure rather than tearing.

The use of fibres as reinforcement in rammed earth walls may be suitable for mitigating earthquake effects related to the sudden failure of walls/structures.

3.35.4 Review of Comments

The following points are relevant:

- With the increase in fibre content in earth blocks, while maintaining a constant cement content in the mix, the density of the composite material will be reduced. This may be utilised effectively and will provide various advantages in design.
- The relationship between dry density and fibre content using the same compaction energy may not yield uniform results due to the resistance offered by the fibres during compaction.
- For the same drying period, the residual moisture content has been observed to increase with the rise in cement content.
- It is recommended to use more than 2.5% cement content of dry soil in the mix to achieve an increase in the strength of blocks.
- With an optimum limit of 0.8% fibre content by weight of dry soil, both compressive and tensile strengths will be increased.
- Significant contributions have been observed in raw soil units by using coconut fibres as reinforcement in rammed earth blocks.
- For strength improvement, the use of 0.8% coconut fibres as reinforcement along with 5%–10% cement for stabilisation is recommended.
- By studying the tensile properties and failure patterns, it can be noted that the material may provide necessary warnings to inhabitants in seismic areas prior to total collapse.

It appears promising to use coconut fibre-reinforced cement-stabilised earth blocks for construction.

Source:

- Raj, Das, Mohammad, and Saha [8]

3.36 STUDY ON EARTH BUILDING BLOCKS REINFORCED WITH JUTE AND BANANA FIBRES

The properties of building blocks have been investigated using jute and banana fibres as reinforcement. Comments have been made on the mechanical, physical, and durability properties of fibre-reinforced building blocks.

3.36.1 Materials

Soil – The soil was sourced from Kallaperambur, Thanjavur district, India, with a maximum dry density of 1,940 kg/m^3 and 10% as optimum moisture content.

The reinforcement of soil was carried out using jute and banana fibres as mentioned.

3.36.2 Properties of Fibres

The properties used in the study are indicated in Table 3.5:

TABLE 3.5
Properties of Fibres

Serial No.	Fibre	Jute Fibre	Banana Fibre
1	Length (mm)	60–70	60–70
2	Tensile strength (MPa)	216–225	259–274
3	Diameter (mm)	0.8–1	0.04–0.10

3.36.3 Approach

The physical, mechanical, and durability properties were assessed by performing tests on density, water absorption by capillary action, compressive strength, linear shrinkage, indirect tensile stress, wear, and erosion.

3.36.4 Block Size

The soil blocks were prepared using dimensions of 215 × 105 × 65 mm with banana and jute fibres as reinforcement.

The fibre content varied from 0.1% to 1.0% by weight. The blocks were prepared using manual pressing with optimum moisture content. The mixing of fibres was done carefully to ensure uniform distribution throughout the blocks.

The blocks were sun-dried for a period of 21 days at an average temperature of 25°C–30°C.

The blocks were cleaned to remove dust before testing.

3.36.5 Physical Properties

The shrinkage assessments of the blocks were conducted by measuring their dimensions before and after drying.

3.36.6 Water Absorption Test

This was conducted according to the BS EN 772-11 code. The relevant IS guidelines may also be followed.

3.36.7 Mechanical Properties

By using a testing machine, the compressive stress of a block was computed.

3.36.8 Durability Tests

The durability properties/tests were conducted following ASTM D559-03. The dry mass reductions of the earth blocks were calculated after completing 12 cycles of strokes on each face of the blocks.

3.36.9 Erosion Tests

This was carried out as per NZS 4298 guidelines.

3.36.9.1 Results and Comments

This may be mentioned as below:

3.36.9.1.1 Physical Properties

Fibre reinforcement will improve the physical properties of reinforced soil/earth blocks. Furthermore, jute fibres performed better than banana fibres. The density of fibre-reinforced earth blocks will be lower than that of unreinforced blocks. The density will decrease with an increase in fibre content in the blocks, as shown in Table 3.6.

It may be noted that at 1% fibre content (i.e., replacement), the decrease in density was observed to be 11% and 16% for JFREB and BFREB, respectively.

where

JFREB – jute fibre-reinforced earth/soil blocks

BFREB – banana fibre-reinforced earth/soil blocks

3.36.9.1.2 Water Absorption

It was observed that water absorption increased with the addition of fibres.

3.36.9.1.3 Linear Shrinkage

Linear shrinkage must be checked and kept within the limits. According to some standards, 3% is recommended as the maximum limit. From this study, it was observed to be less than the upper limit of 3% for both unreinforced and fibre-reinforced earth blocks.

3.36.9.1.4 Compressive Strength

Both JFREB and BFREB showed an increase in compressive strength with an increase in fibre content.

About an 85% increase in compressive strength was observed with 1% jute fibres in the mix.

The presence of fibres may help control the formation of cracks and showed an increase in load-carrying capacity with JFREB.

The addition of fibres may also help control the formation of large cracks and enable the material to behave like a ductile substance.

The tensile strength of BFREB was found to be less than that of JFREB.

The performance for wear and erosion was found to be reasonable with both fibres.

It may be remarked that jute fibres can improve the properties of earth blocks in all cases, covering physical, mechanical, and durability requirements.

It is recommended to use a small quantity of lime or cement in the mix, which may be very useful to meet durability performance.

Source:

- Selsiadevi, Selescadevi, and Varshini [9]

TABLE 3.6
Physical Properties

Fibre (%)	Density, Fibre-reinforced Block (kg/m^3)	
	JF	BF
0	2,040.6	2,040.6
0.25	1,979.49	1,985.18
0.5	1,783.06	1,981.39
0.75	1,754.86	1,955.03
1.0	1,710.19	1,810.50

3.37 FURTHER ON THE USE OF NATURAL FIBRES TO IMPROVE WATER RESISTANCE OF COMPRESSED EARTH BLOCKS

3.37.1 General

The main objective was to study the water resistance characteristics of CEBs using different natural fibres.

3.37.2 Materials

Soil – Two types of soil, i.e., red and brown soil, were collected from Ghana for the experimental work.

Properties of Soil – The details are given in Table 3.7.

3.37.2.1 Fibres

The following three types of fibres were used:

- Coconut husk
- Oil palm fruit fibres
- Sugarcane bagasse

The fibre lengths were 125, 38, and 80 mm for coconut, oil palm, and bagasse fibres, respectively.

3.37.2.2 Specimens

The specimens were 290 × 140 × 100 mm and were prepared using a block-making machine with a constant pressure of 10 MPa. The fibre content was 1% by weight in the mix. The units were dried in the sun for 21 days with an average temperature of 27°C and 72% relative humidity.

3.37.2.3 Tests

Appropriate pressure spray tests (i.e., accelerated erosion) were carried out to assess the resistance of specimens to continuous rainfall situations. The tests were more representative of the actual site conditions related to erosion of earth structures.

3.37.2.4 Comments on Tests

The comments from the tests may be summarised as follows:

- Unreinforced brown specimens could not withstand the total period of 60 minutes of water spraying for erosion. All the units tested failed between 54 and 58 minutes. The red specimens could resist 60 minutes of the water spray test but eroded deeply between 81 and 86 mm. The performance of the red specimens was better in terms of resistance against erosion.
- The inclusion of fibres contributed to lowering the erodibility rate of the blocks with some damage.

TABLE 3.7
Properties of Soil

	Grain Size					
Soil Type	Gravel (>2 mm)	Sand (2–0.063 mm)	Silt (0.063–0.002 mm)	Clay (<0.002 mm)	Optimal moisture content (%)	Maximum Dry Density (kg/m3)
Brown	12	46	28	14	18	1,780
Red	15	39	16	30	19	1,790

- The addition of fibres in the blocks significantly improved their erosion resistance.
- The surface of fibre-reinforced blocks eroded more quickly, in depth, than the internal portion of the material.
- A reduction in depth with increased spraying time was observed. This may indicate that the erosion that took place at the beginning will slow down.
- The addition of fibres in the blocks may not fully protect against erosion problems caused by water, but it will significantly contribute to reducing the rate of erosion.
- The contribution of fibre was more noticeable in the internal parts of the blocks than on the surfaces.

Source:

- Danso [10]

3.38 TEXTILE-REINFORCED EARTH BLOCKS FOR CONSTRUCTION

This study included the addition of natural fibres and recoverable textile components in earth blocks that may contribute to structural resilience for compressed earth units, including better material qualities related to comfort and extended life of buildings. Furthermore, textile and fibre items may be suitable for local environmental conditions, soil, and seismic requirements. The development includes the use of pre-templated items that may serve as formwork to control construction but will remain in place after construction as tensile reinforcement. This will mean the application of carbon-neutral material to satisfy local conditions and will be suitable for providing good interior climate conditions, like an ideal home in the developed/developing world.

Source:

- Widder [11]

3.39 GLASS FIBRE-REINFORCED COMPRESSED STABILISED EARTH BLOCK: A GREEN ALTERNATIVE TO CEMENT CONCRETE BLOCKS [12]

The use of stabilisers such as cement and lime in the mix for compressed earth blocks (CEB) can make the blocks stronger, more durable, and less water absorbent, as stated before.

Furthermore, the inclusion of glass fibres in the mix can improve the strength and crack-resistant properties of the blocks. Compressed Stabilised Earth Block (CSEB) may be considered an alternative green building material compared to traditional bricks or concrete blocks.

Studies have been carried out with glass fibres supported by tests to increase the strength and crack resistance of compressed stabilised earth blocks (CSEB).

For the production of clay bricks, the emission of CO_2 occurs. The release of CO2 for CSEB production is lower.

CO_2 emissions of different products are summarised in Table 3.8.

TABLE 3.8
CO_2 Emissions

Serial No.	Materials	CO_2 Emission (Kg/Tonne)
1	Concrete blocks	143
2	CSEB	22
3	Fire clay bricks	200
4	Aerated concrete blocks	280

3.39.1 Manufacture of Blocks

Materials – For CSEB, the materials would be earth/soil, cement, crushed sand, and lime. The soil has to be of proper quality. If necessary, required modifications/adjustments may have to be made.

Soil proportions are indicated in Table 3.9.

3.39.1.1 Fibres

Twelve millimetre lengths of glass fibres were included as reinforcement in the mix. CSEBGF, i.e., CSEB with glass fibre units, were prepared with 1% of glass fibres relative to the weight of cement.

3.39.1.2 Cement

Portland pozzolana cement of 43 grade was used.

3.39.1.3 Mixing

The ingredients were mixed dry using a pan mixer for uniformity. Water was added later.

3.39.1.4 Moulding and Compaction

Different mould sizes may be used. Common mould sizes may be 220×100×75 mm and 152×165×324 mm. The initial compaction (i.e., about 20%) was done using a table vibrator, then the required pressure was applied to the mix in the mould.

3.39.1.5 Stacking and Curing

Water curing was done for 28 days.

3.39.1.6 Tests and Results

The units were prepared/made ready and tested.

Information on salient items is given in Table 3.10: CSEB proportions are indicated in Table 3.11.

3.39.2 Additional Comments

The following may be mentioned:

- The water absorption for CSEB was within the limit. The reduction of water absorption was due to the presence of the raw material.

TABLE 3.9
Soil Proportions

Serial No.	Ingredients	Percentage
1	Gravel	15
2	Sand	50
3	Clay	20
4	Silt	15

TABLE 3.10
Information on Salient Items

Samples	Average WA (%)	Average Compressive Strength (MPa)	Dry Density (kN/m³)
CSEB	8.2	14.91	20.47–20.81
CSEBGF	7.94	16.11	20.66–20.81

TABLE 3.11
CSEB Proportions

Ingredients	Percentage
Earth	65
Cement	10
Crushed sand	20
Lime	5

- CSEB units took more time to dry compared to fired bricks.
- CSEBGF samples showed little difference in water absorption.
- The compressive strength of CSEB was high, three times more than the locally available fired clay bricks.
- CSEBGF samples indicated an average increase of 8% in compressive strength, which may be considered equal to that of concrete strength.
- The compressive strength will depend on the type of earth and the amount of stabilisers used.
- The surface texture of the blocks will provide good bonding with the mortar.
- The cost of CSEB and CSEBGF blocks may be 25% cheaper than ordinary fired bricks.
- CSEB with glass fibre reinforcement indicated better performance in strength and water absorption.
- The inclusion of glass fibres will provide better crack resistance properties to the blocks.
- A lesser amount of stabilisers in the mix will make the blocks cheaper.
- The blocks with glass fibre reinforcement will be eco-friendly and suitable for green construction.

Source:

- Winsley and Muthukannan [12]

3.40 EFFECTS OF VIBRATIONS ON SOILS [1]

It has been observed that due to vibrations from pile driving, traffic, or machinery operations, the density of sand will increase, resulting in settlement/subsidence of the surface. This may affect the structure or building. It should be noted that vibrations are required for compacting layers of loose sand prior to foundation construction. The effects of vibration on clays will be noticeably less than on sand. The effects of vibrations will require special consideration in the design.

Source:

- Terzaghi and Peck [1]

3.41 REVIEW OF COMPRESSED EARTH BLOCKS (CEB): APPLICATIONS CASE STUDY [6]

3.41.1 Applications of CEB from Projects in West Bengal

Compressed earth blocks (CEB) were produced at the construction site using a hand-operated machine.

Sizes of CEB – 240×240×90 mm and 240×115×90 mm as shown in Figure 3.2:

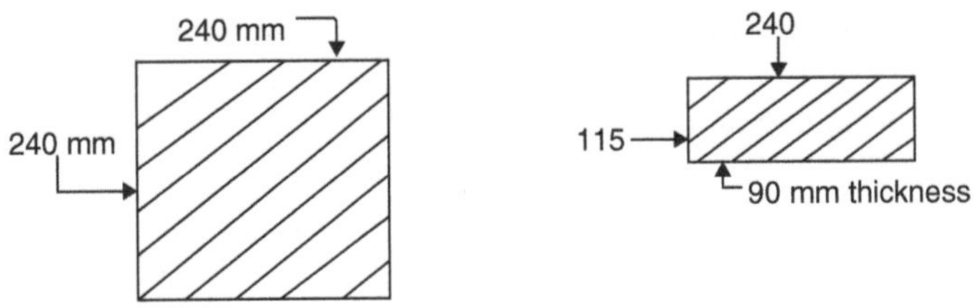

FIGURE 3.2 Sizes of CEB.

3.41.1.1 Strength Achieved

Attempts were made to use locally available soil at the sites. The strength of CEB will depend on the quality of the material used for production, as mentioned. Emphasis was always placed on using locally available soil, and improvements should be carried out accordingly to meet design requirements.

The blocks were tested on-site and at the local laboratory. The strengths of the blocks were low when the bricks/blocks were wet, as expected.

It was possible to produce bricks/blocks at the construction sites with a compressive strength of about 3.5–4.0 N/mm^2 under wet conditions. Locally available burnt bricks had a compressive strength of approximately 2.4 N/mm^2 at a location near one site.

CEB sizes and shapes were uniform; hence, plastering work was often avoided.

Only weaknesses at the corners of some units were observed. This may be due to a lack of application of proper pressure during production.

3.41.2 Applications of CEB from Projects in West Bengal

The following applications may be mentioned:

- **Load Bearing/Filler Walls –** for exposed brick walls
- **Brick Arches –** without plastering
- **Partition Walls –** offices/institutions
- **Brick Domes –** only brick pointing was done
- Decorative/architectural treatments

The performance of CEB units has been found to be satisfactory.

Photographs of ferro-cement and CEB applications in buildings/factories have been presented. CEB blocks/bricks were produced on-site using local people, materials, and hand-operated machines.

Source:

- Ganguly [6]

3.42 RECOMMENDATIONS FOR AN EARTHQUAKE-RESISTANT DESIGN, REPAIR, AND STRENGTHENING MASONRY CONSTRUCTION

The following IS codes may be referred to

- IS-4326:1993, Code of Practice for Earthquake-resistant Design and Construction of Buildings.

- IS-13828:1993, Improving Earthquake Resistance of Low-Strength Masonry Buildings – Guidelines.
- IS-13935:2009, Seismic Evaluation, Repair, and Strengthening of Masonry Buildings – Guidelines.

The recommendations from the above IS codes have been covered briefly in Chapter 4.

3.43 APPLICATIONS AND PHOTOGRAPHS FROM PROJECTS

A few photographs have been presented (Figures 3.3–3.18).

FIGURE 3.3 CEB units and construction. (Source: Ganguly [6]).

FIGURE 3.4 CEB construction. (Source: Ganguly [6]).

FIGURE 3.5 CEB construction. (Source: Ganguly [6]).

FIGURE 3.6 CEB construction. (Source: Ganguly [6]).

FIGURE 3.7 Wall construction, pointed block wall. (Source: Ganguly [6]).

FIGURE 3.8 CEB, corbelling treatment to wall. (Source: Ganguly [6]).

FIGURE 3.9 CEB, arch, and wall. (Source: Ganguly [6]).

FIGURE 3.10 CEB, checking of dimensions of blocks. (Source: Ganguly [6]).

FIGURE 3.11 CEB wall and ferro-cement units. (Source: Ganguly [6]).

FIGURE 3.12 CEB wall and treatment and curved window. (Source: Ganguly [6]).

FIGURE 3.13 CEB arches and domes. (Source: Ganguly [6]).

FIGURE 3.14 CEB dome. (Source: Ganguly [6]).

FIGURE 3.15 CEB dome structure. (Source: Ganguly [6]).

FIGURE 3.16 CEB building. (Source: Ganguly [6]).

FIGURE 3.17 CEB building. (Source: Ganguly [6]).

FIGURE 3.18 CEB building. (Source: Ganguly [6]).

REFERENCES

1. Terzaghi K and Peck R B, *Soil Mechanics in Engineering Practice*, Asia Publishing House, 1962.
2. Teng W C, *Foundation Design*, Prentice-Hall of India (Private) Ltd, 1965, New Delhi.
3. Auroville Building Center, A Training Manual for Technicians and Entrepreneurs, Production and use of Compressed Earth Blocks, Auroville, July 1992.
4. Doat P, Hays A, Houben H, Matuk S and Vitoux F (CRATerre, France), *Building with Earth*. Translated by Asha Puri, The Mud Village Society, New Delhi, India.
5. Wikipedia, The free encyclopaedia, Compressed Earth Block, September 2019, pp. 1–3.
6. Ganguly K, Some Design Issues & Alternative Building Materials - A Brief Review, Innovative World of Building Materials, State of the Art Lecture, 29th National Convention of Architectural Engineers Division, The Institute of Engineers, India, 31.01.2014.
7. IS-1725: 2013 stabilized soil blocks is General building construction – specification.
8. Raj S, Das R, Mohammad S and Saha S, Coconut fibre reinforced cement stabilised rammed earth blocks, *World Journal of Engineering*, March 2017, Vol. 14, pp. 208–215.
9. Selsiadevi S, Selescadevi S and Varshini V, Earth building blocks reinforced with jute and banana fibre, *International Journal of Engineering Research and Technology (IJERT)*, Special Issue-2018, pp. 1–3, ISSN : 2278 -0181.
10. Danso H, Improving Water Resistance of Compressed Earth Blocks Enhanced with Different Natural Fibres, Department of Construction and Wood Technology, University of Education Winneba, Winneba, Ghana, 2017, pp.1–8.
11. Widder L, Earth eco-building: textile reinforced earth block construction - Science Direct, Abstract, September 2017, pp. 757–762.
12. Winsley B J and Muthukannan M, Glass fibre reinforced compressed stabilised earth block - a green alternative to cement concrete blocks, *International Journal of Engineering and Advanced Technology (IJEAT)*, Vol. 9, Issue-1S4, December 2019, pp. 106–109.

4 Adobe, Clay, Fly Ash–Lime Bricks, and Masonry Construction

Various types of construction may be done with different types of blocks/bricks. Some of these will be discussed below:

4.1 ADOBE

Adobe is the method of preparing mud bricks using moulds without compaction and allowing the bricks to dry in the sun as mentioned [1].

The term adobe comes from the Egyptian word 'thobe' meaning brick. This is 'ottob' in Arabic, adobe in Spanish, and toub in French. It is also called sun-dried mud brick and 'banco'.

It may have different shapes such as conical, cylindrical, trapezoidal, etc.

4.1.1 Brick Production in Ancient Times [1]

Brief comments will be made on the items below:

4.1.1.1 Babylon

The raw material was mostly available at the sites. The mix was puddled with feet using large shallow basins. Chopped straw, cut into small pieces, was included in the wet earth to provide more consistency for sun-dried bricks.

Square shapes were used for the production of large bricks. The measurements ranged from 20–40 cm in length by 5–10 cm in width. The most common length was 315 mm, which is larger than Egyptian bricks. The bricks were left in the sun to dry after production.

Different types of mortars were used in some places. It could be simply a clay mortar for the interior of houses or crudely built walls, or a very effective lime-based cement mortar for larger edifices, or a mixture of fly ash and lime.

4.1.1.2 Greece

The use of earth played an important role in Greek architecture, especially as a building element and as a decorative feature. Good quality soil, along with straw and water, was used for making the bricks.

4.1.1.2.1 Types

Three types of bricks may be mentioned for use as follows:

4.1.1.2.1.1 Rectangular Lydian Brick It may measure

500 mm long × 300 mm wide × 80 mm deep
or 450 mm long × 260 mm wide × 100 mm deep
or 390 mm long × 190 mm wide and 100 mm deep

DOI: 10.1201/9781003319108-4

4.1.1.2.1.2 Tetradoron Brick This may measure

300×300×100mm (square size) and was used for private buildings.

4.1.1.2.1.3 Pentadoron Brick This may measure

450×450×80mm (square size) and was used for public buildings.

There were half brick moulds for the above sizes. The depth of square bricks preferred by Greeks was almost constant, i.e., 80–100mm.

Egypt: The use of sun-dried bricks was very common. Different well-known constructions were done using sun-dried bricks.

The following basic steps for production may be mentioned:

- Digging the earth
- Transporting the same
- Drawing water for the mix
- Moulding the bricks
- Removing the moulds in regular lines/manner
- Transporting after initial drying using a type of yoke
- Placing them one on top of the other in small, regular columns with nominal space for air circulation and drying

It may be noted that the walls of Pothom were constructed with sun-dried bricks measuring 440×240×120mm, some of which contained straw or reed fragments in the mix, while others had only silt.

4.1.1.2.2 Soil Selection

The type of soil is critical for the production of the bricks. The soil should be free from large stones, roots, or topsoil. The soil with the following proportions will be suitable for the production of adobe:

- **Sand**: 55%–75%
- **Clay**: 15%–18%
- **Silt**: 10%–28%
- **Organic Matter**: less than 3%

The possibilities with different types of soil and the associated problems that may be encountered at a construction site are mentioned below:

- **Too Much Sand in the Soil:** There may be inadequate cohesion, and the bricks may crumble.
- **Too Much Clay in the Soil:** Cracks may develop in the bricks during drying.
- **Too Much Organic Matter in the Soil:** With decomposition, the characteristics of the soil may change over time, resulting in porosity and poor durability when exposed to water. The proportions must be evaluated properly after conducting tests.

4.1.1.2.3 Quick Testing for Suitability of Soil for the Producing of Adobe

A quick test may be conducted by rolling a sausage of earth in the palm when the soil is in the plastic state. It should not stick to the hands. It must be carefully flattened to form a ribbon as long as possible. The length of the ribbon should be measured from the breaking point.

- The earth will be suitable for adobe when it breaks between 50 and 150mm long.
- When it breaks at more than 15cm, sand has to be added to the soil.
- If it breaks at less than 50mm, clay must be included in the soil.

4.1.1.2.4 Review of Other Items for Adobe Production

The relevant items are given below:

4.1.1.2.4.1 Excavation It is preferable to have the excavated earth from a single location close to the construction site, or it may be mixed with materials brought from other nearby sites. The depth and consistency of the selected material will be important.

The topsoil should be kept separately for possible use later.

The soil may be excavated using a shovel, pick, etc., or by using mechanical diggers.

It should be noted that the volume of loose earth may be about 30% higher than that of the mud bricks.

4.1.1.2.4.2 Sieving Sieves with mesh diameters of 6–12 mm may be used. The finest is meant for stabilised brick production.

The sieving operation is often carried out near the point of excavation.

For reference, it may be mentioned that one person can sieve about 4 m^3 of earth per day.

4.1.1.2.4.3 Preparing the Earth The traditional method may involve a preparatory hydration or soaking approach, i.e., to saturate the clay with water and break down all the small clods of earth. The soaked earth should not be disturbed for 24 hours for adobe production. This will help mix the earth, improve the quality of the bricks, and reduce the likelihood of cracking due to shrinkage.

The prepared earth is placed or piled on a flat surface close to the brick production area, and a crater is dug in the middle of the earth to be filled with water in order to produce a homogeneous mixture in the plastic stage. The work may also be done using a mixture machine.

As a guide and assuming that the mixture is prepared with one-third water, for a daily production of 500 bricks with dimensions of 300×150×100 mm, it may require 650 L of water using a team of four persons.

4.1.1.2.4.4 Stabilisation Attempts Various vegetable or animal fibres have been tried over the years to obtain better performance from adobe. A few of these attempts, in different countries, may be mentioned below:

- **Peru**: Type of grass (Ichu-Fesuca) was used
- **Africa**: Millet husks were used.
- **Trinidad**: Strong, durable fibres from a plant (*Sporobulus indicus*) chopped into pieces were used.
- **Mexico**: Pine needles were applied. The fibres made up 20%–30% of the volume of the bricks, i.e., about 30 g per brick measuring 230×110×70 mm. The fibres were included after water was added to the earth.
- **Iran**: Rice husks, small palm leaves, and goat or camel hairs were used to improve bonding.
- **Egypt**: Hassan Fathy used 20 kg of straw for every 660 bricks measuring 230×170×70 mm, which is equivalent to 1 m^3 of bricks. Vegetable fibres, after mixing with the mud, were allowed to soak for a long time to help decompose non-fibrous materials.
- **Australia**: 56–67 kg of straw was used to produce 1,000 bricks with dimensions of 450×300×100 mm. Alternatively, 4–5 kg of straw may be used per m^3 of brick.
- **USA**: Studies have been carried out to assess the extent of the decomposition of fibres in adobe bricks. It has been observed that bricks over a hundred years old contained dry fibres that were intact. It would be possible to identify the types of species of fibres used in the construction.

Apart from fibres, conventional stabilisers may also be used, such as cement, lime, and bitumen.

The performance of stabilisers will depend on proper and thorough mixing with the material. Stabilisers in powder form are normally mixed into dry earth, while bitumen emulsions are added to the pre-moistened soil.

The discussion on stabilisers has been done in Chapter 3.

4.1.1.2.5 Method of Manual Mixing of Bitumen

The items mentioned below should be noted:

- A given amount of dry earth has to be measured and used, e.g., it may be assumed as 50 wheelbarrow loads.
- The soil must be made wet first and then kept at rest for 24 hours.
- Four wheelbarrow loads of earth should have water added and be mixed adequately to produce a doughy consistency.
- Emulsified bitumen may be poured into the mixture and mixed properly.
- The resulting mix should be combined with the remaining earth and subjected to vigorous blending action.

Good mixing can be observed from its uniform colour. If vegetable fibres are to be used, they should be incorporated after the bitumen mixture to avoid absorption by them.

4.1.1.2.6 Moulds and How to Use Them

The moulds may normally be made of wood but can also be metal. The moulds should be reinforced at the corners. Smooth internal surfaces are preferred to prevent the earth from sticking and to produce bricks with a neat finish. This will also help to clean the moulds easily.

Brick dimensions may vary depending on the requirements. Small moulds may be used, but they will slow down brick production and require more mortar. Larger bricks are expected to make the wall stronger and may be easier to build. However, they will take longer to dry, be heavier to handle, and may crack. A person working alone may find it difficult to handle a mould longer than 800 mm.

The bricks may be square or rectangular. The latter type may be easier to use if they are the same width as the wall. The bonding pattern of rectangular bricks may be more complicated.

4.1.1.2.7 Filling the Moulds

The methods may be

- **Mechanically:** The earth is in a liquid state and poured into the mould.
- **Manually:** The earth is in a plastic state and is placed into the mould.

A simple test may be done to check the water content of the mix related to manual operation. For this test, a V-shaped groove 80 mm deep has to be made in the mix with a sharpened stick to leave a clear mark. The sides of the groove may swell towards each other but should not meet.

4.1.1.2.8 Wall Building

Some general principles may be followed when dealing with sun-dried bricks. The maximum length of wall between two corners should not be more than 6 m. The longer walls should be supported by partition walls, reinforcement, or buttresses every 5 m. The openings should not exceed one-third of the total surface area of the wall and should not span more than 1.2 m. Openings less than 1 m from the corner should be avoided to prevent weakening the corner.

As a general guideline, the ratio of width to height to length of a section of the wall should preferably be 1:8:12 between two corners.

4.1.1.2.9 Mortars

The mortars used in the joints between the bricks should be the same as, or preferably stronger than, the composition of the bricks used.

It should not contain any gravel (sieved through a 3-mm mesh) or straw.

The most common mortars are

- Sand and lime (300 kg/m^3) – for conventional masonry work
- Earth, cement (225 kg/m^3), and bitumen (cut back) 1%
- Earth and bitumen (cutback) 2% – strong but poor adherence
- Sand, cement (400 kg/m^3), and bitumen (emulsion) 3.3%

The thickness of joints may vary between 15 and 25 mm. If it is too thick, it may weaken the wall.

4.1.1.2.10 Protecting the Corners

For the protection of corners against erosion, fired bricks or stones may be considered as replacements for adobe. These will be more resistant materials and can also strengthen the structure.

4.1.1.2.11 Masonry Bonding Patterns

The bricks may be laid in continuous courses so that the walls of the building are constructed simultaneously. In this manner, the stresses would be spread over the whole foundation.

4.1.1.2.12 Reinforcing the Structure

Ring beams and proper foundations are normally provided in the design for the buildings. It may be recommended to reinforce the walls with elements that will work for traction and bending. This is particularly relevant for long walls and where the foundations may be weak and located on unsound ground or where the building is situated in earthquake zones.

The reinforcement can be considered as free-standing, in the form of posts and beams or reinforcing bars placed in the masonry items.

4.1.1.2.13 Reinforcing Items

This may include

4.1.1.2.13.1 Ring Beams

These may be provided in two or three courses below the top of the walls. The ring beam may be used as a continuous band all around the building and will keep the walls intact, preventing them from opening out. It may serve as a member supporting the roof rafters or as a lintel for doors and windows. It can be made of concrete, wood, or a metal lattice. It must be anchored to the walls, especially at the corners, using concrete.

4.1.1.2.13.2 Posts

These are connected to the foundation, ring beam, and complete framework of the house. The posts are located at important places, such as the corners and the middle of long wall sections. The total system will be suitable for resisting horizontal forces.

If reinforced concrete posts are considered, they will be the same width as the walls, making it easier to pour the concrete during wall construction.

If wood is used, the joining work may be complicated and must be completed properly. Metal framework may also be incorporated into the design. The items must be anchored properly into the foundations and ring beams.

4.1.1.2.13.3 Reinforcing the Masonry, Use of Reinforcing Bars The reinforcing bars in the masonry may be placed horizontally or vertically. The horizontal bars may be placed between the layers of bricks every one or two courses. The bars should be thin and fully encased in the mortar. A wide and thin metal lattice arrangement may also be considered. Necessary care should be taken to ensure that the bars are well protected against corrosion.

The vertical reinforcing bars must be properly anchored into the footing and ring beams. Special bricks with holes may be used where the bars are arranged inside the holes or constructed around the bars.

4.1.1.2.13.4 Openings Wooden or reinforced concrete lintels may be used for the openings. They should be placed 200mm on either side of the opening in the brickwork and 500mm for earthquake zones. Arched openings may also be considered for brickwork construction.

Source:

- Building with Earth [1]

4.2 REVIEW OF ADVANTAGES AND DISADVANTAGES

4.2.1 Advantages

- It is in modular brick form.
- The materials will have a low cost.
- It may be used for all types of buildings.
- The construction method will be suitable for do-it-yourself builders.
- It will be a long-lasting material.
- It will be strong and sturdy.
- The adobe construction will have high thermal mass.

4.2.2 Disadvantages

- It may be more time-consuming than some other building methods.
- Steps like moulding, drying, and transporting the bricks will be needed.
- It will require preparing or buying the bricks.
- It will have poor insulation value.
- It will not be monolithic like rammed earth or cob construction methods.

Source:

- Building Your Home with Earth [2]

4.3 THE REVIEW ON SUITABILITY OF SOIL [3]

Comments on the suitability of soil have already been made. Some items will be highlighted from the IS code here.

The following tests may be considered for assessing the suitability of soil as per the IS code:

4.3.1 DRY STRENGTH TEST [3]

For this test, five or six balls of soil, each about 20mm in diameter, must be prepared. After drying for 48hours, each ball may be crushed between the forefinger and the thumb. The soil may be considered to have enough clay if it does not break and may be used for adobe construction. Some control and checks may be necessary for micro-fissures in the mortar due to drying. If some of the balls break, the soil may not be suitable as it does not have enough clay.

4.3.2 Fissure Control Test [3]

At least eight folded units must be prepared with a mixture of soil and sand. In this test, folded units should be prepared with different proportions of soil and coarse sand. The proportions of soil to coarse sand may vary between 1:10 and 1:3 by volume. The units must be checked for the least amount of coarse sand when opened after 48 hours. If they do not show visible fissures in the mortar, this will indicate the correct proportion of the soil/sand mixture for adobe construction. It will also provide the highest strength.

4.3.3 Strength Test of Adobe

Following sun drying for four weeks, the strength of adobe can be assessed qualitatively, and it should be sufficiently strong to support, in bending, the weight of a person weighing 60–70 kg. If it breaks, more clay and fibrous materials may be necessary for the mix.

4.3.4 Compressive Strength (CS) [3]

CS may be computed by testing 100 mm cubes prepared with clay after fully drying the cubes. A minimum strength of 1.2 N/mm^2 will be preferable.

Source:

- Building with Earth [1]
- IS 13827 [3]

4.4 GUIDANCE FROM IS CODE

4.4.1 Block or Adobe Construction [3]

The sizes of the blocks are mentioned in Table 4.1.

Square blocks are preferred for stronger construction due to the presence of fewer vertical joints.

The mud mortar for joining the blocks should be the same soil as that used for making the bricks/blocks. To create blocks with non-shrinkage, straw in the proportion of 1:1 by volume may be used. The wet mix should be allowed to dry for 7 days (minimum 3 days) before use as stated. The last layer of adobe must be moistened properly before laying the mortar. The surface of the wall should be moistened for a few minutes before laying the adobe.

Normal and good construction practices for developing proper bonds in masonry construction should be followed.

These may be summarised as follows:

- All courses should be carefully executed and laid level.
- Perpendicular joints between the walls should be properly designed/prepared to avoid a continuous vertical joint.
- The vertical joints should be discontinued between consecutive courses by overlapping the adobes, and the joints should be filled with mortar adequately.

TABLE 4.1
Sizes of Blocks

Rectangular	380×250×110 mm (overlap of about 125 mm)
Square	380×380×110 mm (overlap of about 190 mm)

The bond details, in adobe walls are given in IS-13827.

Source:

- IS 13827 [3]

4.5 REVIEW OF RECENT DEVELOPMENTS IN FIBRE-REINFORCED ADOBE BRICKS FOR SUSTAINABLE CONSTRUCTION

4.5.1 General Comments

It can be stated that an adobe brick needs improvement against affinity for water, low compressive and tensile strength, and brittle behaviour compared to other traditional construction materials.

These weaknesses can be addressed by using additives or stabilisers in the mix.

Studies have been conducted to include natural and waste materials for the production of the bricks.

It is relevant to study the production methodology, characteristics of the raw materials to be used, and the method of production.

Following this, it may be possible to have sustainable material at a low cost, with a construction method that is relatively simple and affordable.

4.5.2 Sample Preparation for the Study

Adobe bricks may be prepared using soil, water, and additives as stated. The water content in the mix will be important. The methods of making the bricks may include pressing, moulding, and extrusion techniques. The steps mentioned below will be relevant:

- Soil and waste additives
- Characterisation
- Mixing the materials
- Preparation by moulding, pressing or extrusion method
- Drying at room temperature or in the oven

4.5.3 Testing of Items

Different tests have been carried out on adobe bricks to assess their properties suitable for construction purposes. These may include particle size analyses, shrinkage, bulk density (BD), water absorption (WA), CS, flexural strength, thermal conductivity, etc. The properties of the bricks will depend on the quantities of additives included.

The comments from the study may be summarised as follows:

- By using waste additives for the production of adobe bricks, it is possible to enhance the physical, mechanical, and thermal properties of the material.
- Water suction and thermal conductivity tests were relevant for the studies.
- The values of BD, WA, and CS were found to be within the limits for the material considered with waste additives.
- Tests for the commercial production of bricks with waste additives must be conducted for confidence and acceptance.
- Guidance will be necessary for such tests covering all related items.
- Further research using different additives for adobe bricks will be required. The study should cover the technical, scientific, environmental, regulatory, and economic impacts of improved adobe bricks.

- Further work with additives may include date palm fibre, chicken feathers, sugarcane bagasse, waste tea, etc.
- By using waste materials efficiently for adobe bricks, it is possible to provide economic benefits while also protecting the environment.

Source:

- Salih, Osofero, and Imbabi [4]

4.6 COMPRESSED EARTH BLOCKS

The discussion on compressed earth blocks has been covered in Chapter 3.

4.7 COMMON BURNT CLAY BUILDING BRICKS

Common burnt clay bricks are classified based on average CS as per IS Code [5].

Classification – refer to Table 4.2

For higher strengths, IS 2180 may be referred to.

4.7.1 General Quality

The bricks may be either hand-moulded or machine-moulded and shall be prepared using suitable soil. The bricks shall not have cracks or flaws and shall be free from nodules of free lime.

Hand-moulded bricks of 90 or 70 mm height shall have a frog 10–20 mm deep on the flat sides. The shape and size of the frog shall be as per the IS guidelines.

Bricks with a height of 40 mm and those prepared by the extrusion process may not be produced with frogs.

The bricks shall have smooth rectangular faces with sharp corners and be uniform in colour.

Dimensions shall be as per Tables 4.3 and 4.4.

TABLE 4.2
Classification

Class Designation	Average Compressive Strength (N/mm^2)
35–3.5	35–3.5

TABLE 4.3
Modular Size

Length (mm) (L)	Width (mm) (W)	Height (mm) (H)
190	90	90
190	90	40

TABLE 4.4
Non-modular Size

Length (mm) (L)	Width (mm) (W)	Height (mm) (H)
230	110	70
230	110	30

TABLE 4.5
Information

Length (mm) (L)	Width (mm) (W)	Height (mm) (H)
70	110	70⅓ length brick

To achieve proper bond arrangement and use modular dimensions with the brickwork and the non-modular sizes, the following sizes of bricks may be used as given in Table 4.5:

Tolerances – As per IS 1077

Physical Requirements and CS – This shall be as per the guidance of IS 1077.

CS of an individual brick after testing shall not fall below the corresponding class of brick by more than 15%.

4.7.2 Other Information

4.7.2.1 WA

WA tests shall be carried out for bricks as per IS code, and after immersion in cold water for 24 hours, the WA shall not be more than 20% by weight up to Class 12.5 and 15% by weight for higher classes.

4.7.2.2 Efflorescence

The bricks shall be tested for efflorescence rating as per IS code and shall not be more than moderate up to Class 12.5 and slight for higher classes.

Sampling for Conformity – This shall be carried out as per the procedure given in IS 5454.

4.7.2.3 Marking

Each brick must be marked in the frog to indicate the source of manufacture, and the use of a standard mark may be done.

Source:

- IS 1077: 1992 [5]

4.8 BURNT CLAY FLY ASH BUILDING BRICKS

The burnt clay fly ash building bricks shall be used as per the guidance given in IS 13757: 1993 [6]

Some details of such bricks are indicated below:

Classification – The classification shall be from 3.5 to 30. This means that the average CS of the bricks shall not be less than 3.5–30 N/mm^2 respectively.

General Information – The clay and fly ash bricks may be prepared by hand or machine-moulded. Admixtures may be used for production, which will be suitable for fly ash and soils selected for the production. Suitable soils and fly ash have to be selected and used for the production of bricks. The fly ash may be selected for production according to grade-1 or grade-2 as per IS-3812.

The bricks shall be burnt uniformly without cracks and flaws such as black coring, nodules of stone, and must be free from lime and organic matter. The dimensions of the frog, in non-modular bricks, if used, shall be the same as that of the modular size of bricks.

Hand-moulded bricks having 90 or 70 mm height shall be moulded with a frog of 10–20 mm deep on one of its flat sides. The bricks of 40 or 30 mm height, and those produced by the extrusion process, may not be provided with frogs. The shape and size of the frog shall be as per IS guidelines.

The production of bricks must be controlled efficiently. The faces of the bricks must be smooth and rectangular with sharp corners and must be uniform in shape and colour.

4.8.1 Dimensions and Tolerances

Dimensions of clay fly ash building bricks shall be as per Tables 4.6 and 4.7

For achieving proper bond arrangement and using modular dimensions with the brickwork, as well as using the non-modular sizes, the following sizes of bricks are recommended as in Table 4.8.

4.8.2 Tolerances

Modular Size I For necessary tolerances, guidance

Non-Modular Size I may be obtained from IS code.

Twenty (or more, depending on the size of the stack) whole bricks are to be selected at random for necessary checks. Necessary guidelines are available in the code for uniformity and acceptance.

4.8.3 Physical Requirements

These are mentioned below:

- **CS** – The bricks shall be tested as per the IS guidelines. CS from the test results of any individual brick shall not be below the minimum CS specified for the corresponding class of brick by more than 15%. In such cases, if it occurs, the lot shall be checked for the next lower class of brick.
- **WA** – Following the test as per the IS guidance, WA shall not be more than 20% up to Class 12.5 and 15% for higher classes.
- **Efflorescence** – After testing as per IS guidelines, the effect of efflorescence shall not be more than moderate up to class 12.5 and slight for higher classes.
- **Sampling and Criteria for Uniformity** – Shall be as per IS guidelines.

Source:

- IS 13757 [6]

TABLE 4.6
Standard Modular Sizes

Length (mm) (L)	Width (mm) (W)	Height (mm) (H)
190	90	90
190	90	40

TABLE 4.7
Standard Non-modular Sizes

Length (mm) (L)	Width (mm) (W)	Height (mm) (H)
230	110	70
230	110	30

TABLE 4.8
Information

Length (mm) (L)	Width (mm) (W)	Height (mm) (H)
70	110	70⅓ length brick
230	50	70½ width brick

4.9 FLY ASH BRICKS

The bricks may be manufactured using class C or class F fly ash and water. The mixed material for the units is compressed by applying pressure (say 28 MPa) to make the bricks and cured for 24 hours in a 66°C steam bath. It may be toughened by using an air-entraining agent, and the bricks may survive 100 freeze–thaw cycles. The bricks may be referred to as self-cementing types due to the high concentration of calcium oxide in class C fly ash. The bricks may be 20% cheaper than conventional clay bricks.

The basic items are mentioned below [7]:

4.9.1 Raw Materials

The following raw materials, as given in Table 4.9, will be preferred.

The CS of fly ash bricks using the above proportion may vary to some extent. The range may be between 7.5 and 10 MPa. Fly ash bricks may be stronger and lighter than conventional clay bricks.

Advantages

- The dead load will be reduced on the structure due to lightweight.
- They will have higher fire insulation.
- Breakage during use and transport may be reduced due to high strength.
- More area may be covered than with clay bricks using the same number of bricks.
- The seepage of water through the bricks may be reduced due to lower water penetration properties.
- Mortar required for brick joints and plaster work will be reduced to a large extent due to the uniform size of the bricks
- Water soaking for 24 hours may not be required for these bricks. Sprinkling water prior to using the bricks may be adequate.
- Without using a backing coat of lime plaster, gypsum plaster may be applied directly on the bricks.

Disadvantages

- Due to the limitation of sizes, only modular sizes are produced. Larger sizes may have more breakage.
- Mechanical strength may be low, but this may be improved by including marble waste or mortar between blocks.
- The bricks may be suitable for subtropical areas or areas where the climate is warm. They do not absorb heat, which may not be helpful during cold climates.

Source:

- Wikipedia [7]

TABLE 4.9
Raw Materials

Item	Mass
Fly ash	60%
Sand/stone dust	30%
Ordinary Portland cement/(lime + gypsum)	10%
Total	100%

4.10 FLY ASH–LIME BRICKS [8]

Guidance and information on pulverised fuel ash–lime bricks with wet CS less than 30 N/mm² are given in IS 12894. For higher strengths, reference may be made to IS 2180 and IS 1077.

Information regarding dimensions, physical requirements, materials, classification, etc., is mentioned in the IS code for the purpose of using the bricks for buildings.

A few items are referred to from the code.

4.10.1 General Requirements

The bricks should be manufactured to be sound, compact, and uniform in shape when viewed visually. There should not be visible cracks, warpage, or organic matter.

The bricks should be solid and have or not have a frog 10–20 mm deep on one of their flat surfaces as per the IS guidelines. The bricks should have smooth rectangular faces along with sharp corners and maintain uniform colour and shape.

4.10.2 Classes of Pulverised Fuel Ash–Lime Bricks

On the basis of average wet CS, the bricks may be classified into the following classes as mentioned in Table 4.10.

4.10.3 Dimensions and Tolerances

Dimensions (mm) are given in Tables 4.11 and 4.12.

For developing a proper bond arrangement, the modular dimensions of the brickwork, and the use of non-modular sizes, the following sizes of the bricks may be used as shown in Table 4.13:

TABLE 4.10
Classes of Bricks

Class	Average Wet Compressive Strength Not Less Than N/mm²
30–3.5	30–3.5

TABLE 4.11
Modular Sizes

Length (mm) (L)	Width (mm) (W)	Height (mm) (H)
190	90	90
190	90	40

TABLE 4.12
Non-modular Sizes

Length (mm) (L)	Width (mm) (W)	Height (mm) (H)
230	110	70
230	110	30

TABLE 4.13
Information

Length (mm) (L)	Width (mm) (W)	Height (mm) (H)
70	110	70⅓ length brick
230	50	70½ width brick

TABLE 4.14
The Materials

Pulverised fuel ash (fly ash)	Should be as per Grade-1 or Grade-2 of IS 15648
Bottom ash	As replacement of sand should not have more than 12% loss on ignition following the test as per IS 1727
Sand	Harmful materials in sand like clay and silt should be less than 5%
Lime	Lime should be Class-C hydrated lime as per IS 712
Additives	Suitable additives that will not be harmful for durability may be used

TABLE 4.15
Physical Characteristics

Compressive strength	The minimum average wet compressive strength of fly ash–lime bricks should not be less than each class mentioned above following the test as per IS 3495 (Part-1). The wet compressive strength of any individual brick should not be below the minimum wet average compressive strength applicable for the corresponding class of bricks by more than 20%.
Drying shrinkage	The average drying shrinkage of bricks should not exceed 0.15% (from the average of 3 units) when tested as per IS 4139.
Efflorescence test	Should be done as per IS 3495 (Part-3). Efflorescence rating should not be more than moderate up to Class 12.5 and slight for higher classes.
Water absorption	Following the test as per IS 3495 (Part-2), the average water absorption for bricks after immersing, in cold water, for 24 hours should not be more than 20% by mass up to class 12.5 and 15% by mass for higher classes

4.10.4 Tolerances

The limits are given in the code. It shall be within the limits as per 20 bricks.

Materials are mentioned in Table 4.14.

Physical characteristics are indicated in Table 4.15.

For other items such as sampling, criteria for conformity, marking, etc., the necessary guidance may be taken from the IS 12894 code.

Source:

- IS 12894 [8]

4.11 ADVANTAGES OF FLY ASH BRICKS [9]

The following may be important:

- The fly ash bricks may be produced with or without a frog.
- The plaster thickness will be less compared to clay bricks.
- Less mortar may be required as the bricks will be produced with uniform size and shape.
- There will be savings in cement mortar, and the cost will be reduced.
- The bricks will be environment friendly.

4.11.1 More Comments Are

- It will save agricultural land, which is normally used for the manufacturing of clay bricks.
- The operation will be less energy-intensive compared to clay bricks, and the environment will be protected.
- It uses fly ash, which will be available from thermal power stations.
- The manufacture of bricks may be done at construction sites, which will provide numerous advantages.

4.11.2 Review on Main Items for Production

Fly Ash, Sand, Lime/Cement, Water, Pan Mixer, Shaping Machine, Curing Chamber, Storing/testing of bricks, and readiness for applications.

4.11.3 Manufacturing of Fly Ash Bricks

It may be mentioned briefly as follows:

- The fly ash bricks may be machine-made and can be manufactured by hydraulic or vibratory press.
- Ordinary Portland cement may also be used in place of hydrated lime and gypsum.
- The raw materials will include fly ash, which may be available from thermal power stations, hydrated lime, gypsum, locally available sand/stone dust, and water.
- To achieve a semi-dry and uniform mix, the raw materials, in the required proportions, may be mixed in the pan mixer.
- Moulded bricks are air dried for 1 or 2 days in a shed, depending on the weather conditions, and then cured with water for 14–21 days.
- The semi-dry mix may be placed in the moulds of the hydraulic/vibro press.
- The bricks after production may be expected to be sound, compact, and uniform in shape.

Further Comments on fly ash bricks

- It is a well-proven building material and may be used as an alternative to burnt clay bricks.
- IS 12894: 2002 may be referred for fly ash bricks.
- Bricks may be manufactured according to the required strength.
- These bricks have been included/considered in DSR.
- The use of fly ash bricks has also been incorporated in CPWD specifications.

Source:

- NTPC Information [9]

4.12 MACHINE FOR PRODUCTION OF FLY ASH BRICKS [10]

There are different machines available that may be fully automatic for the production of fly ash bricks. The making of bricks from the machine may include items like automatic raw material weighing, auto mixing, batching plant, silos, and discharge of finished bricks through conveyors for fast production. The raw materials required for the production of fly ash bricks will be fly ash, sand, lime, and gypsum or cement as stated.

The bricks may be produced by compressing the raw materials in the mould and pressing hydraulically.

The hydraulic unit may consist of a hydraulic cylinder and a power pack, which may be manufactured in-house. The machine will require less manpower effort, and it may be cost-effective and have the provision for pick and place with an auto-stacking system. The fly ash bricks prepared using such machines will be uniform in shape and size with high CS.

The applications of fly ash bricks may include building construction and infrastructure development, construction of dams, pavements, tanks, canal lining, underwater works, irrigation works, etc.

4.12.1 Further Information on Fly Ash Bricks [10]

The following may be noted:

- The properties will meet the recommendations of IS 4139 and 12894.
- Curing time will be about 8–10 days.
- The maximum CS may be in the range of 8–12 N/mm^2
- WA may vary between 12% and 16% depending on the mix proportions.
- The weight of fly ash bricks may range from 27.5 to 32.5 N/brick, depending on the quantity and quality of sand.

Source:

- AHP Machine [10]

4.13 EXPANDED CLAY AGGREGATE (ECA) BLOCKS [11]

ECA blocks for masonry work are made of ECA, fly ash, and cement. These are manufactured specially with the necessary curing process. The blocks are used for non-load-bearing masonry walls.

These blocks have better fire and chemical resistance, improved durability, reduced shrinkage, versatility, speed of construction, and are economical to use. The use of these units will also be beneficial from an environmental perspective.

The available sizes are

- 600×200×230 mm and
- 600×200×100 mm.

Source:

- Banker [11]

4.14 COMPARISONS OF CLAY BRICKS AND FLY ASH BRICKS [12]

The comparisons between red bricks/clay bricks and fly ash bricks are presented in Table 4.16:

TABLE 4.16
Comparisons

Sl. No.	Properties	Fly Ash Brick	Red Bricks/Clay Bricks
1	Compressive strength	9–10 N/mm^2	3–3.5 N/mm^2
2	Density	1,700–1,850 kg/m^3	1,600–1,750 kg/m^3
3	Dimensional stability	High tolerance	Very low tolerance
4	Absorption	10%–14%	15%–25%
5	Wastage during transit	Less than 2%	Up to 10%

Source: The Constructor [12]

4.15 CONCRETE MASONRY UNITS – SPECIFICATION, IS 2185 (PART-1): 2005

4.15.1 Part-1 Hollow and Solid Concrete Blocks

Brief comments on concrete blocks have been made before.

The code deals with [13]

- Hollow (open and closed cavity) load-bearing concrete blocks.
- Hollow (open and closed cavity) non-load-bearing concrete blocks.
- Solid load-bearing and non-load-bearing concrete blocks.

4.15.1.1 Some Definitions

For items like block, block density, drying shrinkage, face shells, gross area, height, hollow (open or closed cavity) block, length, moisture movement, solid block, webs, width, etc. – IS 2185 (Part-1) may be consulted.

4.15.2 Dimensions and Tolerances

The concrete blocks may be made in different sizes and shapes as per the requirements.

4.15.3 Concrete Blocks

The concrete blocks (open or closed cavity) or solid may be referred to by nominal dimensions.

The nominal dimensions will include the thickness of the mortar joint.

The nominal dimensions of concrete blocks are indicated in Table 4.17.

The blocks may be prepared in half lengths of 200, 250, or 300 mm, considering full height. U-Blocks may be made for bands and lintels.

The variations in the length of units shall not be more than ±5 mm, and for the height and for width, it should not be more than ±3 mm.

4.15.4 Minimum Face Shell and Web Thickness

The dimensions in Table 4.18 are in mm.

TABLE 4.17
Dimensions

Length	400, 500 or 600 mm
Height	200 or 100 mm
Width	50, 75, 100, 150, 200, 250 or 300 mm

TABLE 4.18
Minimum Thickness

Serial No.	Nominal Block Width	Face Shell Thickness, Min	Web Thickness, Min	Total Web Thickness Per Course in Any 200 mm Length of Wall
1	100 or less	25	25	25
2	100–150	25	25	30
3	150–200	30	25	30
4	Over 200	35	30	38

4.15.5 Classification

The bricks shall be of the following grades:

- **Grade A**: These will be for load-bearing type units with a minimum block density of 1,500 kg/m^3. The minimum average CS may be 3.5, 4.5, 5.5, 7.0, 8.5, 10.0, 12.5, and 15 at 28 days.
- **Grade B**: These are also load-bearing units with block densities between 1,100 and 1,500 kg/m^3. The minimum average CS should be between 3.5 and 5 N/mm^2 at 28 days.
- **Grade C** – Solid concrete blocks: These are solid concrete blocks with density not less than 1,800 kg/m^3. The units shall be prepared with a minimum average CS of 4 and 5 N/mm^2.

4.15.6 Materials

The materials for manufacturing the units are cement, aggregates, fly ash, water, admixtures, etc.

- IS 2185 (Part-1) may be referred to.

4.15.7 Manufacture

For mixing, placing and compaction, curing, drying, etc. – IS 2185 (Part-1) may be referred to.

Surface texture and finish – IS 2185 (Part-1) may be referred to.

4.15.8 Physical Requirements

For items covering general information, dimensions, block density, compressive strength, water absorption, drying shrinkage, moisture movement, etc., IS 2185 (Part-1) may be referred to.

Physical requirements are mentioned in Table 4.19.

For other items such as tests, sampling, number of tests, criteria for conformity, manufacturer's certificate, independent tests, cost of testing, and marking, IS 2185 (Part-1) may be referred.

Source:

- IS 2185 (Part-1) [13]

4.16 SPECIFICATION FOR CONCRETE MASONRY UNITS – IS 2185 (PART-2) – 1983

4.16.1 Part-2: Hollow and Lightweight Concrete Blocks

4.16.1.1 Introduction

The concrete masonry construction may be considered economical due to [14]

TABLE 4.19
Physical Requirements

Type	Grade	Density of Block (kg/m^3)	Min. Average Compressive Strength Units (N/mm^2)	Min. Compressive Strength, Individual Units (N/mm^2)
Hollow (open and closed cavity) load-bearing unit	A types	Not less than 1,500	3.5–15	2.8–12
	B types	Less than 1,500 but not less than 1,100	3.5–5	2.8–4
	C types		4–5	3.2–4
Solid load-bearing units		Not less than 1,800		

- The units are large and can be properly sized and shaped, allowing for fast construction.
- Solid load-bearing concrete blocks.
- Solid non-load-bearing concrete blocks.

4.16.1.2 Terminology

The different terms like block, block density, drying shrinkage, face shells, gross area, height, and length may be taken from IS 21815 (Part-2).

4.16.1.3 Scope

The code covers concrete masonry building units that may be used for load-bearing and non-load-bearing wall construction.

These are

- Hollow (may be open or closed cavity) load-bearing concrete blocks.
- Hollow (with open or close cavity) non-load-bearing concrete blocks.

Moisture Movement	]	
Solid Block	]	Reference may be made to IS 2185 (Part-2)
Width	]	
Webs	]	

4.16.1.4 Dimensions and Tolerances

4.16.1.4.1 Concrete Block

The nominal dimensions of concrete blocks are given in Table 4.20 as per the code.

Furthermore, half-length bricks, i.e., 200, 250, or 300 mm, may be prepared for the full-height construction.

The maximum variation in length shall not be more than ±5 mm, and the maximum variation in height and width shall not be more than ±3 mm.

4.16.1.5 Classification

- **Grade A** – These units may be used below or above ground level with a damp-proof course. The exterior walls may or may not be treated with a water protective coating, and the same applies to interior walls.
- **Grade B** – These units may be applied above ground level with a damp proof course, and the exterior walls are treated with a proper protective coating.

Materials – The IS code may be referred to.

4.16.1.5.1 Lightweight Aggregates

The lightweight aggregates shall conform to IS 9142.

TABLE 4.20
Dimensions

Length	400, 500 or 600 mm
Height	100 or 200 mm
Width	50, 75, 100, 150, 200, 250 or 300 mm

TABLE 4.21
Density

Fine aggregates	1,120 kg/m³ (Max)
Coarse aggregates	880 kg/m³ (Max)
Combined aggregates	1,100 kg/m³ (Max)

TABLE 4.22
Physical Requirements

Type and Grade	Min Compressive Strength: Average of 8 Units Min. (N/m^2)	Min Compressive Strength: Individual Units Min. (N/m^2)	Max Average Water Absorption with Oven Dry Mass of Concrete: Less than 1,360 kg/m³	Max Average Water Absorption with Oven Dry Mass of Concrete: Less than 1,600 kg/m³
Hollow, load bearing				
Grade A	7.0	5.5	–	290
Grade B	5.0	4.0	320	–
Hollow, non-load-bearing				
Solid, load-bearing	4.0	3.5	–	–
Grade A	12.5	10.8	–	290
Grade B	8.5	7.0	320	–

Source: IS 2185 (Part-2) [14]

The dry loose BD of lightweight aggregate is given in Table 4.21.

Other Items such as water, admixtures, manufacturing, mixing, placing and compaction, curing, drying, surface texture, drying shrinkage, tests, cost, finish, and marking – IS code may be referred to.

Physical requirements are mentioned in Table 4.22.

4.17 SPECIFICATION FOR CONCRETE MASONRY UNITS, IS 2185 (PART-3)

4.17.1 Part-3 Autoclaved Cellular (Aerated) Concrete Blocks

4.17.1.1 Introduction

The concrete masonry construction may be economical due to [15]-

- The units may be large and properly sized and shaped as mentioned. This will mean rapid construction with fewer joints, which will provide savings in mortar when compared to normal masonry construction. The strength of the wall will be increased.
- Proper plane surface will be available. The use of plaster for unimportant buildings in low rainfall areas may be avoided.

Scope: Part-3 of the code will cover the needs of autoclaved cellular (aerated) concrete blocks with a density of up to 1,000 kg/m³

Terminology: Reference may be made to IS 2185 code, Part-3

4.17.1.2 Dimensions and Tolerances

4.17.1.2.1 Concrete Block

The nominal dimensions are mentioned in Table 4.23.

The blocks may be made half-length, i.e., 200, 250, or 300 mm, to meet the needs for full-height construction.

4.17.1.2.2 Classification

The autoclaved concrete blocks may be prepared into two grades as shown in Table 4.24.

For items such as, materials, different items as per IS 2185, manufacture, surface texture and finish, physical requirements, manufacturing, tests, sampling, number of tests, criteria for conformity, manufacturer's certificate, independent tests, storage, marking, and other items – IS 2185 (Part-3) may be referred to.

Source:

- IS 2185 (Part-3) [15]

4.18 PREPARATION AND USE OF MUD MORTAR IN MASONRY – GUIDE [16]

4.18.1 Selection of Clay for Mud Mortar

The grain size analysis shall be carried out as per the guidance given in IS 2720 (Part-4).

The details are given in Table 4.25.

The total content of clay and silt should not be less than 60% by weight.

TABLE 4.23
Nominal Dimensions

Length	400, 500 or 600 mm
Height	200, 250 or 300 mm
Width	100, 150, 200, or 250

TABLE 4.24
Grades of Blocks

Serial No.	Density in Oven-Dry Condition (kg/m³)	Compressive Strength, Min (N/mm²)	
		Grade 1	Grade 2
1	451–550	2.0	1.5
2	551–650	4.0	3.0
3	651–750	5.0	4.0
4	751–850	6.0	5.0
5	851–1000	7.0	6.0

TABLE 4.25
Details

Clay	18%–22% by weight
Sand	30%–40% by weight
Silt	40%–45% by weight

When the soils have more sand or clay, the same must be blended with suitable plastic or non-plastic materials to bring the mechanical composition within the given limiting values.

4.18.2 Plasticity Index

The plasticity index shall be computed as per IS 2720 (Part-5) for important projects and should range between 12 and 15. For smaller projects, the field tests as per the IS code may be adequate.

4.18.3 Design Considerations

The durability of the mortar will depend on various factors such as setting, hardening, temperature variations, moisture movements in the surrounding areas, changes in its own moisture content, and the accommodation of stress development due to the above factors.

The bond between masonry and mortar is critical and should not be damaged due to volumetric changes and structural movements. If it is damaged, there will be a reduction in the strength of the masonry construction, increasing the chances of seepage of water and other solutions entering the masonry. This may lead to deterioration and affect the resistance against chemical attack under aggressive environmental conditions.

The compressive and bond strength of mud mortar will decrease with an increase in moisture content in the mix. CS may decrease by 0.8 to 1 N/mm^2 when the moisture content is about 6%–10%.

Moisture between 5% and 8% by weight may penetrate into mud mortar even after plastering or pointing on the exposed surfaces of the masonry.

Mud mortar may be used for temporary structures and low-rise structures with adequate lateral stability, as it is likely to be affected by weathering.

The basic stress in masonry with bricks having strengths of 7–10 and 3.5 N/mm^2 shall not be more than 0.2 and 0.1 N/mm^2, respectively.

The thickness of mud mortar shall not exceed 20 mm for brick masonry and 30 mm for stone masonry works.

The masonry may be plastered or pointed with cement or lime mortar on the external faces to eliminate the erosion of mortar from the joints.

Plain concrete or RCC bed blocks may be used on the wall and under concentrated loads for load distribution.

4.18.4 Preparation of Mortar

The soil for the mud mortar shall be prepared and selected carefully to ensure it is free from gravel, coarse sand (i.e., particles more than 2 mm), lime, vegetable matter, and Kankar elements.

Kankar and coarse grit may be removed by wet sieving if required.

The sieved clay must be dried and blended with plastic clay or non-plastic materials that are locally available, such as rice husk, wheat husk, or jute waste, to achieve the required composition.

When wet sieving is not done, the soil may be lightly broken up and wetted. The wet mass shall be tempered underfoot repeatedly, turned over, and thoroughly kneaded. The kneaded mass shall be left for re-tempering for 24 hours, and the soil shall be covered with wet gunny bags.

Immediately before application in masonry, checks must be done for the required consistency and workability. The total moisture content of the prepared mortar shall be within the range of 35%–40% by weight of soil.

Field tests shall be carried out as per the guidance given in IS 13077.

Source:

- IS 13077 [16]

4.19 MODERN FACE OF LIME

Lime may be considered an alternative to cement. Furthermore, lime needs less energy to produce than cement. The following may be considered important [17]:

- To produce a standard type of mortar, one may consider 1:6 standard cement–sand mix or a 1:3 hydraulic lime (NHL 3.5):sand mix. For producing 1 m^3 of mortar, 240 kg of cement (1:6 mix) or 200 kg of hydraulic lime (1:3 mix) may be required. The energy used for the production of 240 kg of cement will be higher than mentioned.
- Low-energy sustainable materials like water reeds, earth, straw, clay, and coppiced timber may be used with lime, as lime breathes and can keep them dry.
- Hydraulic lime produces less CO_2 during the firing process than cement.
- It allows masonry items (i.e., bricks and stones) for the building to be recycled further after completing the service life of the building.
- Lime mortars and plasters may have a better aesthetic appearance than cement.
- Movement joints/expansion joints may be reduced with lime-based mortars and plasters.

Portland cement is a versatile material for various engineering applications. Cement, when used as mortars, plasters, and renders, would not be ideal as it is too hard, too rigid, and too impermeable. There may be more use of hydraulic lime for new buildings in the future. The applications of lime may be a better choice. This comment may be made with confidence after examining the performance of historic/heritage buildings.

Source:

- Pritchett [17]

4.20 THE MYTH IN THE MIX [18]

The ratio 1:3 (lime to sand) is well known for the lime mortar applications related to masonry work.

The use of lime mortar has been prevalent for a long time in different parts of the world. The mix proportions of mortar varied and were largely based on the nature and performance of the predominantly locally available materials. Initially, the uses were with pure non-hydraulic lime applied as putty after mixing with well-graded aggregate.

The common use of the 1:3 (lime : sand) ratio is based on the measurement of voids by volume within a measure of dry sand. The volume of water required to fill all the voids of a measured volume of oven-dried sand is important. This volume will be the minimum volume of lime binder required to produce a good mortar. The quantity has been computed to be one-third of the volume of water to fill the voids, and thus 1:3 may have been determined.

The mix can be volume batched according to the specified ratio using sand and lime.

Source:

- Lynch [18]

4.21 REVIEW OF MORTAR [19]

A few comments may be made on the following:

4.21.1 Materials

The mortar may be prepared with sand using one or more of the following:

- Lime
- Portland cement
    - Ordinary
    - Rapid hardening
    - Blast furnace
    - Sulphate resisting
- Masonry cement
- High-alumina cement
- Plasticisers or other additives
- Water
- Selected fibres

4.21.1.1 Sand

The grading of sand and the proportions of particle sizes will be important for workability, including the ease of spreading the mortar. The necessary guidance may be obtained from IS codes.

4.21.1.2 Lime

The type and quality of lime will be critical. The necessary guidance may be obtained from IS codes.

4.21.1.3 Types of Finish

The following types of finish may be associated with jointing and pointing:

The types are struck flush, square recessed, overhung struck, curved recessed, and struck or weathered.

Source:

- Smith and Plowman [19]

4.22 SPECIAL TYPES OF MORTARS [20–22]

For special types of mortar, in construction, like silicate and resin types of chemical-resistant mortars, respective IS codes may be referred to.

Source:

- IS 4443, 4456 (Part-1) and 4832 (Parts 1 and 2) [20–23]

4.23 CASE STUDY RELATED TO USE OF LIME MORTAR

The analyses of lime mortar for the two structures will be referenced. These structures are Amba Vilas Palace (AVP) and Krishna Raja Sagara Dam (KRS), which are located in India [24].

The above two structures were built in the early 20th century, in the city of Mysore and before the extensive use of Portland cement for the construction in the country.

4.23.1 AVP

After the damage to the original palace due to fire, AVP was constructed in the same location in 1912 and served as the official residence of the royal family.

There are valuable paintings in the palace building, done directly on the plastered walls. The walls were plastered with lime mortar.

4.23.2 KRS Dam

The dam was completed in 1924 and has been used for irrigation in that area.

Lime mortars were used in the construction of both structures.

4.23.3 Study and Tests on Mortars

Mortar samples were collected from the above two structures for study and various tests, including chemical analyses.

The comments from the analyses/tests are as follows:

- It seems that the lime mortar contained about 20% CaO and about 15%–20% MgO, suggesting that the source was dolomite, which was used as the binder.
- The samples from the palace indicated 30% higher alumino ferrites than the mortar collected from the dam.
- The mortar from the palace was found to be about three times stronger and twice as stiff as the mortar from the KRS dam.
- The vegetation on the dam seems to contain 30% albite, while the mortar from the palace had a higher sand content.

Source:

- Bindiganavile, Suresh, and Islam [24]

4.24 DESIGN OF UN-REINFORCED MASONRY RETAINING WALLS [25]

Brief comments will be made regarding design guidance related to un-reinforced masonry retaining walls.

The retaining wall may be subjected to the following different lateral loads:

- Earth pressure
- Hydrostatic pressure
- Surcharge loads

4.24.1 Failure Modes

These may include

- Overturning
- **Spine Failure –** No tension will be allowed in unreinforced retaining walls.
- Sliding

4.24.2 Movement Joints

For concrete block work, vertical movement joints may be provided at 9 m centre to centre with no reinforcement within the bed joints. With bed joint reinforcement, the distance may be increased. De-bonded ties may be considered across the joint to increase shear resistance and strengthen it. Alternatively, the use of piers near the joint may be considered for strengthening the wall at the joint location

4.24.3 Comments

The following points are important:

- The geometry of the unreinforced masonry wall is relevant.
- No tension will be generated within it. Self-weight will be critical.
- Self-weight should account for lateral loads generated by soil pressure.
- The wall may need to be made thicker, and the resultant loads/forces should be placed within the middle third of the wall footing.
- Due to space restrictions, the above requirements may not be satisfied in some cases. In that case, vertical reinforcement may be considered in the wall.

Source:

- O'Regan [25]

4.25 BEARING CAPACITY OF PLAIN AND REINFORCED CONCRETE BLOCKS [26]

Studies were conducted, and test programs were arranged for bearing capacity related to both plain and reinforced concrete blocks. Bearing failure may occur due to the application of concentrated loads on the concrete block.

The items may include

- Anchorage zones of post-tensioned concrete.
- Pile heads.
- Shear keys in composite elements.
- Supports of reinforced concrete deep beams, etc.

Brief details related to the test programme organised are presented below:

4.25.1 Plain Concrete Blocks

The study was carried out to understand

Edge distance effects

- Distance between the load and the nearest edge of the block (edge distance).
- Different combinations of loading area ratios.
- Size effects of the blocks.
- Effect of base friction.

4.25.2 More on Plain Concrete Blocks

The dimensions of the blocks are indicated below. Some basic dimensions for the test specimens are shown in the sketch enclosed (Figure 4.1). The failures of the blocks, as observed, were sudden and explosive. To reduce friction, 2.4 mm PTFE sheets were used at the bottom of some specimens. The blocks had been tested with different dimensions of a & H. In some tests, the dimensions of 'b' were also modified

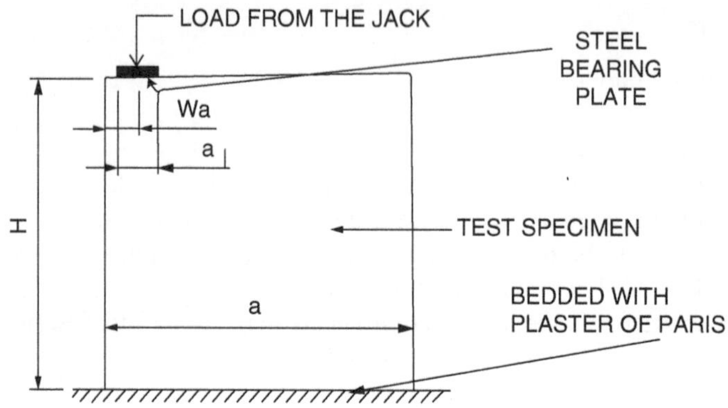

FIGURE 4.1 Loading tests for bearing capacity.

4.25.3 Reinforced Concrete Blocks

The block dimensions are given below. The two top corners were treated with different arrangements of reinforcement for the test specimens. In some tests, the corners were strengthened with steel plates. The failure modes were similar in all RC blocks.

4.25.4 Block Dimensions

The block dimensions for both plain and reinforced concrete blocks were mostly

a = 1,260, 400, 200, 150, 100 mm
b = mostly 100 mm (other dimension of the block)
H = 1,000, 800, 400, 200, 150, 100 mm
Bearing size a_l = 51, 102, 25, 6.4 mm
Edge distance = Varied, 30, 80 – 280 mm (for plain concrete blocks)
= 80 – 330 mm (for reinforced concrete blocks)

The CSs of concrete (f_{cu}) were varied between 45 and 60 N/mm^2.

4.25.5 Comments

The following comments were made after conducting the tests:

- For plain concrete blocks with concentrated loads near the edge ($W_a/05_{al} < 3.5$), failure occurred by shearing the corner. The others failed by splitting into two halves.
- In tests on small-scale models (i.e., a < 400 mm), the bearing capacity decreases with the size of the specimens.
- Plain concrete blocks with a height/width ratio H/a = 0.5 demonstrated a 30% increase in strength compared to specimens with height/width ratios of 1.0 or more.
- Base friction increases the bearing capacity of plain concrete blocks, with more pronounced effects for blocks of smaller height.
- Size effects do not seem to be very critical for larger blocks.
- Based on the test results, a new design equation is proposed for plain concrete blocks that considers both concrete strength and the loading area ratio R (= a/a_l) and inclusion of $\gamma_{m=1.35}$.

Source:

- Cusens, Wang, and Wong [26]

4.26 RECOMMENDATIONS FOR SEISMIC AREAS AND EARTHEN BUILDINGS [3]

The following items are important as per the IS code:

4.26.1 Walls

The height of the walls may be restricted to one-storey plus attic in seismic zones IV and V for adobe buildings. Two-storey heights may be considered in Zone III. For important buildings (i.e., $I > 1.5$), construction should not be done with earthen walls in Zones IV and V and should be limited to one storey in seismic Zone III.

The length of the wall between two consecutive perpendicular walls should not be greater than 10 times the thickness of the wall ‘t’. Furthermore, it should not be more than $64t^2/h$, where ‘h’ is the height of the wall.

For longer walls, strengthening with intermediate vertical buttresses may be considered.

The height of the wall should not be more than eight times the wall thickness.

The widths of an opening, in the walls, should not be more than 1.2 m.

The distance should not be less than 1.2 m between the outside corner of the wall and the opening.

The total width of openings in the wall should not be more than 33⅓% of the total length of the wall in seismic Zone V and 40% in Zones III and IV.

The lengths of bearing for lintels should not be less than 300 mm.

For hand-made walls, tapering upwards will be preferable, with a minimum thickness of 300 mm at the top. A batter of 1:12 may be done with the wider dimension at the bottom.

To increase the seismic performance of the wall, outside pilasters at all corners and junctions may be added

4.26.2 Housing Site

Sites with sandy loose soils, poorly compacted clays, and fill materials may not be suitable against seismic vibrations. Furthermore, sites with a high water table should not be preferred. These recommendations are especially applicable for seismic Zones IV and V.

The site should be located above the high flood level. If required, the ground may be raised accordingly.

4.26.3 Foundations

The widths of strip footings may be as follows:

- One-storey building on firm soil – width should be equal to wall thickness
- 1.5 or 2 storeys on firm soil – 1.5 times the wall thickness
- One storey on soft soil – 1.5 times the wall thickness
- 1.5 or 2 storeys on soft soil – 2 times the wall thicknes

4.26.4 Depth of Foundation

The depth of the foundation shall be at least 400 mm below the existing ground level.

The footing should be constructed with stone or fired brick using cement or lime mortar. Alternatively, it may be constructed with lean cement concrete with plums (cement:sand:gravel:stones in the ratio of 1:4:6:10) or without plums in the ratio of 1:5:10. Lime may be used instead of cement in the proportion of lime:sand:gravel as 1:4:8.

4.26.5 Plinth Masonry

The wall above the foundation and up to the plinth level should be constructed with stone or burnt bricks using cement or lime mortar. Clay mud mortar may be used as a last option.

The height of the plinth should be at least 300 mm above ground level and above the flood water level. Waterproofing treatment will be required at the plinth level and may be achieved using waterproof mud, polyethylene sheets, or heavy black polythene. This should be carried out before starting the construction of superstructure walls. For plinth construction with adobe, the outside face of the wall should be protected against water damage through fascia treatment or plastering. A drain may be provided a little away from the wall to protect it from seepage.

4.26.6 Roof

The roofing structure must be properly tied to the walls and must be light. Trusses may be used for sloping roofs with rafters or A-frames.

The roof covering should also be light, such as sheeting of a suitable type. Heavy roofs with wooden joists and earth topping should be avoided in zones IV and V. Tiled and slate roofs are also to be avoided in these zones.

When thatch is considered for roof covering, it is preferable to make it waterproof and fire resistant by using waterproof mud plaster.

The roof beams, rafters, or trusses should be supported on longitudinal wooden members to distribute the load properly on the walls.

The slope of the overhangs will depend on local climatic conditions. In areas subjected to rain and snow, the walls must be protected by projecting the roof by about 500 mm beyond the walls.

The roof beams or rafters must be designed and placed to avoid being located above the door or window lintels. Otherwise, the lintels must be strengthened by adding additional members accordingly.

4.26.7 Adequate Configuration

The details for adequate configuration are presented in IS 13827. This will be particularly relevant for construction in seismic zones IV and V. Additional strengthening details against seismic effects are also provided in the IS code.

Source:

- IS 13827 [3]

4.27 EARTHQUAKE-RESISTANT DESIGN AND CONSTRUCTION OF BUILDINGS [27]

Guidance may be obtained from IS codes.

A brief review will be made from IS 4326.

4.27.1 Masonry Construction with Rectangular Masonry Units

For the design and construction of masonry walls using rectangular masonry units, guidance may be taken from IS 1905 and IS 2212.

The building categories for earthquake-resistant features are given in Table 4.26.

TABLE 4.26
Building Categories

Importance Factor	Seismic Zones			
	II	III	IV	V
1.0	B	C	D	E
1.5	C	D	E	E

TABLE 4.27
Mortar Mixes

Category of Construction	Proportion of Cement–Lime–Sand
A	M_2 or M_3 or richer
B, C	M_2 or richer
D, E	H_2 or M_1 or richer

4.27.2 Masonry Units

The strength of the masonry units that may be used shall depend on the number of stories of construction and the thickness of the walls. CSs less than 3.5 N/mm^2 for well-burnt bricks or solid concrete blocks shall not be used. The IS code may be referred to for details.

4.27.2.1 Mortar

Recommended mortar mixes are mentioned briefly.

4.27.2.2 Walls

Reinforced masonry walls shall not be constructed at a height greater than 15 m, subject to a maximum of four storeys. The measurement of height shall be considered from mean ground level to the roof or ridge level.

4.27.2.3 Recommended Mortar Mixes

4.27.2.3.1 Mortar Mixes

This may be described as mentioned in Table 4.27.

For the categories and the details of mortar mixes, the above IS code may be referred to.

Masonry Bond – Reference may be made to the IS code.

Openings in Bearing Walls – For dimensions and position of openings, the IS code may be consulted.

4.27.3 Seismic Strengthening Arrangements

Masonry buildings are to be strengthened according to the methods specified for different categories of buildings.

4.27.4 Recommendations

Masonry buildings may be strengthened for different categories as per the recommendations given in the IS code.

The following items are important and shall be executed properly according to the guidance given in the IS code.

- **Lintel Band –** The IS code may be referred.
- **Gable Band –** The IS code may be consulted.
- **Roof Band –** The IS code may be referred to.
- **Section and Reinforcement Band –** The IS code may be referred to.
  Comments/guidelines will be available on the other items as given below:
  - Reinforced brickwork.
  - For full integrity of walls at corners.
- **Plinth Band –** The IS code may be referred to.
- **Category D and E Buildings –** The IS code may be consulted.
- **Vertical Reinforcement**

The vertical reinforcement at the corners and junctions of walls may be critical, particularly for walls up to 340mm (i.e., $1^1/_2$ bricks) thick; the reinforcement must be provided properly according to the details given in the IS code.

The vertical reinforcing bars must be properly embedded/anchored in the masonry foundation and also with the roof slab or roof band to develop adequate tensile resistance. Furthermore, the bars must pass through the lintel bands and floor slabs, including the floor bands provided in all stories. The typical details that are to be followed in brickwork masonry, using rectangular solid units, especially at corners and T-junctions, are provided in the IS 4326 code.

The details will cover

Alternate courses in one-brick walls

- Alternate courses at T-junctions of 1½-brick walls
- Alternate courses at corner junction of 1½-brick walls

The holes/pockets must be filled with concrete/mortar, and the reinforcing bars must be properly anchored as per the design. The sketches are shown in the code.

4.27.5 Framing of Thin Load-Bearing Walls

The load-bearing walls may be made thinner as per the design requirements. The thickness may be reduced to less than 200mm and may be up to 150mm, including the thickness of plastering on both sides. The introduction of reinforced framing columns and collar beams must be done, which will work compositely, having full bond with the walls. The recommended details are shown in the IS 4326 code.

It is recommended to construct the wall first up to four to six courses in height, with toothed gaps left for columns to be filled later. The tooth projection may be about 40mm. The pouring of M15 concrete (1:2:4) may be done against the walls using formwork on both sides of the wall. The reinforcing bars for the columns must be placed in the correct positions all along the wall.

The horizontal length of the walls may be limited to 7m between cross walls, and the storey height to 3m.

The details should include

- Embedded/concealed vertical columns and their locations, along with reinforcement details.
- Location and details of lintel bands.
- Anchoring details.

4.27.6 Reinforcing for Hollow Block Masonry

The relevant items are given below:

- **Horizontal Band –** U-shaped blocks may be applied for construction. The IS code may be referred to.
- **Vertical Reinforcement –** The IS code may be consulted.

The details are indicated in IS 4326 related to hollow block masonry, including the details for horizontal bands and vertical reinforcement in the cavities. The locations of concealed columns will be important.

4.27.7 Floors/Roofs with Small Pre-cast Components

The items are

- Types of pre-cast floors/roofs.
    - Pre-cast reinforced concrete unit, roof/floor.
    - Pre-cast reinforced concrete plank and joist scheme for roof/floor.
    - Pre-cast reinforced concrete cored unit, roof/floor.
    - Pre-fabricated brick panel system for roof/floor.
    - Pre-cast reinforced concrete waffle unit, roof/floor.
- Seismic resistance measures: The IS code may be referred to.

4.27.8 Timber Construction

For the details and information, reference may be made to the IS code.
Source:

- IS 4326:1993 [27]

4.28 IMPROVING EARTHQUAKE RESISTANCE OF LOW-STRENGTH MASONRY BUILDINGS – GUIDELINES [28]

Following the guidance from IS-13828, the main issues for low-strength masonry buildings may be briefly mentioned as follows:

4.28.1 General Principles

The items that will be appropriate for the construction of the buildings may be mentioned as follows:

Lightness – The building should be designed and constructed as lightly as possible to reduce the effects of earthquake forces. However, functional requirements, safety, durability, etc., must be duly taken into account in the process. The roof should also be made as light as possible.

Continuity of Construction – It will be important to connect all parts of the building together so that it can behave as one unit. The floor slabs and the roof should be made continuous. The additions and alterations to the building, whenever required, must be carried out to maintain continuity between the old and new construction as per the design.

Projecting and Suspended Parts – The projecting parts should not be included in the design as much as possible. When the projecting parts are effectively included, they should be properly designed, reinforced, and tied to the main structure. The design should comply with the applicable seismic code.

When stability is ensured by providing counterweights in the form of walls, slabs, etc., checks for overturning should be conducted by increasing the weight of the projecting parts and decreasing the weight of the stabilising mass at the same time, using the vertical seismic coefficients as per the code.

Suspended ceilings should be eliminated. If provided, they should be lightweight, properly framed, and adequately connected. Ceiling plaster may be avoided; if provided, it should be thin.

4.28.1.1 Shape of the Building

To reduce torsion, the building should have a simple rectangular plan and should be symmetrical concerning the mass and rigidity of the building, coinciding with each other. Separate rectangular blocks may be used, especially in seismic zones IV and V.

Small buildings with asymmetric shapes in plan and elevation may be considered. Designing asymmetric buildings for torsion may not be easy.

4.28.1.2 Fire Safety

The buildings should be designed to be fire-resistant as per the IS code.

4.28.2 Special Construction Items

The following will be relevant:

4.28.2.1 Foundations

For foundation design, guidance from the IS code may be followed in general. The same type of soil for the area would be preferred below the building. If required, the buildings may be separated into units and designed accordingly.

Loose fine sand, expansive clays, and soft silt should be avoided. If it is not possible to avoid, the following methods may be considered for foundation design:

- Sand piling/stone columns/ under-reamed piling, etc.
- Soil stabilisation methods.

4.28.2.2 Roofs, Floors, and Pent Roofs

For general guidance, the above IS code may be referred to.

4.28.3 Low-Strength Masonry Construction

For general guidance, the above IS code may be consulted.

Source:

- IS 13828 [28]

4.29 REVIEW AND SEISMIC EVALUATION, REPAIR, AND STRENGTHENING OF MASONRY BUILDINGS

The seismic evaluation and strengthening techniques have been covered in detail in IS codes. For the guidelines, references must be made to IS 4326, 13828, and 13935.

Some brief items have already been mentioned from IS 4326 and 13828.

IS 13935: 2009 will also cover the damageability assessment and improvements for retrofitting/ strengthening work related to upgrading existing masonry buildings against seismic effects. These issues have also been considered in IS 4326 and IS 13828.

A brief overview of the IS 13935 code will be provided below:

4.29.1 General Principles and Concepts [29]

The related issues that may be mentioned are as follows:

4.29.1.1 Non-structural/Architectural Repairing Items

Non-structural architectural components may be easily affected during earthquake effects. The repair work may include one or more of the following:

- Surveying and detecting defects such as cracks and fall of plaster.
- Organising patching work.
- Checking and repairing wiring and electrical conduits.
- Doors and windows require repair work and replacement of glass panes as required.
- Non-structural walls, parapet walls, and smoke chimneys may need to be rebuilt.
- Gas and water pipes, including plumbing lines, should be checked and repaired.
- Roof tiles may need to be rearranged as required.
- Walls may need to be replastered in places, depending on the damage.
- Cracked flooring at ground floor level should be repaired or redone.
- Painting, whitewashing, and redecoration should be done to restore the original condition of the building.

4.29.1.2 Structural Repair/Restoration

The following items may be relevant:

- The structural condition of the building must be examined in detail to determine whether it is suitable for repair work. Furthermore, it must be decided whether continued occupation can be permitted and whether the entire structure or part of it will require demolition if found to be in a risky condition.
- If the structure is deemed reasonable and safe for structural repair, a detailed damage study covering crack patterns, reinforcement bending/yielding, distressed locations, crushed concrete, etc., of the individual structural components must be conducted to assess the residual strengths. Non-destructive tests may be considered if necessary for assessing some structural members.
- Plans and details must be developed for temporary supporting arrangements for the damaged structural members to prevent further distress due to gravity loads during the waiting period for finalisation and repair activities.

After properly assessing the damage to the individual structural elements, supported by a detailed survey, the appropriate repair method must be planned and finalised, and the work carried out properly as suggested below:

- Removal of the damaged portions of the cracked masonry walls and piers may be necessary. These items may be repaired with richer mortar. Non-shrink mortars may also be considered.
- Injection of cement, polymer–cement mixtures, or epoxy materials that are strong in tension may be used for sealing the cracks.
- Inclusion of reinforcing mesh on both faces of the cracked wall, secured to the wall with bolts or spikes if necessary, and appropriately covering it with micro concrete using aggregates of 6 mm or less may be done. Ferro-cement or fibres may also be used.
- Temporary supports or props should be used generously at appropriate locations for safety.
- Damaged reinforced concrete slabs, beams, lintels, etc., may be repaired using polymer mortar or epoxy materials. Micro concrete may also be used for the repair work.
- Suitable fibres may be incorporated into the mortar for the repair process.

4.29.1.3 Seismic Strengthening

The focus will be on increasing the lateral strength of the damaged building while carrying out the repair work.

The following items may be considered:

- Increasing the strength of the structure or building in one or both directions by enlarging the column and wall dimensions or by increasing the number of columns and walls.
- Elements that contribute to the structural weakness should be avoided. Asymmetrical plans for resisting members, abrupt changes in stiffness from one floor to another, large openings in walls without strengthening, and large mass concentrations must be avoided in the design.
- By providing an effective connection between the resisting elements in such a way that the inertia forces generated by the vibration of the building can be transmitted to the other member elements capable of resisting these forces.
- To avoid the risk of brittle failure, it is important to provide proper and adequate reinforcement and connect it efficiently in the resisting members of the structure.

4.29.1.4 Seismic Retrofitting

Many existing buildings may not meet the seismic strength requirements of the present codes due to original design inadequacies of the structural elements, material degradation over time, and changes made during use over the years. A detailed assessment of the existing structural elements will be very relevant and important. The earthquake resistance of existing structures may be improved or upgraded to meet the current code requirements through appropriate seismic retrofitting techniques, as mentioned for seismic strengthening in the code.

4.29.1.5 Strengthening or Retrofitting versus Reconstruction

The replacement of unsafe or damaged buildings and the reconstruction of new ones will require special study.

Reconstruction may be avoided for various reasons. Some of these reasons are mentioned below:

- Higher rebuilding costs may be necessary than strengthening or retrofitting costs.
- Maintaining the functional, social, and cultural environment.
- Preservation of historical architecture.

The relative cost of retrofitting or strengthening to new construction costs will be helpful in making decisions. As a rule of thumb, when the cost of repair and seismic strengthening is less than about 30% of the reconstruction cost, the decision may favour retrofitting.

The updating and strengthening process may require less work time and create less disruption to the living conditions of the population. On the other hand, the reconstruction of the structure and the possibility of modernising the habitat may be preferred by affluent communities.

The cost of retrofitting an existing inadequate building may be as much as 2.5 to 3 times the initial expenditure required to provide seismic-resistant features.

The repair and seismic strengthening of a damaged structure may even be 4 to 6 times as expensive. Therefore, it may be prudent, safe, and cost-effective to design and construct earthquake-resistant buildings from the beginning of the project, i.e., at the initial stage itself, while satisfying the relevant seismic code requirements.

4.29.2 Selection of Materials and Methods

The following may be mentioned:

- **General** – The most common materials for repair or restoration work include cement, steel, various admixtures, and polymer products.

- Shotcreting.
- Use of non-shrink grouts.
- Application of epoxy resin.
- Epoxy mortars.
- Fibre-reinforced plastics.
- Composite materials for wrapping.
- Quick-setting cement mortars.
- Mechanical anchors.
- Other suitable materials.

IS codes and manufacturers may be consulted for detailed information on the above items.

4.29.3 Methods to Restore Original Strength

The items mentioned below will be relevant:

- **General –** All cracks and damages must be identified.
- Repair major cracks and crushed concrete properly.
- Structural repair of minor and medium cracks adequately.

Guidance may be taken from the IS code.

4.29.4 Some More Comments

- The loose materials have to be removed and replaced with the selected material as mentioned before.
- For critical areas with severe damage, the replacement of the member or a portion of it may be necessary.
- If required, additional flexural and shear strengthening with reinforcement may be provided in the repair areas.
- In cases of damage related to walls and floors, diaphragms, steel mesh reinforcement may be used on the outside surface of the member and nailed or bolted to the walls. Finally, it should be treated with plaster or micro-concrete to make it presentable.

4.29.5 Fractured Excessively Yielded and Buckled Reinforcement

With severely damaged reinforced concrete members, it is possible that the reinforcement items are elongated and excessive buckling of bars has occurred. The affected portion of the element may be repaired by replacing the old portion of the steel with new steel bars using butt welding or lap welding techniques. The need for additional stirrups in the elements should be checked and provided if necessary.

Adding additional bars may be done by drilling a hole larger than the bar into the member. The hole may be filled with epoxy, expanding cement, or other equivalent grouting materials. The bar should be forced into the hole and held there until the material sets properly.

4.29.6 Damageability Assessment of Existing Masonry Buildings

The following items should be considered:

- The maximum likely intensity of the earthquake forces.
- Building configuration.

- Building typology.
- The materials and quality of construction and maintenance work carried out during the service life.

The initial and quick assessment of the building may be done through visual inspection and necessary screening.

The consideration of an importance factor of 1.5 as per the code may be applied based on the assessment of the building's damageability. The following actions may be planned according to damageability grades and corresponding retrofitting actions.

The damageability grades and the necessary retrofitting actions are provided in IS 13935. Guidance may be taken from the code.

4.29.7 Assessment of Retrofitting Requirements

The important items include the following:

- Categorisation of buildings
- Special or critically important buildings

The IS code may be referred to for information on the above items.

4.29.8 Strengthening of Existing Walls

The information on the following items may be taken from the IS code:

The steps are as follows:

- Grouting
- Connection between existing stone walls.
- Strengthening with wire mesh on both sides of the wall.
- Pre-stressing the wall
- Making through bond elements in R.R. stone wall
    - Selection of areas where through stones may be installed in horizontal and vertical directions with a distance of about 1 m apart and 500 mm horizontal staggering.
    - The affected stone pieces should be removed gradually and carefully without disturbing other stones in the wall.
    - The plastering should be removed from the surface to expose the stonework.
    - A 75 mm hole should be made in the wall carefully. Bigger holes may not be required.
    - The opposite stone needs to be located on the other face of the wall by gentle tapping.
    - The hole prepared through the wall should be larger on both faces and narrower inside for better detailing.
    - Concrete 1:2:4 may be used to fill half the depth of the hole from both sides. Eight-millimetre mild steel bars with hooks may be placed in the hole, and the hole must be completely filled/sealed.
    - Curing may be done for a minimum period of 10 days by sprinkling water on the exposed surfaces from both sides.

4.29.9 Masonry Arches

The methods of strengthening as recommended in the code are follows:

- Strengthening by using steel tie rods across the springing level or slightly above it by drilling holes and properly grouting/covering the steel bars.

- Relieving the arch loads by inserting beams above it to support the load. Steel channels or I-sections may be used.
- Using ties for the continuous arches to prevent cracking. The use of steel sections and bolting may be adopted.

The sketches are included in the IS 13935 code.

4.29.10 Seismic Belts around the Door/Window Openings

The recommendations are listed below:

The vertical reinforcement in the form of jambs and piers between door and window openings may be provided as suggested below:

- For categories D and E buildings, reinforcement may be provided to resist earthquake forces.
- It will be necessary to restore the strength of the supporting piers in any building category when they are badly damaged.

The **critical items** for the window openings are as follows:

- Ferro-cement and the use of mesh reinforcement
- Seismic belt
- Overlap of mesh

For other details, the IS code may be referred to.

4.29.11 Achieving Integral Box Action

The different items mentioned under this heading will be briefly discussed below:

4.29.11.1 Pre-stressing

For the details, the IS code may be referred to.

4.29.11.2 External Binding

For the details, the IS code may be consulted.

4.29.11.3 Providing Horizontal Seismic Belts

The **items** that should be considered are as follows:

- **Seismic belt** locations
- **Details**/description of **reinforcement** in belt
- **Procedure** for **construction of seismic belt**

The relevant items are as follows:

- Marking the height and width of the belt considering welded mesh and the number of longitudinal wires or mesh size.
- The joints should be exposed 20 mm deep, and these should be raked and cleaned. Water jets may be applied.
- The plaster may be removed using a mechanical cutter for a neat execution.

- Neat cement slurry should be applied to the wall. The plaster to the wall may be done with a 1:3 cement and coarse sand mix after filling the raked joints properly. The wall surface should be finished with a 15 mm thick plaster. The surface should be made rough for better bonding, and a bonding agent may be required for the second layer.
- The mesh should be fixed/anchored to the plastered surface using 150 mm long nails spaced 450 mm apart, effectively tying the mesh to the nails with binding wires.
- The second layer of plaster may be applied above the mesh, providing a thickness of 15 mm.
- An RC belt may be provided on both sides of the wall as per the requirements.
- Curing may be carried out by sprinkling water for a minimum period of 10 days.
- Fixing may be done using anchors instead of nails if necessary.

4.29.12 Vertical Seismic Belt at Corners

The IS code may be referred to for the details.

4.29.13 Providing Vertical Reinforcement at Corners, Junction of Walls

The IS code may be referred to for the details.

4.29.14 Modifications of Roofs or Floors

The following items may be mentioned:

Items such as **slates and roofing tiles** may be brittle and easily damaged. These may be replaced with iron or asbestos sheeting wherever possible. The use of asbestos sheets should be done carefully. Other items are mentioned in IS code.

4.29.15 Stiffening the Flat Wooden Floor/Roof

The use of long planks should be done properly. The thick planks should be nailed at both ends of the joists from below. Metal strips may be used and nailed diagonally. Details with sketches are available in IS code.

4.29.16 Stiffening the Sloping Roof Structure

The rafters should be tied properly as per the details/guidelines given in the IS code.

4.29.17 Inserting New Walls

The following may be noted:

Unsymmetrical buildings may create problems during earthquakes. By separating the parts of the building, the individual units may be made symmetrical. Alternatively, new vertical resisting units may be introduced, such as new masonry or reinforced concrete shear walls internally or buttresses externally. Requirements for cross walls may be examined to provide transverse support to longitudinal walls, especially for long barrack-type buildings.

- The connection details of the new walls with the old structures will be important for performance. The guidelines are available in the IS code for reference.

4.29.18 Strengthening of Foundations

The following may be done:

- The introduction of new load-bearing members may be included in the foundation to relieve the already overloaded foundation/members. Jacking operations may be required in the method/execution.
- Strong elements in the form of reinforced concrete strips may be included, which should be connected to the existing foundation of the building. These elements should be properly introduced and attached to the existing foundation on both sides to work compositely.
- The drainage system of the area should be improved to prevent soil saturation and other problems.
- An apron should be provided around the building to prevent water from soaking the foundation directly and to improve drainage efficiency.

Further information on the **Seismic Strengthening for Bearing Wall Buildings** from IS-13827 has been included in Chapter 5.

Source:

- IS 13935 [29]

4.30 CONCRETE AND MASONRY REFURBISHMENT IN THE WATER INDUSTRY [30]

The parameters covering the physical, chemical, and biological characteristics of water, especially intended for drinking, washing, and cooking, will be important when selecting the repair materials for tanks/reservoirs, etc.

The suitability of non-metallic products should be checked for water tanks/reservoirs meant for human consumption.

4.30.1 Comments for Repair

Polymer and cement-based mortars have been used for repair/restoration work related to corrosion and deterioration of concrete. With these materials, large concrete repair work may be undertaken. The materials are likely to be compatible with the substrate. Reinforcing bars may be strengthened and treated with rust-inhibiting primer or bonding coat before the repair work.

Waterproofing and vapour permeability treatment may be applied to the structure.

Suitable surface coatings should be selected, which will be important for the refurbishment of concrete and brickwork reservoirs after the repair work. Cement-based coatings may be considered, which are normally compatible with the substrate and can resist both positive and negative water pressure, acting as a barrier against water loss or ingress. They can be useful for stopping external contaminants from entering stored water. The coatings are expected to be durable and resistant to abrasion, impact, and chemical attack. The cement-based coatings may also include flexible elastomeric coatings, which will seal the fine cracks in the structures.

Any damage to the substrate must be addressed and repaired. The surface must be cleaned before the application of the coating.

Checks are required on the products to ensure they are suitable for human consumption. Furthermore, checks must be made on the growth of microorganisms and the extraction of substrates that may be harmful to human health, including the extraction of metals.

Technical information on the products must be studied carefully before selecting the materials. The application must be done according to the manufacturer's recommendations.

Source:

- Groves [30]

4.31 MODULAR COORDINATION FOR NORMAL BRICKWORK – RECOMMENDATIONS PART-1 BURNT CLAY BRICKS, IS 13612 [31]

The modular coordination for normal clay brickwork and the recommendations given in the IS code must be referred to. A few items will be briefly presented below:

The main items are as follows:

General consideration, field applications, controlling dimensions of bricks, modular coordination of brickwork using both modular and non-modular bricks, modular brick size (200×100×100mm), non-modular brick size (240×120×80mm), horizontal coordination, vertical coordination, tolerances on bricks, vertical dimensions to the walls, horizontal dimensions of walls, coordinating dimensioning, thickness of walls, location of walls, bond dimensions, and other relevant items.

The above items have been discussed in detail in the code, and necessary guidance may be obtained from the code while designing the structure.

Source:

- IS 13612 (Part-1) [31]

4.32 RECOMMENDATIONS FOR MODULAR COORDINATION IN THE BUILDING INDUSTRY: LOCATION OF STRUCTURAL WALLS AND FLOOR SLABS, IS 13613: 1992 [32]

The different items related to design are covered in the code.

The basic issues are as follows:

General items, field of applications, terminology, design of structures, design details, horizontal location of components, neutral and interrupted planning grids, vertical location of components, continuous modular planning grids, interrupted modular planning grids, design details, modular floor height, non-modular floor height, external walls, spine walls, connections between structural walls and floors, stair, lift, escalator, shaft, door shaft, window shaft openings, columns – locations as per planning grids, beams – locations as per planning grids, floor slabs – side face continuous in planning grids, etc.

The above items have been discussed in the IS code, and the same may be referred to at the time of designing the buildings.

Source:

- IS 13613 [32]

4.33 CASE STUDY [33]

The comments on the design and construction of a masonry arch roof with fly ash–lime and gypsum block masonry will be presented in this case study.

This project is located in Orissa and relates to a health centre and staff quarters project. The commonly used building materials that could be considered were laterite and bricks. These materials may have harmful effects on the environment. Hence, attempts were made to look for alternative materials.

The government of Orissa and the Department for International Development (DFID, UK) decided to use alternative materials for the healthcare infrastructure project. This will involve the construction of walls and curved roofs.

The site was located in a village (Panasapada), 98km from the state capital, Bhubaneswar. Fly ash and gypsum were available for use; hence, it was decided to use fly ash, lime, and gypsum (FALG) blocks for the primary health centre and staff quarters, primarily for walling.

4.33.1 Materials

FALG blocks were prepared by using the following materials:

- Fly ash – 65%
- Lime – 25%
- Gypsum – 10%

The constituent materials were first mixed in an electrically operated rotator-type machine. A steel mould measuring 390 mm × 190 mm × 200 mm was used for production. The moulds were filled with the mixed material, and an electrically driven vibrating plate with a guide was used to press the materials to make 112 mm thick blocks. After de-moulding, the blocks were air-dried for 48 hours as per IS guidelines. Water was sprinkled on the blocks for 5–6 days. The blocks could be used after 7 days.

4.33.2 Properties of FALG Blocks

The important properties are listed below:

- Appearance – grey
- CS – 17.8 N/mm^2
- Size – 390 × 190 × 112 mm
- Drying shrinkage – 0.11%
- Weight – 156 N
- Soundness – Le Chatelier method
- Density – 19.6 KN/m^3
- Boiling water expansion – 2.0 mm

4.33.3 General Comments

FALG blocks were found to be strong in compression; hence, they could be used for members of structural systems subjected to compressive loads, such as domes and vaults. Vaults and domes were constructed by lifting a series of parabolic generator arches on RCC ring beams. Assessing the stability of the independent arches was important.

4.33.4 Uses

Apart from use in walling systems for the primary health centre, the applications included vaults and domes for the roofs of the health centre and staff quarters.

4.33.5 Further Comments

This can be mentioned as a good example of using alternative materials in projects. The construction of roofs with FALG was found to be cheaper than that of concrete. The use of FALG will be preferable for projects as it would reduce the burden of using soil for construction and consume fly ash in the process, which would help reduce fly ash deposits as waste material. The whole approach would be economical and environmentally friendly.

Source:

- Das and Das [33]

4.34 APPLICATIONS AND PHOTOGRAPHS FROM PROJECTS

Some photographs from the projects are shown below for masonry construction (Figures 4.2–4.21):

FIGURE 4.2 Brick masonry boundary wall.

FIGURE 4.3 Brick masonry boundary wall.

FIGURE 4.4 Brick masonry boundary wall.

FIGURE 4.5 Brick masonry wall.

FIGURE 4.6 Brick masonry wall construction.

FIGURE 4.7 Brick masonry boundary wall.

FIGURE 4.8 Brick masonry boundary wall.

FIGURE 4.9 Brick arch construction.

FIGURE 4.10 Brick construction, residential buildings.

FIGURE 4.11 Brick arch construction.

FIGURE 4.12 Brick wall, plastered surface, and window.

FIGURE 4.13 Brick arch construction.

FIGURE 4.14 Brick wall construction.

FIGURE 4.15 Brick arches.

FIGURE 4.16 Brick walls, arches, and timber beams.

FIGURE 4.17 Brick arches.

FIGURE 4.18 Timber beams and cross members.

FIGURE 4.19 Brick walls/arches.

FIGURE 4.20 A traditional masonry building.

FIGURE 4.21 A traditional masonry building.

REFERENCES

1. Doat P, Hays A, Houben H, Matuk S and Vitoux F (CRATerre, France), *Building with Earth*, Translated by Asha Puri, The Mud Village Society, New Delhi, India, 1991.
2. Earth Building Technique - Build Your Home with Earth - This Cob House, pp. 1–22.
3. Bureau of Indian Standards, IS-13827: 1993, Improving Earth Quake Resistance of Earthen Buildings – Guidelines.

4. Salih M M, Osofero A I and Imbabi M S, Critical review of recent development in fibre reinforced adobe bricks for sustainable construction, University of Aberdeen, *Frontiers of Structural and Civil Engineering.* 2020, Vol. 14(4), pp. 839–854. doi: 10.1007/s11709-020-0630-7.
5. Bureau of Indian Standards, IS 1077: 1992, Common Burnt Clay Building Bricks - Specification.
6. Bureau of Indian Standards, IS-13757: 1993, Burnt Clay Fly Ash Building Bricks - Specification.
7. Wikipedia, Fly Ash Brick, pp. 1–3. https://en.wikipedia.org/wiki/Fly_ash_brick. Accessed Oct 02, 2024.
8. Bureau of Indian Standards, IS-12894: 2002, Pulverised Fuel Ash-Lime Bricks - Specification.
9. FLY ASH BRICKS, A Useful & Environment Friendly Building Product, NTPC Information.
10. Fly Ash Brick Making Machines, AHP Machines, Manufacturer, India. https://ahpmachines.com/index.html. Accessed Oct 02, 2024.
11. Banker U J, *Expanded Clay Aggregate (ECA) Construction Masonry Block*, Rivashaa ECO Design Solution Pvt. Ltd.
12. Fly Ash Bricks and Comparison with Clay Brick Composition, Manufacture, The Constructor, pp. 1–7. https://theconstructor.org/building/fly-ash-bricks/5330/. Accessed Oct 02, 2024.
13. Bureau of Indian Standards, IS 2185 (Part 1)-2005, Concrete Masonry Units - Specification, Part 1-Hollow and Solid Concrete Blocks.
14. Bureau of Indian Standards, IS 2185 (Part 2)-1983, Specification for Concrete Masonry Units, Part 2-Hollow and Lightweight Concrete Blocks.
15. Bureau of Indian Standards, IS 2185 (Part 3)–1984, Specification for Concrete Masonry Units, Part-3 Autoclaved Cellular (Aerated).
16. Bureau of Indian Standards, IS 13077:1991, Preparation and use of Mud Mortar in Masonry - Guide.
17. Pritchett I, The modern face of lime, view point, *The Structural Engineer*, 20 April 2004, Vol. 82(8), pp. 13–14.
18. Lynch G, The Myth in the Mix, The 1:3 Ratio of Lime to Sand, The Building Conservation Directory, 2007, pp. 103–104.
19. Smith W F and Plowman J M, Load-bearing and Non-load-bearing Brickwork, The Ibstock Brick & Tile Company Limited, No. 1, Bricks and Mortars, pp. 1–13.
20. Bureau of Indian Standards, IS 4443:1980, Code of practice for use of resin type chemical resistant mortars.
21. Bureau of Indian Standards, IS 4456-1:1967, Methods of test for chemical resistant mortars, Part I: Silicate Type and Resin Type.
22. Bureau of Indian Standards, IS 4832-1:1969, Specification for chemical resistant mortars, Part I: Silicate type.
23. Bureau of Indian Standards, IS 4832-2:1969, Specification for chemical resistant mortars, Part I: Resin type.
24. Bindiganavile V, Suresh N and Md. Islam T, Characterisation of early 20th century lime mortars from Mysore, *The Indian Concrete Journal*, August 2013, Vol. 87, pp. 13–19.
25. O'Regan C, Design of Unreinforced masonry retaining walls, Technical Guide Notes Level 2, No. 18, *The Structural Engineer*, October 2018, Vol. 96, pp. 28–31.
26. Cusens A R, Wang Y B and Wong K K, Bearing capacity of plain and reinforced concrete blocks, *The Structural Engineers*, December 1989, Vol. 67, pp. 407–410.
27. Bureau of Indian Standards, IS-4326:1993, Code of Practice for Earthquake Resistant Design and Construction of Buildings.
28. Bureau of Indian Standards, IS-13828:1993, Improving Earthquake Resistance of Low Strength Masonry Buildings - Guidelines.
29. Bureau of Indian Standards, IS-13935:2009, Seismic Evaluation, Repair and Strengthening of Masonry Buildings - Guidelines.
30. Groves B, Concrete and masonry refurbishment in the water industry, *Concrete*, February 1999, pp. 10–11.
31. Bureau of Indian Standards, IS-13612 (Part-1):1993, Modular Co-ordination for Normal Brickwork - Recommendations. Part-1 Burnt clay bricks.
32. Bureau of Indian Standards, IS-13613:1992, Recommendations for Modular Co-ordination in Building Industry: Location of Structural Walls and Floor Slabs.
33. Das P B and Das P K, The design and construction of a masonry arch roof with fly-ash lime and gypsum block masonry, *The Structural Engineer*, March 2001, Vol. 79, pp. 18–20.

5 Rammed Earth Construction

The earth construction with rammed earth will be briefly discussed in this chapter, with case studies.

5.1 INTRODUCTION [1,2]

Rammed earth is known as Taipa in in Portuguese, Tapial or Tapia in Spanish, 'Pise' (de terra) in French and partly Italian, and Hangtu in Chinese [1]. This construction method covers foundations, walls, and floors of buildings using natural raw materials like earth, lime, chalk, or gravel.

This is a very old method of construction, and many successful applications have been carried out using rammed earth in the past.

Rammed earth technique is a non-combustible, strong, thermally massive, and durable construction method that can be simple and easy to construct. It is labour-intensive and may not be easy to construct building walls, etc., without using machinery such as power tampers. Unstabilised rammed earth may be damaged by water if not protected or maintained properly.

Different types of rammed earth construction may be found in many countries, except Antarctica, and the structures may be subjected to varying temperatures, including wet, semi-arid desert, montane, and tropical regions. Proper design, along with the suitability of adequate soil quality and local climatic conditions, would be relevant for the construction and the structures. These factors must be properly assessed in the design and may be key to deciding its use.

Properly constructed rammed earth walls would function as a single mass [2]. They can last for a long period when the external surfaces are rendered or treated.

The walls may achieve hardness and consistency like soft stone after a long period.

Care must be taken in rehabilitating rammed structures. It is critical to know the materials used originally for the structure and their applications [2]. Further comments on the repair of rammed earth structures are provided later in this chapter.

Source:

- Wikipedia [1]
- Building with Earth [2]

5.2 RAMMED EARTH

5.2.1 History [3]

Man had been using earth for different structures for thousands of years, as mentioned previously. According to the records, Jericho would be the earliest city built with earth. Other structures, such as temples, mosques, and churches, were constructed with mud bricks and rammed earth throughout the ancient Middle East. Egyptian pharaohs constructed cities with rammed earth. In the Far East, the technique was used not only for houses but also for ancient forerunners of the Great Wall in China. The Romans and Phoenicians were responsible for bringing the technology to Europe, where it continued to be used for more than 2,000 years.

In the US, rammed earth construction was popular during the period of 1780–1850, when mass-produced fire bricks and sawed lumber were readily available. It was possible to construct houses easily and quickly with bricks and lumber. Following the shortages experienced after World War I and during the Great Depression, rammed earth construction became popular again for about two decades. Frank Lloyd Wright designed houses made with rammed earth.

DOI: 10.1201/9781003319108-5

After World War II, a modified form of construction using discarded automobile tyres as building blocks was introduced. The tyres were used and rammed full of earth. Research and studies continued to improve the methods and quality of rammed earth construction. The use of pneumatic tampers was considered to enhance efficiency and speed compared to traditional manual methods. Soil compaction could be achieved quickly and efficiently using this process. With time-honoured manual methods, a four-person crew could construct approximately 12–15 m^2 of rammed earth wall per day. In comparison, using power tools, the same crew would be able to build about 91 m^2 of wall per day. For earthquake design, the walls must be properly reinforced and strengthened, integrating carefully with the structure as per the design requirements.

More comments on strengthening will follow later. More comments will follow, in strengthening, later.

5.2.2 Raw Materials [3]

The important material used for rammed earth construction is the earth. There may be five basic types of soil, i.e., gravel, silt, sand, clay, and organic matter, along with dirt that may contain some combinations of these types in a given location, as already highlighted. Some combinations of all the above types may encompass most of the soils available for use. It seems the longest-lasting rammed earth walls are constructed with a combination of 70% sand and 30% clay. The soil for the new building sites must be checked and tested to assess its suitability for construction. Organic materials, if present, must be removed from the soil. Different suitable types of soil may be brought from a source close to the construction site, if possible, and mixed with the existing dirt to create a blend that might be suitable for construction. Cement may be added to the soil to enhance both its strength and resistance to moisture, typically at about ¼ of the ratio required to make concrete.

Steel reinforcing bars may be used in the foundation and in the walls as well. Plywood shuttering may be used to make the forms removable for standard rammed earth construction. About 20 mm thick plywood may be adequate for the job. High-density overlay panels with a thin plastic coating on one side may work well, as the release of the formwork will be easy from the wall after construction. This not only leaves a clean finish on the completed wall but also preserves the form boards in good condition, allowing them to be reused.

Rammed earth tire construction may include rejected automobile tyres, aluminium cans, and cardboard, etc., in addition to compacted soil. Approximately 1,000 tires may be required to construct 610 m^2 wall area in a house.

5.2.3 Design []]

The houses may be designed and constructed with rammed earth to make them the most energy-efficient application for the site. The design may accommodate different climatic conditions, covering humid areas with cold winters. The size and location of windows will be important factors in taking advantage of solar heating in winter and cooling breezes in summer. The houses may be suitably located to take advantage, wherever possible, of hills that may offer protection against storms. Shade trees or trellised vines may provide relief from summer heat but would allow warm sunlight in winter.

Source:

- Vip service [3]

5.3 PROCESS TO BE FOLLOWED [1]

Rammed earth construction is a method of compacting a mixture of damp earth with the required proportions of sand, gravel, clay, and/or stabiliser, which will be placed in the formwork (i.e., an externally supported frame or mould) as mentioned. In the past, lime or animal blood was used for stabilisation. Comments on stabilisation techniques for soil are made in Chapter 3.

5.3.1 Method of Building Technique

The prepared soil mix is placed into the formwork to a depth of 100–250 mm and followed by compaction to bring the volume to about 50% of its original volume. The soil is compacted repeatedly, in batches/courses, to raise the wall gradually to the top of the formwork. The tamping had previously been done with a long ramming pole, and the whole method could be very laborious. The modern method of construction may employ pneumatically powered tampers, which are significantly easier to operate.

After completing the wall, and when it is reasonably strong, the formwork can be removed. After removal, surface treatments may be applied, such as wire brushing, carving, or mould impressions, as per the requirements. It may not be easy to carry out such work later, as the wall may become too hard to work on after about 1 hour. The compressive strength of rammed earth material is expected to increase with curing. Stabilised rammed earth with cement may be cured for the recommended period of 28 days.

In recent times, the walls of rammed earth buildings have been constructed on top of conventional footings or on a reinforced concrete base slab.

The role of the formwork in rammed earth construction is very important. It must be strong, durable, and well braced. The formwork must be clamped on two opposite faces to prevent bulging or deformation that might occur due to high compressive forces. Wooden planks have traditionally been used for formwork; however, plywood or steel may be selected for construction.

Further comments have been made on formwork separately.

Source:

- Wikipedia [1]

5.4 REVIEW OF THE CHARACTERISTICS [1]

The compressive strength of rammed earth will be largely influenced by many factors, such as soil type, particle size distribution, moisture content of the mix, degree of compaction, and the type/amount of stabilisers used in the mix. Well-prepared cement-stabilised rammed earth walls may have strengths of about 10–20 MPa. Using more cement in the mix can affect the permeability of the walls. Normally, it will be less than the amount required for concrete, but the quantity may be more than adequate for sufficiently strong walls for domestic building applications. Observing the ancient structures may give the impression that properly constructed rammed earth structures can survive for thousands of years.

Rammed earth may be reinforced with rebar, bamboo, or wood, and such combinations in construction may be effective against the effects of earthquakes, heavy storms, etc. Unreinforced rammed earth construction may show poor performance against earthquakes and other effects that require resistance against lateral loads. By including cement in the poor soil mixture, the load-bearing capacity of rammed earth buildings will also increase.

Soil may be readily available for the construction work, is inexpensive, and is likely from a sustainable resource. The types of construction may also vary according to the design. Unskilled labour may be considered and can be adequate for most of the building work. The cost of rammed earth construction will be low. However, construction without the use of mechanical tools is likely to be very time-consuming and laborious, as already mentioned. Mechanical tampers and pre-fabricated formwork may only be required for 2–3 days to construct the building walls of about 200–220 m^2 in a house.

Rammed earth will have high thermal mass, similar to brick or concrete construction. This will help moderate daily temperatures and control the need for air conditioning and heating. In colder environments, rammed earth walls may be insulated with styrofoam or similar materials. They must be protected against heavy rain and insulated with vapour barriers.

Rammed earth walls will also regulate humidity when unclad walls containing clay are exposed to internal space.

Untreated rammed earth walls may have a colour and texture similar to that of natural earth.

Moisture-impermeable finishes like cement render should be avoided as they reduce the ability of the walls to de-absorb moisture. This is important for maintaining strength.

The thickness of the walls may vary between 300 and 350 mm. The density of the walls may be suitable for soundproofing. The rammed earth walls are inherently fireproof, resistant to termite damage, and non-toxic.

Source:

- Wikipedia [1]

5.5 FURTHER ON MANUFACTURING PROCESS [3]

Rammed earth construction may be carried out in different ways. It can be done individually, using rammed earth bricks for normal building construction. The rammed earth bricks may also be used for preparing the floors in a rammed earth house built with other methods. However, the standard rammed earth method of construction involves using wood formwork and compacting the prepared soil within the formwork, as mentioned. The general procedure that may be followed will be briefly reviewed below:

5.5.1 Preparation of the Site

The topsoil (25–50 mm) is normally recommended to be removed and stored separately so that it may be reused around the structure when it is finished. Organic matter like weeds and roots should be removed from the soil, and these may be composted for use in landscaping later. When the site is cleared, the layout of the building should be done according to the drawing.

The soil must be excavated for the foundation as per the drawing and design requirements.

Excavation for the floor area should also be done. The depth of excavation should account for environmental factors, including the winter freezing zone, as applicable. A level surface will be preferable for the foundation work.

5.5.2 Laying the Foundation

Depending on the type of soil, the design for the foundation may be completed. A reinforced concrete foundation may be adopted, and the conventional design procedure for the foundation may be followed. A stem wall from the foundation may be used to connect the rammed earth wall above ground level. Depending on the design, the concrete floor may also be accommodated.

5.5.3 Analysing the Soil

Tests are to be carried out to assess the appropriate type of soil for the construction.

The proportions of sand and silt in the soil will be important criteria. Treatments may be required for the soil in some cases. Comments on this issue have been made separately.

5.5.4 Framing the Walls

Typically, wooden formwork may be used to construct the wall to a height of about 0.6 m at a time. The prepared soil may be placed in the formwork as stated. When the mould is filled with soil, it must be properly compacted before it can be removed and reset to construct the next portion of the wall. The wall may also be constructed to its full height. The framework may extend the entire length of the wall or may be provided with gaps. The formwork should be designed accordingly. Necessary gaps/voids in the wall are required to accommodate vertical and horizontal concrete tie members for the applicable seismic design requirements. It may take longer to set, align, and remove the formwork compared to filling and compacting the soil within it. Provisions for grooves or architectural treatment may be made during the wall construction.

The wooden formwork may not be used for rammed earth tyre construction.

5.5.5 Tamping the Soil

Generally, a heavy wooden block with a handle extending upward through the centre may be used to compact the soil. It may be a 100mm square plate with a 25mm diameter pipe welded to it for tamping. A moistened soil layer (approximately 100–150mm) may be placed inside the form, and a worker may drop the tamper from a height of 300–450mm for compaction. This process is repeated to compact the soil. Pneumatic tampers are frequently used nowadays. Manual operations may be required in specific areas, such as near structural members, electrical boxes, or plumbing pipes. After repeated operations with the tamper over the entire surface of the layer, the noise made by the tamper will change from a dull thud to a ringing sound. This condition is achieved when the soil is compacted to half of its original volume. At this stage, another layer of prepared soil is placed, and the tamping operation may be repeated. The wooden formwork will be removed when the tamping activity is completed.

The tamping process may differ when tyres are used as the formwork.

5.5.6 Finishing the Walls

The interior faces of walls may be finished with plaster. If such a finish is not provided, the walls may be treated with a clear penetrating sealant to prevent or control dust accumulation on the surface. The exterior surface must be treated to make it waterproof for certain climatic conditions.

5.5.7 By-product/Waste

Natural resources will be preferred and used efficiently in rammed earth wall construction.

5.5.8 Future/Further Comments

A new version of the rammed earth construction method has been developed called Pneumatically Impacted Stabilised Earth.

In this method, the prepared soil is sprayed under high pressure against one-sided formwork.

Using this method, it would be possible to construct 365.70 m^2 of 45.72 cm (18 in) thick wall per day, which may be 4 times faster than a typical four-person team filling box-type formwork and compacting the earth using power tampers.

Source:

- Vip service [3]

5.6 REVIEW OF RAMMED EARTH CONSTRUCTION FOLLOWING IS CODE [4]

Some items from the code are mentioned below:

Rammed earth construction is known by different names in some countries, as already mentioned.

In this technique of construction, moist soil is poured into long wooden forms for the walls and compacted to obtain the desired density as stated [4–6]. The soil for rammed earth work should contain less clay compared to soil for making adobes. The moisture content should be maintained at a minimum level but should be close to the optimum moisture content as per Proctor compaction tests.

Initial tests may be required to control shrinkage fissures on drying and to assess the quantity of sand required for the clayey soil. Depending on the moisture content, the layering and amount of compaction necessary for construction may be evaluated.

The soil is placed in layers approximately 100 mm thick and fully compacted. Water may be sprinkled on the compacted layer before placing the next 100 mm thick layer. Following this method, the total height of the block may be built up to 500–800 mm. Sufficient water should be sprinkled on the compacted layer to ensure proper adhesion with the new layer.

Higher quality of compaction will provide higher strength up to a certain limit. The compaction may be standardised as mentioned below:

- 50 strokes per 1,000 cm^2 of wall area using a wooden mallet weighing about 8–10 kg [4].

To control fissures, a small amount of straw may be added to the soil, not exceeding one-fourth of the volume of the soil–water mixture [4].

Source:

- IS-13827 [4]
- Ganguly [5]
- Ganguly [6]

5.7 ENVIRONMENTAL EFFECTS AND SUSTAINABILITY [1]

The construction using rammed earth may be considered more sustainable and environmentally friendly than other construction materials that involve cement and construction chemicals. The use of locally available materials would result in low embodied energy and produce minimal waste at the end. The materials used for construction are subsoil, which saves the use of topsoil. The topsoil can be reserved for landscaping and agricultural work. The soil excavated for foundation work may be reused in construction, reducing both transportation costs and energy consumption significantly. Rammed earth can be considered the least environmentally harmful material, and the construction technique is suitable for solid masonry wall construction in buildings.

The formwork for the construction will be of a removable type and can be reused.

Low amounts of cement may be required in the mix, which will help control CO_2 emissions in the atmosphere.

Rammed earth may help reduce ecological impacts related to deforestation and the toxicity of artificial materials associated with conventional construction techniques.

It is important to note that an increased use of cement in the mix, along with transportation and cement production, may significantly alter the overall emissions picture of modern rammed earth construction.

The traditional method of rammed earth construction has resulted in very low greenhouse gas emissions. However, more engineered and innovative designs that involve variations of the rammed earth method (modern techniques) may have the potential to emit large quantities of greenhouse gases into the environment.

Source:

- Wikipedia [1]

5.8 REVIEW OF TRADITIONAL RAMMED EARTH BUILDINGS [2]

5.8.1 Further Review and Comments

The method involves tamping the prepared earth in layers between the formwork to the nominal thickness of stone walls as stated.

5.8.2 Tools Used [2]

Comments are made on the following:

5.8.2.1 Formwork

Each side of the formwork (approximately 900mm high) may be constructed with three or four straight wooden planks with as few knots as possible. The planks may be 3.25 m long, 300mm wide, and 30mm thick. The inside surface should be smooth to avoid lumps of earth sticking to it. They may be held together using four pairs of vertical planks, with two end pairs measuring 270mm wide and the intermediate ones 210mm.

The recommended dimensions are as follows:

Length=2.6m (1.6m min and 4.2m max)
Height=800mm
The formwork should not be heavy to handle during construction.

Further information on formwork is given below;

- **End-board**
  This is also known as a closer or formwork head. Each head may be prepared with two planks held together by two small crosspieces on the outside.
- **Vertical Post**
  The posts or struts may be 1.62m high and are made from sawn squared timber, joists, or rafters. They may extend above the formwork by 500mm.
- **Put-Locks**
  Put-locks or ties may be prepared from hardwood 80mm thick, 95mm wide, and 1.14m long.
- **Wedges**
  The wedges may be inserted into put-locks to secure the posts properly and the side planks on the previous section of the wall.

5.8.3 Soils Used [2]

5.8.3.1 Selection

Not all types of soil will be suitable. The preferred mixture may be specified as follows:

- **Gravel –** 0%–15%
- **Sand –** 40%–50%
- **Silt –** 20%–35%
- **Clay –** 15%–25%

5.8.3.2 Extraction and Preparation

Only the amount of soil needed for the day's work may be prepared. Protection against rain must be provided.

5.8.4 Building the Walls [2]

The following items will be important:

- Foundation work must be completed and levelled.
- The formwork must be fixed in position for the wall. The work may commence from one corner of the building.

- Prior to placing the earth, a layer of mortar should be applied around the edges of the shuttering.
- The corners should be tamped from within the formwork periodically.
- The verticality of the wall must be checked frequently using a plumb line.
- The earth may be placed within the formwork and spread out with the feet. The fill must be compacted with a rammer to achieve a thickness of 80–100 mm.
- Longitudinal strokes should be made with the rammer around the edges of the formwork.
- Criss-cross strokes may be necessary after compacting the earth around the edge of the formwork to ensure that the earth is compacted properly in all directions.
- Slanting strokes of the rammer may be required in the centre.
- After completing the first layer, which must be well compacted, the process should be repeated for the subsequent layers to fill the formwork.
- The formwork may be dismantled.
- The formwork will be re-erected properly with the overlaps from the previous section.
- Before starting to fill the formwork for the next layer, it is recommended to place a mortar layer to enhance the bond with the previously completed section.
- The process may be repeated to complete the full height of the wall.
- The first course should be strong enough to support the weight of the builders and the weight of another course that will be added.
- Grooves may be cut into the previous course for transverse ties, which should be staggered to eliminate the development of vertical cracks.

5.8.5 Vertical and Slanting Joints

There may be two types of joints used between the sections of the same course to enhance the bond. These are given below:

- **Vertical Joints** – The end board will be required. This will provide better rigidity to the formwork. It may be possible to construct a larger volume of wall in one attempt.
- **Slanting Joints** – Normally provided at an angle of 45°. This will improve the bond between two adjacent sections.

5.8.6 Wall Thickness [2]

It should be noted that the ratio of height to wall thickness is often maintained 10:1.

Walls of 500 mm thickness may be constructed up to 7 m. Lesser thickness of walls may be used if required. A thickness of less than 400 mm may not be practical.

5.8.7 Other Associated Items

5.8.7.1 Ring or Tie Beams

The use of ring beams at critical points will be important. This will ensure stability and control the leaning of the loaded building walls.

5.8.7.2 Corners

Various methods of reinforcement may be considered, such as

- Using a triangular fillet of wood
- Reinforcement with stone, brick or other masonry items
- A thin strip of lime mortar may be poured on either side of the corner and between each 100 mm tamped layer.

5.8.7.3 Openings

End boards are used in the shuttering.

5.8.7.4 Gable-End Walls

The wall will be compacted and completed as per the required angle. After drying, grooves may be formed for placing and holding the roof timber members.

5.8.7.5 Roofing

There are different types of roof treatments, including thatched roofs, slates, flat roofs, or Roman tiles, etc.

5.8.7.6 Floors

The floors are laid after completing the walls and placing the floor beams/joists as per the design details.

Source:

- Building with Earth [2]

5.9 ADVANTAGES [2]

There are various advantages, including

First construction; wood will be utilised economically, costs will be lower, it will provide fire resistance, it will be durable and the construction will be strong and solid. With proper design, rammed earth buildings can be visually attractive, thermal insulation will be available, the material can be used as fertiliser after demolition, in most cases local materials and labourers may be utilised, and through different applications, it may be possible to achieve local development and solve local employment issues.

Source:

- Building with Earth [2]

5.10 DISADVANTAGES [1]

The following may be stated:

- Knowledge and experience are important for rammed earth construction.
- Detailed quality checks will be required to control the various aspects of the construction activities; otherwise, this may affect the strength and durability of rammed earth walls.
- The construction of rammed earth is a labour-intensive method and may be expensive in some areas.
- In high-rainfall areas, rammed earth walls will need protection.
- The rain may not wash away a rammed earth wall. However, the rammed earth walls will need overhanging roof protection to keep the walls safe and shielded against erosion in high-rainfall areas.
- Many countries may not have proper guidance for the construction of rammed earth buildings.
- Building authorities, financial institutions and insurers may take more time to approve rammed earth buildings. This may be partly due to a lack of knowledge and scepticism.
- Unlike brickwork, rammed earth walls cannot be easily rectified once constructed.

Source:

- Wikipedia [1]

Repair of Rammed Earth – This has been discussed in case studies later in this chapter.

5.11 COMPOSITE USE OF EARTH WITH OTHER MATERIALS

Earth may be used in conjunction with other structural materials like reed, steel, and reinforced concrete (i.e., the framework) to complete the building. Different uses, as mentioned below, may be involved [2]:

The framework may be designed to support all the building loads, and the earth may be used as an infill material between the load-bearing elements.

The framework and the earthwork may work together to share the loads. Framework elements, which may be lighter, can take the compressive loads and provide rigidity to the structure. The framework members may be located within the thickness of the wall and can act compositely.

The framing elements will also provide bending strength to the structure by utilising connecting items such as flooring, lintels, wooden reinforced flooring, roofing materials, and connections.

5.11.1 Further on Different Items

Details will be presented on some items:

5.11.1.1 Timber Framed Structure

The infill material may be a mixture of earth and vegetable fibres or animal hairs, which may be applied to a wooden lattice arrangement when the material is in a plastic state. The infill material may not fully cover the framework elements; hence the walls will be reasonably thin, and the building will remain light.

5.11.1.2 Earth and Post Buildings

In this case, the structure will consist of wooden posts 100–150mm in diameter (i.e., either rough-cut or squared) placed at 1.5-m intervals in the middle of the wall. The posts may be dug into the ground or provided with footings to prevent rotting.

A series of wall plates may connect the posts at the base and the top to act as ring or tie beams. Apart from earth filling to provide strength, the posts are also braced at the corners to resist lateral loads.

The walls may be 400–500mm thick and can be constructed with a mixture of equal volumes of earth and straw, moistened to achieve a plastic state for construction. The material for constructing the wall may be the same as that used for cob.

5.11.1.3 Horizontal Elements

Higher proportions of straw may be used for the horizontal elements to provide additional strength, especially when subjected to bending stresses. The following items may be mentioned:

5.11.1.3.1 Flooring

The flooring may be constructed using floor joists connected by small sticks that can be placed at 80–100mm intervals. The plastic earth may be used, and the materials should be pressed between the lattice work to completely envelop each stick. By using a rammer, the earth can be tamped firmly to create a smooth surface for the ceiling.

5.11.1.4 Pre-fabricated Elements

Some applications using prefabricated elements have been carried out, as mentioned below:

- **Partition Panels –** The dimensions may be 1,000×250 mm, tongue-and-groove partition units.
- **Door and Window Lintels –** Up to 1.2 m wide.
- **Reinforced Flooring or Roofing Blocks –** The reinforced roofing blocks may measure 700×320×110 mm and are reinforced with two pieces of wood, 30 mm in diameter, which should be prepared by soaking in water or mud and finally coated with a 40 mm layer of earth and fibres.

5.11.2 Earth Roofs

There are various applications that have been executed using earth to construct roofs in the past. The main purpose of a roof is to provide cover and protection against cold, heat, or rain. Stabilised earth was used to improve the durability of the roofs and to reduce maintenance requirements. Brief details of a few types are indicated below:

5.11.2.1 Flat Roof

These applications are very common and represent the simplest way to construct roofs.

Slightly slanting beams may be used to aid drainage, which can be placed on a ring beam made of wood or concrete. Planks or sticks of timber, 30–60 mm, in diameter, can be placed on top. A thick, good-quality paper or roofing felt is then laid over this layer. Earth is then placed in the form of mud, 200–300 mm thick, or rammed in layers to form a layer approximately 70 mm deep. A waterproofing covering is then provided. For additional protection, a further layer of earth or stones may be added on top.

5.11.2.2 Clay Bonded Thatch Panels

Large pre-fabricated thatch panels measuring 900×600 mm have been used. The panels were produced on the ground, placed in position, and fixed properly by punching holes.

5.11.2.3 Mud Brick Vaults

The use of mud brick vaults and domes was common in Iran and the Middle East.

The easiest way to build a vault is to use a form to support the necessary loads during construction. It may be practical to use a removable form where possible. The formwork can be metallic or wooden and should be light and easy to handle during construction.

The bricks are laid in a herringbone pattern for better bonding as per the design of the vault.

5.11.2.4 Domes

The following different methods of construction may be mentioned:

Shuttered Domes – This may be easy to construct by using shuttering for the dome. The formwork should be equal to the size of the room, and arrangements should be made to remove the formwork after construction.

Corbelled Domes – The bricks can be laid in horizontal courses, overlapping each other. The bonding pattern can be created using one of the independent rings in a spiral. The finishing activities at the summit of a circular dome may not be easy. A cone-shaped roof may be preferred.

Source:

- Building with Earth [2]

5.12 PROBLEMS ASSOCIATED WITH THE MATERIAL FOR USE IN RAMMED EARTH CONSTRUCTION

A few items will be mentioned in this connection such as [2]

5.12.1 Choosing a Suitable Soil

The analysis of soil is an important factor for assessment, as mentioned previously. The optimal moisture content of soil is very relevant, as highlighted. The soil intended for rammed earth construction should be dry.

5.12.2 Stabilisation

The different stabilisers that can be used for the soil are discussed in Chapter 3. The common stabilisers for rammed earth constructions include cement and lime.

5.12.3 Compaction

The compaction of soil is a critical aspect of the rammed earth method. Several factors are associated with the choice of rammers, such as weight, area to be compacted (or area of striking), the nature and size of the handle, and the type and shape of the striking head.

A few comments will be made on these factors below:

- **Optimum Weight –** The weight of the striking head should be about 5–9 kg. It may vary depending on the builder.
- **The Handle –** May be made of standard wood or metal. The length can vary between 1.5 and 1.8 m.
- **Head –** May be made of wood or metal. Wooden rammers are easier to prepare and can be constructed on-site as per requirements. Metallic rammers will have smaller heads and can be handled more easily.
- **Area to Be Compacted –** Should be around 6,400 mm^2 (80 mm × 80 mm) and should not exceed 22,500 mm^2 (150 mm × 150 mm).
- **Shape of the Striking Head –** Heads for the rammers may come in different sizes and shapes.
- Round heads may cause less damage to the formwork but may not provide effective compaction around the edges.

Manual rammers can exert a static (average) pressure of 0.1–0. 25 kg/cm^2 for the compaction of earth to a depth of 100 mm. Better results can be obtained with a reduced depth of earth, i.e., 50 or 70 mm. To confirm that the compaction is complete, it is necessary to hear a dull thud sound.

5.12.3.1 Mechanical Compaction

The following factors are associated with the compacting force from the machine:

- Optimum number of strokes available
- The maximum depth of loose earth to be compacted at one time
- Rate of progress required

5.12.3.2 Impact Compacting Machine

The rammers are expected to have a long stroke, a powerful impact, and a reasonable air take. The air pressure may be about 5 kg/cm^2.

The types are as follows:

- **Pneumatic Rammers –** The pneumatic rammers may have a static pressure of 0.45 kg/cm^2. The number of strokes may vary, say up to 400–700/min for thorough compaction.

  The walls constructed using pneumatic rammers will have a uniform dry density compared to manual tamping.

- **Vibrating Rammers –** The static pressure may be achieved in the range of 0.05–0.10 kg/cm^2 and will have a high rate of strokes per minute (i.e., from 500 to 1,000). The depth of compaction may be about 300 mm and the weight will range between 60 and 100 kg.

5.12.4 Foundations

The foundation design has to be carried out properly, considering all the possible loads that may be present. The foundation must be strong enough to carry all the loads and may act as a ring beam with sufficient depth, located below the frost level. It should extend above the ground to form a footing for the rammed earth walls.

The materials for the foundation may include

- Concrete (may be reinforced)
- Highly stabilised soil concrete, compacted or in brick form
- Brick, rubble, cement blocks, etc., masonry

5.12.5 Formwork

Following are the important items that will be relevant:

Rigidity, stability, scaffolding, manoeuvrability, problems of put-locks or transverse size, special forms, correcting the plumb, upper transverse ties, treatments at corners, flexibility, lining, quality of the formwork material has to be checked before use, incorporating pre-fabricated elements, distance between the formwork panels, maintenance and re-use, shuttering of the entire building.

Source:

- Building with Earth [2]

5.13 RENDERS AND PAINTS [2]

The earth houses may be rendered to provide some level of protection and for aesthetic reasons as well. The treatment work on the walls can be expensive depending on the location of the building. Rendering may not always provide a durable solution. Many earth houses have not been treated. It may be possible to eliminate rendering/treatment in some cases.

Rendering to walls may be avoided under the conditions given below:

- Protection for walls against driving rain for low buildings may be provided by sheltering the walls with large roof overhangs.
- A neat exterior finish may be achievable with rammed earth construction by using formwork efficiently, applying joint treatments and simplifying details at corners, gable ends, etc.
- With earth brick masonry, the joints may be made regular.
- Better stabilisation for the exposed walls.

5.13.1 Better Stabilisation for External Face of the Walls

The exterior side of the formwork may be filled with stabilised earth to a width of approximately 100 mm, and the rest may be filled with ordinary earth. Alternatively, 150 mm of stabilised earth may be placed along the length of the external formwork in alternate layers, with bands of more resistant materials provided in the process for the wall.

The main idea of the rendering treatment will be

- To protect the walls against damage from rain and make them waterproof.
- To improve the appearance of the walls and hide defects.

- To improve the appearance by using an attractive colour
- To protect them from knocks and rubbing
- Improving thermal insulation with light renders

5.13.2 Principles [2]

There are various types with different strengths that may be available for rendering.

When the render is too thick, it may resist wall movements such as swelling, shrinkage, and settlement. In such situations, it may crack and come away from the wall.

An untreated earth wall may have untreated earth render, while the stabilised earth wall should be given a coating of stabilised earth.

5.13.3 Different Types of Renders [2]

Various types of renders that may be considered for application include

Cement–earth render, soil render (dagga), fibre-reinforced render, pozzolanic render, lime–earth render, plaster render, sand–cement–lime render, light renders – from industrial products and may be a mixture of cement, lime, small gravel, and other auxiliary products, and balls of earth. Balls of earth, 50–70 mm in diameter, are thrown forcefully against the wall. Other renders – may include bitumen – though bitumen renders may not yield good results.

5.13.4 Preparation of the Wall [2]

Rendering is applied to a dry and stable wall, and the items mentioned below will be relevant:

Brushing to remove dust, chiselling, wire netting, scoring, chicken wire mesh, base coat with holes – a tool is used to poke holes, embedded nails – nails are placed on the surface, scored joints, pottery waste, wooden elements, hollowed-out blocks.

5.13.5 Applying the Renders [2]

Precautions to be taken

Renders are not recommended for application in very cold weather (i.e., below 5°C) or in very hot weather (i.e., above 30°C).

The following suggestions may be useful:

- The condition of the surface to be treated must be examined. Any defects or cracks should be rectified.
- The wall should be thoroughly wet before application.
- Direct sunlight and excessively high winds should be avoided.
- Each façade to be finished in a single day.
- Vertical and horizontal joints should be identified to work in 10–20 m^2 single sections.
- Special attention should be given to the edges of openings as well as door and window frames.
- The renders should not be allowed to dry too quickly.
- Renders may be avoided for footings or horizontal elements.

5.13.6 Applications [2]

The render may consist of three layers of treatment as mentioned below:

The Key or Fixing Layer – This is a fairly liquid mortar and is used with a trowel on the well-prepared wall surfaces. The liquid mortar may be applied by throwing it with a trowel. The thickness may vary between 2 and 4 mm.

Body of the Render – This may be applied 2 or 3 days after the key layer. The application may be done using one or two layers to achieve a thickness of 8–20 mm. The layer may be strong and waterproof, and the surface is smoothed with a straight edge.

Finishing Layer – This may be a decorative layer of 3–6 mm deep. The surface may be treated in different ways as per the requirements. The various types may include

- The surface may be left with a rough cast finish, giving approximately a gravelly texture (i.e., Tyrolien render).
- This is the treatment that should be applied before the render has hardened. The finishes may include rotating brush finish, plastering, brushing, or scoring (ribbed finish).
- A combination of different finishes may be considered (i.e., scoring in some areas) to enhance the surface attractiveness.
- Working on the surface after the render has hardened, finishes may include honeycomb or speckled finish.

Source:

- Building with Earth [2]

5.14 FURTHER ON STABILISERS FOR RENDERING [2]

The stabilisers should be of the same type and proportions as those of the walls. Normally, the common stabilisers are cement, lime, and bitumen. The discussion on stabilisers is covered in Chapter 3.

Brief comments will be made on stabilisers.

5.14.1 Natural Products

The products that may be used as suitable materials for various projects may be applicable to a particular region. The study on the uses of these materials, especially for durability, may not always be available.

- **Agave –** The juice of this plant may be suitable for alcoholic drinks and stabilisers.
- **Cow Dung –** May be used as a binding agent to increase the cohesion of the render.
- **Shea (Karite) –** This is a vegetable fat also known as golam butter.
- **Flour –** After forming a paste, it is added to the soil.
- **Cactus Opuntia –** A plant similar to the Barbary fig tree. Its juice is toxic.
- **Nere –** This is a dark red juice that may be obtained by diluting the powder from the fruit.
- **Euphorbia Latex –** These ligneous, herbaceous plants contain white latex which may be used. Further work on the material may be necessary.
- **Peulh Soap –** It resembles tree gum and should be mixed with brick dust just before being added to the soil.
- Cattle urine and fibres may be used in applications.

Source:

- Building with Earth [2]

5.15 METHODS FOR PROJECTING RENDERS ON THE WALLS

Performance by hand may be improved with the help of different mechanical processes.

These may be mentioned as follows:

- **Pneumatic Spray-Gun with Funnel –** The mortar may be placed into the funnel, and the render will be projected onto the wall using a hand-held spray gun. The spray gun is connected to a pneumatic compressor.
- **Mixing and Applying the Renders –** The mortar will be mixed in the machine, and the material will be driven under pressure and projected onto the surface. It will be possible to operate up to 30 m in height and a distance of 70 m with the help of a pump. A team of three or four persons can complete 150 m^2 of rendering work on the wall per day.
- **Tyrolien Box –** The mortar will be projected onto the walls using a metallic box fitted with a hand-turned paddle.

5.15.1 Further on Paint [2]

This is an important item for the finishing work. Different types of treatments may be considered, following the requirements mentioned below. However, special mention should be made of the treatments using whitewash and banana juice.

- **Banana Juice**

  This may be prepared by chopping the leaves and stalks of the banana tree, cooking them in water, and stirring. It may be strained when it thickens. Following this, it may be used without dilution or mixed with laterite. The treatment (i.e., paint) can perform well for 1–3 years and may be suitable against torrential rain.
- **White Wash**

  The whitewash applications may be very suitable for treating earth walls. It will reflect the sun's rays and is a good antiseptic, resistant to alkali and the influence of bitumen. It may be more readily available than bitumen. Additives may be required to improve its weather resistance, and it may not be durable. Whitewashing may be done twice a year. It may be applied by spraying or with a brush and can be easily dyed.

  Industrial Paint – Below are different types:
  - **Distemper –** May be used for walls. The paint must be completely removed before the application of any other paint.
  - **Latex Paint –** It may yield good results on cement-stabilised soil.
  - **Silicon Paint –** It may not always perform well.
  - **Oil-based Paint –** A first coat may be required to neutralise the alkalinity in the wall and to eliminate bitumen.
  - **Resin-based Paint –** It is generally efficient.
  - **Cement Paint –** It may last 8–10 years on cement-stabilised walls.
  - **Cut-Backs –** It may be applied directly to the walls, turning the colour to grey or black.

  Chemical Products – The products that may be considered are as follows:
  - **Sodium Silicate –** One part of sodium silicate with three parts of water in solution may be prepared and applied with a brush. Two coats may be necessary.
  - **Water Repellent –** These may be silicon-based products that are expensive and may need to be used in large quantities.
  - **Paraffin –** Paraffin may be diluted in benzine and applied/spread at a temperature above 21°C.

  Natural Products – The natural products used for renders can also be used as paint.

Source:

- Building with Earth [2]

5.16 SELECTION BETWEEN RAMMED EARTH, ADOBE, AND COMPRESSED EARTH BLOCKS

5.16.1 (Bricks) [2]

The selection of a particular type of construction in the design will depend on various factors such as technical, economic, climatic, and cultural issues, and may also be associated with knowledge of these factors and their inter-relationships.

Three techniques of construction with earth may be considered: rammed earth, adobe and compressed earth blocks (bricks). The important aspects in this context will be the technological criteria, installation time, information on the performance of the material to be used, and the problems related to the applications experienced. This information will be relevant for the decision-making process.

5.16.2 Soil

The problem may be easily tackled when the soil around the house is adequate. It should be noted that the soil may contain gravel and stones, which will need proper sieving for the use of adobe or compressed earth bricks, as highlighted before. However, the same soil, as it stands, may be used for construction with rammed earth. If the earth at the site location is not suitable, the required quality of soil must be brought from a different place for use. The transportation charges must be considered in the calculations. Alternatively, attempts may be made to improve the quality of the soil by adding stabilisers/admixtures, etc.

For the construction of the building, the mechanical performance of the material will be relevant.

The information presented below is associated with the experiments carried out with various earth wall constructions. The values may be considered as approximate.

The results from the experiments are given in Table 5.1 [2].

TABLE 5.1
The Results

Sl. No.	Type of Wall	Compression (kg/cm^2)	Bending (kg/cm^2) of Wall Section	Penetration (kg)
		Unstabilised Earth		
1	Adobe Bonded with mortar	6.84	0.03	450
2	Rammed earth Volume mass 2 kg/dm^3	6.16	0.028	450
		Stabilised Earth		
3	Adobe Stabilised with 5.6% bitumen emulsion mortar	5.5	0.04	450
4	Rammed earth Stabilised with cement	45.73	0.05	450
5	Compressed bricks Stabilised with cement mortar	57.05	0.05	450

5.16.3 Further on the Items

- **Mortar –** Thin mixed mortar prepared with 200 kg of cement and 125 kg of hydraulic lime per m^3 of sand.
- **Bending –** Two horizontal beams were provided at the top and bottom of the wall. The load was placed horizontally and spread over two metal tubes 1.2 m apart.
- **Penetration –** presented below is associated with the experiments carried out with various earth wall constructions. The values may be considered as approximate.

Walls-Brick – Height 2.45 m, length 1.2 m, thickness 300 mm

Rammed Earth – Height 2.45 m, length 2.4 m, thickness 350 mm
Adobe Bricks (Stabilised and Unstabilised) – dimensions 90×300×120 mm
- Compressive strength 3.5–4.4 N/mm^2 (over a whole brick)

Compressed Bricks – Dimensions -300×260×210 mm
Compressive strength -11 N/mm^2
(over a whole brick)

5.16.4 Comments

The following comments may be made:

- Unstabilised walls, in rammed earth, and adobe bricks indicated similar strengths.
- The inclusion of bitumen tends to weaken the wall.
- Inclusion of cement increases their compressive and bending strengths.
- It may be noted that the walls may be considered suitable for single-storey construction with 400 mm thick walls.

From the study on impermeability of walls against the resistance to rainwater, it may be mentioned that

Cement – Stabilised]
Rammed earth] ----- Good performance
Bitumen – Stabilised]
Adobe bricks]
Stabilised compressed –]
Bricks]
Adobe] ------ Mediocre performance
Rammed earth]

Source:

- Building with Earth [2]

5.17 INFORMATION ON IMPROVING EARTHQUAKE RESISTANCE OF EARTHEN BUILDINGS

For improving earthquake resistance of earthen buildings, reference may be made to IS-13827: 1993. Some comments from the code will be mentioned below [4]:

5.17.1 General Considerations

For the safety of earthen buildings, necessary precautions must be taken, especially against the actions of floodwaters, rain, and earthquakes. The minimum precautions are recommended in the code.

Dry clay is hard and strong in compression and shear; however, the penetration of water into clay can make it soft and weak. The reduction in strength may be as high as 80%–90%. Hence, after construction, the structures must be protected against the ingress of water/moisture into the walls by providing roof projections and waterproofing through mud plastering. The use of stabilisers may be costly for low-cost construction. Attempts may be made to use time-tested stabilised compacted clay blocks or cement-stabilised sandy soil blocks with appropriately strong mortars.

The following items will be relevant to the design:

Height – A single-storey building is recommended for seismic zones IV and V.

Lightness – The building should be designed and constructed to be lightweight after satisfying the design requirements in order to reduce earthquake forces. A lightweight roof may also be considered.

Shape of the Building – The building should be designed as simple, symmetrical, and rectangular in plan. The load-bearing walls of the building may be arranged to run continuously in both directions.

5.17.2 Construction of Earthen Walls

The earthen walls may be constructed in the following manner:

- Hand-made, in layers, using mud lumps may be used to construct the walls as mentioned before.
- This may be constructed by using sun-dried blocks or adobe, which may be cut from hardened soil or prepared in moulds. Compressed earth blocks may also be used.
- Using the rammed earth method, i.e., placing moist soil between the forms for the wall and compacting it manually or mechanically, and laying it in courses as mentioned for constructing the walls.
- The construction may be done by using bamboo, wood, and cane structures or ikra mesh enclosures, followed by mud plaster.

5.17.3 Suitability of the Soil

This item has been discussed in Chapter 3. However, the items mentioned below are important:

- Dry strength test
- Fissure control test
- Adobe strength test
- For compressive strength, a minimum value of 1.2 N/mm^2 will be preferred.

5.17.4 Hand-Formed Layered Construction

This is a primitive method and may be considered the weakest. This issue has been discussed in Chapter 2.

5.17.5 Abode Construction

This item has been discussed in Chapter 4.

5.17.6 Rammed Earth Construction

The rammed earth method and construction have been discussed in this chapter. The discussion is arranged according to the contents.

Source:

- IS-13827 [4]

5.18 FURTHER COMMENTS ON THE DEVELOPMENT OF RAMMED EARTH [2]

Rammed earth construction is well known and still in use in different countries, as stated already. Rammed earth structures over 300 years old can still be identified, continuing to provide shelters for rural people.

According to experts, traces of rammed earth can be found in ancient Roman constructions, and this knowledge has been transferred from generation to generation in the Lyonnais and neighbouring provinces.

Roman earthworks were carried out in the past without the use of straw or fibres. The tamping of earth was done in layers between planks to achieve the normal thickness of stone walls. After proper tamping, the earth binds together to form a homogeneous unit for the walls of the buildings. The earth can be extracted directly from the ground, likely situated near the construction site for use.

It should be noted that properly constructed rammed earth walls form a single unit of mass. When rammed earth walls are coated on the outside with external rendering, they may last for a few centuries.

5.18.1 Further Developments

Two case studies are presented below, which are related to the development of rammed earth construction.

5.18.1.1 Earth Construction in Morocco

5.18.1.1.1 Berber Ksour – Shelter for Local Population

A few items for the constructions are as follows:

- Rammed earth was used for the walls, with adobe bricks applied for the façade.
- The foundation may be 500 mm deep and filled using stone rubble with lime and earth mortar. This may be extended into a footing of 200–500 mm high for setting up the first formwork.
- Simple tools were used, such as a basket, a hoe, and a wooden hammer.
- The formwork could be 600–800 mm high, with a length varying from 1.4 to 1.8 m.
- The walls may be battered from 800 to 1,000 mm at the base to 600 mm for a height of 3 m.
- A two-storey building measuring 12 × 12 m may take about 3 months for construction at the best time of the year.

5.18.1.1.2 Rammed Earth Construction in the ANDES, CORDILLERA:TAPIA

Most of the applications were in the Peruvian environment.

Rammed earth constructions were common in coastal and mountainous areas.

Peasants were involved in rammed earth construction work.

The equipment used may be similar to that used in France.

These may include

- Wooden side panels for marking (1,600 × 550 mm)
- End-board
- Metal rods for ties
- Wooden hammer
- Spade and pick-axe
- A few sacks and a ladder

The common procedure followed included

- The earth would be excavated and collected near the construction site.
- It would be shovelled into sacks.
- It would be poured into the formwork and spread around.
- The person (i.e., builder) may carry it on his back.
- A wooden hammer with handle may be used to compact the earth.

Source:

- Building With Earth [2]

5.19 SURFACE AREA OF WALL THAT MAY BE CONSTRUCTED BY A PERSON IN ONE DAY [2]

The detailed information is given in Table 5.2.

The details may be noted as follows:

Walls – 400 mm thick

- **Rammed Earth** – Erection of shuttering, filling and tamping the earth, and removing the shuttering
- **Compressed Bricks** – Sieving
  Producing the bricks
  Building the wall
- **Adobe Bricks** – Sieving
  Mixing
  Moulding
  Building the wall

The information provided above is based on the output of specialist workers, and it can vary from 1 and 5 between amateur workers and professional builders.

The construction of rammed earth appears to be the fastest; however, experience and knowledge are required for the process.

Other factors such as drying time, the number of people available, and the complexity of the building to be constructed will also be relevant.

TABLE 5.2
Detailed Information

		Item		
Sl. No.	**Description**	**Rammed Earth**	**Compressed Brick**	**Adobe Brick**
1	Surface area of the wall in m^2 that one person can build in one day - In unstabilised earth	3	1.5	2.1
2	- In stabilised earth	2.5	1.2	2
3	Minimum desirable equipment	Shuttering Manual rammer	Sieves Brick press	Sieve Mould Possibly mixer
4	Minimum number of persons on site	2–3	2	1
5	Degree of difficulty	Knowledge will be required	Reasonably easy	Medium

The critical approach would be to opt for a simple solution that aims for durable and high-quality buildings.

Source:

- Building with Earth [2]

5.20 REVIEW OF ADVANTAGES AND DISADVANTAGES

The comments are as follows:

5.20.1 Advantages

- The method of construction is simple and straightforward.
- It is a low-cost material
- It requires minimal water.
- Aesthetically attractive buildings can be constructed using this method.
- It is suitable for load-bearing, free-standing columns.
- It has a long lifespan.
- The structure is strong and sturdy.
- It is environmentally friendly construction.
- It provides high thermal mass.
- Local labourers can be utilised with brief training.

5.20.2 Disadvantages

- It offers poor insulation.
- It is labour-intensive.
- It may not be suitable for vaults and domes.

Source:

- Build your homes with Earth [7]

5.21 SEISMIC STRENGTHENING OF BEARING WALL BUILDINGS

It is important to adequately tie the walls of the buildings with horizontal members for safety. Some comments in this connection have been included from the IS code. These are as follows [4]:

5.21.1 Collar Beam or Horizontal Band

Two continuous horizontal reinforced concrete connecting beams are recommended for proper tying. One should be provided at the lintel level for door and window openings, and the other just below the roof of the wall for seismic zones III, IV, and V. Adequate design for the connection of ties, which must be placed at right angles at the corners and junctions of the walls, must also be ensured to provide the necessary safety.

When the height of the wall does not exceed 2.5 m, the lintel band may be omitted. In that case, the lintels should be connected to the roof band.

The types of bands may be:

- Rough cut unfinished or sawn (70×150 mm) lumber sections, in single pieces, are to be placed as diagonal members for bracing at corners.

- Unfinished rough cut or sawn (50×100 mm or 70×70 mm), in single pieces, are to be placed as diagonal members for bracing at corners.

The details are shown in IS-13827. To lengthen the joints in the elements, proper design must be done using iron straps, and a sufficient number of nails/screws must be used to ensure that the joint will have the same strength as that of the original lumber member.

5.21.2 Pillasters and Buttresses

When pilasters or buttresses are applied at salient corners or T-junctions, the collar beam should also cover the buttresses. The details are shown in IS-13827. By placing diagonal struts at corners, the collar beam can be further stiffened.

5.21.3 Vertical Reinforcement in Walls

In seismic zone V, mesh reinforcement is recommended as per the code. The entire wall must be tied together with mesh made of canes or bamboo, along with the collar beams which may also be constructed with canes or bamboo. The vertical canes should be tied to the horizontal canes and the collar beam at lintel level and the roof beam at the eave level. The details may be seen in IS-13827.

5.21.4 Earthen Construction with Wood or Cane Structures

The information along with details of the bracing and braced frames may be obtained from IS-13827.

5.21.5 Plastering and Painting

Plastering and painting are often carried out to provide protection and enhance the durability of the walls and thatch roof. This treatment is also important for achieving the desired aesthetics.

The details are -

5.21.5.1 Treatment in Dry Areas

The plastering may be done in two layers using natural additives.

The first layer may be 12–15 mm thick, made with a mixture of mud and straw (1:1 by volume) along with a natural additive such as cow dung, which may be used to increase the moisture resistance of the mud. This will also prevent the development of fissures during the drying period. The second layer may be prepared with fine mud. After drying, the surface of this layer should be rubbed with hard, small, rounded pebbles.

5.21.5.2 Treatment in Wet Areas

The surface of the wall may be covered with waterproof mud plaster by following the treatment as mentioned below:

The cut-back material should be prepared by mixing bitumen 80/100 grade and kerosene oil in a ratio of 5:1. For 1.8 kg of cut-back, 1.5 kg of bitumen should be melted and mixed with 300 mL of kerosene oil, and stirred continuously for proper mixing. The mixture will then be blended with 30 L of mud mortar for application. The mixed material will have water-repellent and fire-resistant properties.

For improving the water and fire resistance of thatch roofs, a 20–25 mm thick plaster may be applied to the top surface of the roof and allowed to dry. It may be further treated with a wet mixture of cow dung and waterproof plaster in the ratio of 1:1 and dried again.

The outside walls and thatch roof, after the above treatments, may be painted with water-insoluble paint or treated with a water solution of lime, cement, or gypsum.

5.21.6 Internal Bracing in Earthen Houses

The necessary guidance will be available in IS-13827.

5.21.7 Hold-Fasts to the Walls

Reference may be made to IS-13827 for different treatments.

5.21.8 Other Alternatives and Applications

Guidance may be obtained from IS-13827 for various applications.

Source:

- IS-13827 [4]

5.22 REPAIRING METHODS, RAMMED EARTH STRUCTURES

Prior to attempting rehabilitation work, the information on the original material and its applications will be relevant [2].

These may include

- Selecting a suitable soil
- Compaction
- Stabilisation
- Foundation Provisions
- Form work and possible problems connected with the applications.

A few items in this connection are listed below [8]:

5.22.1 Defects

The defects in earth building may be of two types like

- Problems with surface coatings
- Structural defects

5.22.1.1 Surface Defects

These may include –

Cracks, Peeling, Flakes, Blistering, Boniness, Loss of adhesion

5.22.1.2 Structural Defects

The different types may be

Freeze–thaw heaving of the wall, Water borne erosion of the wall, Nominal erosion at the base of the wall, Structural cracking due to overload and settlement, Bulging, Rat runs and animal holes, Abrasion and damage etc.

5.22.2 Repair Work

The following items may be included:

5.22.2.1 Surface Coating

The defects of surface coating may be repaired locally or entirely if required. Different types of coatings may be available. The repair work may be carried out following good engineering practice for renders and plasters.

5.22.2.2 Structural Defects

These may include the below:

- **Bulging –** Attempts need not be made to realign the leaning or bulged sections of an old building when the wall is standing and performing well. Safety must be ensured. In some cases, the walls may have to be stabilised.
- **Shrinkage Cracking and Spalling –** May be repaired by pointing or filling with wet soil of a similar type (i.e., colour, plasticity, and grading) and matching with the original wall. Colour matching and the development of a proper bond may be critical.
- **Structural Cracking and Under-scour**, may be done by
  - Using pre-cast blocks and repairing with earth mortar
  - Ramming into place new material and using formwork on one face.

5.22.2.3 Renovation of Old Earth Buildings [8]

The relevant items may include

- Historical investigation (drawings, sources from archives)
- Drying out the structure
- Assessing problems to prevent structural failures
- Examining future utilisation possibilities (alternative uses)
- Finalising the schedule of priorities
- Preparing suitable design details
- Ensuring safety during and after execution

5.22.2.3.1 Comments

- The maintenance of rammed earth structures may not differ significantly from that of traditional dwelling buildings. However, a lack of proper maintenance may create more problems in earthen structures than in other building types.
- The repair of earthen buildings may be necessary due to inadequate maintenance, poor dwelling conditions, and substandard construction.
- Careful assessment, including proper planning, may be very important.
- For structural repair work, assistance may be sought from structural engineers with experience in similar problems and structures.
- Execution must be carried out efficiently, with adequate care for the safety of the structure during and after completing the job.

Source:

- Building with Earth [2]
- Maniatidis and Walker [8]

5.23 FURTHER REVIEW AND SUMMARY OF COMMENTS, RAMMED EARTH STRUCTURES [8]

A few comments will be made on the following items:

Codes – Some guidelines may be available from IS code. These are mentioned separately.

A few comments are as follows:

5.23.1 Materials

The items may be mentioned as follows:

- The physical characteristics of soil are very important. These may include grading, colour, and plasticity.
- The important characteristics for rammed earth may include dry density, strength (i.e., compressive, shear, and tensile), shrinkage, durability, thermal properties, and surface finish.
- Standard soil characterisation tests, such as grading, may not be adequate for assessing the suitability of soil for rammed earth construction.
  - The durability of rammed earth may be determined by various accelerated erosion tests. The correlation between test results and field performance will be relevant.
    - The dry density may be determined by Proctor compaction tests.
    - The physical properties are associated with the density of the material.
  - The structural characteristics of rammed earth, in terms of compression and flexure, are not adequately studied. More information may be required.
- The cement stabilisation technique has mostly been used to improve strength and durability. The use of other types of stabilisers, such as lime and natural fibres, has not been widely adopted.
- The proper soil for rammed earth may encompass a wide range of materials, including sand with adequate clay and silt, clayey gravels, clayey silts, and gravel–sand–clay mixtures.

5.23.2 Structural Design

The design should be carried out in a manner that is typically followed for load-bearing masonry walls.

5.23.3 Architectural Design and Detailing

The issues may include

- For external walls, extra thickness or additional insulation may be required. For internal walls, care may be necessary regarding thermal mass and protection against waterborne deterioration.
- Fixings, both structural and non-structural, may be applied directly to rammed earth or by using timber elements placed in the rammed earth during construction. Mechanical raw bolt connections may also be used for stabilised rammed earth.
- Openings may take the form of arches or may be created using timber, steel, and concrete lintels.
- Suspended floors or roofs may be supported on ring beams or wall plates.
- Electrical conduit services may be buried within the rammed earth during construction.
- The rammed earth walls must be protected from waterborne erosion by providing adequate roof overhangs and wall plinths.
- Proper detailing is required to eliminate or control deterioration and to reduce the maintenance costs of rammed earth structures.

5.23.4 Construction

The relevant items are as follows:

- The soil should be properly prepared for construction. Key processes include extracting, screening, pulverising, mixing, and drying.
- The formwork must be sufficiently strong, with adequate stiffness to bear the load during construction.
- For rammed earth construction, the earth should be compacted at its optimum moisture content. Keeping the soil dry during wet weather may prove challenging.
- Sufficient effort will be required to erect the formwork, aligning, striking, removing, and cleaning it.
- Different types of formwork may be employed, including plywood faced with timber or metal soldiers, or the use of metal tie rods.
- Compaction methods may include manual, vibrating, pneumatic, and sheep-foot rollers.
- The uniformity of the soil is important to achieve the proper quality of the finished rammed earth construction.

5.23.5 Foundations

The following items are relevant:

- The rammed earth walls may be constructed on shallow strip footings or stiffened ground slabs.
- A dampproof course should be provided to withstand the stresses from compaction and shrinkage without being damaged.
- Different types of dampproof materials may be used.
- The walls are typically constructed on raised plinths.
- Dampproof treatment should be applied at 150 mm above ground level.
- The footings may be constructed from concrete; however, limecrete and masonry may also be used.
- The ground adjacent to the base of the rammed earth should have a good drainage system, and the walls should be protected against water infiltration.
- Consideration of surface and sub-surface drainage, along with the dampproof course, is crucial for the integrity of rammed earth walls.

5.23.6 Quality Control

The important items may be mentioned as follows:

- Weather conditions will be important during and following construction. The prepared earth should be placed/used at optimum moisture content and protected so that it is not damaged by rainfall during drying.
- Material selection will be critical in order to achieve the proper and finished quality of rammed earth.
- The acceptable tolerance for rammed earth may be greater than that for masonry construction.
- The quality of in-situ rammed earth may be assessed by measuring strength (i.e., cube or cylinder compressive strength), dry density, surface hardness, and finish. Finishing aspects may include colour, texture, and friability.

- There may be disagreements between the client and contractor regarding the finished quality of rammed earth.

5.23.7 Maintenance and Repairs

The salient items are as follows:

- Water is the major concern for rammed earth construction. Maintenance work must ensure that rammed earth walls are protected against waterborne deterioration.
- Protective coatings may be used to improve durability. Options may include renders, paints, lime washes, sodium silicate, and wax treatments.
    - The need for repairs to rammed earth may arise from poor construction, inadequate maintenance, and poor design/detailing.
- As mentioned, defects in rammed earth may include boniness, surface erosion from water, cracking, and abrasion.
- Detailed and careful planning will be necessary for repair work. Structural repairs should be carried out in accordance with the guidance of qualified structural engineers.
- It is essential to ensure the safety of the structure during and after the repair work.

Source:

- Maniatidis and Walker [8]

5.24 CASE STUDIES [9]

The repair works carried out for rammed earth structures are presented below:

5.24.1 Brief Comments on Rammed Earth Structures

The rammed earth method of construction is an ancient technique that developed independently in countries such as China, the Middle East, and Africa, as previously mentioned. This type of construction has dominated in various regions, resulting in elegant structures built in different parts of the world, including India. Some rammed structures may represent many hundreds of years of life, provided they are subject to some maintenance and treatments.

In this case study, two very old rammed earth structures will be referred to. One is located in Spain, and the other in Ladakh, which is situated in the northern part of India.

The discussion of the two cases will focus on the cracks in the structures and the associated remedial works that were carried out for repairs.

5.24.1.1 Case Study-1, Cracks in the Structure, Spain

The structure is located in the north-eastern part of Spain and is a medieval construction that has experienced extensive growth and reconstruction throughout its existence.

Cracks had developed in the structure, and a detailed survey was conducted. From the survey, it was observed that the cracks near the gable end extended to the full wall thickness. It was estimated from the studies that the gable end was slowly moving away from the rest of the structure at a rate of 0.6 mm/year near the roof level.

To rectify this and control the movement, brickwork buttresses could be added to the lower portion, along with other rectification work or reconstruction at the top.

Remedial works were implemented to introduce internal steel beam ties, which were bolted internally to the timber roof trusses and fixed externally using H-shaped plates that would butt against the wall surface.

After the execution, it was noted that following the treatment, the rate of crack growth had not slowed down. It may be that the wall has not yet engaged the ties.

5.24.1.2 Case Study-2, Crack Stitching in the Structure

The stitching of cracks in rammed earth wall construction must be done very carefully to ensure that the method of stitching does not create additional stresses in the wall or lead to further cracks. Furthermore, soft stitching is recommended for cracks using materials similar to those used in the structure.

This study relates to the palace structure located in Ladakh, in the northern part of India. The palace may have been constructed before 1445.

An important structure on the site exhibited two large cracks in the wall. These were easily visible, and one of the cracks extended approximately to the full height of the structure. The cracks may have widened due to the effects of water flow or freeze–thaw cycles. Different cracks were observed on the wall, and one crack extended at a 45° angle from an opening near the bottom corner of the structure. Furthermore, smaller cracks could be seen running nearly parallel to the large vertical crack.

5.24.2 Remedial Work

The remedial work was carried out by introducing buttresses at the base of the sloped wall and placing soft ties across the cracks to maintain the continuity of the structure. The crack stitching technique was adopted, which involved using a mud brick staple across the crack to half the thickness of the wall. Two staples were cut, and hemp matting material was placed inside the cuts after wetting the surface. Sundried bricks with mortar were then placed within the cut to form solid support within the wall.

The soft stitching method of repair work, linked to the staple stitching technique and using the same original type of material for the cracks, may play an important role in repairing cracks in rammed earth construction, particularly in many old structures.

Source:

- Jaquin [9]

5.25 APPLICATIONS AND PHOTOGRAPHS

A few photographs from a project have been included (Figures 5.1–5.3).

FIGURE 5.1 Rammed earth building.

FIGURE 5.2 Rammed earth building.

FIGURE 5.3 Rammed earth building.

REFERENCES

1. Wikipedia, Rammed Earth, pp. 1–7. Wikipedia. https://en.wikipedia.org/wiki/Rammed_earth. Accessed Oct 03, 2024.
2. Doat P, Hays A, Houben H, Matuk S and Vitoux F (CRATerre, France), *Building with Earth*. Translated by Asha Puri, The Mud Village Society, New Delhi, India.
3. Rammed Earth Construction, Vip service for 7/24, Vol. 3, pp. 1/6.
4. Bureau of Indian Standards, IS-13827: 1993, Improving Earth Quake Resistance of Earthen Buildings - Guidelines.
5. Ganguly K K, Some Design Issues & Alternative Building Materials - A Brief Review, Innovative World of Building Materials, State of the Art Lecture, 29th National Convention of Architectural Engineers Division, The Institute of Engineers, India, 31.01.2014.
6. Ganguly K K, Durable Concrete will be more suitable for different applications in projects, *Indian Concrete Journal (ICJ)*, July 2017, Vol. 91, pp. 80–96.
7. Earth Building Techniques - Building Your Home with Earth, This Cob House. pp. 1–22. https://www.thiscobhouse.com/earth-building-techniques-build-home-earth/. Accessed Sep 30, 2024.
8. Maniatidis V and Walker P, *A Review of Rammed Earth Construction*, Natural Building Technology Group, Department of Architecture and Civil Engineering, University of Bath, May 2003.
9. Jaquin P, Study of historic rammed earth structures in Spain and India, *The Structural Engineer*, 22 January 2008, Vol. 86, pp. 26–32.

6 Ferro-Cement

6.1 INTRODUCTION

Ferro-cement is a composite material prepared from cement mortar reinforced with light steel fabric mesh and used as thin sections [1–4].

There are various applications of ferro-cement, which will be presented later in this chapter. Some applications of ferro-cement, however, are already shown in Chapter 1.

It will be interesting to start the chapter by reviewing some of the remarkable applications of ferro-cement.

The work on ferro-cement was pioneered by Prof Pier Luigi Nervi [4] in Italy during the early 1940s. Working in parallel and following similar lines with others, I. L. Lambot constructed a rowing boat in the 1850s, which was in use for several decades [4].

Working independently, Prof Nervi discovered the characteristics of ferro-cement through several tests and constructed different roofing units that represented rational and aesthetically pleasing structural design.

Ferro-cement is related to the special reinforced concrete applications of structural work, where cement mortar and reinforcement in the form of layers of wire mesh and small-diameter steel bars are used to produce stiff structural form, such as folded plates, which are suitable for roofing units, etc. The thickness of these units will be minimal. The ferro-cement material will be homogeneous and may be used for a limited range of loading [4].

The reinforcement for ferro-cement will utilise small diameter wires to uniformly saturate the cross-section of the element. The concrete for ferro-cement may be prepared with cement, water, admixtures, and finely grained aggregate, which will be no more than 5–7 mm in size and normally about 2 mm to make it suitable for more closely spaced mesh [4].

Further comments on the material will follow.

The mesh reinforcement may consist of wires with diameters of about 1 mm or less and with a spacing of 10 mm or less in both directions [4]. A crucial aspect of the design will be the percentage of reinforcement in the cross-section of the element and the effective cover to the reinforcement. The dispersed reinforcement will act both in arresting the cracks and limiting their widths [4]. With smaller mesh diameters, the number of micro cracks formed will increase, resulting in smaller crack widths.

Ferro-cement elements are to be handled carefully to avoid cracking during erection. The fixing and jointing details of the elements must be done properly, considering wind and seismic effects [2,3].

The basic materials may be readily available, and no sophisticated processes or highly skilled manpower will be required for ferro-cement work compared to other materials [4].

Ferro-cement units were constructed and transported from the UK to the USA, providing the main architectural feature of a museum [5].

Ferro-cement units have also been used as structural members in Russia and South America. In China, different types of boats were constructed from ferro-cement, and a complete ferro-cement fishing fleet was built in Cuba [5].

Further work was carried out for sewer relining using sprayed in-situ ferro-cement lining.

A considerable amount of work has been done in the USSR [4] and Eastern Europe on ferro-cement roofing units. The method of production varies from simple in-situ plastering of the reinforcement-cum-formwork to highly mechanised processes such as vibro-bending or vibro-pressing for producing roofing units [4].

DOI: 10.1201/9781003319108-6

Another technique involves winding a ferro-cement sheet onto a circular former to create trough elements [4].

Applications may be made for agricultural and low-cost housing projects. These items may include small roof domes, water tanks, silos, sanitary ware, etc. By using low-skilled labour and inexpensive materials, such applications with ferro-cement are expected to be suitable for developing countries [6]. The composite material of ferro-cement can be produced or manufactured at any location using minimal expertise.

For industrialised countries, the use of ferro-cement may include roofing items such as domes, folded plates, or corrugated sections for medium to large spans [6].

Source:

- Concrete Society [1]
- Ganguly [2]
- Ganguly [3]
- Walkus and Kowalski [4]
- Barfoot [5]
- Jennings [6]

6.2 HISTORICAL BACKGROUND

The history of ferro-cement is quite old and may be over 170 years old [7].

According to ACI, ferro-cement may be described as a form of reinforced concrete with closely spaced multiple layers of mesh and/or small diameter reinforcing bars, fully covered with cement mortar [7].

The concept of ferro-cement is very similar to the Kood type of traditional wall construction. This method involves using bamboo and reeds, which are tightly bound together and filled with a mud mixture combined with cow dung. This method has been used, especially in the rural areas of India. The ferro-cement concept can be seen as a modified/improved version of the Kood technique, incorporating standardised raw materials, controlled construction methods, and reliable structural properties and performance.

In ferro-cement, mesh reinforcement replaces bamboo and reeds, while cement mortar serves as a substitute for the mud and cow dung mixture [7].

The International Ferro-cement Information Centre has been established at the Asian Institute of Technology in Bangkok.

The idea of applying impregnated, closely spaced wire mesh and rich cement mortar was used in Europe in the 1850s [4]. Subsequently, developments occurred, leading to various applications such as domes, boats, roofing units, and opera houses in several countries, including Europe [7]. Applications have also been carried out in India [2,3].

Ferro-cement units possess good strength and resistance to impact and must be constructed with care, using local labour wherever possible. It is a labour-intensive method, as the mortar is typically applied by hand.

Source:

- WRD Hand Book [7]
- Walkus and Kowalski [4]
- Ganguly [2,3]

6.3 HISTORY OF DEVELOPMENT [3,8]

The initial application of reinforced concrete can be considered as ferro-cement, used in a rowing boat built by Lambot in France in 1849/1850. Lambot obtained French and Belgian patents termed 'Ferciment' in 1856. One of his boats remained afloat 100 years later, in 1949, and is currently on display in a museum in France. Much work was done by Prof. Nervi in 1942–1943. Limited applications of ferro-cement construction were recorded between 1888 and 1942. After the World War II, Prof. Nervi demonstrated the use of ferro-cement as a boat-building material by constructing the 165-tonne motor sailor 'Irene,' featuring a 36 mm thick ferro-cement hull.

Furthermore, its weight was about 5% less than that of an equivalent wooden ship.

Prof Nervi also applied the ferro-cement concept to civil engineering structures, including a small warehouse using 30 mm thick corrugated ferro-cement for the roof and walls. Ferro-cement received wide acceptance only in the early 1960s. In 1965, an American-owned ferro-cement yacht was built in New Zealand, the 17 m 'Awanee', which circumnavigated the world without any serious mishap, although it was exposed to gales of 70 knots, collided with an iceberg, and was rammed by a steel-hulled yacht. After this, there was increasing activity with ferro-cement construction throughout the world, including Canada, the USA, Australia, New Zealand, the UK, Mexico, the Soviet Union, Poland, Thailand, India, and other developing countries.

Source:

- Swamy [8]
- Ganguly [3]

6.4 FURTHER INFORMATION [5]

Ferro-cement elements had been constructed and exported to other countries (i.e., from the UK to the USA) using white cement and crushed white dolomite as the main architectural feature for the Menli Museum roof of the building. The units were cast in ferro-cement moulds using white cement, along with crushed dolomite, having a water-to-cement ratio below 0.4. The elements were used compositely with the truss members to support the glass roof. By proper detailing, it was possible to reflect natural light indirectly into the gallery below.

A total of 300 such units were exported, with 210 units having a length of 11.43 m (i.e., 37 ft 6 in) and 84 units with a length of 4.27 m (14 ft).

The application of ferro-cement as supporting formwork had been carried out by Prof Nervi, as mentioned. This was very interesting and demonstrated a clever application of ferro-cement elements. Attempts had been made to use in-situ sprayed ferro-cement lining for sewer relining and pre-cast segmental lining, which was subsequently backgrouted.

In Russia and South America, ferro-cement had been used as structural members. In Cuba and China, ferro-cement had been used for fishing fleets/boats, etc., as mentioned.

Source:

- Barfoot [5]

The applications may be made for agricultural and low-cost housing projects [6].

The items may include small roof domes, water tanks, silos, sanitary ware, etc. The opportunity to use low-skilled labour and cheap materials may be considered suitable for developing countries.

In industrialised countries, the use may be for roofing items such as domes, folded plates, or corrugated sections for medium to large spans, as stated.

The composite material of ferro-cement may be produced at any location to manufacture strong, watertight laminates using minimal expertise.

Source:

- Jennings [6]

Schooners were constructed with ferro-cement hulls.

One was 22.6 m (74 ft) long with a displacement of 80 T. The final completion took place in Malta before sailing.

Another one was 14.6 m (48 ft) long and the displacement of 11.25 T, boat.

Ferro-cement schooner, 22.3 m (73 ft) long and displacing 55.8 T, was constructed and was meant to be used for ocean racing.

The plastering work for the hull was carried out using a stiff, rich mix to allow for slow curing. Proper curing was done using sprinklers.

Ferro-cement was selected as a construction material for the schooners due to its good strength and high-quality damage resistance performance. Furthermore, it is impervious to marine attack, including attacks from teredo worms, and is almost maintenance-free. It is fire-resistant, and any necessary repair work can be done economically.

Source:

- Two ferro-cement schooners [9]

6.5 POTENTIALS OF FERRO-CEMENT [10]

Lambot constructed two boats, one in 1848 and the other in 1849, with 38 mm thickness and disclosed his patent at the Paris exhibition in 1855.

Ferro-cement may be considered the first application mentioned and can be cited as the origin of reinforced concrete.

It was accepted by the Italian Navy in 1943.

Prof. Nervi utilised the potential of ferro-cement by constructing a 165-tonne motor-sailor known as Irene. The thickness of the hull was 35 mm in ferro-cement, slightly less than that of a wooden hull. For marine structures, using ferro-cement can be considered much more cost-effective than steel.

6.5.1 Laminated Ferro-Cement [10]

Laminated ferro-cement can be described as an improved method of construction using highly reinforced, high-quality thin concrete sections.

The method can be applied by spraying high-quality concrete (mortar) against a light mould surface with highly reinforced sections properly embedded in the concrete. Various fibres and admixtures can be used to enhance the quality of the sprayed concrete surface.

This method of laminated concrete will be suitable for producing high-quality reinforced concrete elements that will last for a longer period.

Long life, along with minimal maintenance, will make the material suitable for marine renewable energy structures.

Source:

- Pemberton and Pullan [10]

6.6 MORE COMMENTS

Tenacity of Bolton, constructed as an 18.3 m (60 ft) Samson ketch, has been used as a sailing and adventure training vessel in the UK [11].

For the construction, attempts have been made to use pre-cast ferro-cement panels with insulating fire-resistant coatings for quick building to provide accommodation, warehousing, and facilities as per specific needs.

Ferro-cement can perform well as a ductile material and has the ability to withstand shock loads.

The proper use of ferro-cement will be beneficial and has performed well against earthquakes for the auditorium building at the University of Oaxaca, Mexico.

For damaged structures, effective repair work can be carried out using layers of mesh and mortar to provide adequate ductility to the structure from the repair work.

By using finite element analyses, a shell roof structure was designed and constructed successfully with ferro-cement for a monument in Indonesia.

Source:

- Nedwell [11]

6.7 CONSTRUCTION [7,12,13]

6.7.1 Materials for Ferro-Cement

6.7.1.1 General Items

The materials that are normally used for ferro-cement structures are as follows:

- Rich micro-cement concrete mix/cement mortar
- **Skeletal Steel –** Normally small diameter bars are used
- **Steel Wire Mesh –** May be used in layers
- Admixtures
- Other materials as per the requirements

Different types of cement may be used depending on availability. Good quality sand without impurities will be required. Further comments will follow.

6.7.1.2 Summary of Materials [7]

Rich cement–sand mortar, as the matrix in the former micro concrete.
Skeletal steel in the form of steel bars, welded wire fabrics.
Light steel wire meshes for forming the cages.

All these basic raw materials are commonly used, in practice, for construction related to conventional applications.

Source:

- WRD Hand Book [7]
- Ganguly [12,13]

6.8 FURTHER ON REINFORCING SYSTEM

These may be indicated as follows:

6.8.1 Steel Wire Meshes [7]

The different types of materials that may be considered are as follows:

- Fine wire mesh (woven square mesh/chicken wire mesh, interlocked hexagonal wire mesh)
- Welded wire mesh
- Expanded metal
- Crimped woven wire mesh

6.8.2 Fine Wire Mesh/Chicken Wire Mesh/Interlocked Hexagonal Wire Mesh

The following items may be noted:

- Interlocked woven hexagonal wire meshes (i.e., chicken wire meshes) are interlocked and hexagonal in shape during preparation. The interconnections of chicken wire mesh may not be rigid. Galvanised chicken wire mesh will be available. The sizes of openings may be 13×13 mm, 19×19 mm, and 25×25 mm. The chicken wire mesh may be available in rolls of 15–30 m in length, with widths of 900, 1,200, 1,500 mm.
- The woven square meshes are prepared by weaving using wire weaving machines and may have rectangular opening sizes of 13×13 mm, 19×19 mm, and 25×25 mm as mentioned. The wire may be 18–20 gauges and is usually ungalvanised.
- When square wire is cut, the ends of the wires should be folded back; otherwise, the wire mesh may loosen and separate.

6.8.2.1 Guidance on Using Chicken Wire Mesh

The items that will be relevant are as follows:

- The wires are interlocked in the form of a hexagonal net; geometrically, it is not a rigid framework. Stretching in both directions will be possible. Necessary care should be taken to ensure that while stretching and tying, proper attention is given to maintain the shape of the mesh by applying equal stretches in both directions. Square meshes cannot be stretched and tied tightly. It is easier and recommended to use chicken wire mesh for the applications.
- The chicken wire meshes may be stretched and tightly tied over the weld mesh. For reference, a width of 900 mm of wire mesh may be extended by about 50–75 mm.
- At both edges of the mesh, a straight wire is provided to assist in weaving. The straight wires may be cut only at certain intervals when using the mesh. Otherwise, there may be issues with stretching the mesh if the edges are not restrained. Furthermore, the mesh may bulge at the edges.

6.8.2.2 Comments on Using Weld Mesh [7]

The following points may be important:

- The weld mesh may be available in rolls. When opened, the weld mesh may come out bent and curved. The weld mesh must be straightened out before use.
- The weld mesh should be cut with a weld mesh cutter and not with a chisel and hammer.
- It may be hammered with a light hammer while straightening. The joints should be protected during hammering and should not be damaged or opened.
- The weld mesh should not be cut close to the weld to avoid the risk of wire strands separating. Cutting should be done 5 mm from the weld.
- It may be prudent to complete the skeletal reinforcement first and then tie the weld mesh or chicken wire mesh over it.
- Along the length of the weld mesh roll, small protruding wires are provided at the top and bottom edges. These wires need not be cut, as they will assist in fixing or tying the chicken wire mesh onto it.

6.8.2.3 Comments on Expanded Metal Lath

The expanded metal lath is prepared by slitting thin-gauged steel sheets and then expanding them in the direction perpendicular to the slits. The expanded thin sheets will be diamond-shaped. Considering the fine wire meshes, the shape, size, and structural properties of expanded metal can be more accurately ensured. It should be noted that in research projects, expanded metal lath is often used for assessing the properties of ferro-cement more accurately.

6.8.2.4 Crimped Woven Wires Mesh (Watson Mesh)

The formation of a three-dimensional mesh by Watson was specially developed for boat building. In this technique, crimped creeper wires frictionally lock together in three alternating layers of straight wires. The mesh is formed with a thickness of five wire diameters. Due to the method of frictionally locking the alternative layers, there may be some spring-back action. The mesh can be easily formed into the desired shape. The transverse crimped wires will keep the high-tension wires in place in the longitudinal direction. In this type of mesh, it is possible to have long stretches of straight wires, without twists, crimps, punching, pressing, or welding. This facilitates the creation of a very strong mesh and permits complete flexibility. The system is a three-dimensional mesh with reinforcing members in all three directions.

Source:

- WRD Handbook [7]

6.8.2.5 Summary on Mesh Reinforcement [8]

These may be mentioned as follows:

- Hexagonal wire mesh (chicken wire mesh)
- Square welded wire mesh
- Square woven wire mesh
- Expanded metal lath
- Crimped woven wire mesh (three-dimensional mesh) [7]

Hexagonal-shaped opening meshes may not be very efficient structurally. They are quite flexible and may be suitable for thin sections.

Galvanised wire meshes are also available for use as mentioned.

Source:

- Swamy [8]
- WRD Handbook [7]

6.9 VIEW OF STEEL REINFORCEMENT AND CHICKEN WIRE MESH FROM A PROJECT [13]

The reinforcing arrangement of a ferro-cement element is shown in Figure 6.1.

6.10 CEMENT MORTAR [7]

Mortar/micro concrete generally consists of hydraulic cement, sand, fine aggregates, water, and admixtures.

Typically, the mix includes well-graded fine sand passing the IS 2.36 mm sieve. Considering the size of the mesh and the distance between the mesh layers, small coarse aggregate may be included in the mix. The mortar matrix usually comprises more than 90% of the ferro-cement volume and significantly influences the behaviour of the final material. Care should be taken in selecting the constituent materials and in mixing and placing them properly for the element.

6.10.1 Cement

Ordinary Portland cement of grades 43/53 may be used for ferro-cement work. Sulphate-resisting cement is advisable for coastal areas or for structures exposed to seawater or subjected to acidic industrial waste. If sulphate-resisting cement or admixtures are unavailable, rich cement mortar may be used, and the structure should be treated with a suitable protective coating.

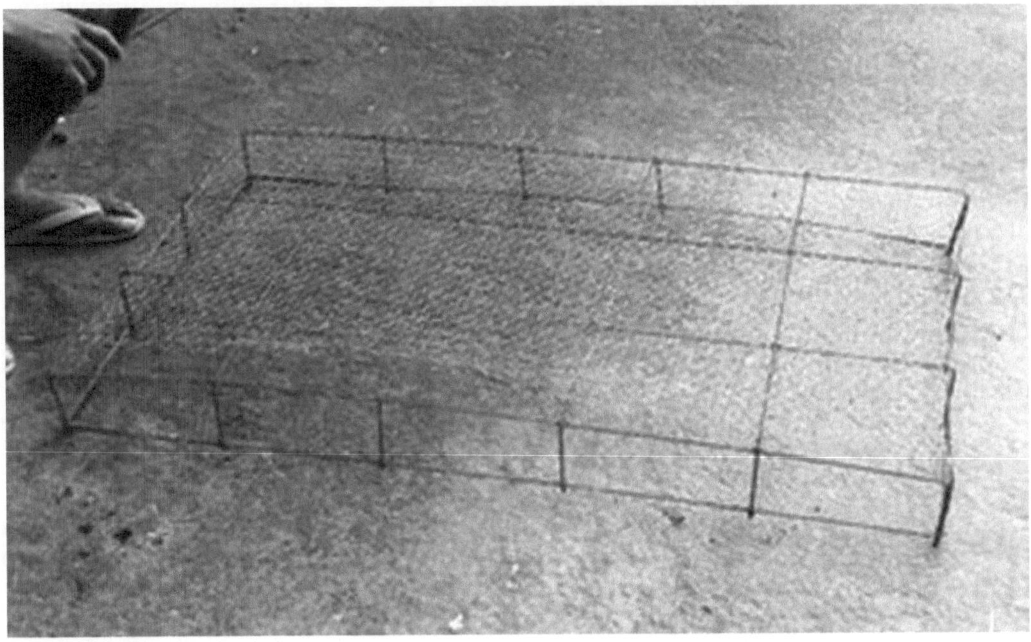

FIGURE 6.1 Reinforcement: small-diameter bars and chicken wire mesh. (Source: Ganguly [13]).

6.10.2 Aggregates

Comments will focus on natural sand.

6.10.2.1 Natural Sand

Sand for applications should be well-graded and washed river sand passing through the 2.36 mm IS sieve. This type is typically used as fine aggregate for ferro-cement work. The maximum size of the aggregates will depend on the size of the mesh openings and the spacing between the mesh layers. Using a 13 mm mesh opening, one-fourth of the opening size may be considered for the aggregate. This indicates that less than a 3.25 mm fine aggregate may be used in the mix. To achieve proper gradation, the fineness modulus of sand should be between 2.4 and 2.5 with a maximum grain size of 1.18 mm. The fineness modulus may be 2.9–3.0 for a maximum grain size of 2.36 mm. Proper control over the grain size will result in minimum water demand, better workability, and high strength.

6.10.3 Alternatives to Natural Sand

Brief comments will be provided on various alternatives to natural sand. The main items are listed below:

6.10.3.1 Crushed Sand or Manufactured Sand [7]

The crushed sand may be considered a proper substitute for natural sand. It may be prepared from quarried rock by reducing its particle size to between 4.75 mm and 150 µm. The crushed sand will differ from stone dust, which is available from stone crushers. The particle size and gradation of crushed sand will be very useful for the mix. Crushed sand has been selected for projects, and by using crushed sand, M40 grade concrete may be produced.

6.10.3.2 Further on Crushed Rock Sand (CRS) [14]

CRS may be prepared by crushing rocks to obtain well-graded aggregates as stated. The aggregates of this type may be more angular and have a rough surface texture than naturally weathered sand.

The rocks used to produce CRS may include sandstone, limestone, diorite, metamorphic siltstone, granite, feldspathic quartzite, etc. Different crushing equipment may be used to produce CRS. The contribution of CRS as filler content in concrete/mortar will be important, as it will reduce environmental impact and provide a cost-to-benefit ratio compared to river sand.

6.10.3.3 Recycled Fine Aggregates (RFA) [14]

Construction and demolition waste may be processed to produce recycled aggregates. These items may include wood, plastic, glass, tiles, mortar, brick, concrete, etc. This type of aggregate may serve as a substitute for natural aggregates and can be used for road construction and landscaping work. The particle sizes can be finer than 4 mm and up to 75 µm or below, and the same may be used for mortar/concrete.

The large-scale recycling of demolition waste and use of recycled aggregates can significantly address disposal problems and also help conserve natural sand.

6.10.4 Industrial By-products [14]

Industrial waste may include items like fly ash, blast furnace slag, coal bottom ash, waste foundry sand (WFS), steel slag, cement kiln dust, ferrochrome, and wood ash, which may be relevant for use in construction and controlled low-strength material.

WFS is high-quality silica sand and is a by-product of ferrous and non-ferrous metal castings. This may be effective for chloride and sulphate resistance properties while simultaneously reducing the workability of the concrete by 20% compared to natural sand. The compressive strength of concrete may be increased by 15%–25% with the use of WFS. A 20% replacement of natural sand may be feasible in some cases for the concrete mix.

Each industrial product may be unique in physical and chemical characteristics. Their behaviour in mortar/concrete must be well assessed before being used as fine aggregates.

6.10.5 Other Materials

Other waste materials, particularly glass, may be recycled and used to replace natural sand in mortar.

The above alternatives to natural sand may be considered in projects for cost-effective and environmentally friendly performance.

6.10.6 Water [7]

The pH value should be about 7. Salt water will not be suitable for the mix; however, chlorinated drinking water may be used.

6.10.7 Admixtures

Different types of admixtures may be considered. The following may be mentioned as basic and common items:

- **Water Reducing Admixture –** This will increase the strength and reduce permeability.
- **Air Entraining Materials –** For resistance to freezing and thawing.
- **Waterproofing Compounds –** For watertight structures.

- Suppression of galvanic action between galvanised steel and cement may be achieved by using chromium trioxide at approximately 300 parts per million (ppm) in the water for the mix.
- Other items as per the requirements.

6.10.8 Fibres

Different fibres may be used in the mix.

Source:

- WRD Handbook [7]
- Singh, Kirthika, and Surya [14]

6.11 MORTAR MIX DESIGN [7]

A rich cement mortar mix is commonly used for ferro-cement construction. The proportions may vary from 1:1.5 to 1:4 by volume, which are often used for ferro-cement work. With an increase in the sand quantity, the water requirement may rise to maintain the same workability. Trial mixes may be necessary to achieve strong and dense mortar that can penetrate easily into the layers of mesh. Factors such as the fineness modulus of sand, water-to-cement ratio, and sand-to-cement ratio for the mix should be assessed properly in advance and applied accordingly. Shrinkage problems may be minimised due to the dispersion of wires/mesh throughout the body of the ferro-cement member. The plasticity of the mortar will be crucial in relation to the method of application.

6.11.1 Guidance for Proportioning Cement Mortar Mix [7]

The following should be noted:

- The stiff consistency of the mortar may be checked by forming a ball with it in the hand. If tossed, it should retain its shape.
- Alternatively, if the trowel is placed in the heap of the mix, it should stand erect.
- To check the silt content in the sand, a simple test may be performed. The material may be mixed with water and placed in a glass jar, then shaken and allowed to settle. The thickness of the layer of silt collected over the sand will roughly indicate the percentage of silt content in the sand.

Source:

- WRD Handbook [7]

6.12 REVIEW OF MORTAR MIX PROPORTION FROM PROJECTS [13]

The information on the mix design is mentioned below:

A cement–sand mixture of 1:2 to 1:2.5 has been used for ferro-cement elements in different projects. Other comments have been made separately.

6.13 MOULDS/FORMWORK FROM PROJECTS [13]

A few typical items are shown below (Figures 6.2–6.4):

FIGURE 6.2 Moulds for ferro-cement units. (Source: Ganguly [13]).

FIGURE 6.3 Moulds for ferro-cement units. (Source: Ganguly [13]).

FIGURE 6.4 Moulds for ferro-cement roofing elements. (Source: Ganguly [13]).

6.13.1 Construction/Steps

The construction activities of ferro-cement may be divided into the following stages:

- Preparation of the mould, specifically for the roofing elements
- Fabrication of the skeletal framing system as per the details
- Fixing wire mesh reinforcement according to the details (the number of layers will be important)
- Preparation of the micro-concrete/cement mortar mix following the trials. Admixtures may be required
- Placement, plastering, and pressing
- Proper curing
- Checking
- Storing/handling carefully
- Despatch

Source:

- Ganguly [13]

6.14 CURVED AND OTHER SHAPES OF FERRO-CEMENT UNITS TRIED IN PROJECTS

For curved and other shapes, refer to Figures 6.5 and 6.6:

6.14.1 Different Shapes of Ferro-Cement Units Tried

CIRCULAR

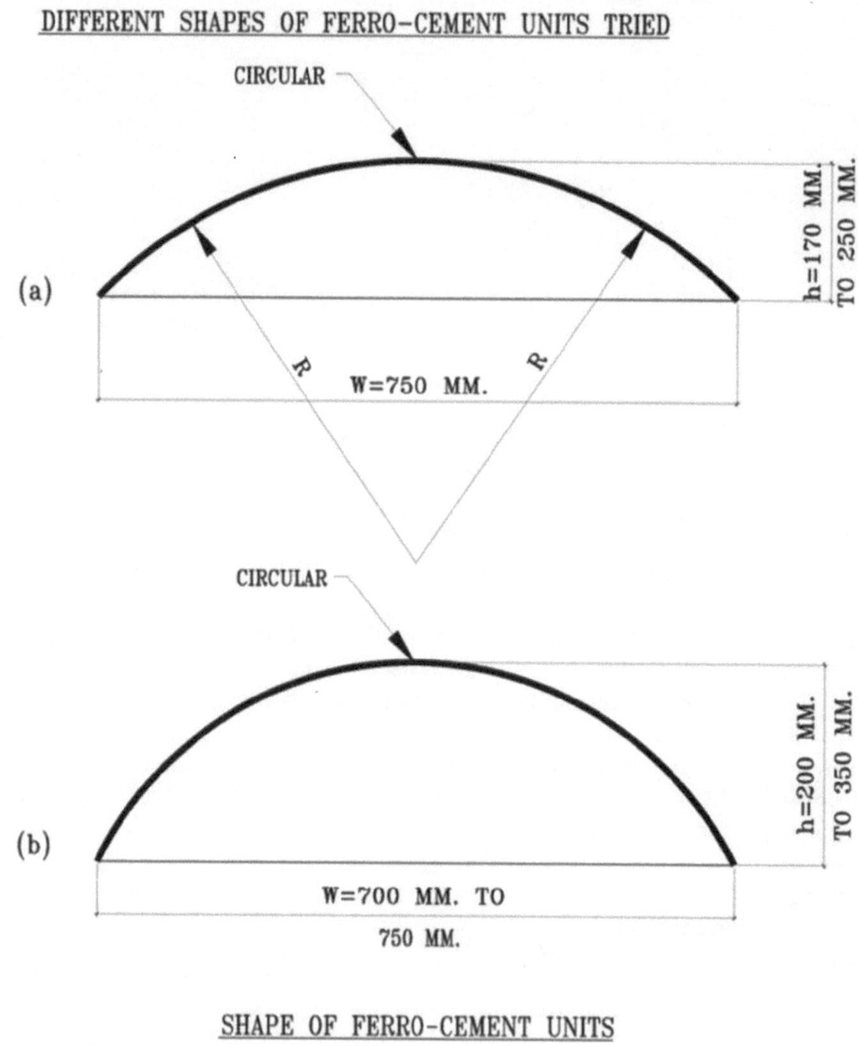

FIGURE 6.5 Shapes of ferro-cement units.

PARABOLIC

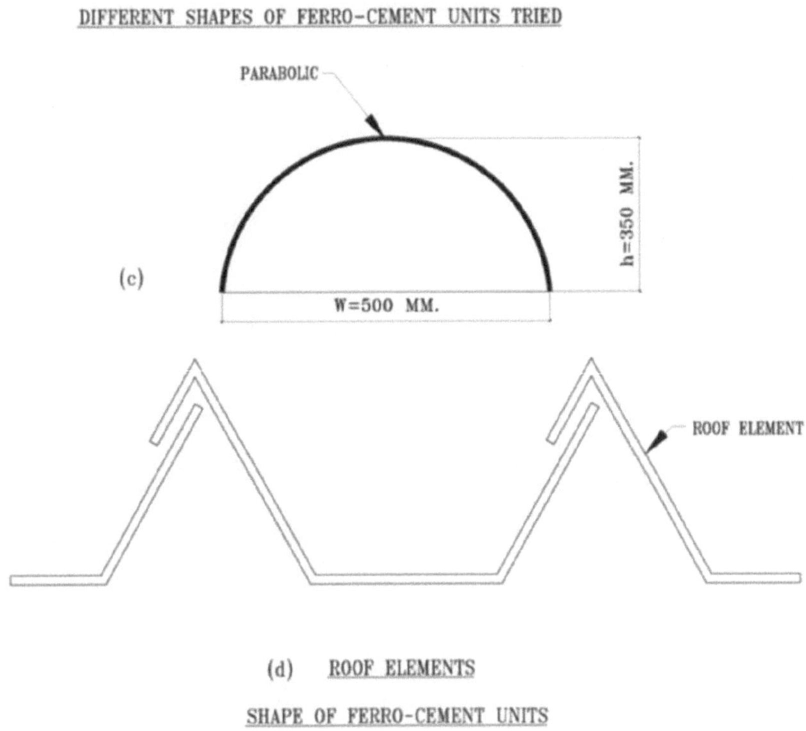

FIGURE 6.6 Roof elements: shapes of ferro-cement units.

The above sections have mostly been used for covering roofs. However, some applications of the units have been made on floors, filled with mud bricks or brick jelly between the units.

Further Details, the dimensions are indicated in millimetres (Figure 6.7).

The units have been used in the roofs.

Source:

- Ganguly [13]

6.15 REVIEW OF REINFORCEMENT DETAILS FROM PROJECTS

6.15.1 General Items

Some basic reinforcement details with mesh used in different projects are indicated below:

6.15.1.1 Reinforcement Arrangement

For structural applications, welded wire mesh 0.5–1.5 mm in diameter and set 12 mm apart may be used (Figure 6.8a). For less structurally demanding situations, 22 gauge 12 mm chicken wire mesh may be more cost-effective (Figure 6.8b).

The cover provided to the outermost layers of mesh is generally very low and normally not greater than 5 mm.

6.16 REINFORCEMENT DETAILS OF CURVED SECTIONS, IN PROJECTS

The applications of curved sections as applied in projects are briefly mentioned below. The shape and reinforcement details are shown in Figure 6.9. These are mostly used in roofs.

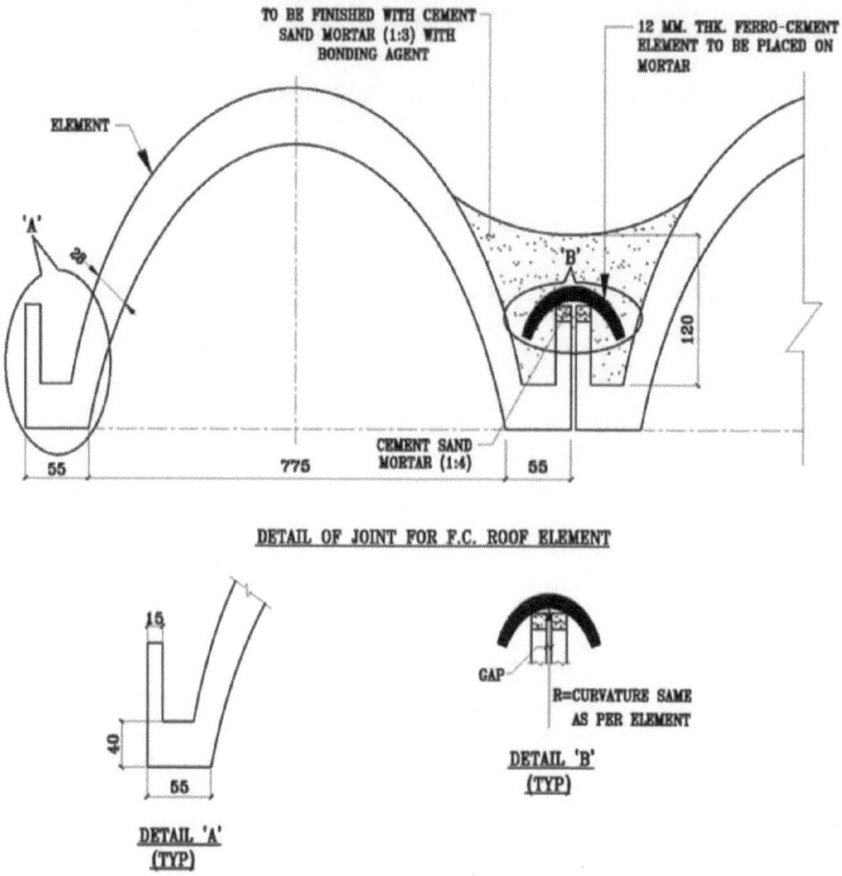

FIGURE 6.7 Details of joints in ferro-cement roof elements.

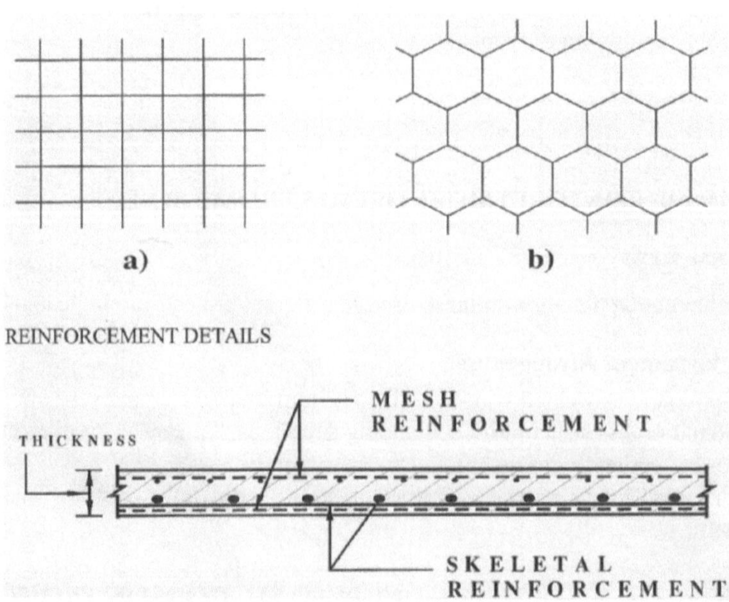

FIGURE 6.8 Reinforcement types. (Source: Ganguly [13]).

6.16.1 Ferro-Cement Details, Curved Section

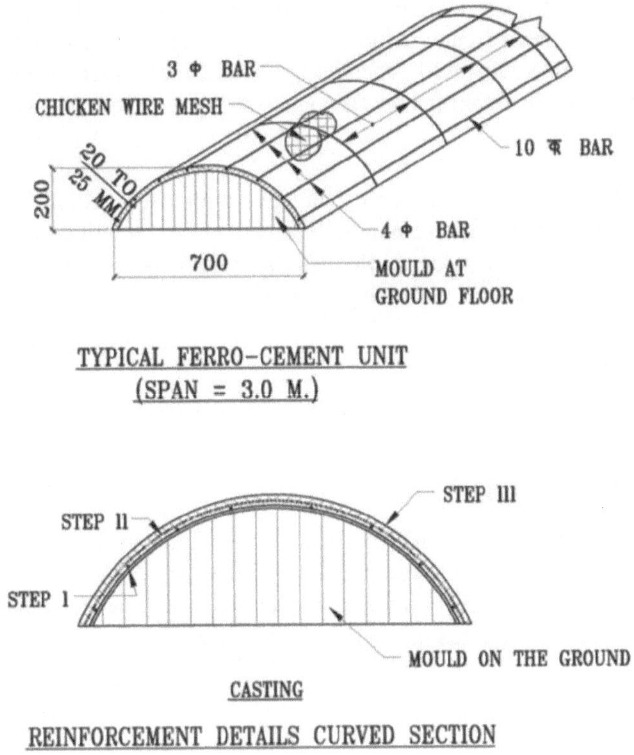

FIGURE 6.9 Casting steps for reinforcement details of curved sections. (Source: Ganguly [13]).

6.17 MECHANICAL PROPERTIES OF FERRO-CEMENT

The material properties of ferro-cement may be obtained from standard books. However, a review will be made on some of the basic items briefly summarised below:

6.17.1 Behaviour under Compression [8]

Cement mortar will control the properties of thin ferro-cement elements in compression. The inclusion of wire mesh in the element will increase the stiffness, but the increase may not be sufficient to be considered in the design. When the reinforcement is very high in the section, the contribution of reinforcement may be taken into account in the design.

6.17.2 Behaviour in Tension [8]

Comments may be made on three stages of behaviour as described below:

6.17.2.1 First Stage

The composite material will behave like a linear elastic solid. The elastic behaviour stage will change with the development of the first crack in the material.

6.17.2.2 Second Stage

This may be referred to as the multiple cracking stage. This stage will start with the development of the first crack in the material and will continue up to the extent when the wire meshes will start to yield. In this stage of loading, the number of cracks will increase with the increase in tensile stress or strain.

However, the widths of the cracks will not increase significantly. With the increase of loads, the distribution will occur among a greater number of cracks instead of widening the existing ones. The number of cracks and their widths will depend on various factors, such as the type and amount of steel. The stage with multiple cracking may be considered as service or working load conditions.

6.17.2.3 Third Stage

This may be referred to as the crack widening stage, with almost no additional cracks developing. At this stage, with loading, the reinforcement will start to yield and will continue up to failure. The number of cracks will remain the same, while the crack widths will increase. In this condition, the behaviour will be controlled by the reinforcement alone.

6.17.3 Fatigue Loading [8]

The fatigue strength of the reinforcement will control the tensile fatigue strength of ferro-cement.

6.17.4 Fracture Strength [8]

The tensile strength of the reinforcement will depend mainly on the volume of reinforcement and the tensile strength of the mesh. Other factors, such as the type of sand, sand-to-cement, and water-to-cement ratios, will have very minor contributions to the tensile strength of ferro-cement. The situation may be different for hexagonal or chicken wire mesh.

6.17.5 Behaviour under Flexure [8]

The behaviour under flexure may be understood by studying the load–deflection curve of a ferro-cement member.

The behaviour may be described as follows:

- Pre-cracking
- Post-cracking
- Post-yielding stage

The evaluation of strength may be assessed in terms of resisting moment, stiffness, and measured for deflection and cracking behaviour for different loading stages.

6.17.5.1 Pre-cracking

Ferro-cement will exhibit high stiffness in the pre-cracking stage. The mortar will provide both compressive and tensile resistance for the composite material. The strengths and stiffness of the elements may be computed using conventional bending methods.

6.17.5.2 Post-cracking

As mentioned, the post-cracking will begin with the development of the first crack and will continue until the extreme tension fibre of the reinforcement starts to yield. The load–deflection curve for the ferro-cement element during this stage may be studied. The element will be cracked in the tension zone; however, the stress in the extreme fibre will remain within the yield strength.

6.17.6 Behaviour under Impact Loading [8]

The impact resistance will increase linearly with the increase in the volume fraction of reinforcement.

It has been observed that welded wire mesh may provide higher impact resistance for the material. It seems that the elements reinforced with chicken wire mesh may provide the lowest impact resistance. Woven wire mesh reinforcement may offer impact strength higher than that of chicken wire meshes, but less than that of welded wire mesh.

6.17.7 Comments on Deflection [15]

It appears that ferro-cement members reinforced with skeletal and mesh reinforcement show higher deflection than those reinforced with mesh only.

6.17.8 Comments on Corrosion [15]

Adequate mortar protection for reinforcement will be important. Insufficient mortar protection for reinforcement may not be adequate to prevent corrosion in harmful environments. The use of a lower water-to-cement ratio along with entrained air can increase resistance against seawater. Protective treatments may be considered in some cases.

6.17.9 Further on Impact Loading [15]

It has been observed that members or panels with a higher specific surface area and strength of meshes may provide better impact resistance.

6.17.10 More on Fatigue [15]

The fatigue strength of ferro-cement is connected with the fatigue properties of reinforcement, including both skeleton and wire mesh. Hexagonal wire mesh can provide more resistance to fatigue compared to welded wire mesh and expanded metal. It seems from tests that ungalvanised mesh reinforcement can provide more fatigue resistance than galvanised mesh.

6.17.11 Creep [15]

Ferro-cement elements demonstrate creep effects under sustained loading on deflection characteristics under pure bending. It has been observed that the composite will exhibit creep characteristics of a metal rather than those of the mortar. For thin elements with more uniformly distributed reinforcement, the behaviour will differ for creep and shrinkage compared to ordinary reinforced concrete.

6.17.12 Performance in Fire [16]

Considering the sections are thin, there may always be a concern about the performance of ferro-cement in fire, as previously mentioned. However, it seems that ferro-cement boats perform better in fire compared to steel boats.

Source:

- Swamy [8]
- Paul and Pama [15]
- ACI 549 [16]

6.18 REVIEW ON DURABILITY OF FERRO-CEMENT [17]

Ferro-cement is made of reinforced composite material, prepared with cement–sand mortar and fine diameter continuous reinforcement as mentioned.

6.18.1 Review of Materials for Construction

Cement, fine aggregates (i.e., sand), water, admixtures, corrosion inhibitors, steel/welded wire fabric, woven wire mesh, protective coatings, sealants, other finishing/decorative items as applicable.

6.18.2 Problems with Durability

6.18.2.1 Resistance against Aggressive Environment

Internal reasons such as

- Cover of the member is insufficient, causing damage to the cover
- Alkali–aggregate reaction
- Permeability of the mortar
- Changes in volume due to differing thermal properties of sand and cement paste

External reasons for damage to cover

- Weathering, differential temperature, fire
- Chemical actions, aggressive environments
- Mechanical abrasion
- Physical defects – cavitation, etc.

6.18.3 Improvement of Durability

The items may include

- Better-quality cement has to be used.
- Well-graded sand as per Zone II with a maximum aggregate size of 2.5 mm must be used.
- The water-to-cement ratio should be kept between 0.35 and 0.4.
- Sand should be cleaned and properly washed.
- Admixtures may be used to increase workability.
- Zinc galvanised wire meshes and bars should be used.
- The use of mechanical vibrations and pressure compaction will aid in the proper placing of the material.
- curing is essential.
- For better permeability and high strength, epoxy mortar or epoxy impregnation of cement–sand mortar may be considered.
- Accelerated curing, such as steam curing or vacuum processing, may be implemented.

Ferro-cement is a low-cost and suitable material that has been used successfully in various countries alongside masonry and conventional reinforced concrete structures.

Source:

- Kaushik [17]

6.19 SUMMARY OF ADVANTAGES [7,18,19]

The following items may be noted:

- Lightweight, homogeneous, and versatile material [7,18] Ferro-cement units are lightweight and homogeneous, making them easy to work with. Any shapes may be created according to requirements.

- The basic raw materials will be easily available in most places [19].
- Maintenance costs will be low [18].
- Long-term performance [18].
- A good construction method is essential, and sufficient care will be required for execution.
- The pieces will remain together even if they leak/damage.
- The construction method can be learned quickly.
- Local labour can be employed after some training.
- Only low labour skills are needed [19]
- Experience in removing the units from the formwork for different shapes will be important.
- **Increase of Bond Strength –** By using small diameter bars and wires, the bond strength between steel and concrete is improved. The strength of concrete/mortar is also high for ferro-cement.
- **Bond Area Increase** [7] – By increasing the bond area, the adhesion between steel and mortar/concrete improves and the units become strong in tension.
- **Dispersion of Steel Wires** [7] – The layers of wire meshes and small diameter bars are well covered with mortar/concrete during production. This enhances the performance of ferro-cement units in flexure, tension, crack, and impact resistance.
- **Crack Control** [7] – Wire meshes are closely spaced and well covered with mortar/concrete. This will help control cracks in the elements.
- **Equal Strength in Both Directions** [7] – The wire meshes are placed evenly in both directions to achieve equal strength in both directions. The units will have resistance against diagonal tension from shear.
- **Formless Construction** [7] – Tied mesh reinforcement and small diameter bars can bond with cement concrete/mortar when pressed and filled into them. The production of ferro-cement may be completed without the use of shuttering/formwork. However, shuttering is often used for optimal performance.
- **Strength through Shape** [7] – Ferro-cement units may be prepared in various shapes to perform for a particular span and meet the standard requirements for slenderness/buckling and general stability.

  The wall thickness for the units may be 25–50 mm as mentioned.
- **High Strength-to-Weight Ratio** [7] – Thin-walled units of ferro-cement are light and possess high strength. The strength-to-weight ratios are high, and the elements are capable of bearing higher loads.
- Can perform well against earthquakes [19].

Source:

- Wikipedia [18]
- WRD Handbook [7]
- The Constructor [19]

6.20 DISADVANTAGES [18,19]

The items may include

- The operation is labour intensive; hence, it may be expensive where labour charges are high.
- There are chances of damaging the reinforcement if air voids are left in the original construction.
- The elements may be damaged by sharp objects [19]
- To remove voids in the wet stage of construction, methods such as vibration, pressurised spraying, or other techniques may be employed [18].

- Treatments for reinforcing bars may be necessary to control rust and cracking.
- Corrosion of reinforcement may develop due to incomplete protection of the metal by the mortar [19]
- Surface treatment may be required for protection and waterproofing.

Source:

- Wikipedia [18]
- The Constructor [19]

6.21 MORE ON THE WEAKNESS OF FERRO-CEMENT ELEMENTS [13]

All ferro-cement members will be thin and, therefore, the protection against fire may not always be very satisfactory. Suitable treatments for fire protection may need to be considered for the elements.

Furthermore, specific attention may be required to meet shear and deflection needs. Proper detailing will be essential to satisfy the requirements for wind and seismic effects.

The treatment for joining the elements, particularly for applications in roofs, may pose problems. Some special care or detailing may be necessary to achieve satisfactory results and control leakage.

Reinforcing bars may be galvanised or coated to increase the durability of the members for some structures. This is particularly important when the cover over the reinforcing bars is minimal. Various protective treatments may also be considered.

Waterproofing treatment may be required on the roof for protection and to achieve better results.

Source:

- Ganguly [13]

6.22 REVIEW OF CASTING OF ELEMENTS [13]

6.22.1 General

The casting of the elements may be done on the floor for many projects as per the steps shown below:

The casting area is pre-selected and prepared on the ground. The moulds are set up on the ground for different requirements/applications. Photographs of the moulds are included.

6.22.2 Mix Proportion

A cement–sand mixture in proportions of 1:2 to 1:2.5 has been used for ferro-cement elements. Other comments have been made separately.

The following construction steps are from a project

- **Step-1** – The mortar bed is first done by hand over the mould on the ground.
- **Step-II** – The reinforcing cage along with wire mesh is placed on top of the material (i.e., the mortar bed from Step 1).
- **Step-III** – The final layer (i.e., mortar/concrete) is applied by hand, and the total thickness is checked and controlled.
- Curing is to be done properly.
- Items are to be stored properly.
- Handling is to be done carefully.

Source:

- Ganguly [13]

6.23 SUMMARY OF MATERIALS USED FOR FERRO-CEMENT MEMBERS

6.23.1 Basic Items for Ferro-Cement [7]

These may be mentioned as follows:

- Skeletal steel in the form of steel bars, welded wire fabrics, etc.
- Steel wire meshes for preparing the cages
- Rich cement mortar for the matrix, commonly referred to as micro concrete
- Finishing
- Curing

6.23.2 Proportioning of Cement Mortar

Generally, rich cement mortars with a mix proportion of 1:1.5 to 1:4 by volume may be used for ferro-cement work. The water requirement may increase with the sand content to maintain consistent workability. Trial mixes may be necessary to achieve a strong, dense mortar consistency that can easily penetrate the layers of mesh provided in the element. The fineness modulus of sand, as well as the water-to-cement and sand-to-cement ratios, should be checked and determined beforehand. The issues of shrinkage may be mitigated to a large extent by placing reinforcement throughout the section of the ferro-cement member. Considering the method of application of the mortar, its plasticity – particularly the placement and pressing – will be very important.

The slump for a stiff cement mortar mix should not exceed 50 mm.

The compressive strength of the cement mortar mix at 28 days should not be less than 35 MPa.

6.23.3 Review on the Guidance for Proportioning and Mixing of Cement Mortars

The following may be noted:

- The stiff consistency of the mortar can be observed by forming a ball of mortar in the hand; when tossed, it should retain its shape.
- Another check would be when a trowel is placed in the heap of the mix; it should be able to stand erect.
- To check the silt content in sand, the material may be mixed with water in a glass jar. After shaking the jar well, allow it to settle. The thickness of the layer of silt that collects above the sand will indicate the approximate percentage of silt content in the sand.

Source:

- WRD Handbook [7]

6.24 COMPARISON OF FERRO-CEMENT AND RCC CONSTRUCTION [7]

The comments listed below in Table 6.1 may be made on some salient items:

6.25 DESIGN GUIDANCE

Some items on design referenced from the ACI code [16].

The design of ferro-cement elements subjected to bending and axial loads may be carried out similarly to conventional reinforced concrete beams or column members, following the limit state method. The relevant guidance for the limit state method of design may be derived from the IS code.

TABLE 6.1
Comparisons

Serial No.	Item	Ferro-Cement	Reinforced Cement Concrete
1	Composition	Almost homogeneous	Nearly heterogeneous
2	Gain of strength	Due to shape of the structure	Due to shape, size, and reinforcement
3	Structural behaviour	Non-rigid	Rigid
4	Thickness	25–50 mm	Minimum 75 mm (say)
5	Material matrix	Rich cement mortar	Cement concrete
6	Reinforcement	Continuous fine wire placed throughout the structural elements	Steel bars > 6 mm and placed distance apart
7	Strength	High tensile strength, better bond, and shear strength	Weak in tension, shear, and bond
8	Tensile strength	80–90 kg/cm^2	4–6 kg/cm^2
9	Strength-to-weight Ratio	45–90	15–50
10	Casting procedure	Filling mortar will be dense and compact, the chances of honeycombing will be less	Formwork will be required, more honeycombing may be present
11	Further on casting/ making	Local labour may be used	Experienced persons will be required
12	Handling	Being thin and delicate, special care will be always required for handling	General care will be necessary
13	Appearance	The elements look very attractive in elevations. Pleasant appearance	Elegant structures may be designed and constructed

Source: WRD Handbook [7]

The members must satisfy both adequate strength and serviceability requirements.

The width and spacing of cracks will be less in ferro-cement elements compared to reinforced concrete under service load conditions [16].

When skeletal reinforcement is used, in the design, care should be taken to ensure that it does not occupy more than 50% of the thickness of the ferro-cement composite [16].

Better results may be expected when the reinforcement is distributed throughout the thickness of the member and by increasing its specific surface [16].

6.25.1 Comments on Mesh Reinforcement [16]

A minimum of two layers of mesh reinforcement will be preferable in ferro-cement elements. Greater advantages may be obtained when more than two layers of mesh reinforcement are provided in the member.

The recommended mesh opening should not exceed the thickness of the ferro-cement element.

6.25.2 Comments on Cover [16]

The cover to the reinforcement should be approximately twice the diameter of the wire mesh or the thickness of other reinforcing bars used in the design.

A smaller cover may be provided when the reinforcement is protected by treatment and will not corrode, with the crack width limited to 0.05 mm.

It has been observed that ferro-cement units less than 25 mm thick performed well with 2 mm cover.

6.25.3 Comments on Crack-Width [16]

For service loads and non-corrosive conditions, the crack width should be less than 0.10 and 0.05 mm for corrosive environments and water-retaining structures.

Source:

- ACI 549 [16]

6.26 REVIEW AND FURTHER COMMENTS ON DESIGN METHODS [20]

The design of ferro-cement members subjected to axial loads and bending may be carried out using

- **Working Stress Method of Design –** For service load conditions

  or
- **Ultimate Strength Method of Design –** For ultimate load conditions

The necessary guidance may be available from the IS code.

6.26.1 Design Assumptions for Ultimate Strength Design

The main items are as follows:

- The maximum strain at the extreme compression fibre will be 0.003.
- The tensile strength of mortar is nominal in flexural strength calculations.
- The compression block may be modified by an equivalent rectangular block.
- The maximum stress in the fibre mesh will be the yield stress of the wires.

6.26.2 Load Factors

Different load factors, as applicable, may be obtained from the IS code.

6.26.3 Relevant Design Parameters

These are listed below:

- The thickness of the elements should not be less than the mesh opening.
- Skeletal reinforcement should not occupy more than 50% of the thickness of the ferro-cement composite as mentioned.
- For non-pre-stressed water-retaining structures, the total volume fraction of steel should not be less than 3.5%. The total specific surface should not be less than 0.16 mm^2/mm^3.
- For improvement, the total steel should be distributed in the element in as many layers as possible.
- The minimum number of mesh layers in the design of the members should be two.

Source:

- Sehgal [20]

6.27 CANAL LINING WITH FERRO-CEMENT [21]

6.27.1 Introduction

The design of canal lining is very important, and ferro-cement can play a significant role. The guidance for the design and selection of the type of lining is provided in IS 13430. The use of ferro-cement as a type of lining has been included in the IS code. It will be necessary to review the guidance quickly in the code before proceeding further with the canal lining.

The critical items are listed below:

- Canal lining can improve flow characteristics and reduce water losses due to seepage.
- The saved water may be used and will be helpful for irrigation purposes.
- A lined canal can also reduce the cross-sectional area of the canal and the land width requirements. Furthermore, it can lower the cost of excavation and related masonry works.
- The lining can enhance command and provide a larger working head for power generation.
- Proper assessment, along with a technical-economic justification for the lining, must be conducted before deciding on the lining.
- Availability of materials, overall costs, and savings in seepage and head, etc., will all be relevant.
- Performance information on different types of lining must be studied and assessed accurately before making a decision on a particular type.
- Comments on the selection of the type of lining are included in the code.
- The purposes of different types of lining are included in the code.

6.27.2 Necessary Information

The items are as follows:

- Capacity
- Longitudinal section along the canal alignment
- Full supply level, bed level, bank level, etc.
- Cross-sections
- Road/railway crossings, drainage crossings, pedestrian/cart tracks, etc.
- Sub-soil water level
- Salt content of the soil
- Availability of skilled/unskilled labour
- Use and availability of construction machinery
- Cattle traffic
- Local and climatic conditions
- Availability of construction materials
    - Cement
    - Building Lime
    - Fly ash as pozzolana
    - Coarse and fine aggregates
    - Stones of proper size and quality
    - Suitable soil for burnt clay bricks/tiles
    - Pulverised fuel–ash–lime/clay fly ash building bricks

6.27.3 Requirements of Lining

The relevant items will be as follows:

Economy, strength and durability, structural stability, easy maintenance and repairability, increased hydraulic efficiency, resistance to erosion, impermeability, ability to control weed growth, resistance for burrowing animals, and moderate flexibility.

6.27.4 Functions of Lining

The necessary items to be considered are as follows:

Seepage control, increase of hydraulic performance, prevention of water logging, better resistance to erosion/abrasion, reduced cross-sectional area, smaller structures, and low maintenance and operation costs.

6.27.5 Different Types of Lining

The following items are as per the code:

6.27.5.1 Rigid Type of Lining

Stone-pitched lining, burnt clay tile or brick lining, pre-cast cement concrete/stone slab lining, stone masonry lining, cement concrete tile lining, in-situ cement/lime concrete lining, burnt bricks or pulverised fuel ash–lime bricks or burnt clay, soil cement/soil cement and fly ash lining, ferro-cement lining, shotcrete lining, and asphaltic cement concrete.

6.27.5.2 Flexible Lining

Geo-membrane, fibre-reinforced plastic tissue as phallic membrane, bituminous or bituminous/asphaltic felt lining, and composite membrane/rubber lining.

6.27.5.3 Combination Lining

Membrane treatment in the bed with brick/tile or concrete lining on the sides.

6.27.6 Selection of Lining

The selection will depend on the type of subgrade, level of the water table, availability of construction materials, climatic conditions, time schedule, speed of construction, and performance of lining in the existing canal and in the surrounding areas.

6.27.7 Canal Cross-Sections

The different types of cross-sections may be listed below:

Trapezoidal, rectangular, parabolic, cup-shaped, polygonal, semi-circular, and other sophisticated shapes.

Parameter for Design of Lined Canals

- Side slopes
  - Inner slopes of lined canal
  - Outer slopes of lined canal
- Freeboard
- Berm
- Dowla
- Bank top width
- Typical cross-sections
- Roadway and drainage
- Cross-section, discharge, and velocity

Under Drainage
Expansive soil

Source:

- IS 10430 [21]

6.28 EXPERIMENTAL WORK WITH FERRO-CEMENT LINING [22]

6.28.1 Testing of Ferro-Cement Canal Lining

The efficient use of canal lining will be very important to save water lost due to seepage and other wastes as mentioned. In this connection, and to improve the use of canal linings, ferro-cement units had been tested at the site for the study. The study was primarily intended for small-sized canals.

The units were fabricated and transported to the site for testing.

6.28.2 Details

Shape – Ferro-cement units were trapezoidal in shape
Dimensions – 900 mm, at the base
1,000 mm, unit length
810 mm, depth
25 mm, thickness
Reinforcement – Skeletal reinforcement, 6 ⌀ @ 260 mm c/centre to centre, longitudinal 6 ⌀ @ 300 mm centre to centre, transverse plus two layers of chicken wire mesh.
Mortar mix = 1:3
Water-to-cement ratio = 0. 5
Normal curing
The **sides were plastered**.

The surface was made proper and the correct thickness was provided.

6.28.3 Tests

The following tests were used:

6.28.3.1 Seepage Loss Test

The test was carried out in a canal. After placing and lining the units in position, water was filled to a depth of 200 mm, and checks were made after 12 hours (i.e., checking after the period from 6 p.m. to 6 a.m.). The test was conducted continuously for 15 days.

Assuming no loss of water due to evaporation, it was found that the level of water remained the same; that is, there was no loss of water from the treatment with lining. It may be noted that ferro-cement lining treatment is seepage proof.

6.28.3.2 Erosion Resistance

Three ferro-cement panels measuring 300×300×25 mm were cast for the test. Skeletal reinforcement using 6 mm diameter bars along with two layers of chicken wire mesh was employed, as mentioned. After proper curing for 28 days, the units were subjected to a jet of water with a velocity of 1.8 m/s for 100 hours.

After the tests, the specimens were sun-dried, and the weight losses were computed.

The percentage of weight loss was about 0.3% (max).

6.28.3.3 Flexural Test

The size of the panels for this test was 600×300×25 mm. The panels had skeletal reinforcement of 6 mm diameter bars along with two layers of chicken wire mesh reinforcement, one layer on each side. The units were subjected to a central concentrated load, and the load–deflection curves were prepared for the specimens. The load–deflection behaviour was linear up to the development of the first crack.

6.28.4 Summary of Comments from the Tests

The comments mentioned below may be relevant:

- There was no loss of water due to seepage from the canal lined with ferro-cement.
- The ferro-cement units subjected to high velocity water showed a greater loss of weight compared to the allowable limit. The mix proportion of the mortar may need to be modified to overcome this problem.
- The units performed well and exhibited a linear relationship following the load–deflection curve until the development of the first crack. After this, the deflection increased rapidly with the increase in load. The cracks were initially fine and expanded quickly with the loads.
- The cost of ferro-cement units has been calculated per metre length and compared with sand–cement block lining and cement concrete lining.
- From the study, it was found that ferro-cement lining units, if adopted, would also be economical.

Source:

- Sekar, Chockalingam, and Sethunarayanan [22]

6.29 HANDLING OF FERRO-CEMENT ELEMENTS [13]

It is always important to handle ferro-cement elements very carefully so that cracks do not develop during handling and execution. It can be easy to overstress the elements during erection operations. A simple erection technique should be aimed for.

Ferro-cement units are tough, durable, and crack-resistant and are expected to perform better than conventional reinforced concrete elements, as already stated.

Source:

- Ganguly [13]

6.30 DESIGN OF FERRO-CEMENT WATER TANKS [23]

Brief comments will be made on the design items.

6.30.1 Introduction

Different shapes for the tanks may be considered. Ferro-cement tanks may be suitable for low income rural areas due to

- Simple skills necessary
- Commonly available materials will be used
- Self-help utilised
- Simple equipment required for execution
- Repetition of formwork and costs may be shared

The cylindrical tanks will be easy to construct. The critical aspect will be the joint between the floor and the wall. The joint must be made strong, and proper reinforcement detailing must be done to avoid leakage.

The conventional design procedure may be followed for the design of the tanks. The tensile load in the element must be carried by the reinforcement alone as provided. IS guidelines may be used for the design.

The items related to the design for the tank will be as follows:

6.30.1.1 Foundation

The floor slab, in smaller tanks, is constructed continuously with the walls. For larger tanks, the floors may be constructed separately from the walls with a separate foundation for the walls.

6.30.1.2 Walls

The walls at the base may have either a sliding or rigid joint. Proper detailing and sealing will be necessary for the sliding joints.

For smaller tanks (i.e., less than 5 m in diameter), the floors are constructed first, and the walls are made continuous with the floor/foundation.

6.30.1.3 Roofs

The roofs may be constructed with lightweight materials for larger tanks. For smaller tanks, they may be made using the wall. For smaller tanks, a cover/lid may be made that will be supported on the walls. A photograph is included to show a ferro-cement tank with a lid.

6.30.1.4 Tank Fittings

The necessary inlet and outlet openings may be provided in the walls as per the requirements. The outlet pipe may be placed a small distance above the floor. The pipe openings may be covered with a layer of fine mesh to control mosquitoes/insects from entering the tanks.

6.30.1.5 Maintenance

Inspection and maintenance will be necessary as per good engineering practice to extend the service life of the tanks.

Source:

- Watt [23]

6.31 GUIDANCE FROM IS CODE [24]

6.31.1 Pre-cast Ferro-Cement Water Tanks

The IS code is meant for ferro-cement water tanks up to a capacity of 10,000 L. The comments will be made on the following items from the IS code:

6.31.1.1 Shape

Different shapes may be considered. For tanks more than 2 m in diameter, a shallow spherical dome may be provided at the base, or suitable fillets may be considered at the junction of the vertical wall and bottom slab.

6.31.1.2 Tolerances

- Length, breadth, height, and diameter up to 1 m – ±5 mm
- Length, breadth, height, and diameter above 1 m – ±10 mm
- Thickness – ±2 mm

6.31.1.3 Materials

- **Cement** – Shall be as per IS 269/IS 1489/IS 455
- **Sand** – Sieved and conforming to Zone-II of IS 383 may be used. The maximum particle size should be less than 1.18 mm. The sand must be clean, inert, and free from organic materials. The contents of silt and clay should be less than 3%.
- **Water** – As per IS 456.

- **Skeletal Steel**
  - Mild steel as per IS 432.
  - Hard-drawn steel shall be as per IS 432 (Part-2)
  - Hard drawn steel wire fabric shall be as per IS 1566

Skeletal steel may be used for the framework of the structure and for placing the mesh in layers. The diameter of skeletal steel reinforcement shall range 3 to 10 mm.

Reinforcement must be free from loose mill scales, coats of paint, loose rust, oil, mud, or other coatings that may harm the bond development with concrete/mortar. The reinforcing bars shall not be visibly pitted by rust.

6.31.1.4 More on Reinforcement

Mild steel wire shall conform to IS 280:1978 for wire mesh. The mesh may have hexagonal, rectangular, or square openings and may be woven or welded. The wires may be galvanised before weaving. An aperture size of 6–20 mm and wire diameter of 0.56–1.25 mm are commonly used. For welded wire mesh, the diameter of the wire shall not be less than 1 mm.

When galvanised wire mesh is used with mild steel bars/wires, chromium trioxide should be added at a concentration of 200–300 ppm to the water used for the preparation of mortar.

6.31.1.5 Admixtures

Admixtures may be used in the mix to reduce the water-to-cement ratio without affecting workability while providing the required strength and durability. Admixtures and integral cement waterproofing compounds, when used, shall comply with their respective IS codes.

6.31.1.6 Design

6.31.1.6.1 Mortar

The minimum compressive strength of cubes with a face area of 50 cm^2, made of cement mortar, shall be 25 N/mm^2. The mix proportion may consist of one part cement to 1.5–2.5 parts sand by mass.

The water-to-cement ratio should be maintained within 0.35–0.45.

The tensile stress in reinforcement for service conditions shall not exceed 200 N/mm^2.

The minimum cross-sectional area of main reinforcement in any one of the two principal directions shall not be less than 1.0% of the gross cross-sectional area of the element.

Laps in wire mesh shall not be less than 100 mm.

Skeletal steel may be placed with spacing not exceeding 300 mm c/c in both directions. The laps in bars shall not be less than 150 mm.

6.31.1.6.2 Wall Thickness

The minimum wall thickness shall be 12 mm for tanks with a capacity of up to 2,000 L. For mechanised or semi-mechanised casting and for tanks with a capacity of up to 1,000 L, the minimum wall thickness may be 15 mm with hand casting. For tanks with larger capacities, the wall thickness may range from 20–40 mm depending on design requirements.

The minimum thickness for a lid or cover shall not be less than 15 mm.

When the bottom slab thickness exceeds 30 mm, the slab should be constructed in ferro-cement only.

6.31.1.6.3 Cover

The minimum clear cover to reinforcement shall be 4 mm.

Source:

- IS 13356 [24]

6.32 BRIEF REVIEW OF FERRO-CEMENT TECHNOLOGY APPLIED FROM PROJECTS

6.32.1 Introduction

Ferro-cement units made of cement, fine aggregates, and reinforcing bars can be used successfully in the building industry. The thickness of the units is small compared to conventional pre-cast concrete sections. Their applications may extend to marine, housing, water resources, sanitation, grain and water storage, biogas structures, and the repair and strengthening of structures.

Ferro-cement units are tough, durable, and crack-resistant, and are expected to perform better than conventional reinforced concrete elements in certain applications. A few photographs are included to cover different applications.

6.32.2 Mortar Mix

The mortar mix should be designed for maximum strength, density, and impermeability, with sufficient workability to minimise voids. A low water-to-cement ratio, not exceeding 0.4–0.45, is generally recommended for hand plastering techniques, combined with a low aggregate/cement ratio, commonly 1.5:1 or 2:1. The aggregate will usually be sharp-grade sand, mixed with ordinary Portland cement, which will generally be adequate. By using suitable admixtures in the mix, it may be possible to achieve the correct type and strength of concrete if required.

Further on reinforcement for structural applications

For structural applications, welded wire mesh 0.5–1.5 mm in diameter and set 12 mm apart can be used. For less structurally demanding situations 22-gauge 12-mm chicken wire mesh may be a more cost-effective solution.

6.32.3 Cover

The cover provided to the outermost layer of the mesh will be very low and generally limited to about 3 mm, and normally not greater than 5 mm.

6.32.4 Casting

The comments on casting have been made separately.

The thickness of the ferro-cement elements will be less than stated. This will mean that there will be little room for error in workmanship, and considerable care in construction will be essential to maintain quality.

Source:

- Ganguly [13]

6.33 SUMMARY OF INFORMATION ON FERRO-CEMENT FROM PROJECTS

Reference may be made to Figure 6.10.

6.34 AN EXPERIMENTAL STUDY OF FERRO-CEMENT BEAMS [25]

Twenty ferro-cement beams were tested under simply supported conditions.

A two-point loading system was applied to the test beams to obtain flexural behaviour/results, i.e., the beams would be under pure bending. The beams were tested using a universal testing machine.

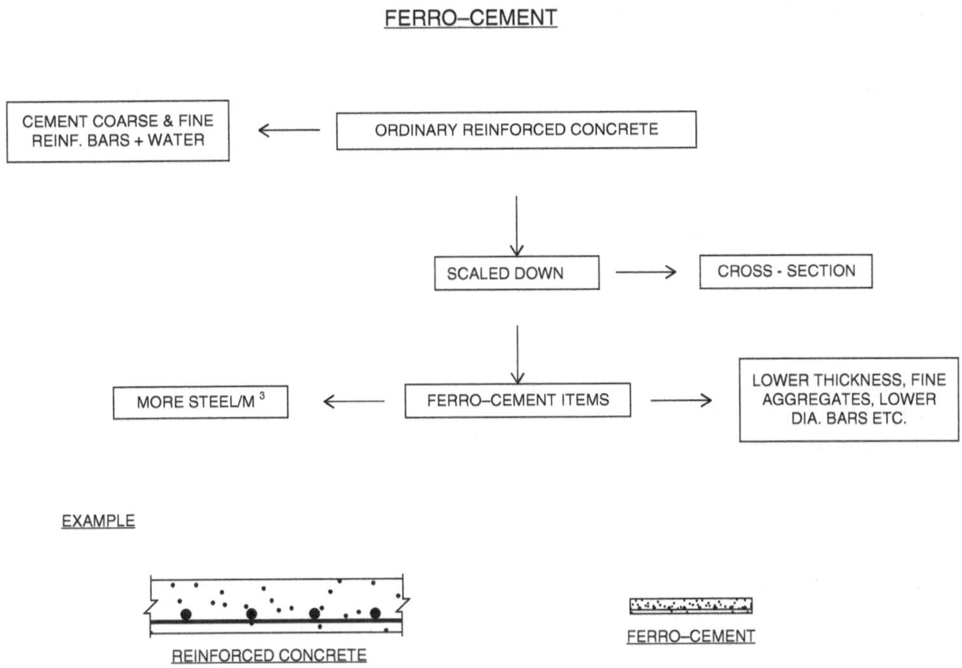

FIGURE 6.10 Ferro-cement. (Source: Ganguly [13]).

Different types of beams were tested, covering I, T, L, U, and box sections. The lengths of the specimens were 3,000 mm, with a thickness of 30 mm for the webs and flanges. The widths of the flanges were 100 mm, and the depth varied from 100 mm and above.

6.34.1 Reinforcement Details

The steel area calculations included the use of both steel bars and welded wire mesh in the beams.

The steel bars were 8 mm in diameter, with wire mesh of 1 mm in diameter and wire mesh openings of 225 mm^2 used.

6.34.2 Materials

- **Cement –** Ordinary Portland cement, 43 grade.
- **Fine Aggregates –** Locally available river sand was used as the fine aggregate.
- **Water –** Clean and fresh water was used for the preparation of mortar and for curing.
- **Skeletal Steel –** HYSD steel grade Fe-415 for 8 mm bars with a tensile strength of N/mm^2
- **Ferro-Cement Reinforcing Mesh –** Galvanised welded steel wire mesh of 1 mm diameter with square grids was used for beam construction. The grid size was 10×10 mm, and the yield strength of the wire mesh was 300 N/mm^2.

6.34.3 Mix Proportion

- **1:2.5 Cement–**sand mix was used

6.34.4 Construction of Ferro-Cement

- Preparation of the skeletal frame work
- Fixing of the wire mesh
- Placing of the mortar/plastering
- Curing

6.34.5 Composite Properties

- **Thickness** – 30 mm
- **Cover to Reinforcement** – 1.5 mm
- **Ultimate Tensile Strength** – 34.5 N/mm^2
- **Composite Strength** – 27.6–68.9 N/mm^2
- **Allowable Tensile Strength** – 10.3 N/mm^2
- **Modulus of Rupture** – 55.1 N/mm^2
- **Cube Strength of Mortar** – 29.9 N/mm^2

6.34.6 Comments from Tests

These may be listed as follows:

- The flexural moment capacity of the ferro-cement beams was found to be within the limits of the calculated ultimate moment value.
- Ferro-cement beams may be used for lintels, floor beams, and temporary structures.
- Problems may be faced with ferro-cement beams regarding deflection. This may be addressed by providing stiffeners at suitable locations.

Source:

- Kanthe and Dhake [25]

6.35 FAILURE OF A FERRO-CEMENT BEAM [26]

The failure of a rectangular ferro-cement beam is presented. The beams were tested with two layers of wire mesh on the surface areas, with skeletal fabricated reinforcement. Cement mortar was applied to the bottom and sides of the beams, while the top was left open. The beams were taken to the site after adequate strength development, placed on supports, filled with structural concrete, and finished with cement mortar and wire mesh layers on the top. The beams were tested for bending, crack width, shear, and deflection. One beam failed in shear.

Comments from tests are as follows:

- The design was adequate, and sufficient reinforcement was provided in the member.
- The shear strength was higher than the applied load.
- Sufficient care should be taken regarding workmanship and proper compaction.
- The water-to-cement ratio should be maintained very low to prevent surplus water in the material, thus minimising corrosion.
- There was separation of concrete from the external cement matrix and the wire mesh layer.
- The shear force was resisted by the cement matrix, wire mesh layers, core concrete, and the shear reinforcement provided.

- The failure occurred sequentially, likely by snapping of the wire mesh, failure of the cement matrix, snapping of the shear reinforcement, and then failure of the core concrete, one after the other.
- The failure progressed gradually.
- Only one beam failed in shear out of 152 beams tested.

It seems the bonding of layers, if applicable, will be important for a ferro-cement member to act compositely against the applied loads.

Source:

- Desai [26]

6.36 COMMON APPLICATIONS

These may include -

Roof and floor construction for various spaces – different stages may be adopted, water tanks of different sizes, doors and windows of different designs including frames, partition walls in a modern office, shelf units, thin slabs used for tables and seating purposes, paving slabs, wall panels to protect embankments of a water body, boundary walls, hand railing for staircases, special architectural details, tiles for various uses, tiles for cable protection, guard rooms/cabins, solar power stations (26 kW) – Sunderban Area, West Bengal.

Source:

- Ganguly [3,13]
- Walkus and Kowalski [4]
- Barfoot [5]
- WRD Handbook [7]

6.37 REPAIR AND MAINTENANCE OF FERRO-CEMENT [16]

Some items on the repair of ferro-cement members from the ACI code will be briefly mentioned below:

6.37.1 Introduction

The repair work for ferro-cement members will be similar to that of conventional reinforced concrete elements. However, a few items related to the repair work for ferro-cement will be reviewed.

Due to variations in temperature and humidity, ferro-cement structures are likely to be affected and damaged.

Maintenance will effectively mean identifying voids and repairing/filling them, replacing the spalled cover concrete, providing suitable protective coatings, and rectifying surface blemishes.

6.37.2 Stain and Blemish Removal

6.37.2.1 General

The stains are not likely to penetrate deeply into the mortar material as the surface will be less porous. The quality of the mortar cover to the reinforcement will be very relevant.

6.37.2.1.1 Formation of Blemishes

This may be mainly associated with poor workmanship, use of faulty materials, uneven evaporation, inadequate curing, etc.

The other reasons may be

- **Cement –** sourced from different suppliers
- **Polythene Sheets –** their use may result in uneven curing
- **Mottling –** presence of calcium chloride in the mix, inadequate curing, etc.
- **Hard Steel Trowelling –** the use may cause rapid drying
- **Water-to-Cement Ratio –** controlling this will be relevant

6.37.2.1.2 Stain Removal

General Comment – This problem may be addressed by thorough flushing and brushing using a stiff bristle brush.

6.37.3 Efflorescence

The treatment may be carried out by breaking into voids and re-plastering. Otherwise, the repair may be done by drilling and injecting non-shrink cementitious grout.

6.37.4 Protective and Surface Treatments

General – A good-quality mortar is expected to have better resistance to weathering.
Coatings – Epoxy and polyurethane materials may be considered.
Hardeners – The use of sodium silicate is often recommended.
Sheathing – Fibreglass laminates had been used in boat hulls.

6.37.5 Repair Work for Damage

General – The repair work must be carried out quickly.

6.37.5.1 Types of Damage

Spalls – Can occur when a portion detaches from the main section due to various reasons such as environmental effects, pressure, or expansion within the affected areas. Spalls are typically caused by corrosion of steel in ferro-cement.

De-laminations – This problem may arise when ferro-cement layers split. This may occur near the neutral axis of the member and under the action of impact loads due to the presence of voids in the interior layers.

Scaling – This may be a local effect and refers to the flaking or peeling of the material near the surface of the mortar.

Cracks and Local Fractures – These may result from changes in temperature, drying shrinkage, and hairline cracks and crazing may occur. Normally, these may not require any repair.

Fire Damage – Ferro-cement boats have performed better in fire compared to steel materials, as previously mentioned.

Evaluation and Testing – The evaluation and the reasons for damage are important for assessing the extent and cause, which may still be active. Different testing methods may be considered.

6.37.6 Surface Preparation

General – To remove the deteriorated or contaminated mortar.

Removal of Deteriorated Concrete – To remove damaged or contaminated mortar. Sound mortar in the mesh need not be removed.

Reinforcement – Loose, scaly, corroded areas are to be cleaned, and the mortar may be removed by sandblasting, water jet, or vacuum technique.
Cleaning – Loose particles and dust must be removed.
Cracks – It may be easier to remove or clean the mortar by carefully hammering on each side of the cracks and re-plastering with the mortar using latex.

6.37.7 Repair Materials

These may include

- Portland cement
- **Polymer Mortars –** Different types are available.
- **Admixtures –** Various types are available for applications according to requirements.
- Other items as applicable

6.37.8 Repair Procedures

The items/procedures include

- Mixing
- Partial depth repair/patches
- Full-depth repair
- Temporary propping as required
- Overlays
- Shotcreting
- Curing

Source:

- ACI 549 [16]

6.38 CASE STUDIES

Two case studies will be included as given below:
These are listed below:

6.38.1 Protection to Embankment of a Water Body by Using Ferro-Cement Elements [3]

The case study presented below is associated with the protection of the embankment of a water body with the help of ferro-cement units.

6.38.2 Introduction

The rehabilitation work on the embankment of a water body has been presented. Detailed inspections were carried out several times for the assessment. Due to the effect of wind, the embankment of the water body started eroding over time. The nature of erosion and extent of damage was not uniform.

6.38.3 Layout

The layout of the water body may be seen in Figures 6.11 and 6.12. It was important to retain the embankment and to maintain a uniform surface profile of the bank to the water body. The rehabilitation work is related to the protection of the embankment and to retaining its uniform smooth

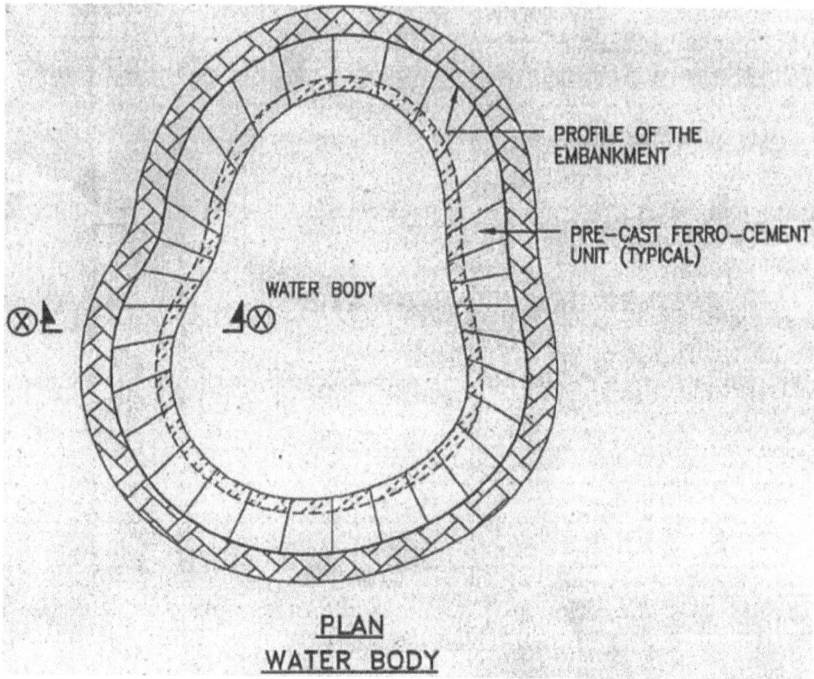

FIGURE 6.11 Plan of water body. (Source: Ganguly [3]).

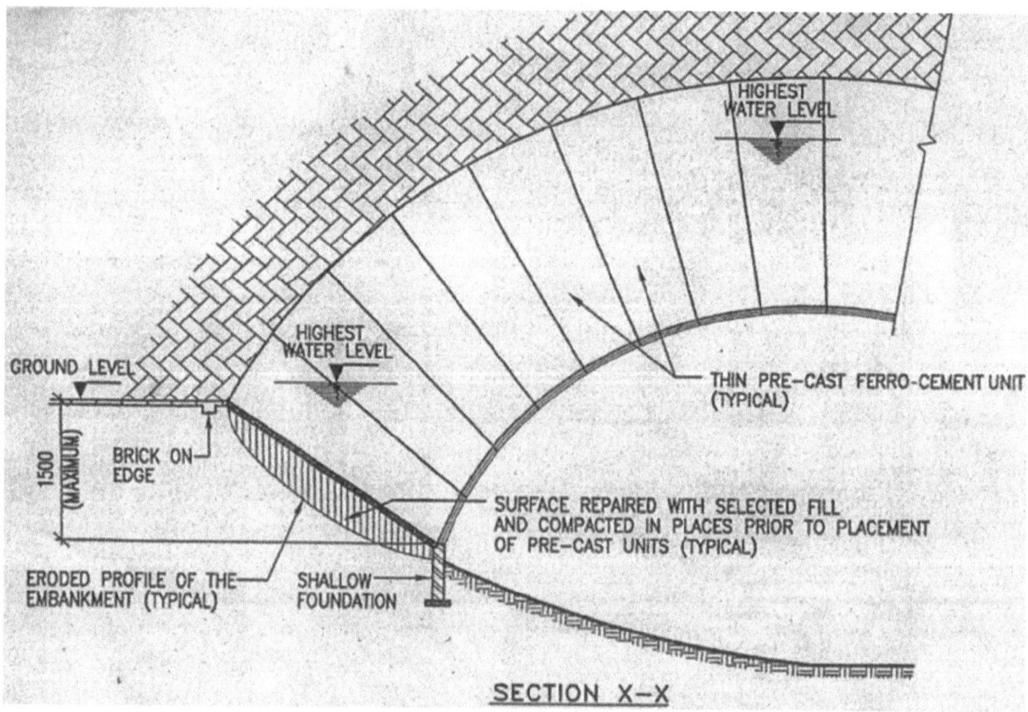

FIGURE 6.12 Water body section. (Source: Ganguly [3]).

profile as assumed initially in the design. For successful repair work, studies of various alternatives had been carried out to arrive at a scheme that would be cost-effective and easy to execute on site. The application of pre-cast ferro-cement units was adopted for the repair work.

6.38.4 Repair Scheme with Pre-cast Ferro-Cement Elements

The scheme with pre-cast ferro-cement elements is presented below:

The pre-cast concrete planks were cast on the banks at three to four locations covering the total area of the water body. The total perimeter of the water body will be about 1.2km. The shallow foundation with brickwork was done first to receive pre-cast ferro-cement units. The thickness of the units was 40–50mm. The pre-cast planks were allowed to slide down from the top of the embankment slowly into their respective positions. In some cases, in-situ concrete had to be used due to site problems. The space between the pre-cast planks and the eroded embankment was filled and compacted with selected fill. Two photographs have been included to present the profile of the embankment. A few photographs have been presented for the details and the embankment after the repair work (Figures 6.11–6.14)

6.38.5 Concluding Comments

A higher thickness was required for in-situ ferro-cement work in some areas. The scheme developed for the protection was found to be cost-effective and very efficient. It was a simple scheme, and the work could be initiated simultaneously from different positions. Placement of pre-cast planks did not pose any problems. The water body had to be blocked in certain areas during the construction of the supporting foundation. The embankment appeared neat and attractive once all the planks were in place. It was possible to maintain the profile of the embankment effectively by positioning pre-cast planks at a slant close to the angle of internal friction of the soil. Even after a couple of years, it can be said that the profile remained reasonably good.

The above procedure may be followed for repairing or retaining existing earth embankments where similar erosion problems exist.

Proper inspection and maintenance work may be required from time to time to extend the service life.

FIGURE 6.13 Embankment after restoration work. (Source: Ganguly [3]).

FIGURE 6.14 Embankment after restoration work. (Source: Ganguly [3]).

6.38.6 Photographs

A few photographs are presented for the protection work. These are shown in Figures 6.11–6.14.

6.38.6.1 Pre-cast Ferro-Cement Shell Roofing for Low-Cost Houses and School Buildings

The case study related to the applications of pre-cast ferro-cement shell roofing for low-cost houses and school buildings will be briefly presented.

The Structural Engineering Research Centre (SERC) in Ghaziabad developed roofing units for large-scale production and use. The units were tested in the laboratory and were found to be suitable for low-cost roof applications.

One of the developments included a pre-cast segmental shell unit, which had been used to cover a large shed, and its performance was observed to be satisfactory without the need for any maintenance work.

The units were produced on-site using local materials and labour.

Normal applications of low-cost materials may include the use of asbestos–cement or galvanised iron sheets.

The disadvantages of these materials include

- Short span
- Noise during rain
- Poor insulation
- Transportation from market points

Ferro-cement elements may be preferred due to

- Aesthetic and insulation qualities
- The casting of the units can be done locally near the site using local labour.
- Coverage of large areas is possible without intermediate supports.
- Transportation of raw materials may be easier.
- Simple erection techniques and assembly.

Salient features of the roofing system

Three types of buildings with different plan sizes may be mentioned.

6.38.6.2 First Category

This is meant for the economically weaker section having plan dimensions of 2.86×8.9 m. For this type, eight pre-cast ferro-cement segmental shell units with an effective span of 2.63 m may be used.

6.38.6.3 Second Category

This may be used for the construction of school buildings in plains. The plan dimensions for the building roof may be 8.59×12.69 m. For this category, 13pre-cast ferro-cement segmental shell units using a combination of effective spans of 5.13 and 3.13 m may be used.

The drainage of rainwater may be achieved by providing an adequate slope.

6.38.6.4 Third Category

This type of school building was meant for construction mostly in hilly areas with exposure to heavy rain and snow. Buildings with a plan size of 9.4×2.42 m have been developed with a double-pitched roof of segmental shells to cover a span of 6.46 m.

6.38.7 Design

The design has been done for a live load of 1 kN/m^2 to cover different spans. The reinforcing arrangement includes skeletal steel bars in longitudinal and transverse directions, along with two layers of wire mesh. The design has been done following the respective IS codes.

6.38.8 Construction

For the purpose of demonstration, two units with spans of 5.3 and 3.3 m were constructed on the ground. The units were supported at both ends and were loaded and tested using sand-filled bags. Loosely filled sand bags were used to manage the curved surface to provide a uniformly distributed load condition for the units.

The units were able to carry a load of 1.5 kN/m^2. There were no cracks and no noticeable deflections after seven days of construction.

6.38.9 Concluding Comments

The ferro-cement shell roofing system designed and developed by the Uttar Pradesh Rural Housing Board (UPRHB) has been found to be a cheaper alternative to the conventional materials that are normally used. By adopting a suitable shape for the basic roofing elements, it may be possible to design a ferro-cement system to cover large areas.

Source:

- Sharma and Murugiah [27]

6.39 REVIEW OF SOME IMPORTANT OVERSEAS APPLICATIONS [3]

These are presented in Table 6.2.

6.40 CONSTRUCTION PHOTOGRAPHS

Various applications such as roofing units, dome structures, water tanks, and pre-cast ferro-cement boundary wall elements have also been constructed in India using ferro-cement. A few photographs are presented to show different applications (Figures 6.15–6.29).

TABLE 6.2
Review of Some Overseas Applications

Serial No.	Item	Comments
1	SYDNEY OPERA HOUSE, exterior [4] by M/s. Ove Arup & Partners	Exterior tile-faced ferro-cement pre-cast units
2	PALLAZETTODELLO SPORT IN ROME [4] By Prof Nervi	Ferro-cement Pre-cast elements as shuttering to the cast-in-situ ribs and topping
3	Hong Kong's first ferro-cement [4] fishing vessel	Hull thickness was about 35 mm
4	Ferro-cement sailing boat in New Zealand [4]	
5	Ferro-cement fishing fleet in Cuba, boats in China [5]	
6	Boats with ferro-cement in the UK [9]	73′-0″ long
7	Grain Silo [6] Mexico City	Hand-plastered and chicken wire mesh
8	Two ferro-cement schooners [9], UK	74′-0″ long
9	White ferro-cement roofing units [3,5], domes, wall units, India	48′-0″ long
10	Sewer re-lining [5], UK	Sprayed in-situ ferro-cement lining and pre-cast segmental lining

Source: Walkus and Kowalski [4]; Barfoot [5]; Jennings [6]; Two ferro-cement cement schooners [9]; Ganguly [3]

FIGURE 6.15 Grain silo structure, Mexico. (Source: Jennings [6]).

FIGURE 6.16 Floating dock. (Source: Jennings [6]).

FIGURE 6.17 Round-shaped structure with an outer layer of ferro-cement. (Source: Ganguly [13]).

FIGURE 6.18 Round-shaped structure with an outer layer of ferro-cement. (Source: Ganguly [13]).

FIGURE 6.19 Ferro-cement elements. (Source: Ganguly [13]).

FIGURE 6.20 Ferro-cement elements in the building. (Source: Ganguly [13]).

FIGURE 6.21 Ferro-cement elements. (Source: Ganguly [13]).

FIGURE 6.22 Ferro-cement units. (Source: Ganguly [13]).

FIGURE 6.23 Ferro-cement roofing units and wall elements. (Source: Ganguly [13]).

FIGURE 6.24 Ferro-cement door. (Source: Ganguly [13]).

FIGURE 6.25 Ferro-cement door, special type. (Source: Ganguly [13]).

FIGURE 6.26 Ferro-cement and CEB combination. (Source: Ganguly [13]).

FIGURE 6.27 Ferro-cement and CEB combination. (Source: Ganguly [13]).

FIGURE 6.28 Ferro-cement and CEB combination. (Source: Ganguly [13]).

FIGURE 6.29 Ferro-cement water tank. (Source: Ganguly [13]).

REFERENCES

1. Concrete @ Your Finger tips, The Concrete Society, pp. 1. https://www.concrete.org.uk/fingertips.asp. Accessed Oct 04, 2024.
2. Ganguly K K, Composite Materials with Particular Reference to Concrete, Geomate- 2017, 21–24 November 2017, TSU city, Mie, Japan.
3. Ganguly K K, *Reinforced Concrete Design, Repairs/Restoration, Strengthening of RC Structures and Concrete Technology*, Scientific International, January 2020.
4. Walkus R and Kowalski T G, Ferro-cement: a survey, *Concrete*, February 1971, Vol. 5, pp. 48–52.
5. Barfoot J, Fine forecast for ferro-cement, *Concrete*, May 1985, Vol 19, pp. 25–26.
6. Jennings P J, Ferro-cement a structural material, *Concrete*, January 1981, pp. 20–24.
7. Maharashtra Engineering Research Institute. Ferro-Cement Technology, WRD Hand Book Chapter No. 1, Govt. of Maharashtra, Water Resource Department, Nashik, pp. 1–13.
8. Swamy R N, *Concrete Technology and Design (Vol. 2), New Reinforced Concretes*, Surrey University Press, 1984.
9. Anon., Two ferro-cement schooners, *Concrete*, March 1973, Vol. 7, pp. 32–33.
10. Pemberton M and Pullan M, Ferro-cement-Have we ignored the potentials of ferro-cement?, *Concrete-Camberley*, July 2012, Vol. 46(7), pp. 42–43.
11. Nedwell P, The use of ferro-cement, *Concrete*, August 2010, pp. 52–54.
12. Ganguly K K, Durable concrete will be more suitable for different applications in projects, *Indian Concrete Journal (ICJ)*, July 2017, Vol. 91, pp. 80–96.
13. Ganguly K K, Some Design Issues & Alternative Building Materials - A Brief Review, Innovative World of Building Materials, State of the Art Lecture, 29th National Convention of Architectural Engineers Division, The Institute of Engineers' India, 31. 01. 2014.
14. Singh S K, Kirthika S K and Surya M, Agenda for use of alternative sand in India, *ICI Journal*, October-December 2018, pp. 34–43.
15. Paul B K and Pama R P, Ferroement, International Ferro-Cement Information Center.
16. Batson G B et al., ACI 549. 1R-88, Guide for the design, construction, and repair of ferro-cement, *ACI Structural Journal*, 1988, Vol. 85(3), pp. 325–351.
17. Kaushik S K, Durability of Ferro-Cement, Institution of Engineers (India), Belapur Local Centre - Workshop on Ferro-Cement Technology, July 1996, pp. 83–101.
18. Wikipedia, Ferro-Cement, pp. 1–4.
19. What is Ferrocement? Applications and Advantages of Ferrocement in Construction, The Constructor, Civil Engineering Home, pp. 1–15. https://theconstructor.org/concrete/ferrocement-in-construction/1156/. Accessed Oct 11, 2024.
20. Sehgal V K, Design Criteria for Ferro-Cement, Institution of Engineers (India), Belapur Local Centre - Workshop on Ferro-Cement Technology, July 1996, pp. 77–81.
21. Bureau of Indian Standards, IS 10430:2000/2009, Criteria for Design of Lined Canals and Guidance for Selection of Type of Lining.
22. Sekar R, Chockalingam S and Sethunarayanan R, Ferro-cement for canal lining, in *Proceedings of the Third International Conference on Ferrocement*, Civil Engineering Department, University of Roorkee, 1988.
23. Watt S B, *Ferro-Cement Water Tanks and Their Construction*, Intermediate Technology Publication, 1978.
24. Bureau of Indian Standards, IS 13356:1992, Pre-cast Ferro-cement Water Tanks Up-To 10,000 Litres Capacity – Specification.
25. Kanthe V N and Dhake P D, An experimental study of ferro-cement beams, The *Indian Concrete Journal*, August 2013, Vol. 87, pp. 55–62.
26. Desai J A, Failure of a ferro-cement beam, in *Proceedings of the Fifth International Symposium on Ferro-Cement*, 1994, pp. 197–202.
27. Sharma P C and Murugiah A, ICI Bulletin, March 1993, pp. 35–39.

7 Fibre-Reinforced Concrete

7.1 DISCUSSION ON FIBRE-REINFORCED CONCRETE MATERIALS

7.1.1 Introduction

It has been observed that the use of composite materials in concrete structures has increased considerably in recent years, covering different applications.

Composite materials have been used by engineers for a long time [1–4]. The technique has been successfully applied in the past by mixing straw to reinforce bricks and using metal reinforcement on the tension face of concrete [3,5]. It has been found that asbestos fibres were used to reinforce clay posts about 5,000 years ago [3].

Early humans initially used wood, rocks, and bones efficiently in their fight for existence against natural and various other types of forces. At that early stage, people used these materials to make weapons, many utility articles, and tools and to build their living spaces. At the beginning, humans used the materials in their original form and gradually learned to use them more efficiently by modifying them. Later, other materials such as vegetable fibres, clay, shells, horns, teeth, skin, and sinews of animals were also tried. In this process, humans learned, and composite materials were developed gradually [4].

Steel reinforcing bars have long been used in concrete to address the tensile stresses in structural concrete elements. The use of different types of fibres, along with reinforcement, has been successfully attempted to overcome the weaknesses of concrete. Polymer composites are multiphase materials produced by combining polymer resins such as polyester, vinyl-ester [1,5], and epoxy (thermo-set) with polyethylene and polypropylene (thermo-plastic), along with fillers and the reinforcing fibres, to create a material with far superior mechanical and chemical properties than the individual base materials combined. The key components of composites include reinforcing fibres, the matrix, and the interface [1,6].

The idea of enhancing the physical properties of concrete by adding fibres originated mainly from developed countries, including the USA. The emphasis was on high strength, low self-weight, and durability. The topic of composite materials for concrete is modern and expansive, making it difficult to cover all areas without omitting some important aspects in the process. Therefore, only brief comments are provided on most items considered relevant in this publication.

By selecting two or more materials with their individual characteristics, it may be possible to achieve advantages through proper combinations [6].

The focus of this publication is on the use of fibre-reinforced concrete (FRC) as a composite material. However, comments on both metallic and non-metallic reinforcement in concrete have also been included, which will be elaborated upon later in this chapter.

The fibres play a crucial role in providing stiffness and strength to the composite material [1]. Load transfer from the matrix to the reinforcement occurs through a proper bond between the matrix and reinforcement substances, such as fillers, which may be used to reduce costs and enhance dimensional stability [6].

The fibres may be continuous or discontinuous in fibre-reinforced composites. Other matrix materials may include polymeric, metallic, and ceramic options [6].

Structural design with composite materials must consider the properties of the composite itself. These properties can be varied to meet specific structural requirements. Properties such as stiffness and thermal expansion may be changed to a large extent depending on the requirements [6].

The main list of fibres that may be used includes the following [1,2].

DOI: 10.1201/9781003319108-7

7.1.2 Asbestos, Glass, Plastic, Aramid, Carbon, Steel, Synthetic Fibres, and Others

Further comments on different fibres have been made separately.

Concrete with fibres is suitable for applications where large amounts of energy need to be absorbed, such as in explosion loading [2]. The composite material will have high tensile strength and will reduce cracking compared to conventional reinforced concrete [2].

Further, fibres will improve the impact strength of concrete and limit the development of cracks, enhancing the strain capacity of the composite material after the cracking of the cement paste matrix [2].

There are various issues related to using different types of fibres. Experience and specialised knowledge will be required for their proper applications [1,2,7,8].

Source:

- Hollay [1]
- Neville [2]
- Mehta and Monterio [3]
- Composite Materials and Structures [4]
- Humphreys [5]
- TIFAC [6]
- Ganguly [7]
- Ganguly [8]

It will be interesting to know what different authors have commented after carrying out studies/applications/tests at different times on the same topic and to examine the developments that took place later. Efforts have been made to collect information from various publications, journals, books, etc. Following this approach, some issues may be mentioned repeatedly.

7.1.3 Broad Classification of Composites

The types may be [9]

Synthetic Composites	–	**Tyre** – Steel or polymer fibres and rubber matrix **Fibreglass** – Chopped fibres and polyester **Fibre-Reinforced Plastic (FRP) Composites** – Panels, beams, columns, etc.
Natural Composites	–	Wood: cellulose fibres and lignin matrix Bone and bamboo, etc.
Hybrid Composites	–	Timber or concrete elements wrapped with polymer composites Timber beam members reinforced with FRP plates on the tension side Sandwich structures (with face sheets and timber core)

Source:

- Qiao [9]

7.1.4 Fibre-Reinforced Concrete [1–3,6,9]

Concrete containing discontinuous and discrete fibres is known as fibre-reinforced concrete (FRC). The concept of adding fibres to the concrete mix is not new. The use of straw and asbestos fibres to strengthen mud bricks and clay posts [3] has been successfully implemented many years ago, as mentioned.

From studies, it was found that the ultimate tensile strengths of concrete with fibres did not increase much compared to normal concrete. However, the tensile strains at rupture did improve [3]. Furthermore, FRC is much tougher and more resistant to impact and fatigue than normal concrete.

Different tests have been conducted on cement concrete composites with various fibres to study and understand their performance under different load conditions. It has been observed that the addition of any type of fibres will reduce workability [3].

The interface region is crucial for the performance of the composite material. Composite materials with weak interfaces have low strength and stiffness but high resistance to fracture [1], while those with strong interfaces possess high strength and stiffness but are very brittle.

FRC has been applied in various building projects to gain benefits related to crack resistance, better concrete installation, and greater impact and abrasion resistance [1–3,10] performances, etc. Engineers have used fibres in ground floor slab construction with the intention of eliminating the commonly used reinforcement with steel bars. The design may be adapted for slabs supported on piles using steel fibre reinforcement [11]. Other applications will include cast-in-situ free suspended elevated slabs in residential and commercial projects, composite floors on steel decking, retaining walls, bridge slabs, etc. It should be noted that the performance of no two fibres will be the same. The effects of using fibres will depend on fibre type, its dimensions, properties, and the dosage of fibres used in concrete, among other factors.

Source:

- Mehta and Monterio [3]
- Hollaway [1]
- Neville [2]
- Ganguly [8,10]
- Vollum [11]

7.1.5 Fibre-Reinforced Cement-Based Materials

Some comments on the initial studies and observations may be presented below:

7.1.5.1 Initial Studies/Observations [12]

In recent years, new man-made fibres and composites have become available for different applications. The initial observations from studies on FRC may be summarised as follows:

- Fibres can provide directional strength and stiffness with ease of arrangement.
- Polymer composites will be suitable for use to meet the requirements.
- The tensile strength of all fibres will be greater than that of the matrix, considering elongation at break.
- The effect of creep will be small but may still be significant.
- Asbestos, glass, and steel can withstand higher temperatures, while plastic will fail at about 100°C.
- Compaction may present difficult problems.
- Pressure vacuum spinning and extrusion appear to be promising techniques.
- The fibres that may be beneficial for composites include polypropylene, asbestos (for tensile strength), steel (for impact strength and ductility at failure), and glass (for tensile and impact strength as well as ductility at failure).
- The advantages of dispersed fibres, apart from strength, may include increased toughness, crack control, and reduction in component thickness. Cost will be an important factor for consideration.
- The uses of FRC may include pipes, roofing items, cladding, flooring, piles, linings, and some asbestos cement products.

- Furthermore, thin-section products may include floor and wall panels, as well as roof shell components.
- Guniting to walls using polypropylene fibres has been used in the tidal section of a river.
- A realistic assessment will include further laboratory work supported by theoretical studies.

Source:

- Materials Technology Divisional Committee [12]

7.1.6 Further Review/Comments on Concrete Reinforced with Fibrous Wire [13]

7.1.6.1 Previous Study/Comments

The use of fibrous material in concrete has been explored in the past, as mentioned. Some initial observations on fibres will be discussed here. In 1910, it was noted from the USA after a study that an increase in the physical properties of concrete might be possible by including cut nails and spikes in the concrete mixes. In 1914, a patent was filed in New York to cover abrasive-resistant and spalling-resistant concrete. It was indicated that, with more closely spaced wire, it might be possible to achieve greater cracking strength.

The following comments may be noted from different tests as initial observations:

- From volume batching to weigh batching was proposed for chopped wire or lightweight aggregate etc.
- Confirmation/clarification would be necessary for 2%, for example, wire in the mix. It should be clearly stated that 2% of the total weight of the wire will refer to the total volume or the partial volume.
- By including plain wire of 0.25 mm (0.01 in) in diameter and 25.4 mm long in the concrete mix, and using a constant water-to-cement ratio, improved cube and flexural strengths of 10.6% and 53%, respectively, were observed.
- The inclusion of chopped wire in the concrete mix reduced workability.
- In concrete containing a volume fraction of 0.3 or more of coarse aggregate, a limitation of the wire content to around 0.01 of the volume fraction may be necessary to control the agglomeration of the fibres.
- Small cross-sectioned flexural specimens provided a greater increase in strength.
- Improvements were observed in impact resistance and thermal conductivity.
- The use of chopped wire increased the resistance of concrete against freeze–thaw cycles.
- Thin sections showed a significant improvement in strength with the inclusion of wire fibre. This was not observed in conventional cube, cylinder, and beam tests.
- The inclusion of steel fibres in the mix would increase both the material and handling costs of the concrete. A proper assessment will be needed to weigh the cost aspects against the advantages.
- Cost savings by reducing the conventional reinforcement mesh may be achievable through the use of steel fibres in concrete.

Source:

- Dixon and Mayfield [13]

7.1.7 Review on Definition and Strength [3]

7.1.7.1 General Consideration

It has been observed that the use of any type of fibres in the concrete mix will reduce workability, as mentioned.

7.1.7.2 Concrete Mixtures

Following ACI committee recommendations, FRC may be specified by fibre content and strength. The flexural strength may be noted for use in paving and compressive strength for structural applications.

- Fibres of different shapes and sizes [1–3] have been tried with concrete.

From studies and observations, it can be commented that the versatility of concrete can be effectively extended by using fibres in concrete, especially overcoming its characteristic brittleness [3].

The review of important items of FRC may consider the following:

Significance, toughening mechanism, materials and mix proportioning, general considerations, concrete mixtures, properties – workability, strength, elastic modulus, creep and drying shrinkage, toughness and impact resistance, durability, different applications, etc.

Source:

- Mehta and Monterio [3]
- Neville [2]

7.1.8 Further Comments and Details [3,14]

Further brief details on FRC may be mentioned as given below:

- **Workability** – Typical concrete or mortar mixes with fibres will have low consistencies. The compactability and placeability of concrete will be better than with low consistency.
- **Strength** – The contribution of fibre reinforcement to concrete is primarily for the improvement of flexural toughness of the material rather than strength. With the increase of fibre length to a limit, both strength and toughness may increase. The increase in strength may be nominal. When flexural strength is the main consideration, fibre reinforcement in the concrete may not be a suitable substitute for conventional reinforcement.
- **Toughness and Impact Resistance** – In addition to the improved flexural toughness of concrete, the fatigue and impact resistance of concrete will be significantly increased. This improvement has been achieved by using crimped steel fibres instead of indented fibres. The use of nylon and polypropylene low-modulus fibres has proven effective for pre-cast concrete elements subjected to high impact.

  It has been observed that the inclusion of fibres in conventionally reinforced beams has increased fatigue life and, under fatigue loading, decreased crack widths.
- **Elastic Modulus, Drying Shrinkage, and Creep** – The use of steel fibres in concrete has minimal effects on modulus of elasticity, drying shrinkage and compressive creep. Flexural creep has been significantly reduced when stiff carbon fibres are used in the mix. Small effects have been observed on tensile creep, which has also been reduced.
- **Durability** – FRC is often produced with high cement content and low water-to-cement ratios. It has been observed that concrete containing steel fibres, when well compacted and properly cured, will provide good durability performance when the fibres are protected with cement paste. In hostile environments where chloride is present, surface rusting may occur, but the fibres in the interior of the concrete will not corrode.

Following tests, it has been noted that the resistance of concrete reinforced with steel fibres against scour from debris in flowing water did not improve [14]. However, fibrous concrete was found to be durable when the loss of material was due to cavitation impacts from large debris [3,14].

Source:

- ACI Report 544.2R [14]
- Mehta and Monterio [3]

7.1.9 Fibre Potential [15]

The potential for using fibres in concrete has been recognised for a long time, and asbestos fibres were initially used, as stated. Other fibres have also been used, as mentioned.

Polypropylene fibres can contribute important benefits to plastic concrete. The dosage rate will be low, and due to the small nominal size of the fibres in the mix, they may help control plastic shrinkage cracking. Micro-fibres in concrete may not provide any significant improvement in post-cracking structural performance. Macro-synthetic fibres (MSFs) can offer better post-cracking strength than micro-fibres, which can help reduce or eliminate the need for fabric reinforcement in some cases.

With the development of high-tensile steel fibres, it is possible to use longer bay sizes in the design, and even to create joint-less and seamless floor slabs. By applying ultra-high tensile strength wire with different anchoring shapes, it will prevent fibres from being pulled out; the fibres will be very useful when the concrete is highly stressed.

The use of other fibres in concrete, such as basalt fibres and Kevlar fibres, will be discussed in their respective sections.

The benefits of using micro-, macro-, and steel fibres will cover different aspects of projects, such as aesthetics, durability, improvements to traditional design, and cost reduction, especially for pre-cast items. The potential of fibres in design and construction appears to be very promising for the future.

Source:

- Armfield and Birkinshaw [15]

7.1.10 Comments on Fibre Materials [3,16,17]

As already mentioned, different types of fibres have been selected and mixed with concrete, and applications have been made. At the initial stage, stainless steel fibres cut from hard-drawn wire were tried [16]. It has been noted that by using 1%–2% of stainless steel fibres by volume, a substantial improvement in the flexural and compressive strength of concrete was achieved. Non-metallic and natural fibres have also been used with concrete.

Following the report from the US Army Engineering Research [17] and Development Laboratories, the comments may be summarised as follows:

- Glass fibres were found to be the strongest but unsuitable for high pH environments of cement slurry.
- The same was true for cotton, rayon acetate, and Dacron polyester.
- Nylon demonstrated a viscosity effect (water absorption), which may limit the amount of fibres that can be used in concrete.
- Nylon, polyethylene, polypropylene, and saran were sufficiently resistant to reinforce concrete.
- Fibres that are subjected to alkaline degradation, such as cotton, rayon acetate, Orlon, glass, and Dacron (polyester), were found to provide little strength to Portland cement products.

Source:

- Beddar [16]
- Mehta and Monterio [3]
- U S Army, Technical Report [17]

7.1.11 Not All Fibres Are the Same [18]

The fibre reinforcement in concrete will act like conventional reinforcement after the concrete has cracked. It will be important to know at what stage the cracking may occur and how to control it. This will depend on the fibre types, dosage, etc., and should be selected properly for the application. Steel and MSFs may control the cracks in hardened concrete. Micro-synthetic and cellulose fibres will also be effective in controlling plastic shrinkage cracking.

Steel and MSFs will increase the toughness, ductility, and energy absorption capacity of hardened concrete. Both types of fibres may be considered to enhance the load-bearing capacities of concrete members. Furthermore, the fibres may help control drying shrinkage cracking. Micro-synthetic fibres may be useful in managing micro-cracking in the plastic state of concrete, as previously mentioned.

7.1.11.1 The Comments on the Fibres

These may be summarised as follows:

- **No Two Fibres Will Be the Same –** A fibre should be selected for a particular application after studying its properties against the requirements.
- The performance of fibres will depend on several factors, including fibre type, its properties, dimensions, dosage, and the performance of the matrix with the fibre.
- Design guides are available. Reference may be made to Technical Report 63, ACI 544.4 R.
- The test reports or certification will be important. FRC may also be tested to verify the basic/important properties expected for the application. This will be particularly important to satisfy the requirements of serviceability and ultimate limit states.

Source:

- Kirby [18]

7.1.12 Further on Fibre Composite Materials

The loss of workability due to the addition of fibres in the mix will be proportional to the volume concentration of fibres in concrete [3].

A few important issues for fibre-reinforced composites are as follows:

7.1.12.1 Requirements of Fibres [1]

The desirable fibre requirements are as follows:

• Fibre should have a high modulus of elasticity.	• Fibre should have high ultimate strength.

- The diameter and surface of fibres should be uniform.
- The variation in strength between individual fibres should be minimal.
- The fibres should be stable and retain their strength during handling and fabrication.

7.1.12.2 Other Qualities [16]

- Good dispersion within the matrix.
- Compatibility with the binder, which should not deteriorate or be compromised in the long term.
- Adaptable modulus of elasticity (sometimes may be higher than that of the matrix).
- Being short, fine and flexible to allow the mix, transporting and placing to be reasonably easy.

- The consideration of price should take into account the proportion of the mix.
- Being strong, yet adequately robust to withstand the mixing operations.

7.1.12.3 Matrix [1]

The following functions will be important:

- To bind together the fibres and to protect their surfaces from damage during fabrication, handling, and service life.
- Chemically compatible with fibres over a long period.
- To disperse the fibres and separate them to avoid the development of cracks and failure of the composite.
- The composite, when under load, should be able to transfer stresses to the fibres efficiently by adhesion and/or friction.

7.1.12.4 Interface [1]

The interface between the fibre and matrix is very important and must have proper and adequate bonding both chemically and physically. The following items and assumptions are critical for the analysis of the composite material using a mathematical model.

- The matrix and fibres behave elastically.
- The material adjacent to the fibre has the same properties as the material in bulk form.
- The bond between fibres and the interface will be adequate, and there will not be any strain discontinuity across the interface.
- The fibres are arranged in a regular or repeating array.

The salient information on the characteristics of different fibres is given separately.

Source:

- Hollaway [1]
- Beddar [16]
- Mehta and Monterio [3]

7.1.13 FRC – Origin of These Species [19]

7.1.13.1 Review of the Composite Material

A few comments on this issue are as follows:

7.1.13.2 Ancient Recipe

It is reported that the use of composite materials began in Egypt more than 2,000 years ago. The applications were associated with mud bricks reinforced with straw fibres. Following this, the use of asbestos fibre cement was observed and, with the continuous improvement in the utilisation of fibres, more sophisticated carbon-based materials were developed.

7.1.13.3 Fibres

FRC is made of cementitious hydrated cement paste with the inclusion of a small amount of steel fibres as mentioned. With the addition of steel fibres, the redistribution of forces occurs, and the formation and extension of cracks will be restricted. Polymer fibres (which may be thinner than human hair) can be used alongside steel fibres to enhance fire resistance. The fibres will melt under heat, providing microscopic tubes within the concrete for the escape of moisture; otherwise, there may be explosive spalling.

7.1.13.4 Real Advantages

Problems related to shrinkage cracking, and to improve impact and fire resistance, may reduce the requirement for conventional steel reinforcing bars in the member, and these issues may be controlled by FRC. Increased joint spacing may be possible in floor slab design due to better control of shrinkage.

The thickness of the floor slab may be reduced, and the cover layer for the fabric may be eliminated.

7.1.13.5 Installations

The fibres may be mixed into concrete during the batching process to ensure uniform mixing. Dosage machines may be used to put the correct amount of fibres into the batching plant. The typical dosage rate for an industrial floor (joint-less) may be around 30–45 kg/m^3.

For tunnel segments, it may be 25–40 kg/m^3.

7.1.13.6 Typical Applications

The applications have been covered separately.

Source:

- Crowther [19]

7.1.14 Review of FRC – Past, Present, and Future [16]

Concrete may be considered a very suitable man-made construction material, consisting of limestone and siliceous conglomerates.

Concrete is strong in compression and weak in tension, and may be considered a relatively brittle material, as stated.

7.1.14.1 Early Beginnings

It may be mentioned that there is evidence for the asbestos fibres that were used to reinforce clay posts about 5,000 years ago. Furthermore, it seems that straw was used for sun-dried bricks, horsehair for plastering, and asbestos fibres for reinforcing Portland cement materials. G. Graham took the first US patent for steel FRC in 1911. One year later, the second US patent was issued to B. Wearkley.

7.1.14.2 Developments up to 1950

The first British patent was taken by William Fickly in 1914. He commented that toughness and abrasion resistance were increased by including pieces of metal, but there were no improvements in tensile and compressive strengths.

7.1.14.3 Recent Developments

Significant developments in fibre concrete technology took place during the 1960s. Different fibres were introduced for various applications. These may be mentioned as follows:

- Natural, vegetable, or mineral fibres
- Steel fibres (different shapes)
- Alkali- or non-alkali-resistant glass fibres
- Polypropylene fibres
- Carbon fibres
- Other fibres (ceramic, nylon, polyester, etc.)

Various fibre types have been tried with concrete. These may be either continuous or discontinuous. The important qualities will be as follows:

- Good adhesion with the matrix
- Compatibility with the binder, which should not be attacked/damaged over time
- Adaptable elastic modulus
- Sufficiently short, fine, and flexible. Mixing, transportation, and placement should be easy.
- Accessible price
- Sufficiently strong and not to be damaged during the mixing process.

Fibres can be very important in the mix and can sufficiently improve the properties of the composite material. Furthermore, the fibres can provide ductility to concrete/cement mortar and absorb a large amount of energy before failure.

The major items for improvement may be mentioned as follows:

- Resistance to crack propagation
- Ability to withstand large deformations
- Tensile, flexural, and impact strength
- Post-crack residual strength

Further Comments – SFRC has been used in various areas and has been found to be suitable where high strength, abrasion-resistant concrete is required for the projects.

Source:

- Beddar [16]

7.1.15 Steel FRC – Breaking New Ground [20]

With the recent development, steel fibres in concrete are being used extensively for the design and construction of ground floor slabs with a reduced volume of concrete, i.e., without increasing the floor depth, and will have additional insulation to meet the design requirements. Some comments have been made regarding ground floor slab design/construction.

System – Insulslab SFRC may provide super-insulated type construction using steel FRC. The system can be arranged to facilitate a waffle type of floor using interlocking expanded polystyrene pods. FRC with steel fibres will be placed on top to finish the floor slab. Steel fibres of about 40 kg/m^3 may be adequate for ground-supported and pile-supported floor slabs.

Alternative Solution – Using SFRC instead of a traditional reinforced concrete slab will significantly improve the installation process. This method will reduce the volume of concrete and the steel requirements in construction, with the advantage of reduced construction time and cost. Furthermore, it will minimise health and safety risks on site by eliminating the need for specialist cutting tools.

Cap of Piles – When piles are required as per the design, the SFRC flooring technique can be suitably used to satisfy the design needs. This construction method may be considered an improved and efficient one.

By combining SFRC with the traditional pile foundation method of construction, it would be possible to save time on site, which would be very important for any project.

Source:

- Gray [20]

7.1.16 Information on Different Fibres

The information on fibres is given in Table 7.1:

Details on Fibres are mentioned in Table 7.1 [16,21]

TABLE 7.1
Details on Fibres

Fibre Type	Density (kg/m³)	Diameter (mm)	Length (mm)	Elasticity Modulus (GPA)
Steel	7,860	0.2–0.5	20–50	200,000
Asbestos (chrysotile)	2,550	0.0002–0.002	5	80,000–150,000
Carbon	1,900	0.5–1.10^{-2}	Variable	400,000
Glass zirconium	2,700	0.005–0.020	40–70	80,000
Sisal	700–800	0.1–0.5	–	–
Polypropylene	910	0.01–0.1	25–75	4,000–8,000

The design may be performed for slabs supported on piles using steel fibre reinforcement. Other applications will include cast-in-situ free suspended elevated slabs in residential and commercial projects, composite floors on steel decking, retaining walls, and bridge slabs.

A few popular types of fibres have been reviewed in this publication.

An attempt has been made to study a few aspects of FRC to a limited extent, including the applications of fibre-reinforced polymers that have been made in recent years. The advantages of each fibre are mentioned specifically and separately. This way, some common items have perhaps been repeated.

Source:

- Beddar [16]
- Swamy [21]

7.1.17 Fibres – The Traditional Method of Reinforcement [22]

There are various types of fibre that may be considered for use in concrete, depending on the project's requirements. The common types that may be considered for concrete reinforcement are as follows:

- Micro-synthetic fibres
- MSFs
- Steel fibres
- Other types for special needs

Micro-synthetic fibres may be used to enhance the properties of concrete at the initial stage, especially concerning early-age drying shrinkage and settlement cracking. Furthermore, they can influence other properties, including impact, shatter, and abrasion resistance of concrete. In addition, micro-synthetic mono-filament fibres can also control explosive spalling in high-performance concrete when exposed to fire.

Steel fibres can be useful for increasing the residual strength and toughness of concrete in the hardened state and can provide post-crack performance of concrete. The use of steel fibres will also improve the impact and shatter resistance properties of concrete and can act as a more ductile material with high load-bearing capacity. These are often considered for industrial floors and sprayed concrete applications.

7.1.17.1 Reinforcement Options

MSFs may be used for external concrete slabs and pre-cast concrete elements, especially to tackle corrosion problems related to steel fibres and steel fabric reinforcement. MSFs can offer similar benefits to steel fibre reinforcement, with the additional advantage of being non-rusting.

7.1.17.2 Brief Comments

Considering various uses and the potential of fibres, it may be possible to have a situation where fibres are used as reinforcement for concrete in a traditional manner. Ready-mix suppliers may be able to provide concrete that includes fibres as per the requirements. The combination and use of micro-synthetic fibres along with steel or MSFs look promising. Mixing fibres will provide the best possible resistance against cracking throughout the service life of concrete structures.

Source:

- Atkinson [22]

7.1.18 Reasons for Using Composite Materials in the Design [6]

Brief comments, as given below, may be made about the benefits:

- The tensile strength of composites may be four to six times greater than that of steel or aluminium.
- The fatigue endurance limit will be more (up to 60% of the ultimate tensile strength).
- Better impact properties and torsional stiffness.
- Compared to aluminium structures, composite materials will be 30%–45% lighter for structures with the same design requirements.
- Embedded energy will be lower compared to other structural materials like steel and aluminium.
- It is less noisy in operation and will transmit lower vibrations than metals.
- Compared to metals, composite materials are more versatile. Changes and improvements can be made to meet design requirements.
- It can provide a long life and is better for fatigue and environmental resistance.
- The materials will have a reduced life cycle cost compared to metals.
- It is beneficial for corrosion resistance and fire performance.
- FRP may be considered suitable for repair, seismic retrofitting, and strengthening of structural members to extend service life.
- It will be suitable for providing better aesthetics to structural elements.
- It can eliminate the use of joints and fasteners in the design.
- Deterioration due to environmental effects will be less. Composites will be suitable for saline environments.
- It can be considered as a suitable material for safe and economic development.

Source:

- TIFAC [6]

7.1.18.1 Summary of Main Reasons for Using Composites

The main reasons for using composites are listed below:

- Reduction of weight
- Flexibility in design
- Corrosion resistance
- Other reasons as stated above

Source:

- Qiao [9]

7.1.19 Further on Advantages of Using Composites [9]

Some of these are listed below:

• High strength-to-weight ratio (specific strength)	• High stiffness-to-weight ratio (specific stiffness)
• Non-corrosive, non-magnetic, non-conductive	• High fatigue life

- High energy absorption properties: acoustic and seismic performance
- Ability to incorporate sensors in the material into monitor and/or correct its performance. Smart construction/composite
- Ease of fabrication of large complicated structural shapes or modules. Modular type of construction.
- Ability to tailor the material for specific applications and to design the material with other existing properties such as UV resistance, flammability, and smoke toxicity.

Source:

- Qiao [9]

7.1.20 Comments on Disadvantages [9]

• Cost of raw materials and fabrication	• Transverse properties may be weak
• Low toughness and weak matrix	• Environmental degradation of matrix may be possible
• Difficulty in connecting	• Difficulty for the analysis

Source: Qiao [9].

7.1.21 Review on Types of Common Fibres That May Be Considered [9,16]

Numerous fibres have been tested, some of which are already mentioned. The common types are listed below [3,16]:

- Natural vegetable or mineral fibres (sisal, jute, bamboo, and asbestos)
- Alkali- or non-alkali-resistant glass fibres
- Polypropylene fibres
- Steel fibres (straight, hooked ends, indented, enlarged end, melt extract fibres, etc.)
- Carbon fibres, aramid fibres
- **Other Fibre Types** – Ceramic, nylon, polyester.

Various tests have been conducted on cement concrete composites with different fibres to study the performance of these materials under various load conditions.

FRC has been applied in numerous building projects to gain benefits related to crack resistance, improved concrete quality, enhanced installation, increased impact and abrasion resistance [3,9], etc. Furthermore, the items that will be improved with different fibres mainly include the following [3,9,16,23]:

- Fracture toughness
- Tensile, flexural, and impact strength
- Post crack residual strength
- Lower embedded energy
- Spalling of concrete
- Less noisy, lower vibration transmission than metal
- Long life
- Resistance to crack propagation
- Ability to sustain increased deformation
- Resistance to fatigue
- Thermal shock resistance
- Strengthening to improve the capacities of the member
- More versatile
- Improved appearance

Source: Mehta and Monterio [3]; Beddar [16]; Qiao [9]; Ganguly [23].

7.1.22 Further Review on Reinforcing System in Concrete and Composite Action

William Boutland Wilkinson of Newcastle upon Tyne was the first to [24] suggest the use of iron as reinforcement for concrete in 1854, considering the weakness of concrete in tension. In 1892, Francois Hennebique [24], a French contractor, obtained a British patent for his system. His patent was based on plain round bars with fish-tailed ends and stirrups made of flat strips of mild steel. During this time, the construction was generally known as "Ferro-Concrete". Rules were established later in 1894 for design calculations based on the modular ratio concept.

We have come a long way since then, and the quality of reinforcing materials has improved significantly in recent years. Stainless steel bars have also been used in design and construction. The current options for designers regarding reinforcing bars for concrete members may be mentioned as follows:

- **Mild Steel Bars –** Normally not widely used [25]
- Galvanised Bars [8]
- Coated Bars [8]
- **FRP Bars –** FRP is particularly beneficial in highly corrosive environments and electromagnetic areas for applications [8].
- Cold Twisted High-Strength Bars [25]
- Stainless Steel [8]

Stainless steel may be used in areas directly at risk of chloride-induced corrosion. The comments on stainless steel in concrete are included later.

Source: Bussell [24]; IS 456 [25]; Ganguly [8]

7.1.23 Comments on the Hybrid Reinforcement [26]

The use of hybrid reinforcement in design and construction, particularly the application of steel fibres, may be preferred. Applications have been made in industrial floors and pre-cast tunnel segments as stated. Ground-supported floor slabs may experience considerable water pressure at times due to the rise of the water table. The pressure can be particularly high when the structure is located close to a canal, river, or the sea.

This issue can be handled efficiently by incorporating fibres along with the standard reinforcement in the concrete slab. A reduction in traditional reinforcement within the slab may be possible, which can also help control crack development simultaneously. This approach can be described as the conversion of the classical type of reinforcement into hybrid reinforcement by adding steel fibres.

7.1.23.1 Crack Control

The inclusion of hybrid reinforcement in design to control crack development should be considered. The aim should be to limit crack width to some extent. It appears that the use of Dramix 4D fibres may be effective in limiting crack widths to between 0.1 and 0.3 mm. This is crucial for creating a liquid-tight and durable structure. The steel fibres will be evenly distributed throughout the entire concrete mass.

The use of hybrid reinforcement can be effective in design, as mentioned. The application of hybrid reinforcement using fibres may lead to more cracks; however, these cracks will be smaller, finer, and water will not be able to pass through.

More applications have been carried out in walls, ground-supported slabs, walls in underground basements, and car parks, etc.

It is interesting to note that hybrid reinforcement is gaining recognition and being applied in design. This will mean that traditional reinforcement will be considered along with steel fibres in the design. In order to control crack widths, more traditional reinforcement may often be used in the design, which will be more time-consuming, expensive, and may not be easy to place.

By using steel fibres in concrete for design and construction, better performance may be expected.

Source:

- Vanheuverzwijn [26]

7.1.24 Comments on the Perfect Union

Some observations from studies will be made on the following items:

The cracking process with fabric reinforcement only [27]

The items are listed below:

7.1.24.1 Initial State

The cracks will develop in the weakest area of the stressed member. At this stage, the fabric reinforcement will take the stress, and there will not be any contribution from concrete, with the deformation of steel being maximum. Away from this zone, the stresses will be on the concrete. This will mean that the deformation of steel will decrease while the deformation of concrete will increase. At the anchoring length of reinforcement, both concrete and steel will exhibit deformation. The section is not cracked, and there is no slippage between concrete and reinforcement fabric.

7.1.24.2 Final State

In this state, more cracks develop, and the transferred force exceeds the tensile strength of concrete. With more cracks, less anchoring length is left between two cracks. The cracks cannot form when no anchoring length is available, as the increased tensile stress in concrete cannot develop. With further deformations, new cracks will not develop, and the existing cracks will widen. This condition can be described as the final cracking state.

The minimum distance between cracks will be the anchoring length, and the maximum distance is two times the anchoring length.

7.1.24.3 Steel Fibres and Cracking

The items mentioned below are relevant:

7.1.24.4 The Effects of Steel Fibres on Cracking

The fibres will bridge each crack, and the stresses are reduced at the bottom of the crack. The fibres may cross the cracks at various points and may also be bent. The frictional forces may be locally increased by the compressive forces in areas parallel to the cracks. Secondary cracks may develop due to the tensile forces perpendicular to the cracks. In steel FRC, this may be typical throughout the cracked section.

7.1.24.5 The Effects of Fibres on the Distribution of Cracks

There will be more cracks in the member, and their widths will be smaller. This will mean a decrease in permeability, i.e., better performance in service life.

7.1.24.6 RC with Steel Fibres Only

With normal dosage, post-cracking strength will be less than the strength before cracking. With high-performance fibres and higher dosage, the post-cracking resistance moment may be greater than the resistance moment of the non-cracked section.

7.1.24.7 Crack Control Using Hybrid Reinforcement

By using traditional reinforcement, the stress and tension in the steel will be reduced due to the presence of fibres. The relevant factors will be anchoring length, the distance between the cracks, and the width of the cracks.

7.1.24.8 With Steel Fibres

- Post-cracking stress is not zero.
- It will bridge the fully cracked section, including the portion of unreinforced concrete cover.
- Less energy will be available to open other cracks.

Some of the stress may be transmitted through the cracks by the fibres. A portion of the elastic energy stored will be released with the cracking.

7.1.24.9 Cracking of RC with Hybrid Reinforcement

7.1.24.9.1 Initial State

The inclusion of fibres, along with fabric reinforcement in the member, will reduce cracking energy and the anchoring length.

7.1.24.9.2 Final State

A reduction in the maximum distance between cracks will occur. There will be a larger number of cracks with smaller openings.

7.1.24.10 Further Comments

The items mentioned below will be relevant for using traditional reinforcement along with steel fibres in the section:

- Reduction in crack openings may be possible.
- Savings in time and money.
- A reduction of traditional reinforcement in the member/section.

Source:

- Camiola [27]

7.1.25 Use of FRC in Composite Metal Deck (CMD) Applications

7.1.25.1 Introduction [28]

The CMD applications using light-gauge steel decking and concrete have already been mentioned.

At ultimate limit state, the composite action between the metal deck and concrete topping will support the loads.

In the conventional method of construction, a single layer of fabric reinforcement would be provided towards the top surface of the slab to control cracking. The minimum amount of reinforcement in the slab will be based on the relevant code and the slab thickness used in the design.

During a fire, the steel deck at the bottom of the slab will lose strength quickly, and the bending resistance of the slab will depend on the fabric and the concrete.

From various applications, it was observed that fibres, if used properly, can provide the necessary resistance to the slab for the loads.

Fire tests were conducted on CMD samples for studies and assessment. Satisfactory results were obtained from the tests, and the applications may be extended to multi-deck structures/constructions.

7.1.25.2 Benefits from Applications

Fibre–deck composite floor decking will provide various benefits apart from good engineering performance compared to traditional reinforcement. The relevant items from traditional construction that may be omitted include delivery, lifting, storage, and placement of fabric reinforcement, etc. The omission of these items will be especially advantageous in multi-storey constructions.

It is noted that after applying the fibre deck solution with FRC in projects, about 20% time savings would be possible.

It may be stated that FRC is expected to provide significant benefits when considered for the fibre–deck composite floor decking system.

Source:

- Evans [28]

7.1.26 Non-metallic Reinforcement for Concrete [29]

The use of non-metallic reinforcement in concrete will help improve durability since the material will not corrode.

Fibre-reinforced plastics for reinforced concrete structures may be known as FRPRCS.

FRP bars have been developed that resemble steel bars with similar properties.

Carbon and aramid fibre strips for FRP, etc., have been developed, which can be bonded to structural elements to strengthen members for retrofitting work. By following this treatment, additional loads may be accommodated on the members.

In this context, carbon fibre-reinforced polymer (CFRP) has contributed significantly.

Fibre composite materials for strengthening and retrofitting work have been discussed in detail elsewhere.

Source:

- Burgoyne [29]

7.1.27 Review of Fibre Reinforcement – The Non-metallic Way [30]

A few comments will be made as given below:

The use of fibres in concrete has increased recently to replace steel bars and fabrics, as already mentioned. Applications can be found in various areas, as described in Section 7.7. There are several advantages to using fibres in concrete, such as resistance to tensile stresses, control of drying shrinkage cracks, and resistance to impact. Developments have also taken place in the use of structural synthetic fibres.

Various types of fibres, starting from horse and animal hair to goat hair, have already been tried with a cement matrix. Later, asbestos fibres and glass fibres were attempted for reinforcement, as stated.

Asbestos fibres could not be used due to health and safety considerations. It was also found that alkali-resistant glass containing a proportion of zirconium oxide would be suitable for use with concrete. More comments on glass fibres will be included later.

7.1.27.1 Polypropylene Fibres

Recently, monofilament polypropylene fibres have been used for floor screeds. The material is flexible and tough, capable of controlling plastic and drying shrinkage cracks when used with concrete. Attempts have been made to replace steel reinforcement in ground floor slabs, paved areas, etc., by using fibres in the concrete.

The important items for using non-metallic reinforcing fibres are given below:

7.1.27.2 Configuration

The shape of the fibres is important. It has been observed that round-shaped fibres may provide limited strain resistance. Embossed and twisted fibres may offer increased bond strength and reduce the chances of fibre pull-out. Embossed and deformed heavy-duty fibres may provide similar post-cracking performance to that of steel reinforcement.

7.1.27.3 Chemistry

The fibres should be able to withstand the highly alkaline environment of concrete and should resist various chemicals to which the concrete structure may be exposed.

7.1.27.4 Contents

The dosage rate of fibres is very relevant in concrete. The fibres may be oriented randomly in the mix. The dosage must be appropriate to ensure that there is a sufficient amount of aligned fibres. The surface areas of the fibres should be wetted to ensure contact with the cement paste. The typical rate may be 3–8 kg/m^3 with a specific gravity of 0.91. This represents about 0.3%–0.7% of the volume of concrete.

7.1.27.5 Correct Length

The aspect ratio of fibres (i.e., fibre length: thickness) is a key factor for satisfactory performance. Longer fibre lengths with the proper texture will be more effective in providing anchorage within the cement matrix than shorter fibres. With adequate fibre length, it may allow for the development of shear force at a stress equal to fibre strength.

The applications and development of synthetic fibres will be suitable for use in floors, slabs, and bridge decks, and will also control plastic shrinkage cracking. Furthermore, they will increase impact resistance and tensile strength of the element in a more homogeneous manner than steel. Furthermore, the fibres are likely to eliminate loose and broken materials on the floors and runways, providing satisfactory and long-term safe performance.

Source:

- Jowett [30]

7.1.28 Fibre-Reinforced Using Plastic Reinforcement for Concrete [31]

To replace steel bars, developments have occurred for alternative reinforcement in concrete to address corrosion in aggressive environments. Stainless or coated reinforcing bars have been trialled. Furthermore, developments include fibre-reinforced plastic (FRP), using glass, carbon, or aramid fibres along with suitable resin to form a rod or grid.

By using proper resin and fibres, the material will be highly suitable and more durable.

7.1.28.1 Bar Sizes and Shapes

The maximum diameter of bars is about 25 mm; however, 32 mm diameter bars have also been produced.

Most of the bars are circular in cross-section; however, square and rectangular shapes are also available.

The common method of producing FRP reinforcing rods is through the pultrusion process. This mainly involves pulling the uncured material through a heated die. A secondary process is required to ensure the appropriate surface for developing a proper bond. Thermo-setting resins are used in the production process, which do not allow the rods to be bent into the required shapes after full curing. Hence, special arrangements are needed for shaping.

7.1.28.2 Material Properties

This is connected with the amount and type of fibres and the resin used to meet the requirements for durability. FRP thermo-set items will have no plasticity and will respond linearly to failure.

The mechanical properties of glass and CFRP reinforcement are given in Table 7.2.

7.1.28.3 Further Comments

A few relevant comments are as listed below:

7.1.28.3.1 Design

- Failure may occur due to the compression of concrete.
- Crack widths may not be important for durability.
- The deflections will be greater than those of equivalent members with steel reinforcement.
- Fire design will need special considerations.

Further comments have been added in items 7.3 and 7.4.

7.1.28.3.2 Durability

Glass fibres and some resins may degrade due to high alkalinity. Applications using AR glass or carbon and aramid fibres will be important, along with suitable resins.

7.1.28.3.3 Costs

Comments on cost have been included in item 7.3.

Source:

- Clarke [31]

7.1.29 Next-Generation of Fibre Reinforcement for External Applications

The comments are as follows:

Polypropylene micro-reinforcement fibres

Polypropylene micro-reinforcement fibres have been used in external concrete to enhance performance against freeze–thaw cycles, improve homogeneity, aid early crack control, and provide better impact resistance, using a dosage rate of 0.9 kg/m^3. Further comments on this fibre have been made separately.

TABLE 7.2
Mechanical Properties

Item, Material	Tensile Strength (N/mm^2)	Stiffness (N/mm^2)
Carbon fibre reinforcement	1,500	150
Glass fibre reinforcement	1,000	45

Design demands

The demand for using fibres in design is increasing, especially for low maintenance related to heavy-duty pavements, container handling, lorry parking areas, etc. Polypropylene micro-fibres may not be able to improve the flexural toughness of concrete. Steel fibres have been considered for such applications.

Control of early age plastic shrinkage

Micro-fibres can help control plastic shrinkage and settlement cracking at the initial stage and improve resistance to freeze-thaw damage, as mentioned. However, macro-fibres can enhance flexural toughness and may provide better post-cracking performance, as observed from tests.

Cost-effective alternative

The combination of MSFs and micro-synthetic fibres may be used to reduce construction time and costs. The slab thickness may be reduced, and the elimination of normal steel fabric reinforcement will be possible. This type of concrete reinforcement will be suitable for a wide range of applications, such as

External hard standings, airport taxi ways, lorry parks, roadways, driveways, ports/dock areas, footpaths/walkways/bike paths, golf cart tracks, boat ramps, and others.

7.1.29.1 Typical Dosage Rates

The following may be mentioned:

- Ground bearing slabs, driveways, parking areas, and for higher load applications – 4.5–7 kg/m^3
- Footpaths, walkways, and other normal road applications – 2–4 kg/m^3

Further Comments – The inclusion of synthetic fibres may be considered as a replacement for steel wire fabric, steel fibres, and air-entraining admixtures.

More applications covering long-term performance will be important.

Source:

- Atkinson [32]

7.1.30 Review of Properties of Fibre-Reinforced Concrete (FRC)

FRC is a comparatively new material, and its use is increasing in recent years, as stated.

The contributions of fibres in concrete may be summarised as follows:

They will reduce the water and air voids, increase the tensile strength of concrete, reduce bleeding in fresh concrete, likely provide higher flexural strength compared to the strength offered by reinforcing bars, control the growth of cracks under load, increase the durability of the concrete, and some fibres will enhance impact and abrasion resistance.

The fibres that may be mentioned in this connection are steel, glass, synthetic, and natural fibres. More comments on fibres have been presented in item 7.2.

Source:

- Kaur and Talwar [33]

7.1.31 Factors Influencing the Workability of Steel FRC, Part-1

Some comments from initial studies/observations are as follows:

The workability of the concrete mix depends on various parameters like the water-to-cement ratio, fine aggregate-to-coarse aggregate ratio, and aggregate-to-cement ratio.

Experimental studies have been carried out to assess the workability of various mixes with different proportions of fibre content, aggregate content, water content, etc. The contributions of fibre type on the workability are also relevant.

7.1.31.1 Materials

The following materials have been used for the studies:

- **Cement** – Ordinary Portland cement
- **Aggregates**
  - **Fine Aggregate –** Sea dredged fine sand corresponding to zone-3.
  - **Coarse Aggregate –** Crushed granite, 10 mm maximum size.
- **Fibres** – Three types of steel fibres (i.e., crimped, round, and duo-form) with different diameters and lengths were used. The aspect ratios (l/d) varied from 57 to 100.

7.1.31.2 Mix Proportions

Four types of tests were conducted.

For series-1, sand-to-cement ratio was 1.5 with varying water-to-cement ratios, coarse aggregate, and fibre content.
For series-2, a constant aggregate-to-cement ratio of 4.5 with fine aggregate was used alongside different fibre contents and water-to-cement ratios.
For series-3, an aggregate-to-cement ratio of 3 and a water-to-cement ratio of 0.5 were considered.
For series-4, cement, sand, and aggregates were used in the ratio of 1:2.5:2.

7.1.31.3 Mixing

Uniform distribution and mixing of fibres are important. Three stages of mixing were performed due to the small quantities of materials. The materials were weighed and hand mixing was carried out on a non-porous surface in three stages.

These are given below:

Stage 1 – Coarse aggregates, sand, and cement were mixed thoroughly.
Stage II – Water was added gradually.
Stage III – The fibres were added by sprinkling small quantities at a time using a sieve.

7.1.31.4 Assessment of Workability

Slump and Vebe consistometer tests were conducted to assess workability.

7.1.31.5 Workability Test Results and Comments

- With an aggregate content of 53% in the mix, Vebe time decreased with the increase of the water-to-cement ratio, as expected.
- Ball effects of fibres were observed in the mix with the increase of the water-to-cement ratio. There was a tendency for balling to occur at lower volume fractions of fibres.
- To achieve a workable mix with a large volume fraction of fibres, it may be necessary to increase the sand content of the mix to a higher value.

Source:

- Narayanan and Kareem-Palanjian [34]

7.1.32 Further on the Factors Influencing the Workability of Steel FRC, Part-2

Further comments may be included.

7.1.32.1 Introduction

The effects of fine aggregate and the percentage of fibres in the mix have been studied and discussed. It has been observed that the optimum fibre content increases linearly with the increase of fine aggregate. It should be noted that water-to-cement ratio and aggregate-to-cement ratio practically had no influence on the optimum fibre content of the mix.

Furthermore, a linear relationship may exist between Vebe time and fibre content, provided that balling does not occur.

It has also been noted that fibre distribution in stiff mixes was uniform when the mixes were adequately vibrated to achieve full compaction. In mixes with high workability, the fibres showed a tendency to agglomerate under vibration. Prolonged mixing resulted in balling effects, even with moderate fibre content.

7.1.32.2 Comparing Strength Results

An increase in strength was observed with 1% fibre content while varying the water-to-cement ratio, compared to the equivalent specimens with no fibre content. The rate of increase in strength with fibre content would be steeper when the water-to-cement ratio is lower.

7.1.32.3 Additional Comments

- Increasing the water-to-cement ratio in a mix and using a high coarse aggregate content may result in a reduction of the optimum fibre content and will increase the chances of balling. However, workability may be improved.
- There is a maximum percentage of fibres that can be used in a mix with specified proportions and water-to-cement ratio that will not cause balling. This may be referred to as the optimum fibre content of the mix.
- For a given workability, an increase in sand content in the mix will result in an improvement in the optimum fibre content.
- The compressive strengths of FRC were found to be greater than the corresponding concrete without fibres.
- Concrete with a low water-to-cement ratio showed a larger increase in strength with the same fibre content.
- Concrete with a higher aggregate-to-cement ratio showed greater increases in strength due to the inclusion of fibres.

Source:

- Narayanan and Kareem-Palanjian [35]

7.1.33 Other Types of Fibres

The brief comments on other known fibres are as follows:

7.1.33.1 Basalt Fibres

Comments on basalt fibres have been made separately.

7.1.33.2 Fibreglass

Comments on fibreglass have been made separately.

7.1.33.3 Geotextiles

Polyester and PP non-woven materials are used for geo-bags, landfills, water canals, etc.

Source:

- Arrowtex Structural Strengthening Materials [36]

7.1.33.4 Silica Fibres [37]

Silica fibres are made of sodium silicate (water glass). They are used for heat protection and may serve as a substitute for asbestos. The fibres can be used for making wet webs, filter lining, and reinforcing material.

Source:

- Wikipedia [37]

7.1.33.5 Kevlar [38]

Kevlar is a type of para-aramid synthetic fibre. It was initially used as a replacement for steel in racing tyres. It can be spun into ropes or fabric sheets, which may be used as components in composite materials.

Source:

- Wikipedia [38]

7.1.33.6 Nomex [39]

It is an aromatic polyamide (aramid) fibre and is associated with nylon. It may be used in military and industrial applications and is considered effective for flame resistance and thermal protection. It is suitable for resisting hazards connected with electric arcs and flash fires involving intense heat and flame.

Some important properties are given below:

Resistant to insects and fungi, effective for flame resistance and thermal protection, fibres will not melt or drip, high elongation, abrasion resistant, cut resistant, resistant to chemicals and industrial solvents, etc.

Source:

- CSR [39]

7.1.33.7 Quartz Fibres [40]

The quartz fibres may be prepared by extruding molten quartz through a stream of hot air. The fibres were first made in France in 1838 and were used for the construction of delicate balances. The fibres have good heat resistance and high dielectric strength, with diameters ranging from 0.7 to 10 µm, and are non-flammable.

Source:

- Quartz fibre – CAMEO [40]

7.1.33.8 Human Hair

The reported applications are less.

7.1.33.9 Animal Hair

Animal hair was used for some applications, including its use in mud and lime plaster to prevent cracking [33].

7.1.33.10 Sheep's Wool [33]

Some applications have been made with sheep's wool, one of which is mentioned in item 7.7.

7.1.33.11 Horsehair [33]

Some applications have been made with horsehair, one of which is mentioned in item 7.7.

Source:

- Kaur and Talwar [33]

7.1.33.12 ADFIL Fibres [41]

7.1.33.12.1 Fibrin XT

These are mono-filament polypropylene fibres that may be used in concrete mixes to control plastic shrinkage and plastic settlement cracks. Furthermore, they are expected to enhance the surface properties and durability of hardened cementitious items. Mono-filament polypropylene fibres are discussed further in a separate section.

Common applications may include

Internal floor slabs, external hard surface, pre-cast concrete, sea defence work, bridges, road pavements, shotcreting/guniting, runways, water-retaining structures, agricultural areas, etc.

Source:

- ADFIL Construction Fibres, Fibrin XT [41]

7.1.34 Load–Deflection Behaviour of FRC

The typical load–deflection curves of plain cement concrete and FRC are shown in Figure 7.1 [3]. It can be observed that plain cement concrete fails suddenly once the deflection corresponding to ultimate strength is exceeded. FRC, however, continues to sustain loads at deflections considerably exceeding the fracture deflection of plain concrete. In other words, FRC specimens do not break immediately after the first crack as plain cement concrete does. The mechanism of multiple cracking continues until the fibres fail or local de-bonding occurs, leading to fibre pull-out. These are illustrated in the figures 7.1–7.3.

7.1.35 Load–Deflection Curve

The interface region will be very important for the performance of the composite material as mentioned. Composite materials that have weak interfaces exhibit low strength and stiffness but high resistance to fracture [1]; those with strong interfaces demonstrate high strength and stiffness but are very brittle.

FRC has been applied in various building projects to gain benefits related to crack resistance, better concrete quality, better installation, enhanced impact and abrasion resistance, as mentioned [1,3,21].

Engineers have implemented fibres in ground floor slab construction with the intention of eliminating the commonly used reinforcement with steel. The design may be adapted for slabs supported on piles using steel fibre reinforcement [3,21].

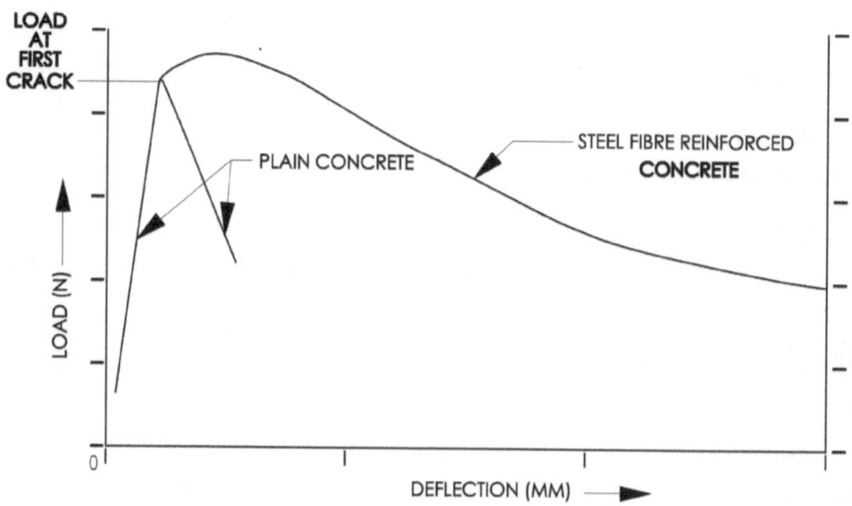

FIGURE 7.1 Load–deflection curve.

Other applications include cast-in-situ free suspended elevated slabs in residential and commercial projects, composite floors on steel decking, retaining walls, and bridge slabs.

The effects of using fibres will depend on the fibre type, its dimensions, properties, and the dosage of fibres used in concrete, etc.

Fibre composites can be grouped into high volume (i.e., 5%–15%), moderate volume (1%–5%), or low volume (i.e., less than 1%) for different applications.

The advantages of each fibre are detailed specifically and separately, which may have resulted in some common items being repeated. A few popular types of fibres have been reviewed in this publication.

Source:

- Mehta [3]
- Swamy [21]

7.1.35.1 Further Comments from Tests

Various tests have been conducted on cement concrete composites with different fibres to study the performance of these materials under varying load conditions. It has been found that the principal constituents influencing the strength and stiffness of the composites are the reinforcing fibres, the matrix, and the interface.

7.1.35.2 Post-cracking Behaviour in Tension

The behaviour pattern is shown in Figures 7.2 and 7.3.

Comments on Figure 7.3

a. The composite carries decreasing load with fibre pull-out at the crack.
b. The composite carries an increasing load after matrix cracking.
c. Idealised stress–strain curve of fibre cement or concrete.

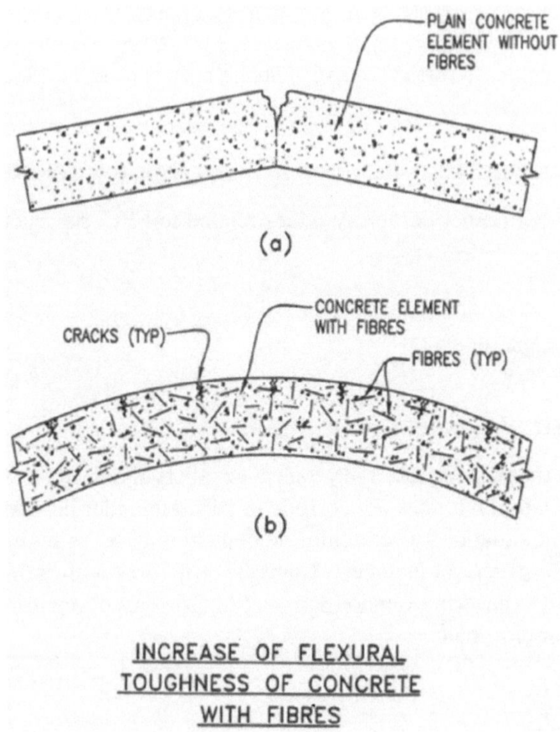

FIGURE 7.2 Increase of flexural toughness of concrete with fibres. (Source: Mehta [3]).

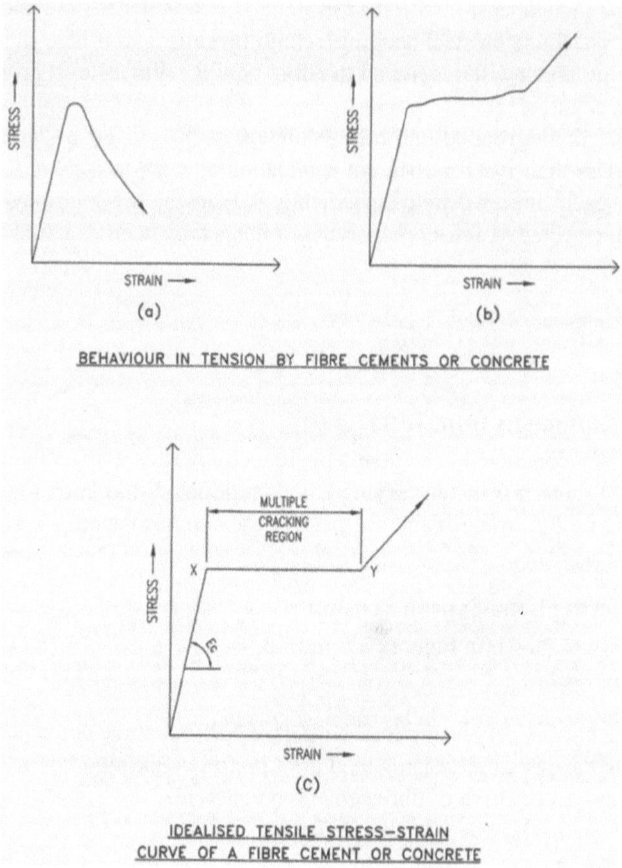

FIGURE 7.3 Idealised tensile stress–strain curve of fibre cement or concrete. (Source: Swamy [21]).

7.1.36 Durability of Steel Fibre-Reinforced Self-compacted Concrete

The above item, along with other durability issues related to FRC, has been covered separately in this chapter.

Source:

- Frazao, Camoes and others [42]

7.1.37 Stainless Steel in Concrete [43]

- Stainless steel is increasingly used in projects to satisfy the need for durability, longevity, and structural reliability. It does not corrode in concrete under normal conditions and can withstand difficult situations where chloride content may be as high as 8% by weight of cement within the structural members. Concrete structures with stainless steel reinforcement are expected to have an extended service life. The cost of repair/restoration work will also be reduced or eliminated.

Source:

- Guildhall Project [43]

7.1.38 Further on Stainless Steel [44]

Reinforced concrete structures may be affected by corrosion of carbon steel bars due to carbonation of the concrete cover or by the ingress of chloride. The use of stainless steel bars may be considered a suitable alternative solution.

The initial cost of stainless steel in a project may be high. However, the decision must be made carefully, considering the following:

- It may be selected for projects that are at high risk of corrosion. Furthermore, it may be difficult to organise remedial works related to some problems that may become necessary later.
- The most suitable grade should be selected for the requirement. Normally, this may be 1.43xx steel.
- Stainless steel has inherent resistance to chloride-induced corrosion in concrete. This should be taken into account in the design to satisfy durability requirements.

The items that are relevant to using stainless steel and may be considered for design changes are mentioned below:

- **Cover to Reinforcement –** Nominal cover may be reduced irrespective of concrete grade.
- **Crack Width –** Crack widths may be increased to 0.4 mm for structural members with stainless steel reinforcement.
- **Saline Treatment –** May not be required.
- **Nominal Steel –** May be reviewed and modified for using stainless steel with $f_y > 500$ N/mm^2 for beams, while no changes may be necessary for columns and walls.

Using stainless steel may be economical for areas directly at risk of chloride-induced corrosion, as stated.

The areas of applications are given in Table 7.3.

7.1.39 Stainless Steel in Swimming Pool Buildings [45]

7.1.39.1 Introduction

Stainless steels are iron-based alloys containing a minimum of 10.5% chromium. The corrosion resistance is provided by a thin surface layer of chromium-rich oxide.

Stainless steels have been used for swimming pool applications. Important factors for selecting stainless steels include hygiene, ease of cleaning, good resistance to corrosion, ease of fabrication, and recyclability.

TABLE 7.3
Areas of Applications

Types of Structures	Location at Risk
Marine and offshore	Inter-tidal and splash zones
Bridge super-structures	Parapet edge beams, bearing areas
Car parks	Entry ramps and slabs
Bridge substructures	All surfaces exposed to salt spray or inter tidal and splash zones

Source: Gedge [44].

Both structural and architectural applications cover pool liners, handrails, structural components, ladders, diving structures, decorative items, fasteners, furniture, water treatment, and ventilation items.

7.1.39.2 Further Comments

Following experimental investigations, it has been noted that the appropriate grades of stainless steel can serve as a suitable and excellent material for both new and existing swimming pools.

The grades of stainless steel that are normally used in architecture and building construction are 1.4301 and 1.4307 (304 types) and 1.4401 and 1.4404 (316 types). These are the grades suited for swimming pool environments and non-structural applications.

Types 304 and 316 stainless steels have performed well in non-structural pool water and poolside equipment.

Stainless steel will be an excellent material for use in swimming pools related to outdoor and indoor applications and to avoid stress corrosion cracking.

Source:

- Baddoo and Cutler [45]

7.1.40 Sustainable and High Performance of Concrete Structures with Stainless Steel [7,46,47]

For sustainable and high-performance concrete structures, the use of stainless steel bars in concrete members can be considered favourably as mentioned. The stainless steel bars may be used as the first layer [46] of reinforcement for the structural members. The initiation of corrosion in steel reinforcement may occur early and within the expected lifespan of concrete structures, depending on the environment. Furthermore, localised corrosion may propagate beyond the point of the first crack. With the presence of chloride and the initiation of corrosion, the relative rate of diffusion will be guided by the gradient between surface concentration and the first layer of reinforcement. With a thinner cover layer of concrete over the reinforcement, there will be a greater degree of spalling, which will lead to faster structural degradation. The use of stainless steel will facilitate the construction of durable concrete structures with low maintenance requirements, ensuring that the structure remains sustainable and in service for a longer period of time.

In a marine environment where corrosion is more aggressive, the use of stainless steel bars may be justified over carbon steel, considering the reduced maintenance and extended service life of structures [46,47]. Stainless steel bars may be used for roads, bridges, or similar exposed structures where they are subjected to de-icing salts containing chlorides in extreme cold weather.

Stainless steel of Grade-304 (1.4301) will be suitable for most atmospheric environments. Grade-316 (1.4436) may be recommended for marine and corrosive environments, along with newer duplex grades of stainless steel [47].

Stainless steel bars have been used for refurbishment and repair work on structures, in tunnels, and on boat slipways, etc. [47].

The use of stainless steel bars in a marine environment is recommended to protect the structure and its components against corrosion. In highway projects, the application of stainless steel bars will also be justified where chloride attack is experienced and the consideration of life cycle cost is important [47].

Source:

- James and Skovsgaard [46]
- Wells [47]
- Ganguly [7]

7.1.41 Fibres in Ultra-High-Performance Concrete [48]

7.1.41.1 Introduction

Initial development

The development of ultra-high-performance concrete (UHPC) began in 1978. From the studies, it was understood that the ultra-high-strength binder (i.e., cement and ultra-fine fillers) would be brittle like glass and normal cement paste.

The ultra-high-strength cement paste must be mixed with sand, aggregate, and reinforcement for better performance. By using very strong aggregates, the fracture energy of the material can be improved. With a greater volume and dimension of strong aggregates in the mix, the brittleness of UHPC may be reduced. The applications of UHPC may not be very suitable for pre-cast items because of rheological issues.

After conducting tests, it was found that incorporating fibres can enhance the performance of ultra-high-performance concrete. Furthermore, using ultra-high-strength binders can produce a very dense microstructure in UHPC.

To provide a high degree of ductility to UHPC, it is necessary to incorporate high-energy-absorbing reinforcement (i.e., bars, fibres, and combinations of both).

Including high-strength steel fibres can alter the brittleness of the material and make it more ductile. The following parameters will be important: fibre content, fibre geometry (covering length, diameter in relation to the maximum particle size of UHPC), bond between fibres and the matrix, fibre orientation, and the stiffness of the fibres.

7.1.41.2 Types of Fibres Used in UHPC

The selection of fibres will have a significant influence on the composition, strength, ductility, and cost of the finished material.

Many types of fibres are available, including carbon, basalt, and different synthetic fibres. Three types of fibres have been used in UHPC: polyvinyl alcohol (PVA), polypropylene, and steel. Different fibres yield varying results for the fresh and hardened states of UHPC. The lengths of the fibres ranged from 6 to 20 mm, and the fibre diameters varied from 0.009 to 0.5 mm.

Steel fibres are economical and can contribute significantly to the properties of both the fresh and final states of concrete. Polypropylene fibres can be used for non-structural elements to control plastic shrinkage cracking, sprayed concrete, and toppings along with steel fibres. Such applications provide some degree of fire resistance. PVA fibres are suitable for smaller elements requiring high flexural strengths.

Steel fibres for UHPC can be straight and cold-drawn. Carbon steel fibres may have a brass coating. The quantity of steel fibres in UHPC is relevant, with short fibres and micro-fibres numbering between 81,000 and 675,000 per kg.

7.1.41.3 Cocktails and Hybrids

It may be possible to use a fibre blend/cocktail system in the design. This may involve a mixture of different types of fibres, such as steel and polypropylene, to improve fire resistance. Steel fibres of different lengths (i.e., short and long) and diameters may be used. Studies were carried out in 1986 with UHPC, combining very short, thin and strong steel fibres with reinforcing bars. Short fibres were very efficient in bridging micro-cracks, while long fibres were effective in bridging macro-cracks.

The combination improved strain hardening and also improved the strength and toughness of UHPC.

7.1.41.4 Selection of Fibres

The selection and dosage of fibres will be very important, as they will influence performance. Stainless steel fibres for UHPC are a good option, although the fibres may be expensive. Special fibre shapes may be used, such as hooked, corrugated, crimped, or paddled. These special fibre shapes will affect the dosage of steel fibres. The typical percentage of steel fibres in UHPC may be 1%–3% (V/V). For special applications, higher dosages may be used.

7.1.41.5 Typical Applications

A few examples may be mentioned as follows:

- **TGV station, France**
  10,000 m^2 of roof structure was involved.
  The structural arrangement consisted of UHPC shells with embedded glass panes.
  Each unit was 18 m long, 2.7 m wide, and 3 m deep, with a wall thickness of 50 mm.
- **University of Southern Denmark**
  A perforated façade was designed using 292 elements. Each unit measured 5×3 m with 65 mm thick. The façade could reduce direct sunlight by up to 50%.
- **Steel through Truss Bridge, France**
  Lightweight UHPC pre-fabricated slabs were used for the deck.
- **River Bridge, Czeck Republic**
  36-m-long footbridge was constructed using pre-fabricated elements made from UHPC.
- **National Road, Switzerland**
  A 5-km national road, including bridges, was involved. The use of a UHPC overlay was applied.

Source:

- Buitelaar and Muller [48]

7.1.42 Brief Comments on Polymer Composites

Often, polymer matrix composites comprise a thermo-setting resin matrix and fibrous reinforcement. For certain special applications, thermo-plastic resins have been used. In other cases, composites may have mineral filler reinforcement, either alone or in combination with different fibrous types [6].

Cellular items, such as foam and honeycombs, have been tried as reinforcement to provide stiffness along with ultra-light weight.

The reinforcements that have been used include steel, glass, carbon, and aramid fibres. These materials are available in various forms, such as continuous, woven, non-woven, chopped, and multi-axial. Combinations of these materials have been tried and used [6].

Careful selection of the fibres will be important for specific applications to satisfy different requirements.

Experience and skill will be essential to use FRC in the design.

It may be possible to achieve the best results using composites along with traditional materials. Further comments on this item have been made separately.

Source:

- TIFAC [6]
- Hollay [1]
- Humphreys [5]

7.1.43 Basalt Reinforcement [49]

Basalt reinforcement has been attempted in concrete beams as an alternative to steel bars. The bars are made of basalt from naturally occurring volcanic rocks, which are crushed to powder. The powder is then heated to its melting point (about 1,450°C) and extruded through a fine nozzle to produce thin continuous strands with diameters of 9–13 μm. The basalt strands are very flexible at this stage and may resemble hair.

The strands are bundled and bound together using a polymeric compound to create long straight bars. Normally, extra basalt strands are wrapped around the bars transversely to form spiral ribbing. This ribbing provides containment, rigidity, and helps to improve the mechanical anchorage between the concrete and the bar. The basalt bars can be of conventional reinforcement sizes and are expected to be used as reinforcement for concrete.

Concrete beams with basalt reinforcement have been tested for flexural behaviour. The basalt reinforcing bars are lightweight, possess high tensile strength, and are resistant to heat and fire.

The important properties of basalt reinforcement are listed below:

Tensile Strength – The tensile strength of basalt bars is twice that of high yield steel reinforcing bars. To meet the strength requirement, only half the required reinforcement will be necessary when using basalt bars.

Light Weight – The weight of basalt bars is one-quarter that of steel, making them easier to handle on site, and reducing transportation costs.

Oxidation – Basalt reinforcement is free from oxidation. This leads to better durability, and the concrete will not discolour, which is often experienced resulted from the rusting of steel reinforcing bars.

Residual Strength – The loss of strength under fire is greatly reduced compared to steel reinforcing bars. This provides a significant advantage for concrete structures using basalt reinforcement, which may be susceptible to fire damage. The loss of strength results from the breakdown of the polymeric compound used in the production of the bars. The coefficient of thermal expansion of basalt reinforcing bars and concrete is similar.

Energy Consumption – The energy consumption required to produce these bars is lower. The energy needed to manufacture basalt reinforcement is about one-third of that necessary to produce steel reinforcement per kilogram.

Cost – The cost of basalt reinforcement is half that of steel, i.e., per metre length of a similar size of steel bars.

Elastic Modulus – Young's modulus of basalt bars is about one-quarter of that of reinforcing steel (i.e., 50 GPa compared to 200 GPa for steel).

Ductility – The elongation at failure would be about one-tenth that of steel. The mode of failure tends to be brittle, with hardly any ductility. The basalt reinforced beams will not be very efficient at absorbing energy, which will be important in earthquake zones.

Deflection – The deflections are expected to be larger for the loaded members reinforced with basalt reinforcing bars. To handle this problem, four times the amount of basalt reinforcement will need to be used for the members to achieve the same stiffness as steel reinforcement.

Constructability – The basalt reinforcing bars are manufactured as straight bars and cannot be bent or shaped later. This will create problems in the reinforcement detailing of the elements. Steel reinforcement may be lapped to basalt bars near the ends of the beams.

Poisson's Ratio – This value is not easily available. An approximate value may be about 4. For most structural materials, Poisson's ratio can vary from 0.1 to 0.3.

The possible uses of basalt reinforcement will be for

- Post-tensioned slabs
- Basalt fibres for making sheets, which may be used for the repair work on damaged concrete members.
- A hybrid system using basalt reinforcement and steel bars, where basalt reinforcement is for strength and steel bars are for ductility.

Basalt will not be damaged in an alkaline environment and is not as expensive as carbon fibres. It has superior strength compared to many polymer-based fibres (FRP). The potential of basalt appears impressive.

Further comments on basalt fibres and reinforcement have been made in items 7.2 and 7.6 separately.

Source:

- Gohnert, Gool, and Benjamin [49]

7.1.44 Special Applications of Basalt Fibre-Reinforced Polymer Bars [50]

Basalt fibre-reinforced polymers (BFRP) have been used as rods for the connections of timber structures instead of steel bars. These are inorganic fibre materials extruded from volcanic basalt rock deposits, as stated. The investigations/tests were conducted using the bars. From the tests, it is reported that the pull-out capacity of the bars increased approximately linearly with the bonded length of the whole diameter. The adhesive used was a two-part epoxy thixotropic gap-filling material. The failure mode was mainly shear failure of the timber close to the joint, resulting in the pull-out of the bars.

Further comments on this are given in item 7.6 of this chapter.

Source:

- Yeboah, Taylor, McPolin, and Gilfillan [50]

7.1.45 Fibre Composite Material Solution for Aggressive Environments

The use of micro-polypropylene and MSFs can play an important role in aggressive environments.

The applications of micro-fibres in concrete will reduce plastic shrinkage and settlement cracking, as stated before. This will help the concrete perform better in resisting impact and abrasion in the hardened state. The inclusion of 0.91 kg/m^3 of monofilament micro-fibres will help bridge the aggregates, provide better mechanical interlocking between them, limit the rate of settlement, and resist segregation. Settlement is connected to cracking, and different rates of settlement will be controlled by improved mechanical interlocking of the aggregates in the matrix.

By using micro-fibres, it will be possible to control the amount of bleed and increase the tensile strain capacity of the surface of the concrete by a factor of about three. The reduction of laitance will provide a more durable concrete slab.

7.1.45.1 Impact Resistance

From tests, it can be noted that the micro-fibre concrete showed an improvement in impact resistance over plain concrete of about 30% at 7 days. At 28 days, the improvement may reach about 90%.

7.1.45.2 Abrasion Resistance

From the study, it can be reported that an improvement in abrasion resistance of about 86% over plain concrete may be achieved with micro-fibre reinforced concrete.

Hence, for areas such as unloading and sorting areas of a waste transfer station, the application of 0.91 kg/m^3 micro-fibres in concrete may provide a cost-effective alternative solution for the design of the concrete.

For ground-bearing slabs, the use of micro-synthetic fibres may be considered instead of welded fabric, which will be made from polypropylene and will be chemically inert. Furthermore, it will be resistant to liquids and leaching and will not be subjected to corrosion, hence preventing spalling of the concrete.

The application of macro-synthetic fibres, like Durus S400, may eliminate the need for cover to reinforcement, allowing for a potential reduction in the depth of the slab in the design. The full depth may still be considered in the design.

Source:

- Evans [51]

7.1.46 Explosive Design with Fibre-Reinforced Cement Composites [52]

7.1.46.1 Introduction

FRC can withstand shock waves generated by high-velocity impacts. The tests were carried out on square-section bars for assessment.

The fibres used in the tests can be listed as below:

- E-glass fibre
- Steel fibre
- Polypropylene fibre
- Carbon fibre

Further information regarding the study has been presented in item 7.6 separately.

Source:

- Raouf, Al-Hassain, and Simpson [52]

7.1.47 Some Common Applications of Fibres in Concrete [1,3,8,10,21]

Some of the common applications may be for

Highway and airfield pavements [21], hydraulic structures [21], fibrous shotcreting [3,21], pre-cast applications [21], structural applications [21], external cladding [21], industrial floors, noise barriers [21], second-phase layers on top of concrete planks, ground-bearing general slabs/rafts, composite slab on steel decking, marine concrete [21], retaining walls, bridge slabs, tunnel linings [3], strengthening work to structures, parking garages, taxi way [3], underground tunnel linings and caverns, rock slope stabilisation, and surface protection [3].

The discussion on different applications with fibres has been included in detail later, in item 7.7 of this chapter.

Source:

- Mehta [3]
- Hollaway [1]
- Swamy [21]
- Ganguly [8,10]

7.1.48 Comments on Time to Call in Reinforcement [53]

7.1.48.1 General Comments

7.1.48.1.1 Fibres for Floor Slab Construction

By using micro-, macro-, and steel fibres, it may be more helpful to meet the design requirements for FRC, as already stated.

7.1.48.1.2 Proper Ingredients for the Work

By introducing fibres into the concrete mix, it can contribute to various areas, such as reducing the chances of plastic shrinkage, control cracks, reducing the cost of construction, and providing better structural performance.

Some of the critical and relevant items for using fibres may be mentioned as follows:

- Reduction of the requirement for plant and equipment for the work.
- The operation will reduce the involvement of labour.
- Problems with trips will be reduced or eliminated.
- Discharge at the point of installation.
- The requirement for large areas for storing steel reinforcement bars on site will be reduced.

7.1.49 Comments on Case Study

7.1.49.1 Case Study

7.1.49.1.1 Floor Slab Construction

FRC was used for the construction of a ground floor in a warehouse project in the UK, where the floor slab was supported on piles. The concrete slab had to be strong enough to support a large number of high-rise racking systems. There were space restrictions, making it difficult for vehicles delivering concrete to access and move around the building. The time schedule for the project was also quite tight.

The following items may be mentioned as relevant for FRC design and construction:

- FRC was discharged from the wagon directly to the point of the pour, and pumping operations were avoided.
- FRC met the requirement for residual strength in the piled floor design/system.
- It was possible to eliminate the need for traditional fabric reinforcement by using FRC.
- FRC controlled the risk of cracking, resulting in a robust floor slab.

7.1.49.2 Further Comments

The items below may be noted:

- It was possible to provide a glass-like finish for the slab as per the requirements.
- The steel fibres provided the necessary strength through efficient distribution throughout the concrete slab.
- Construction could be carried out in restricted spaces with proper planning.

Furthermore, it was possible to complete the work within the scheduled timeframe and budget.

Source:

- Taurah [53]

7.2 INFORMATION/DISCUSSION ON DIFFERENT FIBRES AND COMMENTS

7.2.1 Types of Fibres

Some common types of fibres will be discussed below:

7.2.1.1 Introduction

FRC with steel fibres is well established and generally more resistant to impact than traditionally reinforced members. Its advantages include controlling crack width, increasing load-bearing capacity, redistributing stress, saving time, and improving ductility [54]. Furthermore, tests indicate that the performance of the slab with steel fibres is influenced by the geometry, tensile strength, and aspect ratio. Such concrete demonstrates effective crack width control and will resist the development of cracks into the macro phase [7,54].

Common applications of FRC have often involved steel fibres. The fibres are used as replacements for conventional steel fabric reinforcement to control early thermal contraction and drying shrinkage cracking. Polypropylene fibres have been used to increase fire resistance. The fibres redistribute the forces within the concrete, restraining the mechanisms of forming and extending cracks as noted. Typical applications include fibre reinforced industrial ground floor slabs, suspended ground floor slabs on piles, tunnel linings, and pavements [7,55].

Source:

- Greenhalg [54]
- Ganguly [7]
- ACIFC [55]

It is interesting to note what different authors have commented after carrying out studies/applications/tests on the same topic at various times and to examine the later developments. Efforts have been made to collect information from different books and publications/journals. Following this approach, some issues may be mentioned repeatedly.

7.2.1.2 Steel Fibres

7.2.1.2.1 Types [55]

The method of classifying steel fibres is briefly mentioned:

Four types can be described as follows:

- Type-1 – Drawn wire
- Type-1 – Slit sheet
- Type-1 – Melt extract
- Type-1 – Others

Properties – These may be described as below:

Cross-Section	Deformations/Types
• Round • Crescent • Flat • Any other types	• Straight, smooth surface • Wavy • Round with end paddles • Round with end buttons • Crimped (round, flat, or any section) • Polygonal twisted
Length – 19–60 mm **Young's Modulus** = 205 kN/mm^2	**Aspect Ratio** – 30–100 (Length/diameter)

Tensile strength = 345–1,700 N/mm^2

Ductility – Able to bend through	180° without rupture

7.2.2 Influencing Criteria

The following items will be relevant:

Anchorage mechanism (e.g., straight or deformed shape end cones or hooked ends)

- Dosage (kg/m^3)
- Fibre length and diameter, i.e., aspect ratio
- Fibre count – fibre size and number of fibres per kilogram of fibre
- Tensile strength
- Ductility and bond

7.2.2.1 Durability [55]

The fibres are protected against corrosion by the alkalinity of the cement paste in the mix. Bursting stresses associated with corrosion in large diameter bars in concrete will not occur with fibre reinforcement. If the fibres corrode, the small volume of each fibre and the discontinuous nature of the fibres will not create cracking problems like those found in conventional concrete with corroded steel reinforcement. There may be some corrosion only at the surface, which will not be harmful [55]. Generally, there will not be minimum cover requirements.

7.2.2.2 Panel Sizes and Joint Centres in Ground Floors with Steel Fibres

By using steel fibres, it is possible to construct joint-less floors. Up to 50 m×50 m is size [55] may be constructed with no intermediate joints. The typical spacing of joints may be within the range of 6–12 m, and the aspect ratio of each panel should not normally exceed 1 : 1.5.

Careful design and detailing will be required to transfer loads at the joints, especially for larger panels.

Source:

- Greenhalg [54]
- ACIFC [55]
- Ganguly [7]

7.2.3 Comments from Other Applications with Steel Fibres

7.2.3.1 Ground Floor Slabs

The steel fibres in concrete slabs have various applications, and in many cases, the fibres have been able to reduce the use of traditional steel reinforcing bars and fabric to a large extent. The details provided below will be relevant [56]:

- Applied in pile-supported industrial floors, where the ground could not provide adequate bearing capacity.
- Freely suspended cast in-situ elevated slabs in residential and commercial uses.
- Ground-supported general rafts as foundations, e.g., schools, hospitals, shopping malls, and clad-rack warehouses.
- Composite floors on steel decking.
- A second top layer over concrete planks and void formers.
- Bridge slabs.
- Retaining walls.

Steel fibres, in reinforced concrete applications, as mentioned above may have a dosage rate of 30–100 kg/m^3 [56], depending on the type of fibre and the applications. To meet workability requirements, 1.3 mm diameter fibres with high dosages of 80 and 100 kg/m^3 may be used. With 1 mm diameter fibres, the dosage rate may be limited to 50 kg/m^3 [56]. Smaller diameter fibres, i.e., those with a diameter of 0.6 mm or less, are generally not used as they may saturate the concrete at lower dosage rates.

It has been observed that slabs with steel fibres are able to resist negative and positive bending moments, along with shear, shrinkage, and punching stresses. Continuity and starter bars are generally provided in construction as per conventional procedures. The fibre length may be 2.5–3 times the size of the aggregates, allowing the fibres to overlap and fully bridge all the large aggregate particles [56].

The aggregate grading should be uniform, with 20 mm as the maximum size of aggregate and a higher content of 12 mm to facilitate the accommodation of steel fibres between the aggregate particles. The gravel-to-sand ratio may be 0.9 : 1.0, with a minimum of 475–500 kg/m^3 of fines, in the mix, smaller than 200, including 320–350 kg/m^3 of cementitious material [56].

Steel fibres have a significant impact on the application of FRC, as mentioned. Round steel fibres with diameters in the range of 0.25–0.75 mm have been used. Flat steel fibres that are used [5] may have cross-sections varying from 0.15 to 0.4 mm in thickness and from 0.25 to 0.9 mm in width.

Crimped and deformed steel fibres are available either in full length or crimped at the ends only [3].

Steel FRC was selected for the tunnel lining [57]. It was observed that fibres with a high aspect ratio (>65–80) would be more effective in the long term with a lower dosage rate.

A higher dosage rate would be required with a lower aspect ratio (<45–55) to meet performance requirements.

The applications of steel fibre reinforced spray concrete were carried out for rock support [58]. The steel fibres had lengths of 20–40mm and an aspect ratio of 50–100. It was observed that a higher aspect ratio and volume concentration resulted in better performance of hardened concrete; however, it became more difficult to mix, pump, and spray. The dosage of steel fibres may vary; however, 40 kg/m^3 may be considered a normal figure for rock support work, and the value may vary within the range of 20–80 20–80 kg/m^3.

MSFs may be an alternative to steel fibres where movement of the substrate is expected, as well as for applications in wet and aggressive environments like mines and marine slope stabilisation work [58].

MSFs may have a diameter of >0.3 mm and be of similar length to steel fibres, with a different anchoring system. By using a higher fibre aspect ratio and volume concentration, the performance of hardened concrete can be improved. The dosage rates are typically in the range of 2–10 kg/m^3. With a higher dosage, considering fibre reinforced sprayed concrete (FRSC), it is important to account for rebound [58].

Generally, fibres are introduced at a regulated rate into a ready-mixed concrete truck at the batching plant, revolving at maximum speed, or directly into the back of the truck mixer. Five minutes of mixing in a truck mixer will ensure even distribution without clustering [59].

Tests were conducted to assess the adequacy of the punching shear capacity of slabs with steel fibres (hook-ended) with an aspect ratio of 60–75. By using an effective depth of 0.75 h (h = depth of slab), the results were found to be conservative [60].

Steel fibres, with or without steel reinforcing bars, are being used extensively in ground slab construction supported on piles, as mentioned previously. Guidance from the manufacturers may be necessary for fibre dosage (generally 40–50 kg/m^3) and type to ensure adequate load-carrying capacity and control of cracks [60].

Steel fibre-reinforced concrete (SFRC) slabs have been tested [61] successfully. The addition of steel fibres to concrete is intended to improve serviceability limit state performance through enhanced crack control and limited deflection. Reasonable agreement between elastic theory and actual test results was observed for the test samples. Furthermore, there is potential for the development of design methods based on yield line theory and a simplified plastic model.

Steel fibres were also selected for a suspended two-way spanning concrete slab for a car park in Belfast supported on piles [62]. The critical items for the design were as follows:

- Pile centres and size.
- The slab must be able to withstand heavy imposed loads.
- Location of formed day contraction joints.
- No requirements for fibre suppressants and special techniques.

Further comments on this will follow in item 7.6.

The crack widths may be more critical than deflections in pile-supported SFRC floor slabs [11].

It has been reported from tests [63] that high-strength concrete corbels with high tensile steel reinforcement exhibited significant inelastic deformation and ductility at failure when steel fibres were incorporated in both under-reinforced and over-reinforced specimens. The compressive strength of concrete ranged between 43 and 84 N/mm^2, and the shear span-to-depth ratios ranged between 0.11 and 1.1. An increase in load capacity with steel fibre reinforcement of up to 35% has been observed in the investigation. Further details are provided later in this chapter.

For explosion and impact resistance, ultra-high-performance FRC has been studied [64]. The developed material features a low water-to-binder ratio with high cement content and a high content of super-plasticiser. Micro-silica and fine silica sand (<400 μm) are used in the mix for the enhanced homogeneity. High dosages of short small diameter steel fibres were used to improve ductility. Further details will follow in this chapter.

For the extension of the Barcelona (Spain) metro [65], construction was carried out using pre-cast tunnel lining segments with steel reinforcement fibres. It is noted, after studying the site trials in the laboratory, that 60 kg/m^3 of steel fibres in the concrete for the segments could satisfy the design requirements without using conventional steel bars as cage reinforcement.

Considering various applications, steel fibres do not replace main reinforcement in columns and beams. Often, designers provide nominal steel at salient points, typically at 0.1% of the gross [66] cross-sectional area of the concrete. This 0.1% Ac may be referred to as "anti-crack" steel. However, it might be possible to replace some secondary reinforcement with steel fibres at a regulated rate where the secondary reinforcement is used as "anti-crack" steel. Nonetheless, conventional reinforcements must be provided around the edges of the floor slab, stairwell, lift shaft penetrations, etc., as good engineering practice.

Mixed Design [56] – The concrete mix design should consider proper aggregate grading, cement content, and a water-to-cement ratio not exceeding 0.55. super-plasticisers may be required to create flowing concrete. Information on slump prior to adding super-plasticiser will be required, as it relates to the strength of concrete to meet other requirements.

7.2.3.2 Advantages of Steel Fibres in Concrete [3,7,56]

The items mentioned below are relevant:

Higher quality concrete elements can eliminate installation mistakes related to steel and providing the correct cover; etc., in the member.

- Reduced and controlled concrete shrinkage.
- Improved durability.
- Improved impact and fatigue resistance.
- High resistance against spalling.
- Low crack widths under service conditions.
- Increased flexural strength in all three directions.
- Material savings may be possible due to reduction in thickness in most cases.
- Labour savings as cutting, bending, and placing of steel reinforcement will be avoided.
- Labour savings during installation as the concrete may be of a self-levelling type and will not need poker vibration.
- Time savings will be possible, in the project, which will help in maintaining the target schedule.
- The concrete will have a more professional appearance and may be totally smooth against smooth formwork.
- Better personnel safety, as risks related to handling traditional steel reinforcement will be eliminated.
- Concrete can be easily pumped.
- No handling of steel reinforcement by cranes.
- Less noise during construction.
- Better site management may be expected

Source:

- Destree [56]
- Mehta and Monterio [3]
- Ganguly [7]
- Swannell [58]

- Greenhalgh [57]
- Better concrete with fibre mesh [59]
- Beckett-1 and 2 [60]
- Ward [61]
- Cameron [62]
- Vollum [11]
- Magnet and Halabi [63]
- Barnett and Tyas [64]
- Berge [65]
- Viney [66]

7.2.4 SFRC Suspended Slab in Flexure

Steel FRC (SFRC) is being used [61] for ground-bearing slabs and spread concrete tunnel linings. The use of steel fibres in the mix aims to improve the serviceability limit state by achieving better crack control and limited deflection. Steel fibres will increase the ductility and tensile strength of the concrete mix.

7.2.4.1 Tests/Experimental Procedure and Results

The tests were carried out on concrete slabs with different fibre contents. Information on the tests is given in item 7.6.

Source:

- Ward [61]

7.2.5 More on Steel Fibres Reinforcement for Suspended Slabs

The application of steel fibres may be suitable for suspended slabs.

The geometry of fibres is relevant for the performance of steel fibre reinforced slabs, as straight fibres may provide only nominal bond stress with the concrete. The deformed steel fibres are expected to offer better bond with concrete and will help to redistribute the stresses efficiently.

7.2.5.1 Test Programme

Tests were carried out using

- **Tabix Plus** – 1-mm-diameter crimped 60-mm-long steel fibres having a compressive strength of 1,500 MPa.
- **Twincone** – 1-mm-diameter 54-mm-long straight steel fibres with conical ends having a compressive strength of 1,100 MPa.

Both the fibres provided full anchorage with nominal slippage within the concrete.

In both cases, the dosage rates were 40, 45, and 50 kg/m^3.

7.2.5.2 Comments

For successful performance, sound design and efficient execution will be critical. Blast machines may be used for proper mixing of fibres in the concrete.

The advantages of using steel fibres in the design for flooring.

These may be mentioned as follows:

- Pumping of concrete may not be required.
- The use of mesh reinforcement and re-bars may be avoided.

- Execution may be carried out using a laser-screed machine for ground-supported slabs.
- It may be possible to execute 3,000 m^2/day.
- Better for shrinkage and crack control.
- Economical and cost-saving.
- Faster mixing of fibres may be achieved by using the blast machine.
- The design may be conducted using the yield-line method.

Source:

- Destree [67]

7.2.6 Further Comments for Using Steel Fibres [56]

Steel fibres in concrete slabs have various applications, and in many cases, the use of steel fibres was able to reduce traditional steel reinforcing bars and fabric to a large extent, as already stated. The comments on applications mentioned below will be relevant:

- They may be applied in pile-supported industrial floors where ground conditions are poor to provide adequate bearing capacity.
- Freestanding, cast-in-situ elevated slabs for residential and commercial uses.
- Ground-supported general rafts as foundations.
- Schools, hospitals, shopping malls, clad-rack warehouses, office buildings, etc.
- Composite floors on steel decking.
- A second top layer over the concrete planks and void formers.
- Bridge slabs.
- Retaining walls.

Further applications have been presented in item 7.7.

Source:

- Destree [56]

7.2.7 Additional Comments for Concrete Floors with Fibres [68]

There are several items to consider for the design and construction of concrete ground floor slabs to ensure good engineering practice, as mentioned below:

- The specification should adequately cover the details of the requirements.
- A suitable and experienced contractor has to be selected for the job, considering their experience and resources.
- The floor must be designed to carry the required loads, taking into account the adequate factors of safety.
- Design drawings are to be reviewed, covering the selected materials, joints, and isolation joints, etc. The details of the joints must be provided properly.
- Waterproofing, site conditions, and sub-slab conditions, etc., are to be reviewed.
- The smooth supply and control of the concrete will be important, and the execution work must be supervised continuously.
- There may be some problems with joint-less floors compared to saw-cut floors; this must be managed properly. However, there are many advantages to joint-less floors.

Source:

- Williamson [68]

7.2.7.1 Review and More on the Contributions of Steel Fibres [68]

Apart from increasing the load-bearing capacity of the slab, the use of steel fibres in the mix will also improve the mechanical properties such as

Impact resistance, ductility, fatigue resistance, durability, good resistance against spalling, reduced crack widths under service load conditions, and flexural strength in all directions.

Some problems may arise from using steel fibres in concrete floor slabs. These can be overcome by careful selection of suitable materials, thorough checking of details, and applying independent and experienced supervision. The benefits of using fibres in concrete will outweigh these challenges.

Source:

- Williamson [68]

7.2.8 Steel Fibres for Steel Decks [66]

Concrete slabs, whether pre-cast or in-situ, are typically placed between the beams of a steel-framed building. The use of steel FRC has also been implemented in the design of composite floors on steel decks.

Typically, the top reinforcement in the concrete slab serves as anti-crack reinforcement for the slab. It may be possible to eliminate this secondary reinforcement by using steel fibres.

However, conventional steel bars will still be required around the edges of the floor slab, at stairwells, lift slab penetrations, and other cut-outs to ensure good detailing, as mentioned. Fire tests were conducted on the floor panel across the beams, specifically to cover 60 and 90-minute fire ratings.

The application of steel fibres in concrete on steel decks supported by steel beams will have a significant impact, particularly when working at heights and handling fibre reinforcement during construction.

Further work will be necessary to ensure safety and economical building design and construction using fibre-reinforced composite floors.

Source:

- Viney [66]

7.2.9 Steel Fibre-Reinforced Pile Supported Car Park, Belfast

The use of steel fibre for the development of 1,200 m^2 of retail space, alongside 4,800 m^2 for heavy goods access roads and a car park, took place in Belfast. A reinforced concrete pile-supported slab was finalised for the car park. The project was associated with car parking as previously mentioned. Considering the peat layer beneath the site, a pile-supported slab was adapted for super-structure the and internal floor slab. The design considered a suspended two-way spanning slab for construction using laser screed.

The critical items for the project are given below:

- Bekaert design method may be used.
- Heavy goods vehicle loading in external areas
- Traditional flexible road construction for external areas
- Pile size and centres.
- The ground floor would be subjected to an imposed load of 100 kN/m^2.
- The position of formed day contraction joints.
- No other joints will be necessary.

- Bakert's design is applicable to inherent strength with steel fibres for the concrete slab which will be able to span in two directions between the beams. This will be an integral part of the slab design.
- The beams will span between the piles and will be reinforced with pre-fabricated mesh strips placed before the laser screed.
- The system eliminates the need for enlargement of pile heads for punching shear.
- The concrete slab will be supported on pile heads
- Furthermore, it will eliminate the requirement for ties between pile heads and the slab.
- 6,000 m^2 of concrete floor slab was completed within five working days, satisfying the requirements of a tight construction schedule.

The application and the method of concrete flooring indicated savings in cost and time.
Source:

- Cameron [62]

7.2.10 Review and Summary of Steel FRC (SFRC) Floor Slabs

The review on some specific items is made below [69]:

7.2.10.1 General

- Micro-cracks will be intercepted, and their propagation will be controlled with the help of better tensile strength.
- It will increase the initial cracking strength.
- It will increase the flexural strength, hence a thinner slab thickness may be possible in the design.
- It will eliminate the need for reinforcement bars/fabric placement in the slab.
- Fibres will be dispersed throughout the slab isotopically, and there will not be a weak plane for cracks.
- SFRC slabs can withstand the applied loads well. The dosage of fibres may vary from 12.5 kg/m^3 to 100 kg/m^3.

7.2.10.2 Comments on Saving of Time and Money

- Fabric reinforcement can be eliminated entirely from the slab, resulting in savings on materials and labour.
- The reduction in slab thickness, as mentioned, will mean savings in cost for the concrete and its placement.
- Construction will be simpler.
- Joint spacing may be wider, saving costs for joint forming and maintenance.
- The speed of construction will improve, reducing both time and construction costs.

7.2.10.3 Benefits of Using Fibres

These may be mentioned as follows:

The risk of cracking will be reduced; joints should be stronger; spalling of joint edges will be minimised; high impact resistance can be expected; fatigue endurance will be better; service life will be extended; and maintenance costs are expected to be low.

7.2.10.4 Review on Advantages

These may include

- Steel FRC will provide durable concrete with improved flexural and fatigue strength, as well as better abrasion, spalling, and impact resistance.
- It will be a better design alternative to traditional concrete with steel fibres.
- The random distribution of steel fibres in concrete can create a stress-free condition throughout, managing the development of micro-cracks and thus enhancing the performance of the concrete.
- Possible reduction of slab thickness as stated and elimination of conventional reinforcement will lead to improved productivity. Steel fibres in concrete will result in cost savings, reduced volume of material, rapid construction and decreased labour costs.

7.2.10.5 Review of Disadvantages

The items may include

- The fibres may not float on the surface of the finished floor. The damaged concrete surface may reveal both aggregates and fibres, which may not be aesthetically pleasing. However, the structural soundness will be maintained.
- In some cases involving structural elements, the cost and economy of the fibre alternative may not be achieved in the design.
- Strict control of concrete waste must be enforced and kept to a minimum. Wasted concrete will also lead to wastage of fibres.

Source:

- CANZAC [69]

7.2.11 Sprayed Steel Fibre Concrete [70]

Applications of sprayed steel fibre concrete have been carried out in various projects. Due to the spraying action, the fibres may align parallel to the target surface in a random manner in two dimensions. The sprayed concrete may change from a brittle to a pseudo-ductile material.

It is important to assess the properties after the spraying applications. These will be associated with flexural strength, impact resistance, post-cracking toughness, and the direction of the applied stresses with respect to the fibres.

7.2.11.1 Toughness

Post-cracking resistance will be crucial, especially in cases where the movements of the substrate must be contained after spraying.

The toughness index may be computed using the load/deflection curve from flexural tests.

$$\textbf{Toughness Index} = A_{fibrous}/A_{plain}$$

The toughness index will increase quickly with the fibre content.

7.2.11.2 Splitting Strength

The Splitting strength can be evaluated through cylinder splitting or point load tests. This will give information on the splitting strength of the core.

The point load index I_s may be computed from

$$I_s = P/D^2$$

where

I_s = point load index
P = failure load
D = length of the core

The point load index will increase linearly with fibre content, similar to that of flexural strength.

7.2.11.3 Quality Control

It is advisable to conduct tests on panels of sprayed concrete at the beginning of the project to establish the relationship between the point load index and flexural strength.

7.2.11.4 Comments

Sprayed fibre concrete will produce anisotropic material, with fibres predominantly located parallel to the sprayed surface, resulting in a more efficient reinforcement system.

Source:

- Robins [70]

7.2.12 Flexural Properties of Steel FRC

The performance of concrete, particularly the properties of flexural strength, crack control, ductility, etc., was studied using different steel fibres.

Tests were conducted on fibre-reinforced elements to study the increased properties of concrete.

7.2.12.1 Observations from Tests Are [71]

The comments from the tests have been arranged separately.

The following basic comments from tests had been reported:

- The increase in flexural modulus observed due to the inclusion of fibres was in the order of 5%–10%.
- The shaped fibres indicated improved strength, strain and elastic properties compared to the straight round fibres.
- The brittle type of failure of plain concrete was transformed into ductile type failure by the presence of steel fibres.

Further comments from the tests will follow in item 7.6.

Source:

- Swamy and Al-Noori [71]

7.2.13 More on Flexural Strength and Toughness

7.2.13.1 FRC [72]

The recommendations have been prepared for computing flexural strength and toughness parameters of FRC. Tests were carried out on specimens and the procedure has been included.

The computation of flexural strengths and the brief procedure have been mentioned in item 7.6.

Source:

- ICI Technical Committee Recommendation [72]

7.2.14 Fracture of Latex Modified Steel FRC in Plane Shear

The application of fibres in concrete as shear reinforcement to avoid brittle fracture has been [73] done and mentioned before. The mechanical and fracture properties may be improved with latex-modified steel FRC. The latex-modified concrete is likely to be stronger than conventional concrete. It may be possible to make brittle concrete more homogeneous and isotropic by using discontinuous fibres.

The tests were carried out on specimens under pure shear.

7.2.14.1 Test Method and Details

It may be noted that steel fibres were found to be effective in improving the shear strength and shear toughness in SFRC and LMSFRC specimens.

Styrene butadiene rubber (SBR) latex polymer and super-plasticisers were used for the concrete mix.

The test results and comments have been included later in item 7.6.

It may be noted that steel fibres were found to be effective in improving the shear strength and shear toughness in SFRC and LMSFRC specimens.

Source:

- Rao and Rao [73]

7.2.15 Steel FRC for Sewer Pipes [74]

The studies were conducted to use steel fibres in concrete for different structural applications as stated.

The comments would be related to the use of steel fibres for pipes.

The main items regarding the use of steel fibres for such applications may be mentioned as follows:

- The structural performance under loads, particularly for crack widths and their distribution.
- The durability aspects of the pipes under normal operative conditions.

It is noted that steel FRC pipes are expected to sustain higher loads before visible cracks can be seen compared to plain concrete pipes. Furthermore, the FRC pipes would fail in a ductile manner and not suddenly as with plain concrete pipes.

Source:

- Perkins [74]

7.2.16 Compactibility of Steel Fibres [75]

The placement and compatibility of steel fibres in concrete are associated with fibres and coarse aggregates. The use of thin and long-shaped fibres may create interlocking, and some water adsorption may take place at large surface areas of the fibres. This may cause drying of the mix.

The coarse aggregates in the mix may create inter-particle friction with the fibres, leading to distortion and agglomeration of fibres during the mixing process.

7.2.16.1 Test Programme for Compactibility

For the tests, the cement-to-sand ratio was maintained at 1.0:2.5, with a water-to-cement ratio of 0.55. The concrete mix was prepared with Ordinary Portland cement, washed pit sand, and graded crushed gravel with a maximum size of 10 mm.

Initially, fibres with dimensions of 0.254 mm × 25.4 mm (i.e., an aspect ratio of 100) were used. This was followed by tests in which the aspect ratio was varied from 14 to 333 14 to 333 (0.15–1.82 mm diameter × 12.5–50 mm long) with different volume fractions to examine various fibre-aggregate interactions.

It is noted that for the same extent of compactibility, the amount of fibres in the mix is important to avoid the balling up of fibres. The effects of fibre balling can be significantly reduced when the aggregate content is increased from 10% to 20% in the mix.

The components related to the volume fraction of coarse aggregates, the volume of fibres, and the geometry of the fibres are all important for the compactibility properties of the fresh fibre concrete. This will also influence the flexural strength of the concrete material in the hardened state.

Source:

- Mangat and Swamy [75]

7.2.17 Micro-silica Steel-FRC Flooring

The use of micro-silica concrete is common in flooring where high impact resistance, strength, durability, and wear resistance are required. Micro-silica is a by-product in the manufacturing of ferrosilicon alloys.

These requirements are particularly suitable for industrial floor applications where the floors are subjected to a high degree of wear and tear, such as in waste transfer stations and scrap metal pens.

The role of micro-silica in concrete is to produce a finer pore structure within the concrete, as the silica fume particles are smaller than the cement particles. Furthermore, silica fume will speed up the hydration process, and the resulting concrete will therefore be of high strength, durable, and have reduced permeability.

For an industrial joint-less floor, high tensile steel fibres were used in the concrete along with micro-silica.

The details of the fibres are given below:

7.2.17.1 Steel Fibres

Fibre length = 40–60 mm
Fibre diameter = 1 mm
Fibre dosage = 40 kg/m^3
Maximum aggregate size in concrete = 20–25

Trials were conducted to finalise the mix design.

M 60 grade of concrete using 10% of micro-silica in the mix was used.

A tough, durable, and joint-less floor slab was produced at a lower cost.

Similar applications of micro-silica, along with steel fibres, may be conducted for other industrial floors.

Source:

- Martin [76]

7.2.18 More on Pile Supported Concrete Slabs with Fibres [77]

It has already been mentioned that concrete with steel fibres for pile-supported concrete slabs will be beneficial, especially for improving performance in terms of post-cracking tensile strength, impact resistance, and toughness.

Tests were conducted on the pile-supported concrete slabs with 40 kg/m3 of steel fibres (Dramix). The wire-drawn fibres, with a diameter of 0.75 mm and a length of 60 mm (i.e., the aspect ratio of 80), were used for M 45 grade concrete. The slabs were loaded in stages, and measurements were taken for deformations, cracks, and crack widths at different stages and recorded.

The test results indicated agreement with the theoretical calculations using yield-line theory for the suspended slabs.

Source:

- Thooft [77]

7.2.19 Stainless Steel Fibrous Concrete [78]

Stainless steel fibrous concrete was used for tunnel support related to an aqueduct project in the UK. The fibrous concrete dry mix was used for the project, to which 25 mm stainless steel fibres were added. Water was added to the nozzle of the spray gun. The work started with using 6 mm aggregate in the mix, which was rebounding off the rock surface. Later, a dry mix using a 3:1 proportion of sulphate-resisting cement, sand, and 3% stainless steel fibres by dry weight, along with a suitable accelerator, was finalised. Initially, steel mesh measuring 80 mm square and 10-gauge wire was tried. Fixing the mesh was time-consuming and labour-intensive. Furthermore, it was found that using steel mesh did not increase the strength of the concrete.

Source:

- Barfoot [78]

7.2.20 More on Stainless Steel Fibres [79]

The stainless steel fibres mentioned have been used with concrete. The fibres may be available in different shapes and cut lengths as per the requirements. The stainless steel fibres are manufactured under strict production control, and checks are performed at regular intervals to maintain quality. The stainless steel fibres are expected to be suitable for high-temperature performance, provide better dispersion, and be easier to handle for extended life.

Source:

- Fiberzone [79]

7.2.21 Glass FRC (GFRC) [80]

7.2.21.1 GRC Developments

Various alternatives have been studied with glass fabrics/fibres reinforcement to match fibre orientation with structural requirements.

Different approaches have been tried to improve GRC properties.

Some of these include

- Treatment of fibre surfaces, optimisation of packing characteristics, fibre bundle size, and testing techniques based on nanotechnology.
- By including super-critical carbon dioxide to accelerate carbonation. Following the treatment, it was possible to increase toughness, matrix strength, and bond strength of unaged samples. It was observed that carbonated samples could retain significant toughness after ageing, while the control samples were completely degraded.
- High fluidity and lightweight pre-mix GRC using new fibres, super-plasticisers, and silica fume.
- Incorporation into renders.
- Coating with fluorinated resin.

Source:

- GRCA Congress [80]

7.2.22 Further on Glass FRC [81]

Glass FRC is a cementitious matrix formulation for high performance. The proportions of a typical mix may be as shown below:

7.2.22.1 The Details

7.2.22.1.1 Mix Proportion

- Cement (Portland) – 50 parts by weight
- Sand (siliceous), 0–1 mm – 50 parts by weight
- Water-to-cement ratio – 0.3–0.35
- Alkali-resistant (AR) glass – 2.5%–5.0% of the mix
- Super-plasticiser – 2 parts by weight of cement

The other items are as follows:

Acrylic Polymer – May be used to simplify the curing process, reducing in-service shrinkage and increasing durability.

Accelerators, waterproofing agents, flow aids, etc. – May be included with different types of cement.

7.2.22.1.2 Types of Manufacturing Process

The types may be mentioned as follows:

- **Hand-Spraying** – In this method, glass fibre roving will be fed to a gun for the application. Fibres will be chopped to the required length and sprayed with unreinforced mortar (slurry) into a mould. The reinforcement quantity will be about 5% (by weight) and 32–37 mm long fibres. This technique may be applied to large areas like facade panels of about 30 m^2.
- **Premix (Cast or Sprayed)** – In this case, 12-mm chopped fibres are mixed into the matrix at the final mixing stage, and the mix is sprayed or poured into the mould.

Mechanical Properties of GRC are indicated in Table 7.4.

TABLE 7.4
Mechanical Properties

Property	Hand Spray-up	Cast Premix
Dry bulk density (tonnes/m^3)	1.9–2.1	1.9–2.0
Compressive strength (MPa)	50–80	40–60
Young's modulus (GPa)	10–20	10–20
Impact strength (kl/m^2)	10–25	10–15
Poisson's ratio	0.24	0.24
Four-point Bending		
Elastic limit (MPa)	7–11	5–8
Modulus of rupture (MPa)	20–30	10–14

(Continued)

TABLE 7.4 (*Continued*)
Mechanical Properties

Property	Hand Spray-up	Cast Premix
Direct Tension		
Bend-over point (MPa)	5–7	4–6
Ultimate tensile strength (MPa)	8–11	4–7
Strain to failure (%)	0.6–1.2	0.2–0.6
Shear		
In plane (MPa)	8–11	4–7
Interlaminar (MPa)	3–5	N.A.

7.2.22.1.3 Other Properties

- **Freeze–Thaw Resistance –** 4% and 23% fall in strain after 25 and 200 cycles respectively, with no reduction in modulus of rupture.
- **Coefficient of Expansion –** $10^{-6/0}$C
- Non-flammable with no polymer.

7.2.22.1.4 Limitations

The following items may be mentioned:

- Loss of ductility after 15–20 years' ageing
- **Manufacturing Costs –** It will be labour-intensive for the premix manufacturing process and hand spraying.
- Non-optimisation of mechanical properties when the fibres are not properly mixed or randomly dispersed into the mix.

7.2.22.1.5 Improvement for Ductility

Two areas have been studied for enhancing ductility as follows:

- Adding meta-kaolin to the matrix and reducing some cement content to replace it with lime. PFA may be used but may not be very effective.
- Using cement that produces less lime.

7.2.22.1.6 Positioning Reinforcement

The following may be noted for better performance:

- Correctly oriented, i.e., aligned with the stresses in the components.
- Positioned in the optimum location within the components, i.e., under maximum tensile stress areas occurring during tensile or flexural loading.

There may be the following types of positioned reinforcement:

- Meshes or nets
- Mats (chopped strand)
- Unidirectional layers (UD)

The summary may focus on the following areas for development:

- Experimenting with new types of reinforcement, such as mesh or mats.
- The loss of mechanical properties may be addressed by using metakaolin or a special cement as mentioned. cement as mentioned.

Source:

- Muller [81]

7.2.23 Further Details on Fibres

The main types of fibres used with concrete are steel, carbon, AR glass fibre, and polypropylene. The important properties of some fibres are outlined below [82]:

The properties are detailed in Table 7.5.

7.2.24 More on Specifying GRC [83]

Glass FRC (GRC) has been successfully applied in various building projects and is increasingly recognised by engineers and contractors. GRC is a cement-based composite material reinforced with special alkali-resistant glass fibres, which are thoroughly mixed with the mortar of cement and fine aggregates, as stated. The final material will exhibit high tensile, flexural, and impact-resisting properties. A key issue will be the performance of the mix design, especially regarding the glass fibre content.

GRC is generally associated with its reference to flexural strength (known as the modulus of rupture, MOR) [83]. GRC can be grouped into three types [83] as Grade-8, Grade-10, and Grade-18. Each grade numerically represents the characteristic MOR value in MPa. The specification covering manufacture, curing, and testing is crucial for ensuring the quality of GRC. The continuous filament process for glass fibres may be used for producing thermal insulation. Alkali-resistant glass compositions may vary but will typically contain zirconium oxide at a level of 15%–20%. Levels below 15% may exhibit poor chemical resistance, with no additional advantages compared to levels above 20%. Glass fibre has an elastic modulus that is about three times that of concrete. This enhances the stability of the cracking stress of the cementitious mix, suppressing micro-cracks and ensuring reliable cracking stress. The elastic limit (stress) is known as the limit of proportionality (LOP).

Requirements for selecting a GRC Product

The following items may be relevant:

- The proper specifications must be followed for the project.
- The design and the grade of GRC to be used should be consistent.

TABLE 7.5
Properties

Property	Concrete	Steel Fibres	AR Glass Fibre	Polypropylene
Elastic modulus (Mpa)	20,000	190,000–205,000	72,000	3,500
Tensile strength (MPa)	Varies	490–980	1,700	550–700
Density (kg/m^3)	2.0–2.5	7.83	2.68	0.9
Fibre diameter (μm)	NA	13–1,000	14	>18

Source: Ferry [82].

- The design should be conducted efficiently.
- The GRC manufacturer should be experienced and equipped to carry out the required tests.
- The history of successful projects/work may be checked if possible.

Source:

- Morrison [83]

7.2.25 Additional Comments on GRC

For quality control, routine testing will be required. A series [84] of wet and dry state tests has been developed and used for an extended period. GRC is normally constructed in thin sections, typically 12 mm thick, depending on the application. The properties of GRC are measured against flexural strength testing on thin samples and the performance contributed by the fibre reinforcement, which may be evident.

For sprayed GRC, typical specification requirements include a characteristic MOR value of 18 MPa [83] and an limit of proportionality (LOP) value of 7 MPa [84].

Source:

- Morrison [84]

GRC cladding was used at Hays Wharf Galaria [85] as a single skin panel with a fascia that would match the stone and masonry dressings of the original building. Steel sub-frames were used to support the GRC cladding and cornices and were given two coats of bituminous paint for protection against corrosion. Stone-like finishes were required on GRC elements to meet the relevant requirements. The thickness of GRC members varied, but was not less than 10 mm; however, the stiffening ribs were 35 mm. After thorough cleaning, smooth finishes were achieved by rubbing down with a fine abrasive pad. The total area of GRC cladding was more than 4,000 m^2, with 2,750 individual units.

Tests [86] were carried out on GRC structural elements to study limit state behaviour. Tests were conducted on elements 28 days and 2 years old. It was noticed that instability of equilibrium under compression was less dangerous than predicted. Furthermore, creep under long-term load occurred but was much less than predicted.

Source:

- Spence [85]
- Domingos and Finzi [86]

7.2.26 Improving the Appearance of Concrete Buildings and Quality of Urban Development by Using e-GRC

The material e-GRC has contributed significantly to the applications of concrete, especially for the external surfaces of buildings. Some comments will be made on this.

The main items are listed as follows:

- Photocatalysis
- Cleaner and brighter concrete surfaces
- Reduction of pollution
- Self-cleaning
- Comments

The basic information is described as follows:

The versatile e-GRC [87] is suitable for economical production of cleaner and brighter facades of buildings and contributes towards a reduction of pollution in the urban environment. The photoactive titanium dioxide (TiO_2) has been found to be a more powerful photocatalytic agent when the particles are reduced to nano-size. TiO_2 is chemically stable, making it suitable for applications related to photocatalysis in construction.

By including TiO_2 in glass FRC, the most efficient and economical way to achieve the benefits of photocatalysis will be provided. TiO_2 contributes positively to environmental quality.

The applications of e-GRC may be mentioned as [87]-

- Cladding panels and façade items.
- Roofing tiles/slates.
- Permanent formwork and form liners for the outdoor environment.
- Motorway and railway sound barriers, safety barriers, and screens.
- Street furniture.

The summary of principal benefits [87] of using e-GRC for construction may be described as follows:

- A much more aesthetically pleasing appearance, primarily due to the brightness of exposed concrete structures, along with the flexibility in shapes and sizes.
- Better air quality, especially in congested urban areas.

e-GRC is a versatile material that can be used in various forms to meet specific architectural requirements and accommodate special details in a building. It is worth noting that e-GRC will change the perception of concrete from a grey, dirty material to structures with brighter, more attractive appearances. Regular checks on the test results for MOR data will be important to compare the values used in the design. e-GRC products must be designed properly by a competent designer with adequate fixing details and arrangements to ensure satisfactory long-term performance, especially under adverse weather conditions.

Source:

- Bartos [87]

7.2.27 Further on GRC for Houses [88]

GRC has been used in the past in various house building activities to replace traditional natural and cast stone. A brief review will suggest the following items in favour of GRC as a composite material.

7.2.27.1 Durability

Handling natural and cast stone must be done carefully at site to avoid damage, especially the protection of edges. This issue can be handled efficiently by using high-quality GRC elements. GRC elements will contain some alkali-resistant glass fibres and will have high tensile and impact strength. With modern finishing methods, GRC items can be manufactured to closely resemble natural and cast stone in appearance.

7.2.27.2 Door Surroundings and Canopies

These items can be manufactured using GRC. GRC surrounds may be constructed on site after completing the masonry work. Other GRC items may include roof tiles, ornamental decorations, and fire surrounds.

7.2.27.3 Weight

GRC elements will have less weight compared to natural stone. GRC products can be produced as thin elements that satisfy design requirements.

7.2.27.4 Proper Selection of Suppliers

Proper manufacturing methods using the correct specifications are essential for GRC products.

The costs of repair work can be significantly reduced by using GRC composite materials.

Source:

- Faulding [88]

7.2.28 Summary of Glass FRC (GFRC) [89]

It is a cement-based composite prepared using cement, sand, admixtures, and alkali-resistant glass fibres, as mentioned before. GRC items are generally thin (8–20 mm) compared to concrete materials.

The following information will be useful:

Light weight, easy to handle with lower erection and transport costs. Due to its lightweight and attractive appearance, GRC can also be used for restoration works, including heritage buildings. It reduces loads in the building, which means less stress on the structure and foundation. Construction costs will be lower, maintenance requirements will be reduced, and expenses will be minimal. The material has low permeability, provides good performance against freeze/thaw attacks, and offers improved fire resistance.

The uses may cover items such as decorative mouldings, sunscreen panels, acoustic barriers, street furniture, site drainage items, sewer liners, roofing slates, and permanent formwork.

Alkali-resistant glass fibres are non-hazardous and can be used as a substitute for asbestos fibres in many applications.

Source:

- Glass-FRC Association GRCA [89]

7.2.29 Thirty Years of GRC Permanent Formwork [90]

When it is difficult to remove the formwork, it may be a more economical option to leave the shuttering in place permanently. GRC permanent formwork has long been used for bridge construction. The following items are relevant for GRC permanent formwork:

It can provide the required level of corrosion protection to steel bars with the minimum cover as per the design.

7.2.29.1 GRC Mix Proportions

- Cement content in GRC may be more than 800 kg/m^3.
- Water/cement content is typically less than 0.4.
  A low level of carbonation is expected when compared with conventional concrete.
- Decorative permanent formwork may be used, as follows:
  Exposed aggregate finishes can be easily manufactured in factory conditions.
  A white finish may be produced by using white cement.
- Site efficiencies may be improved through interaction with the designer/client.
- Long-term durability for buildings/structures may be achieved with the use of GRC.

Source:

- Ferry [90]

7.2.30 Permanent Formwork for Composite Bridge Decks [91]

There are two types of permanent formwork as listed below for the composite decks:

7.2.30.1 Participating Structurally

This type is designed to support the loads of the wet concrete and construction loads. Ultimately, it behaves as a permanent item and works compositely with the completed unit to carry the applied loads.

7.2.30.2 Non-participating Structurally

This category of formwork carries only the load of the wet concrete and construction loads.

7.2.30.3 Advantages of Permanent Formwork

The items to be mentioned are as follows:

- The need for falsework may be eliminated.
- Standardisation in design/construction may be done.
- The requirements for site skill levels may be reduced.
- The fixing time may be faster and could eliminate the time for stripping.

This will allow early access to the floor for carrying out various activities.

The additional advantages in bridge projects may include

- It will provide a safe working platform early in the deck. This will be greatly appreciated when the bridge must span a rail track.
- It will avoid stripping of formwork in confined and difficult spaces, especially when working at heights.
- GRC panels may be used as flat sheets for spans of up to approximately 800 mm and profile sheets for spans of up to 1.2 m. A temporary support system may be considered for greater spans with profile sheets.
- Both glass fibre reinforced plastics (GRP) and glass fibre reinforced cement (GRC) may be used for non-participating types of formwork.
- GRC may be considered for spans of up to 1.2 m as an economical choice for permanent formwork where closely spaced beams will be available for the applications.

Source:

- McClelland [91]

7.2.31 Flexural Behaviour of Steel and GFRP Reinforced Concrete Beams

The experimental programme was conducted on concrete beams to study the behaviour of glass fibre reinforced polymer (GFRP) used along with steel as reinforcement, particularly to observe the performance at the serviceability stage. Simply supported beams were tested by applying a load at the mid-span.

7.2.31.1 Observations from Tests

The comments mentioned below are very relevant:

- The type of reinforcement is important for the behaviour at the service load stage.
- There was less cracking in steel reinforced beams compared to GFRP reinforced beams for the same applied load.
- GFRP reinforced beams deflected considerably more than steel reinforced beams for the same applied load.

- With higher concrete compressive strength, the deflections were lower for a particular load for both types of beams. This was more noticeable in GFRP reinforced beams.
- The tension stiffening effect may be observed in GFRP beams due to increased concrete tensile strength.
- Increased cracking in GFRP beams affected the bond between concrete and reinforcement compared to steel reinforced beams.
- The service load may be considered as 70% of the failure load.
- GFRP reinforced concrete beams experienced deflections equivalent to span/101 (approximately) under service load conditions.

7.2.31.2 Ultimate Strengths

The capacity of GFRP beams was greater than that of steel reinforced beams due to the higher flexural strength of GFRP reinforcement. The failure mode may be compared to an over-reinforced concrete beam [42].

It may be summarised from tests that GFRP-reinforced beams had higher failure loads, reinforcement strain, deflections, and cracking compared to similar steel-reinforced concrete beams. The deflections at service load in GFRP-reinforced beams were higher than the limiting deflection value of span/350, whereas the deflections of beams reinforced with steel were less than span/500.

Source:

- Rooney and Taylor [92]

7.2.32 The Influence of Glass, Polypropylene and Steel Fibres on the Physical Properties of Concrete [93]

The study was carried out to assess the influence of three types of fibres – glass, polypropylene, and steel – on the physical properties of concrete. In this connection, four dosage rates were considered: 1, 5, 10, and 20 kg/m^3 of concrete, to examine the following properties:

- **Workability –** slump test
- Abrasion resistance
- **Crack Formation –** from visual examination
- **Flexural Toughness –** first-crack strength
- Impact resistance
- Flexural strength/modulus of rupture

The characteristic strength of concrete was 35 N/mm^2, and a slump of 50 mm was considered for the study.

The following comments will be important [93]:

- Concrete workability was least affected by glass fibres and more affected by polypropylene fibres.
- All fibres were effective in compensating for shrinkage cracks.
- All fibres increased the abrasion resistance of concrete up to 10 kg/m^3.
- The flexural strength increased with the addition of glass fibres.
- The increase in abrasion resistance with glass fibres was more consistent.
- All fibres increased the impact resistance of concrete.
- At lower dosages (i.e., 1 and 5 kg/m3), glass fibres were less effective in providing toughness.
- At dosages of 10 kg/m^3 and above, glass fibres were found to perform best in terms of absolute toughness, first-crack strength, flexural strength, and first-crack toughness.

Source:

- Sadegzadeh, Kettle, and Vassou [93]

7.2.33 Carbon Fibres [1]

Carbon fibres have been used successfully in concrete. These fibres are produced by using controlled pyrolysis and cyclisation of some organic fibre precursors. The common types are precursor polyacrylonitrile (PAN) and precursor mesophase pitch (MPP).

The first type is a synthetic fibre, and the second is manufactured from the destructive distillation of coal. The sequence of the procedure includes stabilisation, carbonisation, graphitisation, and the necessary surface treatment.

Three different types of carbon fibre may be referred to from heat treatment temperatures. These may be classified as follows:

Type-1: This is the stiffest carbon fibre and requires high heat treatment.
Type-2: This is the strongest carbon fibre, which has been carbonised at a temperature that provides the greatest tensile strength.
Type-3: The lowest heat treatment will be required for this type. This will be the cheapest carbon fibre. The stiffness will be less than that of the previous types, but the strength will remain good. The carbon fibres may be used with a polymer or a cement matrix.

Source:

- Hollaway [1]

7.2.34 Further on Carbon Fibres

Short pitch-based carbon fibres may be used in concrete. The following points may be noted from a mix design:

7.2.34.1 Details

The quantity may be 0.5% by weight of cement, 0.189 volume % (i.e., vol %) of concrete, with a dispersant, chemical agents, and silica fume in the concrete. The fine and coarse aggregates were included in the concrete mix.

The following comments may be made:

Flexural strength increased by 85%, compressive strength increased by 22%, optimum water-to-cement ratio was 0.5, flexural toughness increased by 205%, the slump was 102 mm, the increase of material price was 39%, and the air content was 6%. Durability related to freeze–thaw was increased.

The aggregate size had nominal effects on the aforementioned properties. The minimum carbon fibre content required for the increase in flexural strength was 0.1 vol %. The fibre length decreased from 12 mm before mixing to 7 mm after mixing. The drying shrinkage and electrical resistivity were reduced by 90% and 83%, respectively.

Source:

- Chen and Chung [94]

7.2.35 Aramid Fibres

The aramid fibres are man-made materials of organic type, having a high degree of crystallinity, and these fibres may be used for the composites. Two types of stiffness grades are normally

TABLE 7.6
Mechanical Properties

Material	Fibre	Elastic Modulus (GPa)	Tensile Strength (MPa)	Density (g/cm^3)	Maximum Temperature (°C)
Aramid	49	125	2,760	1.44	180
	29	83	2,750	1.44	180

Source: Hollaway [1].

available: one with a modulus of elasticity of 60 GPa and the other with a modulus of elasticity of 130 GPa. The higher fibre modulus may be used for polymer composites.

The fibres are produced by an extrusion and spinning technique. Different types of aramid fibres may be available.

The mechanical properties of fibres [1]:

A few basic items are mentioned below in Table 7.6:

7.2.36 Synthetic/Non-metallic Fibres

The synthetic fibres are prepared from various thermoplastic polymers. The relevant ones include polyolefins [1] (polypropylene and polyethylene), polyamide (nylon), and polyester. The polyolefins are considered for preparing cement/mortar composites. This, along with other types, may also be used for the requirements of geo-synthetics.

Source:

- Hollaway [1]

The use of structural synthetic fibres will control plastic shrinkage cracking in concrete floor slabs and bridge decks and will demonstrate better performance against impact resistance and tensile strength in a more homogeneous manner than steel. This will ensure long-term safe and satisfactory performance, as already highlighted [30].

It has been noted that polypropylene mono-filament fibres have been used in concrete floor slabs to prevent plastic shrinkage and drying shrinkage cracks. However, due to the low E-modulus of the fibres, there are limitations for such applications. Research workers have been trying to manufacture fibres that will have better effects on tensile strength, impact resistance, and E-modulus without affecting other properties of concrete. In this connection, non-metallic reinforcing fibres will be preferred to eliminate steel reinforcement in pre-cast concrete units, piles, manhole rings, ground floor slabs, and paved areas.

The following items will be important in relation to non-metallic fibres [30].

Configuration.
Chemistry
Contents.
Correct Length

The above items have been covered separately before in this chapter.

Source:

- Jowett [30]

7.2.36.1 Additional Comments on Improving Performance of Concrete in Fires

It has been noted that the use of polypropylene fibres in concrete will increase the permeability [95] of heated concrete under fire, releasing the built-up pressure and preventing explosive spalling.

Hence, the performance of concrete in fires can be improved by using these fibres. The diameter of fibres may vary from 5 to 30 μm. Furthermore, it has been found that the greater the quantity of individual fibres present in the concrete, the higher the number of channels and voids that will be generated during the fire. This will help to release the pressure and will effectively protect the structure.

Source:

- Kitchen [95]

7.2.36.2 Further on Fire Tests

The following comments may be made:

The tests were conducted using both mono-filament and fibrillated fibres with a dosage of 2 kg/m^3. The concrete sections were heated up to 1,300°C using furnaces. It was noted that concrete test panels with mono-filament fibres showed greater explosive spalling resistance compared to the concrete containing fibrillated fibres [96].

Source:

- Kitchen [96]

7.2.37 MSFs

Fibre mesh polypropylene fibres will inhibit the formation of micro-cracks, which can create weaknesses within the hardened concrete.

With MSFs, attempts have been made to replace steel fabric in concrete foundations, tilt-up walls, slabs, and roadways. The use of MSFs has been shown to be worthwhile in various applications. Attempts have been made to replace steel reinforcement with MSFs in projects [97] as stated.

It had been noted from the study/research [98] that the use of MSFs in concrete was efficient under seismic conditions due to the three-dimensional distribution nature of the reinforcement. Furthermore, structural elements performed more ductile, and the concrete with these synthetic fibres would be able to absorb more energy.

STRUX 90/40 synthetic macro-fibres (flat fibres 40 mm long and cross-sectional aspect ratio = 90) have been used successfully over metal decking in various projects [97]. It was found that the combination of synthetic macro-fibres with metal decking could reduce construction time and was cheaper to use than more traditional methods.

Strux 90/40 synthetic macro-FRC [97] has been used in Blackpool's promenade. Such applications of synthetic fibres are potentially very effective in aggressive marine/coastal areas.

Source:

- Ridout [98]
- Perry [97]

7.2.38 MSFs for Macro-Size Projects [99]

MSFs have been used in mining areas. There are also other applications related to cases with ground movements and where the risk of corrosion is high.

The comments on the uses may be mentioned as follows:

- High-speed railway tunnel
- Devil's slide road tunnel
- Hindhead road tunnel
- Hydro-electric power station

7.2.38.1 Further Comments

MSFs have been used in various projects both as primary and secondary linings. The fibres are expected to perform very well technically compared to steel fibres. For a design life of 50–100 years, MSFs could be considered a good option and a viable alternative.

Source:

- Ridout and Martin [99]

7.2.38.2 More Comments [58]

MSFs may be used with the same length as steel fibres, with [58] a different anchoring system for improved bond. A higher dosage may be considered for fibre reinforced sprayed concrete (FRSC).

Careful assessment will be required to decide on the particular type to be used for a project considering different aspects/requirements.

Source:

- Swannell [58]

7.2.39 Micro-Polypropylene Fibres

The addition of specially engineered mono filament polypropylene fibres has been found to be beneficial specially for improving early tensile strain capacity of the concrete after placing. Attempts, in the past, had been done to use concrete with air-entraining agent (AEA). The use of AEA, in concrete may not be always satisfactory.

The following will be relevant [100]:

Cement type, source, and fineness, concrete and ambient temperatures, mix proportions, admixture type, dosage, transport and mixing of concrete, concrete workability, placing, compacting, and finishing the concrete.

It will be important to ensure that the specifications are not only satisfactory at the time of testing but also include the specified amount of entrained air in the concrete, so that protection against freezing and thawing is not lost after concreting.

It is commented [32] that the combination of macro- and micro-synthetic fibres will provide a significant advantage in both design and performance for external applications. This type of design may have different applications such as

Lorry parking areas, external hardstandings, airport taxiways, roadways, driveways, port/dock areas, walkways/footpaths/bike paths, boat ramps, golf cart tracks, etc.

The dosage rate may be as follows for the applications:

- Walkways, footpaths, and other areas with minimal load applications – 2–4 kg/cum.
- Ground bearing slabs, driveways, parking areas, and higher load applications – 4.5–7 kg/m^3 [32].

The combination of MSFs and micro-synthetic fibres in concrete, along with the effects of three-dimensional reinforcement, will inhibit the development of plastic and settlement cracks and improve resistance to impact, shatter, abrasion, and freeze/thaw cycles. This approach may be considered a replacement for steel wire fabric, steel fibres, and air-entraining admixtures in concrete [32].

Source:

- Kitchen [100]
- Atkinson [32]

7.2.40 How Do Polypropylene Fibres Reduce Explosive Spalling?

Comments are made as follows:

It may be mentioned that at 160°C, polypropylene fibres in concrete will start melting, resulting in a reduction of the volume of individual fibres. With the increase in heat, the fibres will degrade and will begin to ignite at about 360°C.

They alter the constituent materials, and the remnants of the fibres occupy only 5% of the voids. The gases generated will be burnt away by the fire or dissipate into the atmosphere.

The remaining voids will provide channels created by the fibres and create escape routes for the water vapour. The vapour pressure will be released, and explosive spalling may be avoided.

7.2.40.1 Fire Tests Channel Tunnel Rail Link (CTRL)

Some comments are as follows:

- Monofilament polypropylene fibres may be used in the concrete.
- The use of lightweight aggregates is not recommended for high-strength impermeable concrete, especially for tunnel lining.
- Further trials/tests may be required to finalise the concrete mix design.

To reduce explosive spalling, it will be necessary to consider compressive strength, load level, and the use and quantity of steel reinforcement required for the project.

Further comments on fire tests have been provided separately.

Source:

- Kitchen [96]

7.2.41 Further on Fire Protection to Concrete Linings in Tunnels [101]

The study was carried out in the Netherlands to assess the use of polypropylene fibres in concrete for the passive fire protection of concrete linings in tunnels. The application of polypropylene fibres in concrete may be considered an alternative approach for passive protection.

The tests were conducted to meet the requirements of M 50 concrete grade, using a water-to-cement ratio below 0.4.

The critical requirement for passive protection in a fire is to prevent concrete spalling.

Initial comments from the test are as follows:

- The explosive spalling will not stop once it starts. This is likely to continue until the failure of the structure, which may occur in less than 15 minutes.
- At the concrete surface temperature over 200°C and with a heating rate of 2°C per minute, spalling may take place.

For passive fire protection, polypropylene fibres may be used in concrete. Concrete containing polypropylene fibres will be less sensitive to spalling. The fibres will melt during the fire and will provide an escape route for the steam pressure, thereby reducing the likelihood of spalling.

Source:

- Varley [101]

7.2.42 More on Improving Concrete Performance in Fires [95]

The following information is relevant for using polypropylene fibres in concrete:

- The fibres have been developed specifically to melt in concrete at about 160°C, as mentioned.
- It has been observed that the greater the quantity of individual fibres present in concrete, the higher the number of voids that will be created during the fire.
- It has also been noted that the extrusion process will ensure that the fibre is not degraded and will provide effective protection for the life of the concrete structures.
- Further studies may allow lower volumes of fibres to be used without compromising the effectiveness of the passive fire protection system.
- It should be noted that the use of polypropylene fibres in concrete projects, both above and below ground, will play an important role in the performance benefits of the materials, particularly in increasing the permeability of heated concrete and preventing explosive spalling.
- The use of polypropylene fibres in projects is expected to increase in the future.

Source:

- Kitchen [95]

7.2.43 Fibre–Mesh Polypropylene Fibres

Fibre–mesh polypropylene fibres (fibre–mesh PPF) are manufactured using olefin fibres prepared from polymers or co-polymers of propylene. Polypropylene is produced by melt spinning the molten polymer, followed by stretching and orientating the fibre molecules. The molten or semi-soft propylene is extruded to make flat sheets or filaments. Two types of PPF can be produced.

By using fibre–mesh, it is possible to increase the tensile strain capacity of concrete during the setting period. This will inhibit the formation of macro-cracks within the hardened concrete and avoid the development of plastic cracks on the surface. Fibre–mesh PPF in concrete will also control the evaporation of water from the surface but cannot be considered a substitute for good curing.

The other advantages of using 1% by volume of fibre–mesh PPF include resistance to bleeding, impact, plastic shrinkage cracking, freeze–thaw cycles, and abrasion. Furthermore, it can reduce permeability, enhance residual strength, provide fire resistance, and offer anti-microbial protection.

Studies report that using 1% by volume can improve shear strength, which would be beneficial in designing slab–column junctions.

Source:

- Better concrete with fibre mesh [59]

7.2.44 Further on Using Polypropylene FRC

The following items will be briefly discussed:

7.2.44.1 Health and Safety Issues

Considering health and safety issues, the use of polypropylene FRC may be preferred over steel fabric and reinforcing bars.

The benefits of using polypropylene FRC can be categorised into the following three important areas:

- Construction
- Maintenance
- Demolition

The relative density of polypropylene fibres is less than that of steel, and the risk of handling on site will be greatly reduced.

7.2.44.2 Applications in Flooring

The use of polypropylene fibres in floor applications is increasing, particularly in the retail sector.

In terms of sustainability, the carbon footprint of one tonne of polypropylene fibres is less than that of one tonne of steel.

The future use of polypropylene fibres in projects appears to be very promising.

Source:

- A kitchen [102]

7.2.45 Polypropylene Fibres Control Explosive Spalling in HPC

There are various issues that can be associated with the spalling of concrete. It has been observed that factors such as concrete permeability, moisture content, presence of reinforcement, and rate of heating are all very relevant to concrete spalling.

Primarily, two types of concrete spalling may be mentioned in relation to fire:

- Gradual spalling
- Explosive spalling

7.2.45.1 Gradual Spalling

This type is associated with a situation where the deterioration of concrete occurs slowly. The rate of temperature rise in the concrete is gradual. The moisture within the concrete will have time to escape from the sides exposed to heat, resulting in less pressure build-up.

7.2.45.2 Explosive Spalling

In this condition, the concrete is heated very quickly. As temperature rises, the moisture in the concrete pores transforms into vapour. This vapour increases the pore pressure, which cannot escape from the concrete. As vapour pressure builds up, it may exceed the tensile capacity of the concrete, leading to explosive spalling.

7.2.45.3 Fire Tests

Various tests have been carried out to assess concrete spalling.

7.2.45.4 Practical Considerations

Studies indicate that structures with HPC can be protected from explosive spalling through the inclusion of monofilament polypropylene fibres in the concrete mix. It is crucial for engineers to gain confidence in using these fibres and to specify them for various structures, including high-rise buildings and tunnel construction. It should be noted that 18-μm fibres can entrap more air in the concrete than larger diameters. Increased air content in concrete can lead to a loss of strength and increased permeability. By using monofilament polypropylene fibres in concrete, it is possible to protect the lives of those attempting to escape a fire and those fighting it.

Source:

- Atkinson [103]

7.2.46 Sprayed Concrete Containing Polypropylene Fibres

Sprayed concrete containing polypropylene fibres has been used in various projects, including applications in tunnels.

The advantages are briefly summarised below:

- Less rebound.
- Reduced leakage.
- The steelwork within the concrete is protected.
- Increased strength and toughness.
- Good adhesion.
- Improved concrete workability.
- More consistent concrete mix for the project.
- Low permeability for construction.

Source:

- Kitchen [96]

7.2.47 Polyester Fibres

It may be stated that various applications have already been carried out in the past using polyester fibres in cement concrete. However, it has been observed that polyester fibre may not be suitable due to its degradation in the alkaline environment of concrete. This is due to the presence of calcium hydroxide in Portland cement material.

Source:

- US report [17]

7.2.48 Basalt Fibres

7.2.48.1 Introduction

Attempts have been made to produce basalt fibres in the USA in 1923. Significant development work on this has been carried out after World War II.

Basalt is a material prepared from extremely fine fibres of basalt and consists of the minerals plagioclase, olivine, and pyroxene. The material can be compared with fibreglass and will provide better physico-mechanical properties than fibreglass. It will be cheaper than carbon fibre. The material has been used for fireproof textiles in the automotive and aerospace engineering industries.

7.2.48.2 Manufacturing

The production of basalt continuous fibres (BCF) is associated with melting, homogenisation of basalt, and fibre extraction. It is a one-stage manufacturing process, and the basalt may be heated only once.

More work on BCF has been carried out using cold technologies with low energy costs.

7.2.48.3 Further Information

Basalt fibres can be prepared from a single crushed basalt after carefully selecting the source. Basalt with high acidity (i.e., more than 46% silica content) and low iron content is preferable for

TABLE 7.7
Properties

Property	Value
Tensile strength	2.8–3.1 GPa
Elongation at break	3.15%
Elastic modulus	85–87 GPa
Density	2.67 g/cm^3

the manufacture of fibres. No other materials are included during production. The basalt is washed properly and then melted.

The temperature needed to melt crushed and washed basalt would be approximately 1,500°C.

The molten rock is then extruded through small nozzles to create continuous filaments of basalt fibres.

The filament diameter of basalt fibres may vary from 10 to 20 μm. This will be above the respiratory limit of 5 μm, making basalt fibres a possible replacement for asbestos fibres.

Basalt fibres will have a high elastic modulus and high specific strength, which will be three times greater than that of steel. The properties are mentioned in Table 7.7.

The important applications of continuous basalt fibres will be for the production of basalt reinforcing bars, which may replace traditional steel bars in construction work.

Source:

- Wikipedia [104]

7.2.49 Summary and Comments [105]

The following items may be mentioned:

Inert mineral fibres, resistance to corrosion, exceptional properties, improved flexural strength of concrete, improved tensile strength of concrete, and better toughness for concrete.

Research work/study with basalt FRC has been carried out using the percentage volume fraction of fibres as 0.5 (i.e., 13.0 kg/m^3), 0.75 (i.e., 19.5 kg/m^3), and 1.00 (i.e., 26 kg/m^3). The increases in compressive strength, splitting tensile strength, and flexural strength were observed to be in the range of 26.79%, 42.71%, and 44.06% respectively, using a dosage of 0.5% of basalt fibres when compared with control concrete. Furthermore, the basalt fibres were found to be of an amorphous and hydrophilic type.

Source:

- Singh [105]

7.2.50 Natural Materials and Fibres

7.2.50.1 Introduction [4,21]

It may be mentioned that stones, wood, and clay were initially used by humans mainly for constructing structures/accommodation [4]. Straw from grass plants, fibrous leaves, etc. were used as natural fibres for roofing material. At the early stage, people used stone axes, spears with wooden handles, wooden bows, fishing nets prepared with vegetable fibres, etc. There were attempts to improve and obtain efficient materials with better properties and performance. Developments occurred over time, and composite materials were introduced in stages.

7.2.50.2 Natural Fibres [21]

Natural FRC will be a special type of concrete consisting of small diameter discontinuous, discrete natural fibres of various origins randomly mixed in a cementitious matrix. The uniform dispersal of fibres in a cementitious mix will distribute stresses and increase resistance to cracking, shock loads, and impact, whilst also improving the ductility of the matrix and enhancing energy absorption of the composite, as mentioned previously. These types of construction materials with natural fibres also possess good acoustic and thermal properties [21].

The basic requirements of natural fibres, as already mentioned for other fibres, will be to provide high tensile strength and elastic modulus, proper chemical and geometric stability, satisfactory bond at the interface with the matrix, and good durability.

A few natural fibres that have been tried by engineers and researchers are mentioned briefly below:

Coconut Fibres	–	The outer covering of fibrous material from a mature coconut fruit is used. The fibres are normally between 150 and 350 mm long. They contain lignin, cellulose, tannin, pectin and other water-soluble materials. The physical properties of coconut fibres may vary from region to region.
Sugarcane Bagasse Fibres	–	Bagasse is the fibrous residue obtained from sugar production after extracting the juice from the cane stalks. Sugarcane belongs to the grass family. It can grow up to approximately 6 m high with a diameter of 6 cm, depending on the species and cultivation area.
Sisal Fibres	–	These are the strongest natural fibres for applications. They can be used for gypsum plaster sheets and are suitable for producing high-quality FRC.
Bamboo Fibres	–	Bamboo fibres are strong in tension but have a low modulus of elasticity and high water absorption capacity. They are grown in tropical and sub-tropical areas. Bamboo belongs to the grass family and can grow to about 15 m high, with a diameter ranging from 25 to 100 mm Comments on bamboo have been made separately
Palm Fibres	–	The fibres may be extracted from decomposed palm trees. These fibres are often brittle, having low tensile strength and modulus of elasticity, along with high water absorption capacity.
Flax Fibres	–	Flax fibres are strong and durable. They are obtained from a blue-flowered plant.
Jute Fibres	–	Jute fibres are extracted from mature jute plants after submerging them in water for about 4 weeks, allowing the bark to decompose. The fibres are then exposed, followed by manual stripping from the stems, washing, and sun drying. Various types of jute fibres with different properties may exist. Jute is available in Bangladesh, India, Thailand, and China.
Wood Fibres	–	Wood fibres are generally obtained from softwood trees such as pine, larch, and spruce. These fibres are strong and durable.
Vegetable Fibres	–	The potential fibres for reinforcement have been found to be akwara, water reed, elephant grass, plantain, and musamba. The geometry of these fibres may vary; they can be circular, rectangular, or elliptical in cross-section and tapering along the length. The equivalent diameter may vary between 1 and 4 mm.

7.2.50.3 Further Comments

Apart from the ability to sustain loads, natural FRC must be durable to meet the aforementioned requirements. Limited information on durability for natural FRC is available. However, the success of natural fibre reinforced composites depends more on the durability of the material against the environment to which they are exposed than on their strength properties. The limited dimensional stability of natural fibres due to changes in moisture may present durability problems [21].

The large-scale application of natural FRC for structural works is rather limited [21]. It seems only a few special applications have been undertaken where natural FRC has formed an outer layer to other materials. This outer layer can serve as a permanent strong and tough covering over a weak core and can help provide architectural and ornamental features required in the design.

The greatest advantage of natural fibre cement composites is that such fibres are available in various forms [21] and in abundance, freely, in many countries where they can be used for housing, especially for low-cost construction as an alternative building material.

Source:

- Swamy [21]
- Composite materials and composites [4]

7.2.51 Flexural Behaviour of the Composite

Several theories can be employed for evaluating the flexural tensile strength of fibre-reinforced cement composites. The simplified stress distribution [21] is shown in Figure 7.4.

7.2.51.1 Comments

Uniform stress distribution in the tensile zone

The neutral axis has been assumed at 1/4th the depth of the compression face.

$$\text{Moment of resistance} \quad MR = f \cdot (13/32) \cdot D^2$$

where

f = uniform stress in the tensile zone
D = depth of the section

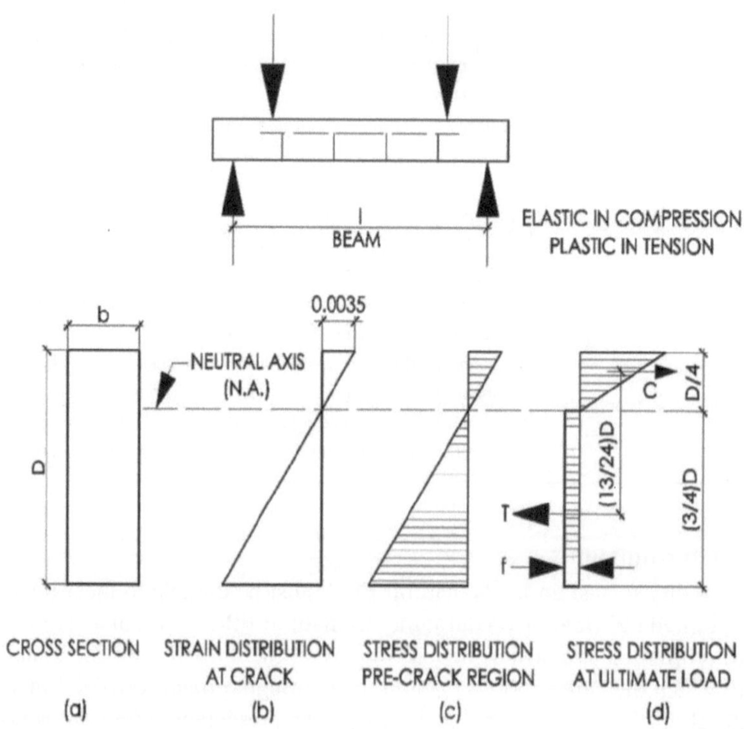

FIGURE 7.4 Stress/strain distribution of the composite.

Alternatively, a stress block [11] beyond the elastic range may be assumed for computing the moment of resistance of the concrete element. This stress block will be more appropriate if the design moments are calculated based on yield line analysis.

Source:

- Swamy [21]
- Vollum [11]

7.2.52 Further Review on Different Types of Fibres

Some applications of different fibres have been presented later in this chapter.

The common types are as follows:

- **SFRC** – steel fibre-reinforced concrete
- **SNFRC** – synthetic fibre-reinforced concrete
- **GFRC** – glass fibre-reinforced concrete
- **NFRC** – natural fibre-reinforced concrete

7.2.52.1 Other Developments of FRC

This may be mentioned as follows:

High-fibre volume micro-fibre system
Polymer concrete
Compact reinforced composites

Source:

- Kaur and Talwar [33]

7.2.53 Comparison between Macro-Synthetic and Steel FRC [106]

The comparison between two fibres has been studied specifically in respect to limiting crack widths. In concrete, control is generally achieved by providing conventional steel reinforcement using reinforcing bars or fabric. Both macro-synthetic and steel fibres can alternatively perform the same function.

It has been observed through applications that MSFs with sufficient dosage rates can provide equal or better control than nominal fabric reinforcement and can improve the flexural capacity of ground-bearing floor concrete. The typical dosage rate may be 2–9 kg/m^3. The fibres may be 15–60 mm long and have equivalent diameters of 0.4–1 mm.

MSFs have been used in sprayed concrete for ground support and in composite metal decking arrangements, apart from ground-bearing floor slabs. Similar performance has been noted with steel fibres when the proper type and dosage are used.

7.2.53.1 Properties of Macro-Synthetic and Steel Fibres

The properties, and hence the performance, of both fibre types can vary considerably among different suppliers. Proper comparisons must be made after considering the variations in the properties of the two fibres.

7.2.53.1.1 Flexural Toughness

Considering the same fibre volume as crimped MSFs and the higher modulus of Strux 90/40 fibres, Strux showed better load-carrying characteristics. The flexural toughness of Strux was found to be closer to the behaviour of FRC with 39 kg of 50 mm long fibres than to crimped macro-synthetic polypropylene FRC.

7.2.53.1.2 Creep Resistance

It was noted that specimens reinforced with high modulus MSFs showed creep behaviour similar to that of steel fibres. However, low modulus fibres showed a wide range of creep coefficients.

7.2.53.2 Further Comments

It is worth mentioning that FRC with MSFs featuring high-end properties, like Strux 90/40, has demonstrated similar flexural toughness and creep resistance performance when compared with specimens of concrete containing steel fibres.

Source:

- Hathaway [106]

7.2.54 Polypropylene and Steel Fibre Combination [107]

In this study, polypropylene and steel fibres have been considered to assess the properties and benefits of combining the two products. This type of hybrid system may be used for all types of concrete applications but will be more suitable for ground-supported slabs.

The fibres are combined in the proportion of 0.1% polypropylene fibres and 0.25%–1% steel fibres by volume. The polypropylene fibres, at a rate of 0.9 kg/m^3, are added to the batching plant either in a pan mixer, in the drum, or alternatively in the truck mixer. The steel fibres, at a rate of about 25 kg/m^3, are then added in the batching plant or on-site, either manually or through a loading gantry, or via mechanical arrangements using conveyors, hoists, or automated dispensers.

The fibres will be quickly distributed throughout the mix without balling or clustering.

7.2.54.1 Fibres Increase the Tensile/System

Polypropylene fibres increase the tensile strain capacity of the material and help concrete to resist micro-cracks and capillaries, thereby reducing surface permeability, as mentioned.

Steel fibres are effective when the concrete has hardened.

The mix design will be important for compressive, flexural, and tensile strength for the hybrid concrete for the development of the first crack, rather than the dosages of polyurethane or steel fibres in the mix. However, post-crack performance (i.e., strength and durability) will be significantly improved with the increase of steel fibres.

Concrete will be protected from explosive spalling when high-strength concrete is used with fibres in high-rise constructions, sprayed constructions, tunnel linings, etc., as demonstrated in fire tests.

Source:

- Carr [107]

7.2.55 Fibres in Floor Screeds

Polypropylene fibres have been used for screeds and mortars.

They may replace concrete reinforcement and wire mesh in the screeds. There are various applications.

A proportion of 1:3 for the mix is intended for heavy and frequent traffic, 1:4 may be used for medium to light traffic, and 1:5 mixing may be done for light traffic.

The following items are relevant for screeds:

- Well-graded sharp sand
- Good-quality cement
- 60 mm long polypropylene fibres and 60 μm in diameter
- Suitable admixtures as per requirements
- Adequate water in the mix to produce a semi-dry consistency for easy placing and proper compaction.

7.2.55.1 Testing In-Situ Screed

The testing of screeds may be conducted by dropping a load.

Further information will follow in item 7.7 of this chapter.

7.2.55.2 Review and Comments

The following comments may be made regarding the screeds with fibres and the advantages over common cement–sand screeds containing crack control reinforcement:

Better pumping characteristics; placement and compaction will be easier; three-dimensional effects for diffusing the cracks; no lapping, cutting, and placing of mesh; better impact resistance; expected reduction in plastic shrinkage cracking; elimination of fibre corrosion; and nominal wastage.

Source:

- A kitchen [108]

7.2.56 Review and Summary of FRC [109,110]

7.2.56.1 Introduction

FRC is a composite material prepared with a mixture of concrete and discontinuous, discrete fibres, which are uniformly dispersed into the mix to achieve various advantages. There are different types of fibres that may be used with the concrete [110].

The fibres are small pieces of reinforcing material with different characteristic properties. They can be described by their aspect ratio (i.e., the length divided by their diameter). The aspect ratio may vary from approximately 30 to 150 [109,110].

7.2.56.2 Types of Fibres [110]

These include

Steel – normally used; glass fibres – discussed separately; carbon fibres – discussed separately; stainless steel fibres – mentioned separately; polypropylene – discussed separately; nylon fibres – can improve impact strength and have a low modulus of elasticity; basalt fibres – mentioned separately; asbestos fibres – mentioned separately; coir fibres – organic or natural fibres; and natural fibres – discussed separately.

7.2.56.3 Important Items for the Properties of FRC

7.2.56.3.1 Relative Fibre Stiffness of Matrix [110]

The modulus of elasticity of the cement matrix should be less than that of the fibres.

The bond between the cement matrix and fibres will be important and should be strong for high tensile strength of the composite and also for the transfer of stresses.

7.2.56.3.2 Aspect Ratio of the Fibres [109,110]

This is an important item for FRC. It has been observed that the relationship between the strength and the aspect ratio of the composite is linear up to the aspect ratio of 75. Above an aspect ratio of 75, the toughness and strength will be reduced.

7.2.56.3.3 Volume of Fibres [109,110]

It has been observed that the strength and toughness of the fibre composite depend on the volume of the fibres used. The relationship is nearly linear. This means that with an increase in the volume of fibres in the mix, the toughness and strength of the composite will also increase. However, a large volume of fibres may cause segregation of the concrete/mortar.

7.2.56.3.4 Workability and Compaction of Concrete [109]

Fibres will decrease the workability of concrete, as mentioned. With the introduction of fibres into the mix, it may be difficult to achieve adequate compaction. Non-uniform distribution of fibres may also lead to poor workability.

7.2.56.3.5 Orientation of Fibres [109]

The orientation of fibres will be random within the material, which differs from the arrangement of conventional reinforcement. Conventional reinforcement is placed in the directions required based on specific assumptions.

7.2.56.3.6 Size of Coarse Aggregates [109]

The size of aggregates may be restricted to 10 mm.

7.2.56.3.7 Mixing of the FRC [109]

The mixing and preparation of the FRC must be done carefully to avoid segregation and balling of fibres, ensuring the mixing is carried out uniformly.

7.2.56.3.8 Necessity of FRC [110]

The main benefits may include

- Increased tensile strength of concrete
- Reduced air/water voids
- Increased durability of concrete
- Some fibres exhibit good resistance to creep. Orientation and volume of fibres will be important for creep performance.
- Small, closely spaced, and uniformly dispersed fibres can act as crack arresters, improving the static and dynamic properties of the composite.

Other benefits are mentioned separately.

Source:

- Cement Concrete.org [109]
- Constructor [110]

7.2.57 Applications of FRC

The applications of FRC have been covered in detail later, in item 7.7 of this chapter.

A few may be mentioned as follows:

Cladding items for buildings, door and window frames, permanent and temporary formwork, pressure pipes, sun breakers, bus shelters, decorative grills, staircases, wall panels, sanitary items, manhole covers, etc.

Source:

- Mehta and Monterio [3]
- Swamy [21]
- Cement concrete.org [109]
- Ganguly [7]

7.2.58 Current Developments in FRC

The following items may be mentioned [109]:

- High-fibre volume micro-fibre system
- Compact reinforced composite (CRC)
- Slurry infiltrated fibre concrete (SIFCON)

7.2.58.1 Effects of Fibres in Concrete [110]

The comments have been made separately.

Source:

- Constructor [110]
- Cement Concrete.org [109]

7.2.59 Fibres in Ultra-high-Performance Concrete

7.2.59.1 Initial Development

The development took place in the year 1978. It was discovered from studies that the ultra-high-strength binder (i.e., cement and ultra-fine fillers) would be brittle like glass and the normal cement base.

Like normal cement, it has to be combined with sand, aggregate and reinforcement for better performance. By using very strong aggregates, the fracture energy can be improved, and with greater volumes and dimensions of strong aggregates, the brittleness of ultra-high-performance concrete (UHPC) will be reduced. The material may not be very suitable for most pre-cast applications because of rheological issues.

Following the tests, it was found that by using fine fibres, the performance, i.e., the mechanical fixation, can be improved within the very dense micro-structure of ultra-high-strength binders. This will exceed that of conventional concrete mortar.

To provide a high degree of ductility to UHPC, it will be necessary to incorporate high energy-absorbing reinforcement. This may include fibres, bars, or a combination of both.

By including high strength steel fibres, the brittleness of the material will be reduced, making it more ductile. The relevant factors in this connection may include fibre geometry (i.e., fibre diameter in relation to the maximum particle size of UHPC), fibre length, fibre content, orientation, stiffness of fibres, and the bond between fibre and the matrix.

7.2.59.2 Types of Fibre

The selection of fibres will be relevant and will significantly impact the composition of the material, strength, ductility, and cost of the final product.

There may be many types of fibres available, such as carbon, basalt, and other synthetic fibres. Three types of fibres have been used: PVA, polypropylene, and steel, and the results were analysed. The lengths of fibres for UHPC varied from 6 to 20 mm, and the fibre diameters ranged from 0.009 to 0.5 mm.

Steel fibres may be used, as they could be economical and contribute significantly to most properties of fresh and final concrete. Polypropylene fibres may be utilised in non-structural elements to control plastic shrinkage. Furthermore, UHPC sprayed concrete with steel fibres may be selected for toppings to provide some degree of fire resistance. PVA fibres may be used for smaller elements and to achieve high flexural strengths.

The steel fibres may be straight and cold drawn. The carbon steel fibres may have a brass coating.

The quantity of steel fibres in UHPC, covering micro-fibres and short fibres, may range from 81,000 to 67,500 per kg.

7.2.59.3 Cocktails and Hybrids

It may be possible to design fibre blends or cocktails, mixing different fibres like polypropylene and steel to improve fire resistance. Studies on steel fibres of different lengths and diameters were conducted in 1986 with UHPC, combining both long and short steel fibres. The short fibres are suitable for bridging micro-cracks, while the long fibres are effective in bridging macro-cracks.

The combination may improve strain hardening and is also expected to enhance the strength and toughness of UHPC.

7.2.59.4 Applications

There are various applications. A few may be mentioned as follows:

- **University of Southern Denmark** – application of pre-fabricated perforated UHPC façade elements.
- **TGV Station, France** – use of UHPC shells and glass panes.
- **Bridge over Loire River, France** – use of lightweight UHPC pre-fabricated slabs.
- **Brunnen, Switzerland** – sealing and waterproofing, using UHPC overlay.
- **Lubina River Bridge, Czech Republic** – constructed with pre-fabricated UHPC elements.

Source:

- Buitelaar and Muller [48]

7.2.60 Further on Hybrid Reinforcement in Design

7.2.60.1 General

Steel fibres have been used for industrial floors, pre-cast tunnel segments, etc., as mentioned. In recent years, more applications have been carried out for ground-supported slabs, basements, and walls in underground car parks.

For underground construction, the designer must address the issue of water pressure. The contribution of steel fibres, along with a reduced quantity of traditional reinforcement, may significantly control cracking in concrete.

The use of hybrid reinforcement with steel fibres may also be suitable for underground works in order to manage cracking.

7.2.60.2 Control of Cracks

For crack control, the designer may decide to use hybrid reinforcement with steel fibres in the design. This approach may limit the crack widths to between 0.1 and 0.3 mm and to produce watertight and durable construction.

The fibres should have the highest flexural tensile strength. The fibres, after mixing, should be well distributed throughout the concrete mass.

The hybrid reinforcement may generate more cracks; however, these cracks are likely to be small and finer. With finer cracks, water may not be able to pass through the concrete.

7.2.60.3 Demand for Fibres

It appears that designers are leaning towards using more hybrid reinforcement with steel fibres in their designs to control crack width. Otherwise, a greater quantity of traditional steel may need to be used for the structural members to manage crack widths, which could be more expensive in the long run.

Source:

- Vanheuverzwijn [26]

7.2.61 Durability of FRC [21,55,111]

For durability, the corrosion of steel fibres and the performance of polymeric fibres in an alkaline environment are the main concerns. Polypropylene fibre is the only polymeric fibre that has been tested for field applications over an extended period. The performance of polypropylene fibres in an alkaline environment has been found to be satisfactory; however, they provide inferior thermal and fire resistance.

Corrosion of steel fibres due to carbonation is reported to occur only at the surface, and this does not affect the performance of the remaining fibres. Surface staining was observed within the first month. It has been noted that chloride ion penetration does not affect fibre corrosion. Carbonation depth was also measured to be less in a few limited tests. In most cases, it was observed that corrosion of fibres was restricted to the carbonation depth. Furthermore, since the failure of FRC often results from fibre pull-out rather than fibre fracture, the unutilised and uncorroded fibre strength should be adequate for durability [21]. Hence, a small reduction in diameter due to corrosion may be acceptable, provided that interfacial bond strength remains unaffected [21]. From various applications with steel fibres, it seems that fibre corrosion does not pose a significant problem affecting durability. The composite material is environmentally friendly and suitable for green building design. This is the key advantage of FRP composite materials.

The fibres are protected against corrosion by the alkalinity of cement paste, as stated in the mix [55]. Bursting stresses associated with corrosion in large diameter bars in traditional concrete structures will not occur with fibre reinforcement. Even when the fibres corrode, the small volume of each fibre will be minimal, and the discontinuous nature of the fibres will not create cracking issues like conventional concrete with corroded steel reinforcement. There may be some surface corrosion associated with partly embedded or loose fibres, but this will not be harmful. Generally, there is no minimum cover requirement [55].

Concrete structures may be protected by using a protective coating.

Source:

- ACIFC [55]
- Swamy [21]
- Witerberg [111]

7.2.62 Further on Durability of FRC [3,21]

FRC is prepared with a low water-to-cement ratio and high cement content, making it very durable. The fibres are well protected within the concrete. In exposure to chloride, surface rusting may occur, but the fibres inside the concrete will remain well protected. Long-term tests on FRC exposed to de-icing salts indicated that only nominal corrosion of fibres may take place, with no significant adverse effects.

The comments on various common fibres are briefly mentioned below:

7.2.62.1 Steel Fibres [21]

Steel fibres will be protected from corrosion by the alkalinity of concrete. Carbonation problems in concrete may initiate corrosion of the fibres, and the situation may worsen when the concrete is cracked. Air entraining admixtures may be used under freeze–thaw cycle conditions.

7.2.62.2 Glass Fibres [3]

As mentioned, ordinary glass fibres are not suitable for use in Portland cement mortars/concrete. Alkali resistant glass fibres and zirconia show better durability in alkaline environments. Comments on glass fibres have been presented separately.

7.2.62.3 Stainless Steel Fibres [21]

Stainless steel fibres have also been used to achieve durable performance. Comments on stainless steel fibres have been presented separately.

7.2.62.4 Polypropylene Fibres [21]

Polypropylene fibres may be damaged by ultraviolet radiation and thermal oxidation. The cement matrix can mitigate the effects of ultraviolet radiation. For thermal oxidation, suitable stabilisers may be employed to enhance service life.

7.2.62.5 Natural Fibres [3]

Various natural fibres, such as cotton and wool, along with many synthetic polymers, may not be suitable and could exhibit weaknesses in terms of durability under the alkaline conditions of cement paste.

7.2.62.6 Comments on Bamboo [21]

Bamboo may be attacked by insects such as borers, termites, and rot fungus. Chemical preservation methods may be used for bamboo members. Bamboo can be treated with a solution of arsenic pentoxide, copper sulphate, and sodium dichromate.

For fire protection, the treatment may consist of ammonium phosphate, copper sulphate, boric acid, zinc chloride, and sodium dichromate.

Source:

- Mehta and Monterio [3]
- Swamy [21]

7.2.62.7 More on Durability

- The performance of high-quality polypropylene fibres has been found to be good during exposure in crack widths of up to 5 mm.
- High-quality polypropylene MSFs will not be affected by corrosion and will maintain their performance quality irrespective of exposure conditions.
- Uncracked SFRC concrete elements will not be affected by corrosion. The performance will deteriorate once they are cracked due to corrosion.
- Steel fibre corrosion may result from the loss of fibre diameter at the crack opening. The corrosion effect will increase with the width of the crack and decrease with the distance from the crack face.
- Crack widths greater than 0.1 mm will enhance deterioration due to corrosion. The crack width will be responsible for the duration from the start of corrosion.
- Corrosion of steel fibre reinforcement may lead to a transition from ductile pull-out failure to brittle snapping failure due to the rapid loss of toughness. This may be referred to as corrosion-induced brittle failure.

Source:

- Witerberg [111]

7.2.63 Durability of Steel Fibre-Reinforced Self-Compacting Concrete [42]

Studies have been carried out on the durability of steel fibre-reinforced self-compacting concrete (SFRSCC), specifically to find out information on corrosion resistance properties. It is particularly relevant to know whether corrosion to steel fibres contributes, in this type of concrete, to cracking and ultimately spalling of the surrounding concrete.

Experimental work with SFRSCC and self-compacting concrete specimens was conducted to study and assess the mechanical properties and durability indicators.

The relevant information is as follows:

- Self-compacting requirements were maintained, and 60 kg/m^3 of steel fibres were added to SCC
- The inclusion of steel fibres did not alter the durability indicators
- It has been noted that SCC is very effective with steel fibres, particularly in increasing post-cracking flexural resistance and energy absorption.
- Furthermore, the use of fibres did not appreciably affect the self-compacting requirements or the durability indicators of SCC.

Source:

- Frazao, Camoes, and others [42]

7.2.64 Fibres Reduce Marine Concrete Permeability [112]

The marine concrete will be exposed to not only the chemical attack from seawater but also the structures will be subjected to continuous assaults from the waves and the abrasive actions of wave-born sand and shingle with the ebb and flow of the tide. The impact of these abrasive actions on the concrete can be very serious, resulting in spalling of concrete and corrosion of structural steel as applicable, which may ultimately lead to structural failures.

By using fibres in the mix, the permeability, bleeding, and segregation of the concrete can be reduced. Furthermore, the concrete may be made more durable by reducing plastic shrinkage cracking and preventing the chances of attack from chlorides.

To protect the structures against the aggressive marine environment, the concrete must be very durable and have low permeability. It has been observed that by incorporating monofilament polypropylene fibres in the mix, such durable concrete may be produced, making it suitable for this purpose.

When monofilament polypropylene fibres are used, they can control bleed water migration within the concrete and reduce the chances of segregation of fine cement and sand particles. This process provides more efficient hydration of the cement and leads to a tough and durable concrete surface through proper and improved bonding of the cement matrix. The abrasion resistance of the material will be enhanced due to a lower water-to-cement ratio in the mix.

Steel reinforcing bars are typically used in the design of structures. The concrete is supposed to protect the reinforcing bars. With continuous impact from the waves, along with the abrasive actions of sand and shingle, and supported by freeze/thaw cycles (as applicable), the concrete may spall. Consequently, the protective layer around the reinforcing bars may be damaged and removed, exposing the reinforcing bars to corrosion, which may eventually lead to structural failures.

In some projects, Fibrin XT was used at 0.91 kg/m^3 in the mix. The fibres improved the durability of the concrete and could also replace the air-entraining agents in the mix. By using Fibrin fibres, it may be possible to increase resistance to abrasion and impact by 40% and 400% respectively, while also enhancing concrete durability against chloride attack.

Multi-length fibres with a diameter of less than 30 μm may be used for the concrete.

Fibrin may be produced using an innovative extrusion process with special blends of chemicals that will produce polypropylene beads which can be stretched to diameters within the range of about 5–30 μm.

Monofilament polypropylene fibres may be used in the mix for cost benefits and to achieve durability of concrete in various marine environments.

Source:

- Kitchen [112]

7.2.65 Special Applications with Fibres

Two items will be mentioned in this connection:

7.2.65.1 Translucent Concrete [113]

Translucent concrete is also known as light transmitting concrete and is prepared with concrete building material embedded with optical elements such as optical fibres.

This material may be used as facade cladding for internal walls.

Translucent concrete was first mentioned in a Canadian patent in 1935. Its development largely took place with the introduction of optical glass fibres in the mix [113].

In 2003, transparent concrete was successfully produced by adding 4%–5% glass fibres (by volume) to concrete [114].

7.2.65.1.1 Materials Required for the Manufacture of Transparent Concrete [114]

The materials for production will be similar to those of traditional concrete, and the requirement will be to use fine aggregates along with optical fibres (4%–5% by volume) in the mix. The optical fibres may be plastic or glass.

For the transmission of light, the thickness of the optical fibres may vary from 2 μm to 2 mm, which is almost equal to the diameter of human hair.

This material will be suitable for green technology.

7.2.65.1.2 Types of Optical Fibres [114]

These are as follows:

- Single-mode fibre
- Multimode fibre
- Multimode step-index fibre
- Multimode graded-index fibre

Single-mode fibres can propagate only one mode at a time, whereas multimode fibres can propagate hundreds of light modes simultaneously.

7.2.65.1.3 Review of Manufacturing Process [114]

The items are as follows:

- The manufacturing process will be similar to that of traditional concrete as mentioned.
- It will consist of 96% concrete and 4% optical fibre.
- Quick-setting cement may be used along with stone dust and sand as fine aggregates.
- The fibres will blend well with the concrete due to small size.
- Craft clay may be used to provide a base for the optical fibres.
- The finished items may be used for exterior and interior walls, illuminated pavements, or in art/aesthetic designs.
- It must be ensured that sufficient natural light is available when working with this material. The wall mounting arrangement should provide adequate lighting to achieve uniform illumination across the entire surface.

7.2.65.1.4 Applications [114]

The following may be mentioned:

- Illumination of the interior of buildings.
- Flooring surfaces illuminated from below, which may shine in different colours after sunset.
- Exterior wall surfaces.
- Sidewalks with lighting below the surface.
- Aesthetically attractive items, including floors and furniture.
- Facades, interior walls, dividing walls, etc.
- Colours may be used for different varieties.
- Subways may be illuminated.

7.2.65.1.5 Advantages [114]

- It is a good material for bringing sunlight into the interior of buildings, reducing power consumption during daylight hours.
- It can be visually appealing and aesthetically appealing.
- It may be more effective in cold climates.

- It is environmentally friendly.
- Illumination of a space may be achieved using natural light.
- Various creative designs can be implemented with the material.

7.2.65.1.6 Disadvantages [114]

- Skilled supervision and labour will be required for casting related to the manufacturing process.
- The cost of using optical fibres will be high. While applying plastic fibres will be cheaper, light transmission will decrease.

Source:

- Wikipedia [113]
- Gahrana, Raj, Chouhan, and Krishnia [114]

7.2.65.2 Bendable Concrete [115]

Bendable concrete, as a material, can be stronger and more durable. It may also be referred to as flexible concrete or engineered cementitious concrete (ECC).

It can be described as an ultra-ductile fibre-reinforced cementitious composite material with high ductility and good control over crack widths.

ECC may have a strain capacity of more than 3% and act like a ductile metal material.

7.2.65.2.1 Composition of Bendable Concrete

Bendable concrete will contain all the ingredients of traditional concrete, excluding coarse aggregates, and will be reinforced with polymer fibres. This means the concrete will consist of cement, sand, water, fibres, and admixtures.

Polyvinyl alcohol fibres covered by a very thin (i.e., nano-metre thick) slick coating are used to prevent the fibres from rupturing.

ECC will deform more than normal concrete without fracturing.

7.2.65.2.2 Method of Working

The main points are as follows:

- With the increase in load, cracking will be prevented due to the strong bond between the concrete and PVA fibres.
- The ingredients of PCC will work together to share the load.

7.2.65.2.3 Advantages

Some advantages include

- The ability to bend like a metal.
- Not as brittle as glass.
- Self-healing properties.
- Better resistance to cracking.
- Greater strength and durability than conventional concrete.
- No harmful gas emissions.
- The flexible concrete will be about 20%–40% lighter.
- The requirement for steel reinforcement may be reduced.
- Suitable for pre-cast concrete work.
- Will reduce costs by using flexible concrete.

7.2.65.2.4 Disadvantages

These are listed below:

- The initial cost will be high compared to conventional concrete.
- It will depend on the quality of the material used.
- It will require skilled labour for construction.
- Compressive strength may be lower than that of conventional concrete.
- It will need special types of materials which may be difficult to obtain in some areas.

7.2.65.2.5 Applications

- It may be suitable for earthquake conditions as it may not break down due to vibrations.
- Concrete canvas may be made with flexible concrete.
- It may be used in military areas.

7.2.65.2.6 Special Comments

- The tensile capacity of flexible concrete may be about 10–15 MPa and the compressive strength up to 70 MPa.
- It has self-healing properties.
- It is light and can be 500 times more flexible than normal concrete.

Source:

- Prasad [115]

7.2.66 Synthetic Fibres for Port and Dock Applications, Marine Structures

The supporting concrete slabs with sufficient reinforcement can deteriorate during use related to loading and unloading operations. The damage to the slab from the corrosion of reinforcement may be a common problem that can lead to temporary closure of the affected areas. This may be an important issue for delays to loading/unloading operations and disruption to tide control ship movements.

Following the damage, the concrete pavement will be subjected to the ingress of chloride ions from seawater, de-icing salts, etc. Due to this, the corrosion of steel fabric reinforcement in the concrete will occur, resulting in surface spalling and reduced service life of the concrete.

As a solution, the conventional steel fabric reinforcement in concrete may be replaced by synthetic macro-fibres to avoid the risk of corrosion.

The application of synthetic fibres will be easier and will reduce construction time, which is important under tidal conditions.

The advantages of avoiding steel bars/fabric in concrete construction have already been discussed/mentioned.

The use of these fibres will also reduce health and safety hazards in construction.

The inclusion of micro-synthetic fibres in the concrete mix can increase the impact and abrasion resistance of the concrete.

The applications of synthetic fibres in marine structures have been detailed in item 7.7 of this chapter.

Source:

- Boggust [116]

7.2.67 Further Comments on Concrete Floor Slabs Supported on Ground

7.2.67.1 Use of Macro-Polypropylene Fibres in Ground Bearing Concrete Floors

Ground floor slabs had been designed with steel reinforcement and steel fibres to control cracks and hold the concrete together. Macro-polypropylene fibres can also be used for similar applications in designing ground-bearing concrete floors. The fibres can help hold the concrete together, redistribute cracks if they occur, improve performance at joints within the concrete, and assist in spanning over any soft patches beneath the concrete slab.

The following are the advantages over traditional methods:

- No cost for handling and placing reinforcing bars.
- No reinforcing bars to be removed, and part load changes for the requirement of extra bars will be eliminated.
- No storing, handling, and lifting of reinforcing bars on site.
- No damage, corrosion, or theft of bars from the site.
- No cutting of bars, trip hazards, and a lower health and safety risk.
- Costs for chairs and tie wire will be reduced.
- No third-party accidents or liability.
- The concrete pour sequences will be simpler and quicker without the presence of reinforcing bars or fabrics.
- No problems related to reinforcing bars or fabrics placed incorrectly in the concrete slab.
- Cost savings will be possible on materials.
- It will reduce the carbon footprint by using polypropylene fibres compared to steel reinforcing bars.

In one application, 360 kg/m^3 of cement content with 4 kg/m^3 of Adfil's Durus S300 macro-polypropylene fibres were used for the floor.

Source:

- Brookfield [117]

7.2.68 The Structural Performance [7]

The performance of the composite materials is expected to be better, considering their high specific stiffness and strength. The low self-weight of the material will provide other benefits, such as reduced structural dimensions and lower costs for transport and erection. It seems that the performance of FRP materials, especially for strengthening work, is satisfactory when executed properly. However, regular checks and monitoring for condition assessment will be required. It will be important to observe the performance over a further period of time to gain more confidence.

Source:

- Ganguly [7]

7.2.68.1 Failure Investigations and the Forensic Process [118]

Forensic engineering to investigate structural failure is becoming increasingly important and is considered different from designing and constructing new structures. Forensic engineering is associated with detective skills as opposed to design skills, which will be the key requirements for the investigator. Causation investigation will be the prime focus of the forensic approach. Without carrying out the forensic investigation, the causes of a structure's failure may not be concluded.

Source:

- Brady [118]
- Ganguly [7]

7.2.69 Fibres Used in Concrete Floors

7.2.69.1 A Case Study

7.2.69.1.1 General

The case study was associated with voids in the concrete slab of a large warehouse. The problem was identified as needing to accommodate loads from the racking system on the floor.

7.2.69.1.2 Potential Defects

The floor slab was constructed with FRC, and the voids were present as a result of the clustering of fibres during construction. The ultrasonic pulse velocity technique may be suitably used to identify the voids in FRC slabs.

7.2.69.1.3 Identifying Presence of Defects

Cores may be extracted from the affected areas of the slab to identify the presence of voids. Ground-penetrating radar (GPR) could be used for the investigation. The work started with the use of GPR, but the application was not found to be successful.

7.2.69.1.4 Limitations of GPR

Reinforced concrete members with layers of reinforcement may present challenges for GPR. Only limited information may be obtained regarding void detection. The complications may be greater for slabs with densely packed steel fibres.

7.2.69.1.5 Ultrasonic Pulse Technology

The use of a wireless inspection technique with multi-channel ultrasonic pulse echo technology was adopted. The ultrasonic technology (tomography) emits ultrasonic pulses into concrete, with about 40%–50% of the pulse being reflected or attenuated due to the presence of metallic objects. The pulses that encounter an air gap will be fully reflected and provide information. The remaining pulses (i.e., about 50%) will penetrate deeper into the concrete.

7.2.69.1.6 Survey/Inspection

The survey work is very important, and some information may be gathered regarding the voids initially by drilling, as stated.

The survey may indicate the following:

- Good-quality concrete
- Concrete with small isolated voids
- Concrete with large voids

Tomography may detect voids, but the actual volume of voids and their connectivity cannot be properly identified.

However, the method may be suitable for checking the quality of the resin injection repair method adopted for the slab (member).

7.2.69.2 Comments

Judgement and experience are crucial for using modern technology and applying non-destructive tests.

Source:

- Thomas [119]

7.2.70 Comments on Creep

The tests on creep are mentioned separately later in item 7.6 of this chapter.

7.2.71 Codes and Standards for Steel FRC

For guidance, steel fibre concrete work may be divided into

- Design
- Production
- Testing

7.2.71.1 Design

Steel FRC has been used for industrial floors and tunnels, as mentioned.

TR 34 may be referred to for the design of industrial floor slabs, along with EC2 for applications related to structures.

7.2.71.2 Production and Product Standard

British standards may be consulted. BS EN 14650 may be used for the production of pre-cast units using SFRC.

7.2.71.3 Testing of SFRC

For testing, BS EN 14721 and BS EN 14488 may be used.

Source:

- Viney [120]

7.2.72 Testing of Materials

Different tests and results have been presented separately in this chapter.

7.3 FIBRE-REINFORCED POLYMERS

7.3.1 Introduction

7.3.1.1 Initial Study

Suitable fibres are being considered in composite materials for construction and applied to structures as reinforcement. These may be glass, carbon, or aramid, as already mentioned. They all possess high ultimate strengths and stiffness and exhibit a linear elastic response to ultimate load [121].

The fibres may be difficult to handle and place in concrete and may be combined with resins to prepare composite rods, grids, plates, etc.

There are different types of resins available for composites. Factors such as durability, manufacturing process, and cost will be critical in the choice. Thermo-setting resins are typically selected for applications; however, thermo-plastic resins may also be used.

The common process will be the pultrusion method, which is suitable for producing different types of structural shapes [121].

The fibres may be supplied in the form of continuous rovings. They will be drawn off carefully and in a controlled manner through a resin bath, which will impregnate the fibre bundle. This will be followed by pulling through a die, which will consolidate the fibre–resin combination and create different shapes. The die may be heated to set and cure the resin, allowing the formed composite to be drawn off by suitable reciprocating clamps or a tension device. High strength and stiffness can be achieved from the process as a high proportion of fibres is placed in the cross-section. To achieve a smooth surface, a secondary process may be employed to improve the bond, such as over-winding with additional fibres or applying a sand layer.

The fibre composite materials are lighter and stronger than steel, and with proper resin and fibre combinations, the composite materials can be very durable. However, the composite material may not exhibit any yield at ultimate load.

Carbon, glass, and aramid fibre composites have been used for various applications, including bridges.

The important items are as follows:

- Development of suitable manufacturing processes.
- Suitable resin and fibre selection.
- Study to determine the durability of FRP rods exposed to aggressive environments directly or placed in concrete.
- Testing of structural behaviour and analytical methods.
- Techno-economic feasibility studies.
- Development of suitable design guidance.
- Case studies of trial structures and composites.
- Cost of fibre composite materials.

As already stated, fibre composite materials may be used as reinforcement and for strengthening concrete and other structures.

Source:

- Clarke and Waldron [121]

7.3.2 Fibre-Reinforced Polymers (FRPs)

This is basically a combination of fibrous reinforcement held in a polymer matrix to form plates, tubes, rods, and other structural profiles as mentioned [122]. The fibres may be carbon, aramid, or glass, and the resins may be epoxy, vinyl ester, polyester, or phenolic.

These may be used as follows [123]:

- Composite plates made from fibres and epoxy resins, fixed with epoxies to the soffits of slabs and beams.
- Preformed fibre composite shells, normally tailor-made and bonded around columns.
- Sheet materials impregnated with resin and wrapped around columns or other members to be strengthened.
- Fibre composite strips or rods that may be placed into grooves prepared in the concrete for near-surface-mounted (NSM) reinforcement.
- Pre-formed shear straps bonded to the soffits and sides of the beams to create external stirrups.
- Trials have been conducted to strengthen concrete beams using fibre-reinforced polymers (FRPs) [122]. The popularity of using bonded fibre composites has increased in recent years to strengthen concrete/steel structures.

Such composites can be successfully applied to beams, slabs, columns, and other parts of the structure.

FRP materials are easy to handle due to their low weight compared to traditional steel plates. FRPs have good fire resistance as they are poor heat conductors, which is advantageous over metals. To increase fire resistance, additives can be mixed with the resins, or intumescent [122] paints may be applied to the surface as a protective coating. Attempts have been made to bond either unstressed or pre-stressed CFRP plates to the underside or the bottom flange of beams. CFRP has higher tensile strength than steel, and they are also non-corroding. Alternatively, reinforcing rods may be embedded as NSM FRP reinforcement [122] at the bottom flange of the beam. This technique increases the tensile stress capacity of the lower part of the beam.

CFRP plates can be bonded to the sides of the beam near the supports to enhance the shear capacity. The reinforcement may be applied as cross-wrapped or in bands, with the tensile capacity of the carbon, aramid, or glass fibres acting as confinement [122].

Research has revealed that FRP systems can sufficiently increase the stiffness and load-carrying capacity of concrete structures beyond the range achievable with conventional methods. It is needless to say that the design and installation of such applications must be carried out at both the design office and site by experienced personnel in order to ensure the long-term durability and performance of the strengthened structural elements.

When the composite and adhesive have been correctly specified, FRP materials will require little or no maintenance during their service life. Resin properties must be tailored to satisfy different requirements for wetting ability, viscosity, and gel and curing times. Some resins may need protection from UV radiation [124]. The fibre composites should, however, be inspected for signs of crazing, cracking, or de-lamination, which would indicate some level of deterioration [125]. The composite should be checked for local damage from impact or abrasion [125].

FRP wrapping can increase the bending and shear strength of an element by a factor of around three [124].

The performance of an FRP strengthening system will depend on the quality of workmanship for both the preparation of the concrete surface and the installation of the fibre composite material [123,125].

Source:

- Halliwell [122]
- Clark [123]
- Sadka [124]
- Clark [125]

7.3.3 Further on Fibre-Reinforced Polymers (FRP)

The popularity of using bonded fibre composites has increased in recent years to strengthen concrete and steel structures.

Such composites may be applied successfully to beams, slabs, columns, and other parts of the structure. The material is essentially a combination of fibrous reinforcement held in a polymer matrix to form plates, tubes, rods, and other structural profiles.

FRP materials are easy to handle due to their low weight compared to traditional steel plates.

Concrete bridges are being widely strengthened by suitable fibre composite materials. Fibre composites may be considered an efficient and cost-effective means of strengthening concrete bridges. The major advantages include the speed of installation, leading to short possession time [125].

Tests [124] conducted at the Transport Research Laboratory on half-scale column specimens with varying amounts of axial and hoop fibre wrapping enhancements in strength observed in the following areas [124]:

- Concrete strength and strain capacity – due to confinement by hoop fibres
- Additional compressive reinforcement provided by axial fibres
- Extra tensile reinforcement provided by axial fibres

Workmanship will be important at all stages of the installation process.

Inspection should occur at all stages of the strengthening process [125]. Immediately after the installation of the FRP, a full inspection should be carried out to examine the "as-installed" condition. All strengthened structures should be subjected to routine visual inspections, with more detailed inspections carried out from time to time. For buildings, the portion of the area strengthened may be inspected.

NSM FRP bars, as a complementary item to externally bonded FRP laminates, have been used [126] to strengthen a bridge for a CFRP system; NSM tapes and/or externally bonded laminates in combination may be used for retrofitting work.

The presence of moisture may be the main reason for the deterioration of the FRP strengthening system [5]. Hence, gutters, drains, etc. should be kept clear of debris so that rainwater does not accumulate. The surface should be kept clean [125].

Source:

- Clarke [123]
- Clarke [125]
- Sadka [124]
- Stone, Tumialan, Nanni, and Parretti [126]

7.3.4 More on Polymer Composites as Construction Materials

As stated before, polymer materials are well known for their lightness, resistance to corrosion, and ease of application. Fibres may be combined with the polymers to produce composites with better properties, which can be used for structural elements to meet different requirements and satisfy various purposes in the construction industry.

The development of polymer composites began in 1940 and started with military and aerospace applications. With the advancement of these materials, composites have been used for construction purposes, such as load-bearing infill panels and structural members.

Pre-fabricated units for such purposes can be made using glass-reinforced polyester, and the units can be connected to provide rigid forms that meet the requirements. Glass-reinforced plastics have been used for various items, such as pressure pipes, roofs, and tank liners.

For design involving polymer composite materials, greater knowledge and understanding of the materials will be necessary. Knowledge of different items related to the anisotropic behaviour of the material, as well as its elasticity, micro-mechanical properties, strength, and stability, will be required. These properties may be affected by the manufacturing method, loading patterns over time, and exposure to environmental conditions. The design process may be interactive, involving the constituent materials used and an understanding of the manufacturing methods and the design of the composite materials ultimately.

Polymer composite materials have been used in civil and structural engineering, covering bridge design, bridge repair, structural strengthening, mooring cables, and other structural components.

Source:

- Polymer Composites as Construction Materials [127]

7.3.5 Simplified Stress Block

The simplified stress block is shown in Figure 7.5. The simplified stress block can be used for pile-supported slabs when the design moments are computed using the yield line method of analysis.

Source:

- Vollum [11]

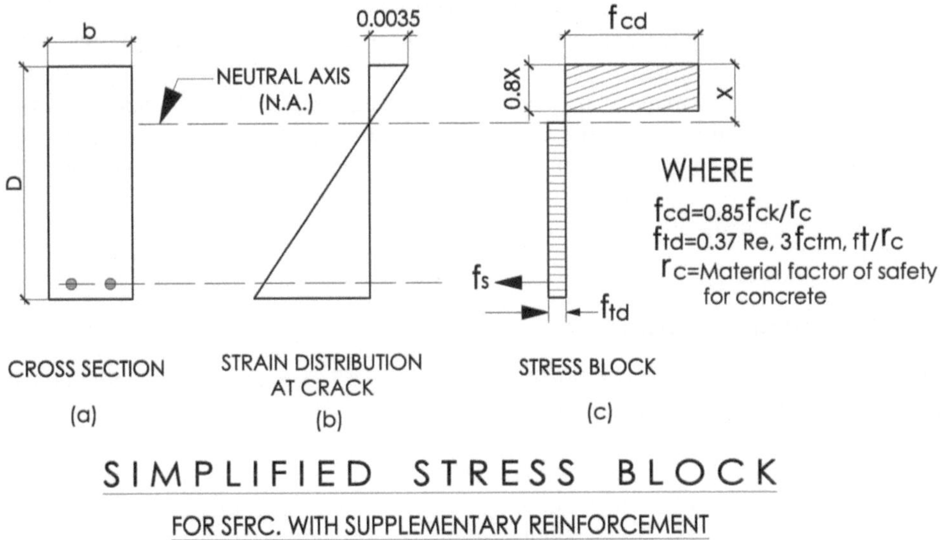

FIGURE 7.5 Simplified stress block for SFRC with supplementray reinforcement.

7.3.6 FRP Reinforcement and Development

The development and interest in the applications of fibre-reinforced polymers in projects have been ongoing. They can be considered an efficient alternative to traditional reinforcement types and their uses. The development of structural FRP reinforcement has taken place in advanced countries.

It had been used efficiently for the repair and strengthening of existing structures. The material has good properties and flexibility and may be considered suitable for various applications. There have been many applications carried out on different types of structures, allowing for the strengthening and improvement of the load-bearing capacities of structural members, including reinforcement work to meet seismic requirements.

It seems that FRP reinforcement will play a major role as an externally bonded reinforcement system.

It may be used as an external bonding element for structural elements, and guidelines are available.

Source:

- Matthys and Taerwe [128]

7.3.7 Advantages of FRP

The application of FRP reinforcement was tested in a project where the structural elements had very little cover. By using FRP reinforcement, this issue can be resolved as mentioned. The re-bars would weigh one fourth of that of steel, allowing for additional bars to be used externally to increase strength. The bars can be placed externally anywhere as per convenience or requirement. The bars may be disturbed and displaced during concreting and compaction, but this is unlikely to have significant effects on the appearance of the concrete. The affected surfaces may be sanded or ground smooth locally and appropriately repaired. This approach does not significantly increase the chances of corrosion to the reinforcement.

FRP bars have been used around masonry openings for windows, doors, and arrow slits. Stitching work can be carried out to stabilise the walls by carefully drilling into the stone/masonry around the openings to about 600 mm and placing FRP bars in position. This will be followed by filling or injecting the holes with polyester resin, with the bars inserted into the holes and held in place until cured. The bars will also help to stabilise the walls and act as dowels for the external stone treatment around the openings.

The bars are non-reactive to magnetic fields and radio frequencies. The applications of FRP bars are suitable and beneficial in electromagnetic contexts. Other applications may include magnetic resonance imaging rooms in hospitals, structures with high-voltage lines, and transformer vaults.

Source:

- Jeanes [129]

7.3.8 NSM FRP Reinforcement

NSM FRP composites are anisotropic, and the materials possess good tensile strength in the direction of the fibres. They are elastic up to failure and do not yield. They are corrosion resistant and provide good weather-resistant characteristics compared to other construction materials.

The applications of NSM FRP bars may be used as a complementary item to externally bonded FRP laminates. This NSM FRP technique may be used for strengthening the flexural and shear strengths of weak RC elements. The technique was used to strengthen a bridge with a CFRP system, NSM tapes, and externally bonded laminates in combination for the retrofitting work.

Additional information on the applications has been covered separately in item 7.7.

Source:

- Stone, Tumialan, Nanni and Parretti [126]

7.3.9 Concrete Strengthening with Fibre Composites

Due to revised load requirements, existing structures may need to be strengthened to carry additional loads. There are different types of fibre composites that can be used for strengthening structures. Here, some comments are made on CFRP composites and wraps for strengthening.

Further information will be given in item 7.4 of this chapter.

Source:

- Hackett [130]

7.3.10 Review of Polymer Materials

7.3.10.1 Introduction

The applications of fibre-reinforced polymer materials may be used to strengthen structural members and to prevent structural collapse, as indicated. This will replace the traditional use of steel and special concrete jacketing methods.

By using the materials properly, it is possible to increase the axial, shear, and flexural capacities of the affected members and to carry additional loads if required. In some cases, the weaknesses in the design may be addressed accordingly.

Traditional strengthening/retrofitting methods are time-consuming and labour-intensive. These methods also increase the cross-sectional area of the member. FRP materials possess good mechanical properties, corrosion resistance, better durability [131], lightweight, easy application, reduced construction time, efficiency, and low life cycle cost.

Different FRP materials are available for applications.

The composite materials may include [132]

- CFRP
- Glass FRP (GFRP)

- Aramid FRP (AFRP)
- Steel FRP (SFRP)

Information on wrapping using SFRP is not very common.

Source:

- Parvin and Brighton [131]
- Abdelrahman and El-Hacha [132]

7.3.11 More on Fibre-Reinforced Polymer Composites for Construction

7.3.11.1 General Comments

In recent years, we have observed extensive applications of fibre-reinforced polymer composites in various projects.

In strengthening and retrofitting projects, FRP composites have been used success fully. The advantages of using FRP composites have already been mentioned briefly.

Initially, glass fibres embedded in polymeric resins were used for FRP materials. The properties related to lightweight, high stiffness, high strength, lower cost, and environmental resistance allowed polymer composite materials to be produced that were superior to the constituent materials alone. With these developments, it became possible to create structures that are safe, economical, and functional, meeting the requirements of society.

7.3.11.1.1 Structural Shapes

Different structural shapes/profiles have been manufactured for various uses and constructions, especially related to bridge super-structures and buildings.

7.3.11.1.2 Pultrusion Process

The product will be economical and of consistent quality for the structural shapes when using this process. This will be a continuous and automatic process. The system consists of pulling reinforcing fibres and fibre fabrics that are resin-impregnated through a heated curing die. Single- or multi-celled closed section profiles, along with open sections, may be produced.

7.3.11.1.3 Fibre-Reinforced Polymer Structural Shapes

Large-sized pultruded FRP shapes larger than 150×150 mm have been produced for buildings and bridge super-structures. Standard profiles are also available for smaller structures.

7.3.11.1.4 Development in Pultruded Structural Shapes

Pultruded shapes, covering both standard profiles and those made to specific requirements, have been applied in building and housing construction applications.

7.3.11.1.5 Applications

A few applications include

Highway bridge decks, sandwich construction, and pultruded shapes using adhesives.

7.3.11.1.6 Cost Comparison

FRP materials tend to be more costly, as mentioned. However, considering the long-term performance, including maintenance and the costs of necessary repair work during the service life, the overall picture may differ.

7.3.11.1.7 Comments on Deflection and Cracking

The deflections and crack widths are greater in FRP-reinforced concrete beams and slabs compared to steel-reinforced concrete members such as beams and slabs. This is primarily due to the lower value of the elastic modulus of FRP.

The crack pattern will be similar to that of steel-reinforced sections under lower load conditions. However, as loads increase beyond the service condition, crack spacing may decrease and crack width may increase compared to steel-reinforced members.

It should be noted that the creep and shrinkage performance of FRP-reinforced elements will be similar to that of steel-reinforced members.

7.3.11.1.8 Other Types of Strengthening of Reinforced Concrete Structures

- **Flexural Strengthening –** Comments have been given separately.
- **Confinement –** Comments have been given separately
- **Shear Strengthening –** Comments have been given separately

7.3.11.1.9 Masonry Construction

The composites have been applied in old masonry structures for the purpose of strengthening.

Often, epoxy-bonded FRP strips are placed on the surface of masonry structures in the direction and location corresponding to the principal stresses.

The execution of FRP strips or sheets may be carried out in a manner similar to that for RC members.

The FRP composites may be used as confinement reinforcement for masonry as unbonded strips that can be made with matching colours. Transparent glass FRP fabric or bonded FRP sheets may also be used.

7.3.11.1.10 Timber and Metals

For strengthening, FRP strips or sheets have been used for timber beams and columns. This may be applicable for increasing stiffness, strength, and ductility.

7.3.11.1.11 More Comments

The important differences between FRP materials and conventional materials include higher strength, stiffness, and linear–elastic behaviour up to failure. Other considerations may involve the thermal expansion coefficient, heat and fire resistance, and water absorption, which should be considered in the design.

Training for engineers, construction personnel, inspectors, and owners is necessary to cover the relevant issues of FRP technology for successful and effective utilisation.

It should be noted that with the expected moderate decrease in material prices, their use may be limited to those applications where the unique properties of FRP materials are critically important and required.

Source:

- Bakis, Bank, Brown, Cosenza, Davalos, Lesco, Machida, Rizkalla, and Triantafillou [133]

7.3.12 Review of Special Strengthening Methods with FRP

These may be summarised as below:

> Wrapping and curing, using pre-stressed strips, applying fusion bonded pin-loaded straps, pre-fabricated shapes, mechanically fastened FRP strips, and placement inside slits.

It seems that FRP methods will play an important role in strengthening/retrofitting projects involving bridges, buildings, historical, and other renowned structures.

Source:

- Bakis, Bank, Brown, Cosenza, Davalos, Lesco, Machida, Rizkalla, and Triantafillou [133]

7.3.13 Strengthening of Columns

7.3.13.1 Introduction

Structural deterioration occurs during service life due to various reasons, including exposure to saltwater, freeze–thaw cycles, and de-icing chemicals.

Structures may need to be strengthened or retrofitted to meet the revised requirements of codes and standards.

The strengthening of RC structures with FRP composites is quite common and well-known, as already mentioned. Conventional retrofitting techniques, involving concrete and steel jacketing, are labour-intensive and time-consuming processes. These methods also increase the cross-sectional areas of the structural members, especially for columns.

FRP materials are highly efficient, possessing good mechanical properties, being lightweight, corrosion-resistant, durable, easy to apply, requiring less time for construction, and offering reduced life cycle costs.

7.3.13.2 Strengthening

The applications of FRP compositions may be very useful for strengthening columns. This may include FRP wrapping, FRP spraying, and FRP encasement. The columns may be strengthened to have additional axial, shear, and flexural capacities to meet different needs. The requirements may also cover seismic detailing, eccentric loads, accidental impacts, and provide protection against corrosion.

7.3.13.3 FRP Confinement of Columns

The confinement may involve wrapping the columns with FRP sheets, in-situ cured sheets, or pre-fabricated jacketing, with the fibres positioned in the circumferential direction. The confinement will be less efficient for rectangular and square columns compared to circular columns.

Further information on the confinement of columns will be included in item 7.7of this chapter.

7.3.13.4 Further Comments/Observations for FRP Composites Using Experimental Information

These may be mentioned from the experimental studies:

- It seems that the angle of the hoop plies and the stacking sequence in the wrapping system will be important. This may provide different types of ductility and strength for the strengthened columns with the same thickness of FRP wrapping.
- More applications and studies were carried out on FRP-confined axially loaded cylinders and short columns. The scale effects should be checked for designing full-size columns. The size effects of columns will also need to be assessed.
- Further studies and applications of hoop and ply combinations in the wrapping arrangements for prismatic columns may be required. This will primarily aim to avoid or delay premature fracture of the corners.
- The corner stress concentrations may be changed by modifying the shapes from square to circular and rectangular to elliptical columns. This will improve axial load capacity and ductility. The effects of shape modifications on performance will need to be studied.
- FRP wrap stiffness will be critical for the column jacket design. The jacket must be adequately stiff against low axial strain in the column.
- Columns confined with FRP sheets that have lower strength will benefit more from increased compression load capacity.

- When handling eccentrically loaded columns, a lower factor for enhancement should be considered in the design for FRP-wrapped concrete columns.
- Seismic damage may be controlled or prevented by using unidirectional fibre composite sheets in the longitudinal direction to improve flexural capacity. Additionally, wrapping the columns in the lateral direction may be necessary, as this will provide better ductility and energy absorption capacity.
- Corrosion-damaged concrete columns may be repaired with FRP. This will improve strength and ductility and will also reduce the rate of corrosion reaction in the member.
- For impact loading, the concrete columns will need to be adequately strengthened to have sufficient capacity for energy absorption.

Source:

- Parvin and Brighton [131]

7.3.14 Technique of Wrapping [134]

Carbon fibre fabric, when epoxy-bonded and placed on the external surfaces of concrete members, can provide significant strength to the structural elements without adding much extra weight.

Carbon fibres have high tensile strength and are lightweight, as stated. The composite material is called fibre-reinforced plastic (FRP). FRP wraps may be used for any size or shape of structural elements to be strengthened and are easy to execute. The traditional method of strengthening by adding concrete or steel may not be suitable due to time constraints and increased workload.

The system manufacturers would match the fibres with the resins in order to achieve the required increase in strength. Carbon fibres are often used for FRP composites.

Other fibres that may be considered for the same purpose are glass, aramid (Kevlar), or basalt. Carbon is expensive, and when high strength is not required for a particular application, glass or basalt may be preferred primarily for cost reasons. Basalt may be favoured as a low-cost fibre that exhibits high abrasion resistance and does not wick. Glass fibres may wick and absorb moisture and chemicals, leading to potential failure.

FRP strengthening may be 30%–50% cheaper compared to traditional strengthening methods due to its simplicity and ease of installation.

Source:

- Palmer [134]

7.3.14.1 The Comments on Applications, Carbon Fibre Wrap [135]

The applications for strengthening may include the confinement of columns, strengthening of weaker structural elements, natural stone and timber structures, irregularly shaped structural substrates, masonry construction, etc.

These may be summarised as follows:

- Confinement
- Seismic strengthening
- Shear strengthening
- Weak substrate strengthening

Source:

- Sika Information [135]

7.3.15 Further on Strengthening of Reinforced Concrete Structures with Fibre-Reinforced Polymers

The different types that may be covered are as follows:

7.3.15.1 Flexural Strengthening [133]

This may be carried out by placing the composite material with epoxy bonding in the tension zone, parallel to the principal stress direction. Special care may be necessary to check the linear elastic type of FRP materials and the required bond between FRP and concrete.

The possible failure modes may be

- Yielding of steel followed by FRP fracture
- Crushing of concrete (compressive failure)
- Yielding of steel followed by compressive crushing of concrete, with FRP remaining intact
- Peeling off of FRP at cut-off points or terminations, following the shear failure of concrete
- Initiation of peeling off far from the ends following inclined shear cracks in concrete.
- De-bonding at the FRP–concrete interface may occur in locations with uneven concrete surfaces or due to faulty bonding. This type of failure may be avoided through proper quality control checks.
- Information on wrapping using SFRP is not very common.

The failure mode related to the yielding of steel and crushing of compressive concrete may be desirable.

The slip at the concrete–FRP interface may not be considered in the design.

7.3.15.2 Confinement and Strengthening

Applications involving the confinement or wrapping of weak RC columns in seismic areas using FRP materials have been carried out. An efficient method of confinement increases the rotational capacity in the plastic hinge areas and can control the de-bonding of internal reinforcement in lapping areas.

Confinement may be executed for RC columns using FRP materials in the form of pre-fabricated jackets or in-situ cured sheets, maintaining the principal fibre in the circumferential direction. By adopting the confinement method, the axial load capacity can be increased, as stated.

In circular columns, FRP wrap would control the lateral expansion of the concrete.

The following items are relevant for models with FRP-confined concrete:

- The circumferential stiffness of FRP.
- Following micro-mechanics, the composite action of FRP concrete columns and FRP concrete interaction will be important.
- Control of the dilation tendency of concrete through continuous restraint provided by the FRP.
- The confinement is less efficient for RC columns with a rectangular cross-section. In such cases, confining stresses will concentrate on the four corners of the cross-section, as stated.

7.3.15.3 Shear Strengthening

The items mentioned below are important:

- Shear strengthening may be achieved by bonding FRP materials with epoxy, with the fibres placed parallel to the principal tensile stresses.
- The efficiency of strengthening through external FRP reinforcement may be linked to the mode of failure. This may occur either by peeling off through the concrete or by FRP tensile fracture at stresses that may be lower than the FRP tensile strength. Other factors such as

bond condition, available anchorage length, the type of attachment at the FRP termination point, axial rigidity of FRP, and the strength of concrete, etc., will all be relevant.
- Depending on the requirements and accessibility, the strengthening may be a partial or full wrapping system for the elements.

In some cases, the actual mechanism may involve a combination of peeling off in some areas and fracture in others.

It is not easy or clear to assess the load to be carried by FRP at the ultimate limit state for shear related to reinforced concrete members.

Source:

- Bakis, Bank, Brown, Cosenza, Davalos, Lesco, Machida, Rizkalla, and Triantafillou [133]

7.3.16 Strengthening with Carbon Fibre-Reinforced Plates

Carbon fibre-reinforced plates have been used for bridge repair. Considering their low weight, high strength, excellent handling flexibility, varying elastic moduli, high resistance to corrosion, and excellent fatigue, creep, and fire resistance characteristics, carbon fibre-reinforced plates have been selected for the repair work of a bridge.

The bridge is located in Skipton, UK. It features pre-stressed concrete [136] hollow edge beams that were damaged and in need of strengthening.

CFRP plates have been utilised for shear strengthening of beams [137].

A large number of beams were tested to study the performance of carbon fibre-reinforced polymer (CFRP) plates in increasing the shear resistance of the reinforced concrete beams. The beams were designed to fail in shear.

The surface preparation is crucial for developing a bond, and thus, necessary guidance will be required from the manufacturers.

The following surface preparation and treatment have been completed:

- Grit blasting and cleaning the surface with compressed air.
- Wire brushing and hand-abrading with a carborundum stone.

The application of CFRP strips to shear spans significantly increased the strength [137].

Further comments have been made in item 7.6 separately.

CFRP plates have been successfully used to strengthen a live bridge [138]. The bridge was constructed with two main steel edge girders simply supported on the abutments at each end. There were cross beams connecting the longitudinal girders and a concrete deck over permanent pre-cast deck soffit panels.

Ultra-high modulus (UHM) CFRP plates were used on the outer face of the bottom flange of all cross beams.

Sikadur-30 was also selected as an adhesive for fixing CFRP plates after treating the steel surface with grit blasting and applying a suitable primer.

Further information on strengthening has been covered separately in item 7.6.

CFRP plates were applied to strengthen a 65 m tall fertiliser prill tower [139] in Qatar.

Different options for strengthening as listed here:

- Steel plate bonding in hoop direction (jacketing)
- CFRP plate bonding
- Adding ten concrete rings spaced uniformly along the height

In accordance with CFRP requirement, two 160×1.8-mm plates per metre were provided for strengthening throughout the tower. A specially formulated adhesive suitable for the Middle East was used. Other options, as mentioned above, were also considered; however, the CFRP option was ultimately selected as it would have the least impact on the operations of the tower.

Source:

- Lane, Fashole-Luke, and Skwarski [136]
- Walker [137]
- Dodds [138]
- Luke, Godman, and Hamnevoll [139]

7.3.17 Further on Designing with Carbon Fibre Composites

7.3.17.1 The Comments on the Design Guidelines

The following may be relevant:

- Joints
- Consideration of peeling stresses
- Thickness
- Design should be maximised
- Surface area

Items such as de-lamination, galvanic corrosion, jointing and fatigue may be considered the most relevant and critical.

Further tests may be carried out for better understanding.

Additional information is provided in item 7.4.

Source:

- Uusitalo [140]

7.3.18 Polymer Composites and Concrete Slabs

A concrete slab was designed with CFRP corrugated sheets as permanent formwork and tension reinforcement. The slab was simply supported in the direction of the corrugation [141]. It was tested for a point load at the centre of the slab, spreading over an area. The CFRP sheet was made from high-strength carbon fibre fabric in epoxy resin.

The neutral axis was computed to be 19 mm below the top surface of the concrete based on the analysis.

Adequate bond between the CFRP and concrete was achieved by using a proper adhesive, which was applied to the CFRP before placing the concrete.

The slab's performance at the serviceability limit state was also satisfactory.

Further information on the test may be seen later.

Source:

- Szczerbicki [141]

7.3.19 The Behaviour of Polymer Materials for Concrete Repair and the Issues Responsible for Selection

7.3.19.1 General Comments

Polymer materials are used for the repair of concrete structures and may be applied by casting or trowelling on the surface. The repair material can be applied to the prepared base concrete, which should be suitable and sound.

Polymer materials can be formulated to provide a wide range of properties, from brittle or ductile, porous to impermeable, and skid resistant to water shedding, among others.

The critical items of the material may include

- High strength
- Low viscosity (i.e., for injection to cracks)
- Low permeability
- Chemical/mechanical resistance
- Free flowing
- Surface penetration (i.e., for protective coatings)
- Shelf life, pot life
- Water shedding (i.e., for removal of contaminants)

7.3.19.2 Other Issues

Thermal and mechanical properties may include

- Elastic modulus
- Bond strength
- Effects of temperature and humidity
- Creep and shrinkage

Different combinations of the properties may be possible with values required for a particular application.

It should be noted that the performance/behaviour of the materials may be affected by environmental conditions during the curing period and service life.

7.3.19.3 The Curing Agents

The main items can be divided into

- Aliphatic amines
- Polyamides/amido-amines
- Cyclo-aliphatic amines
- Aromatic amines

7.3.19.4 Concrete Repair

To control the deterioration of concrete for various reasons, the repairing material should satisfy the following conditions:

- The material should have low permeability to prevent the penetration of water, chemicals, or carbonation, which can damage the reinforcing bars through corrosion.
- It should provide extra stress-carrying capacity, considering that the compressive strength of the structures may have been lost.
- The material should be aesthetically acceptable, especially in relation to restoration work involving spalling and cracking.
- For abrasive conditions, the material should possess additional wear resistance properties.
- The material may be stress-carrying for structural applications or for cosmetic purposes. Cosmetic applications may focus on abrasion resistance, permeability, or the finished look.
- The price of the material should be competitive.
- Application of the material should not be complicated.
- The service life after the repair/strengthening work should be largely trouble-free. However, routine inspection and assessment will be required, and nominal work may be expected in some cases.

For structural applications, a low creep material may be preferred for the expected environmental range. For cosmetic applications, a higher creep material may be advantageousin certain environmental conditions.

More information may be required on rapid expansions and contractions in which creep may not play an important role, such as in thermal shock.

Source:

- Plum [142]

7.3.20 High Performance Polymer Fibres

High-performance polymer (HPP) fibres have been used for the design and construction of ground-supported slabs. They have high resistance to acidic and alkaline conditions. Tests indicated, at the initial stage, that [143] crimped HPP fibres would require much greater energy to pull out compared to a straight fibre of similar length and diameter.

- High-performance fibres are inorganic fibres and are generally meant for high [144]:
- Tensile strength and modulus.
- Resistance to chemical agents that normally degrade conventional fibres.
- Resistance to heat and flame.

Some high-performance fibres are also used for their thermal resistance. They do not catch fire and provide resistance against heat. These may be obtained by [144]

- Using additives
- Special spinning processes
- Surface modification applying chemical finishes

For protection against fire, the fabrics constructed with these fibres may be considered [144].

HPP fibres are used in pre-cast housing panels and industrial floor slabs, especially in external [143] slabs where surface fibres will not corrode.

Source:

- Carr [143]
- NPTEL [144]

7.3.21 Foamed Polymers

Foamed polymer may be prepared by adding a blowing agent [1] to the chemical formulations, which causes the material to expand. This increases its volume many times by the formation of small cells.

There are various applications. It is lightweight and has good thermal insulation properties. It may also be used for packaging and sound absorption.

The applications are often for electrical/electronic components.

Source:

- Hollaway [1]

7.3.22 More on CFRP Plates as Shear Strengthening for Beams

The investigation was carried out to assess the efficiency of various CFRP configuration methods to increase the shear resistance of beams.

Test Programme – A total of 38 beams were considered, which were tested until failure in shear. Strengthening was done using CFRP methods along with other factors.

Variables were as follows:

- Percentage of main (bottom) reinforcement
- Different configurations of CFRP strengthening
- Link spacing in the shear span

SIKADUR-30 adhesive bonding material was used for the concrete, irrespective of grading method and surface treatment.

The application of two horizontal CFRP strips over the shear span and tension zone may be an efficient method for shear strengthening. A better increase in shear strength could be achieved by bonding plates over the total depth and shear span.

The use of SIKAWRAP 103C on the soffit and sides of the shear span could significantly increase the shear strength.

Source:

- Walker [137]

7.3.23 Comments on the Differences between FRP-Reinforced and Steel-Reinforced Concrete Structures [31]

The differences between FRP-reinforced and steel-reinforced concrete structures may be summarised as follows:

7.3.23.1 General Comments on Design

- FRP will have high strength and low stiffness. Failure may primarily result from the compression of the concrete rather than the rupture of the reinforcement.
- The anchorage lengths and lap lengths will depend on the type of bar used and are related to bond stress.
- Crack widths may be restricted for aesthetic and watertightness reasons, although this may not be necessary for durability requirements.
- Partial safety factors will be higher for FRP reinforcement than for steel reinforcement due to material variability. Furthermore, there may be uncertainties regarding long-term performance.
- Deflections are expected to be greater than those for equivalent steel-reinforced elements.
- FRP bars possess lower compressive strengths compared to their tensile capacities. Traditional design methods for columns may not be applicable.
- Fire safety will be a significant consideration in the design.
- The glass transition temperature of the resin will be critical. If this threshold is exceeded during a fire, the fibres will not act compositely. Therefore, it is important to control the rise in temperature at the surface of the bar.

7.3.23.2 Costs

FRP bars will be more expensive than standard high yield steel bars. The high initial cost may be offset by considering the overall costs of structures in aggressive environmental conditions, incorporating factors such as performance, inspection, and maintenance. The durability of FRP bars in concrete will be a crucial factor.

7.3.23.3 Further Study/Work

The following points should be considered:

- Materials should be produced with a standard range of properties.
- Suitable reinforcement shapes should be developed.
- A deeper understanding of the behaviour of the structures in fire is needed.
- Production must ensure consistent quality.
- Design approaches that are most suitable for the materials need to be identified.

Source:

- Clark [31]

7.3.24 Cost and Ductility Effectiveness of Concrete Columns Strengthened with CFRP and GFRP Sheets

7.3.24.1 Introduction

Reinforced concrete columns have been strengthened by efficiently wrapping fibre-reinforced polymers (FRP) in many projects. The typical applications for confinement in columns involve the use of CFRP and glass FRP (GFRP) sheets. The use of steel FRP (SFRP) for strengthening is relatively limited.

This study focuses on the effectiveness of concrete columns wrapped with SFRP sheets, considering factors such as concrete strength, the number of SFRP layers, and the size and slenderness effects of the columns.

The experimental study indicates that columns wrapped with SFRP performed better under normal room temperature, freeze–thaw cycles, humidity, and prolonged exposure to high temperatures compared to columns wrapped with conventional CFRP sheets.

7.3.24.2 Experimental Study

For the experimental study, importance was placed on the following areas:

Type 1 – to investigate the cost and ductility effectiveness of large-scale concrete columns.
Type 2 – to study the cost and effectiveness of columns with varying slenderness ratios.

In Type-1, 18 columns with dimensions of 300 mm in diameter and 1,200 mm height were considered.

The columns were divided into various groups, such as unwrapped, CFRP-wrapped columns, SFRP-wrapped columns, and columns with longitudinal reinforcement, with different exposure conditions and varying fibre orientations of FRP sheets. For the durability study, inner longitudinal FRP was introduced specifically to regulate crack distribution and reduce stress concentration.

The second type of tests included 34 unreinforced concrete specimens having a diameter of 150 mm, with heights varying from 300 mm to 900 mm, which were divided into groups. Each group contained six specimens; three were left unwrapped and treated as control specimens, while three were wrapped with one layer of CFRP or SFRP sheets.

It may be noted that the columns were wrapped in such a manner that the fibre directions of the SFRP- and CFRP-wrapped columns were similar for the purpose of comparison. The wrapping of concrete columns with one layer of SFRP sheets had an equivalent stiffness to columns wrapped with one layer of CFRP sheets.

The columns wrapped with CFRP and SFRP sheets were tested under uniaxial compression loading.

7.3.24.3 Comments on Cost-Effectiveness

7.3.24.3.1 Cost-Effectiveness

The issue of cost-effectiveness will relate to the strength gain against the total costs involved in the construction and FRP wrapping of the concrete columns.

Cost-effectiveness may be considered in assessing the efficiency of SFRP sheets as enhancement materials for concrete.

The cost-effectiveness parameters may be expressed by the strength efficiency scale, defined as the percentage increase in strength for FRP-confined concrete columns divided by the total cost of the FRP-confined concrete columns.

The cost of SFRP may be 56% of the cost of CFRP sheets. Labour costs for SFRP wrapping may be higher than for wrapping with CFRP sheets. The total cost for SFRP columns may be twice the labour cost of CFRP-wrapped columns.

7.3.24.4 Large Columns

From the analysis, it can be noted that columns wrapped with SFRP may have a higher cost-effectiveness parameter when compared with columns exposed to environmental conditions.

For determining the effective strengthening of concrete columns with two layers of SFRP sheets, the results were compared with RC columns wrapped using only one layer of SFRP sheet. The strengthening of concrete columns with two layers of SFRP sheets, with the fibres in the longitudinal direction and one sheet with fibres in the transverse direction, indicated insignificant effects in terms of strength enhancement when compared with reinforced concrete columns wrapped with one layer of SFRP sheet.

7.3.24.5 Smaller Scale Columns

For small-scale samples, it should be noted that the size effect may be an important parameter to consider in relation to the cost-effectiveness approach. The size of the specimen will be relevant, and it may affect the applied confinement pressure on the concrete. With the increase in the size of the samples, the confinement pressure decreases, leading to a lower increase in strength.

7.3.24.6 More and Summary of Comments

The following may be noted:

- The contribution of different parameters, such as concrete strength, the number of SFRP layers, size, and slenderness of SFRP-wrapped concrete columns was studied analytically.
- Following the study on cost, it has been observed that the columns wrapped with SFRP showed better performance than the columns wrapped with CFRP sheets.
- A large increase with different fibre orientations may be possible for strength and ductility.
- Cost considerations may be relevant when deciding on the performance of FRP sheets related to confinement and concrete columns.
- With the consideration and inclusion of cost, the overall benefit may be nominal.
- The optimum design may be developed using cost-effective and ductility-effective analyses for some cases using FRP strengthening configurations.
- Comparing the cost-effectiveness parameters determined experimentally with those determined analytically indicated good agreement.

Source:

- Abdelrahman and El-Hacha [132]

7.3.25 Durability of FRP Composite Materials

The following points may be made:

- The issue related to environmental–mechanical durability will be better than that of conventional materials.
- By selecting constituent materials, it may be possible to design a material that will have better long-term performance. This will provide good resistance covering the influence of moisture, chemical attack, dynamic loading, ultraviolet radiation, and freeze–thaw cycles, thus improving the properties against aging.
- Fibre composite materials have been used for a short period. The assessment for durability may not yet be entirely known.
- FRP composites have been applied in infrastructural projects, and they have been appreciated to a large extent.
- Long-term durability performance may be required for different applications covering a period of 75 to 100 years.
- The important items for durability and the necessary information may include alkaline solution, moisture effects, physical deterioration, and creep.

Source:

- Humphreys [5]

7.3.26 Durability Properties of Strain Hardening Fibre-Reinforced Cementitious Composites

The requirements for durability will be very relevant and critical for structures, especially in aggressive environments. Compared to normal concrete and ordinary FRC, strain hardening FRC (SHFRCC) can resist the migration of harmful substances into concrete and may enhance the durability of reinforced concrete.

Ordinary reinforced concrete structures are susceptible to damage under service loads and difficult environmental conditions. Cracks may develop due to the deterioration of the structure and may later initiate the corrosion process. SHFRCC will be suitable to satisfy the tight crack width limitations of different codes for the requirements of durability of RCC structures.

SHFRCC will be suitable for meeting the strict crack width limitations of various codes regarding the durability requirements of RCC structures.

Source:

- Uddin Ahmed and Mihashi [145]

7.3.27 Fatigue in Composite Materials

Owing to the anisotropic characteristics of composite materials, failure may be complicated under static and fatigue loading. Composites may exhibit good resistance when the fibres are unidirectional and follow a linear relationship to failure. When composites contain off-axis plies, different damage mechanisms may occur, and the response will be non-linear.

The following basic failure mechanisms may occur:

- Matrix crack
- Fibre breakage

- De-lamination
- Fibre breakage and interface de-bonding

The type and extent of damage may vary and depend on

- Material properties
- Stacking sequence
- Types of fatigue

Source:

- Hollaway [1]

7.3.28 Other Applications

These may include

7.3.28.1 General

FRP pipes – water supply and sewage pipes, skylights and dome lights, doors and windows, coffer moulds, internal partitions, FRP water storage tanks, glass fibre-reinforced polyester sheets, pre-fabricated/re-locatable structures, fibreglass-reinforced plastics pipes, etc.

Some of the advantages in building constructions are as follows:

Easy to manufacture, non-combustible, impermeable to the environment, suitable for air conditioning and ventilation, resistant to corrosion, thermal and sound insulation, permits natural light, precise tolerances may be maintained, mechanically strong, easy to handle, transport, and install due to lightweight, allows good flow capacity and surge pressure due to a smooth interior surface, low pumping costs and reduced frictional loss for water supply, easy to repair and maintain.

7.3.28.2 Structural Uses

May be considered suitable for

Better properties in flexure and tension. The structure will be strong and durable, may be compared with steel, aluminium or PVC cross-section profiles, will have greater resistance to corrosion, and can be fabricated in different shapes. It can be bolted or riveted, and also allows for translucent roofing.

Other different applications may include

Ladder, floor gratings, roof truss work, window/door frames, staircase railings, conduit pipes, cable-carrying trays, etc.

Source:

- Civil Engineering and Construction Review [146]

7.3.29 Further Comments on Cost

The cost of the material will be connected to short-term and long-term costs.

Short-Term Costs – These costs include design, construction, and installation.
Long-Term Costs – These will include maintenance, repair, modification, demolition, and disposal.

For the analysis, these may be classified as direct costs like materials and manufacturing, and indirect costs covering traffic interruption, resale value, depreciation, and environmental issues and their impact.

7.3.29.1 Short Term Cost of Composites

Initially, fibre composite materials will be costly compared with conventional construction materials. The cost may be high due to the expensive raw materials, possibly because of the use of improved imported materials and the processing costs for production.

It should be noted that with increased use of composite materials in different applications, costs are expected to decrease.

7.3.29.2 Fabrication Cost

FRP materials will be associated with fabrication, which may be costly.

7.3.29.3 Cost of Fibre Composite Materials

Different considerations may be included to determine the actual cost of the composite materials. The method may be based on life cycle costs, including initial costs, maintenance costs, replacement costs, operating costs, refurbishment costs, and disposal costs. Other items like taxes, depreciation, and additional management costs are also involved. The process of determining the cost may not be easy and may include various elements to the extent possible for comparison with conventional materials.

The method must consider the advantages of existing versus new materials in structures, based on similar performance covering minimum service life, strength, stiffness, and finally, the impact on the environment.

Source:

- Humphreys [5]

7.3.30 Inspecting, Installing, and Monitoring of FRP-Strengthened Concrete Structures

7.3.30.1 General Comments

The use of externally bonded fibre composites with carbon, glass, or aramids, along with suitable resin, may be applied for strengthening structures as previously mentioned.

The main advantages would be cost, speed of execution, and the reduction of contract duration. The performance with such strengthening methods has been reported to be satisfactory.

7.3.30.2 Types of Inspection

After installation of FRP, a detailed and thorough inspection may be necessary to record the as-installed condition of the material. This may be used as a reference for other inspections that may follow later. Routine visual inspections, as well as regular and more detailed inspections, may be carried out for all strengthened structural areas. For building projects, the strengthened portions of the area may be inspected.

7.3.30.3 Inspection of the Strengthened Structures

Regular inspections are important for concrete and the areas that have been strengthened. This may be critical for structures like bridges and in coastal environments.

7.3.30.4 Deterioration of the Composite Materials

The important items of deterioration for FRP materials may include

- Damage from fire
- Mechanical damage

Efforts may be made at the design stage to minimise these effects to the extent possible.

Mechanical damage may occur from impacts, such as soffits of beams for bridges being struck by over-height vehicles and damage to bridge piers.

To take precautions, suitable plates or markings indicating the locations of FRP should be fixed close to the composite areas. Drilling through the composites for fixing items should be avoided wherever possible.

7.3.30.5 Visual Inspection of FRP Materials

FRP materials normally need very little maintenance during their service life, provided the adhesive and composite have been properly executed according to the correct specifications. It may be prudent to check or inspect the FRP composite for cracking, crazing, or de-lamination, if possible, to assess the degree of deterioration. The composites should be inspected for local damage or abrasion. Visual inspection is very important and may play a critical role, especially when carried out by an experienced person.

7.3.30.6 Inspection and Routine Maintenance

Moisture is considered the main reason for the deterioration of the FRP strengthening method. Hence, it is necessary to clear gutters, drains, etc., and maintain them free of debris. The type of solvent used for cleaning must be checked to ensure it does not cause any damage. Water jetting or grit blasting should be avoided to prevent damage. Steam cleaning is also not recommended, as it may soften the adhesives in the process.

7.3.30.7 In-Situ Tests for Inspection for FRP

There are limited methods available for testing FRP after installation, other than gently tapping the area of the structure with a hammer or coin. This method is based on the change of sound in different bonded areas of the work. The application of this technique is highly dependent on the experience and skill of the operator. Other methods may include thermography and acoustic emission.

It may be efficient to place and bond additional samples of fibre composite materials away from the strengthening areas for comparison. These should also be inspected and tested.

7.3.30.8 Repair

For local repairs, plate overlapping may be done. For larger damage, it may be necessary to replace the deteriorated material covering a larger area. It is critical to ensure that the peripheral material is undamaged and adequately bonded. The concrete surface to be repaired must be properly prepared again for installing further FRP materials. The execution will be important.

7.3.30.9 Summary of Comments

- The FRP strengthening method, when executed according to proper specifications and installation, can be expected to be durable.
- The success of the method depends on the quality of workmanship.
- The work should be executed by experienced personnel with proper knowledge of the method to be used.
- Tests should be carried out during installation to assess the quality of materials and workmanship.
- The requirements for monitoring and inspection arrangements should be discussed and agreed upon by the owner.
- It may be prudent to keep dummy specimens that may be required for subsequent testing.
- Arrangements for access to the structure and the required areas must be provided for inspection/monitoring.

Source:

- Clarke [125]

7.3.31 Further Review and Summary of Composite Materials

These may be mentioned as follows:

- The composite materials will contain two important components: fibres and resins.
- These two components are combined to achieve the properties of the final product.
- The fibres are the essential elements of the composite structure that provide the mechanical properties
- Depending on the requirements, different fibres such as carbon, glass, and aramid may be used to develop the matrix of the structure.
- There are different types of resins available, each with various properties.
- The main types of resins are thermo-setting and thermo-plastic resins.
- The main types of resins are thermo-setting and thermo-plastic resins.
- Common types of thermo-setting resins include vinyl esters, epoxies, and phenolics.

7.3.31.1 Different Uses

The applications of FRP composites will be discussed in this chapter.

However, some comments are given below:

- Sandwich structures may be selected appropriately to meet the requirements.
- Integration along with the optimisation of materials takes advantage of the properties that can be effectively utilised for composite structures.
- Core materials for this may include plastic foams, honeycomb materials, and timber.

Source:

- Fy-Composites, Structural Composite [147]

7.3.32 Widening a Bridge and Stitching the New Section to the Existing One

The bridge in the UK was widened and connected to the existing one with minimal disruptions during construction.

The relevant items for bridge widening are mentioned below:

- A composite structure was considered for the new section.
- Fabricated steel beams along with a reinforced concrete deck were used.
- The concrete was poured over glass FRC permanent formwork.
- The work was executed with reduced traffic disruptions.

Further information on this case study is covered separately in item 7.7.

Source:

- Chong [148]

7.3.33 Creep on Fibre-Reinforced Corrugated Concrete Roof Sheeting

NETCEM sheets were tested against loading, specifically for deflection, to assess the long-term effects. NETCEM sheets consist of two layers of a continuous open network of fabricated polypropylene film in a cement-based matrix. These corrugated composite sheets may be used on roofs as an alternative to asbestos–cement materials.

7.3.33.1 Details of the Tests

Eight corrugated sheets were tested considering snow loads for a period of 100 days.

The sheets were loaded under simply supported conditions for a span of 1,380 mm, taking into account the spacing of purlins. The loading was done using oven-dried sand, which was spread uniformly over the surface, and the loads were increased in stages.

It was observed that the increase in deflection over time did not pose problems for the loading at room temperature.

It can be noted that such sheets may not present issues with increasing deflection over time, even under conditions where the cracking stresses in the matrix are slightly higher.

Source:

- Marsh [149]

7.3.34 Case Studies

A few case studies are presented below. These structures had been strengthened using the fibre-reinforced polymer (FRP) wrapping technique.

7.3.34.1 Bridges in Chengdu, China

The bridges were located in mountainous regions. Two arch bridges had to be strengthened to improve their shear and flexural capacities. The span of the simply supported bridges was 120 m over a deep chasm. Due to increasing traffic loads, the bridges had to be upgraded. The conventional strengthening system was found to be unsuitable considering the site conditions. The Tyfo SEH Fibrwrap system was finally selected for strengthening. A light-duty gondola system was used. The repair work was completed in 2 months.

7.3.34.2 Fire Damaged Factory, Indonesia

The structure was damaged during a fire inside the factory. From visual inspection, it was found that spalling and cracking of concrete had occurred in the beams, slabs, and columns.

After carrying out a detailed study, it was observed that the concrete structure was badly damaged and would need strengthening.

Tyfo SCH carbon and SEH glass types were used to increase the flexural and shear capacity of the damaged structure.

The repair work was finished in 15 days.

7.3.34.3 Bridge in the USA

FRP was used to strengthen the bridge in Rhode Island using the Tyfo SCH carbon fibre wrap method. Shear strengthening was also performed for the overstressed sections. The FRP system was applied to the internal and external faces of the concrete girder. It was possible to use the lightweight Tyfo system, and the material was placed by hand in the areas requiring proper strengthening.

7.3.34.4 Telecommunication Towers in Brazil

Tyfo SCH carbon system and WEB bi-directional glass system were used for strengthening.

7.3.34.5 Waterfront Structure in Korea

The existing piers had to be strengthened to carry extra loads. The beams and slabs were strengthened using the Tyfo Fibrwrap SEH E-glass system.

Source:

- Donnelly [150]

7.3.35 Further Case Study

The application of an alternative material in a bridge will be presented below:

7.3.35.1 Dawlish Station Foot Bridge, Plastic Bridge in UK [151]

With the increasing demand for the development of alternative construction materials, engineers are constantly trying to find different and suitable items that will be useful and environmentally friendly. As a case study, the construction of a bridge is reported here.

By using an alternative material, a footbridge has been constructed. The steel bridge referred to here, having a span of 17.5 m and linking railway platforms, was damaged over the years primarily due to a saltwater environment. This bridge was originally designed by Brunel in 1830. The original bridge was replaced using the same design in 1937. Considering the age and importance of the bridge, repairs or replacement work had to follow the same form. This bridge has now been replaced using FRP as an alternative design. The FRP bridge is aesthetically identical to the original steel structure, but it is sufficiently lighter and more durable. Because of space constraints, the use of a large crane was not possible on-site, as the bridge is located near a small village. The bridge is made of two main girders, 1.66 m deep I girders, spanning 17.5 m across the railway tracks. The 15-tonne steel structure was replaced by a 5-tonne FRP bridge.

Each girder is formed from foam-cored shear webs, moulded by film infusion using epoxy resin and biaxial glass fibre reinforcement, capped top and bottom with pultruded angles and plates to form the flanges.

The deck is formed from lightweight composite pultruded panels with a skin thickness of 3 mm. The only non-FRP areas are stainless steel bolts through the panels. All the primary connections are bonded, but the bolts are used to provide an additional degree of confidence.

Source:

- Smale and Smith [151]

7.4 DESIGN GUIDANCE AND STRENGTHENING STRUCTURES WITH FIBRE COMPOSITE MATERIALS

7.4.1 Introduction

Fibre-reinforced polymers (FRP) have been used for a long time, as already mentioned. These materials are being applied in the construction of structural elements. Some comments on FRP materials have been made in item 7.3. The areas related to strengthening specially related to concrete structures, will be discussed in this section.

7.4.2 More on Advantages

- Lightweight
- Simple and fast installation.
- No need for temporary support for the structure.
- Materials may be available in long lengths and can be cut as per requirements.
- The flexibility of the materials also helps in installation.
- The materials can accommodate irregular surface shapes.
- No requirements for laps and joints.
- Can be installed behind existing services.
- Overlapping may be necessary when strengthening is required in two directions.

- Simple execution is very important for bridges, avoiding lane closure issues and possession times on major highways and over railway lines.
- The weight of the structure and its dimensions are not significantly affected by the strengthening work.
- This is especially relevant in areas with limited clearances for structures, such as tunnels, bridges, and other constructions.

7.4.3 Disadvantages

- Risk of fire damage
- De-lamination
- Vandalism
- Accidental damage
- Soffits of bridges and reinforcement over roads may be hit or damaged by overweight vehicles.
- Some protection arrangements or systems may need to be considered to avoid this.

7.4.4 Design Methods

The following will be relevant:

- Fibre composite materials will have a stress-strain response that is linear to the ultimate load with no yielding.
- The traditional method of design may be adopted for members in bending.
- Elastic analysis may be performed with no re-distribution.
- Additional checks will be necessary to eliminate peeling failure at the ends of the laminate and de-bonding from the surface of concrete.
- If failure occurs, it will be in the outer layer of the concrete surface.
- To control de-bonding, the strain in FRP should not exceed 0.8% when the applied load is uniformly distributed. The strain should not exceed 0.6% when high shear forces and bending moments are present.
- The longitudinal shear stress in the concrete at the ultimate stage may be limited to 0.8 N/ mm^2.
- An anchorage length of 500 mm may be adequate
- Wrapping of circular columns may be carried out to increase axial load, bending, and shear capacities.
- FRP strips may be considered for strengthening members. The strips can be used as external stirrups, applying normal design methods. The strain in FRP should be controlled within the limit of 0.4%.

7.4.5 Applications

Some applications with FRPs have been covered separately in item 7.7.

Only a few items will be mentioned here as follows:

- Columns have been repaired in multi-storey buildings, car parks, etc.
- Many bridges, including small bridges, have been strengthened using FRP to increase their load-carrying capacity to meet revised requirements. The supporting columns have also been strengthened.
- Various other structures, including cooling towers, tunnels, lighthouses, and chimneys, have been repaired.

7.4.6 Workmanship and Installation

The installation of FRP material must be carried out properly according to the manufacturer's instructions to achieve the expected long-term performance. Adequate guidance on the correct selection of materials, proper preparation of the concrete surface, adhesive, correct application of the composite, and proper curing of the adhesive is important.

The work must be executed by a qualified contractor with adequate knowledge of similar applications and trained staff.

7.4.7 Inspection and Maintenance

Routine and regular inspections are necessary for important structures. A team of specialists may be involved in planning, preparation, and controlling this work. It may be prudent to install additional materials that can be removed later from the structural elements for testing. The information and details should be properly recorded for future use.

Source:

- Clarke [152]

7.4.8 External Bonded Plates for Strengthening

7.4.8.1 Initial Comments/Observations

The bonding of steel plates to concrete structures may be considered for strengthening work as stated. Conventional methods of strengthening and crack control may be time-consuming and expensive.

Furthermore, plate bonding has a nominal effect on headroom, is less costly, and maintenance may be easier

Generally, there may not be any need for structural propping during the application of plate bonding. Hence, the strengthened section can support additional loads from non-permanent sources. The initial guidance for design is briefly presented below.

Following the design using the working stress method, the increased flexural capacity of the section must be computed for different conditions:

- The stresses of concrete and steel reinforcing bars in the unstrengthened section for the dead loads.
- Computation of concrete, reinforcement, and external plate stresses in the strengthened section for additional loads (i.e., live or imposed)
- Checking the ultimate moment of resistance of the strengthened section, which must be sufficient for the factored ultimate loads.
- Adding the stresses of strengthened and unstrengthened sections and ensuring the values do not exceed the permissible limits for the total design loads.
- Checking shear stresses in the adhesive layer.
- The cover concrete may be weaker in shear than the adhesive or the interfaces.
- Plates with a width-to-thickness ratio of at least 40 should be used to reduce horizontal shear.
- To control nominal elastic shear stresses, a suitable factor of safety must be used.
- Bolting may be considered at the plate ends to control peel failure.

7.4.8.2 Comments on Materials

These may include

- Steel should be used, considering it to be adequately strong.
- Normally, two-part epoxies should be used. The initial shear strength will be greater than that of concrete.

- There are different types of epoxies available. The materials may need to be modified with various additives and hardeners.
- Polyamine hardeners may be more effective for moisture resistance compared to polyamide products. Furthermore, they may provide better creep resistance than hardeners based on polysulphides.
- Proper surface preparation for bonding may be challenging for stainless steel.
- Epoxy adhesives are thermo-setting polymers and may exhibit reduced physical and mechanical properties at high temperatures.
- Temperature may also affect the rate of curing.
- Manufacturers can produce products suitable for particular environmental conditions.
- Freshly mixed products must be considered and applied.
- For overhead and vertical surface applications, it is important to ensure that bonding is properly completed before hardening starts.

7.4.8.3 Proper Bonding

- The work must be carried out by trained personnel who understand and appreciate the results of good workmanship.
- Concrete elements with high concentrations of chloride ions should not be strengthened using this method.
- Surface preparation will be critical for developing a good bond.
- The plates may be de-greased and grit blasted. They may be lightly grit blasted again prior to the application of the adhesive.
- Plastic spacers may be used to achieve uniform glue line thickness.
- The surface of the concrete must be sound. Grit blasting may be performed to remove surface laitance from the concrete and expose coarse aggregates.
- Excessive mixing of adhesives must be avoided to eliminate air pockets.
- An epoxy primer may be necessary for achieving better durability.
- Pressure should be applied, especially for applications to soffits and vertical walls, until the resin is fully cured.
- Bonding plates at ambient temperatures below 10°C may not be desirable without the use of an auxiliary heating system.
- The exposed surfaces of the steel plates may receive protective treatment.
- Extra steel plates bonded to the vertical surfaces of the beams may be used to improve shear resistance.

7.4.8.4 Further on Applications

Bonded plate applications have been carried out to strengthen structures in different projects. Only a few are mentioned below:

- Bridges
- Multi-storey office buildings, columns
- School buildings
- Roof beams in police car parks
- Shops
- Roof structures, railway stations

Source:

- Mays and Calder [153]

7.4.9 Strengthening Structures with Carbon Fibres

7.4.9.1 General Comments

The strengthening of buildings and bridges by applying CFRP plates has been previously stated. The replacement of members or structures would be very expensive and may present other disruptive problems in many cases.

The application of external steel plate bonding is an alternative solution that has been used in the past. However, there may be some durability concerns. The use of CFRP plates is an attractive solution for strengthening work.

The bonding of CFRP plates needs a good quality substrate, which can be prepared by grit blasting or scrabbling and must be dust-free using a vacuum process. The adhesive should be applied to both the plate and the substrate.

CFRP plates had been used to strengthen different structures instead of steel plate bonding, considering quick applications and saving time and cost. Further uses had been made for timber structures, concrete floor slabs, and rectifying deflection problems of structural members, as well as shoring up masonry walls.

A completed project with CFRP

In a Bridge Project – A single CFRP plate of 120 × 1.2 mm thickness was attached to the underside of the bridge to restore the lost flexural capacity. A specially prepared adhesive was used for bonding. The bridge was kept open during the period of repair work.

7.4.9.2 Tests Carried Out

Tests in the laboratory and on-site were conducted to study the flexural strength, creep, fatigue, and durability of beams.

The results indicated an increase in strengths in the range of 24%–80% over an unplated beam.

By developing a stressing method on-site and using a stressed middle plate, both strength and stiffness increased to a large extent.

7.4.9.3 Application Technique

Applications have primarily been done on concrete structures. Following a similar technique, it may also be used for timber, wrought iron, cast iron, and steel structures. However, the use of proper materials, correct adhesives, and efficient applications will have significant implications for the lifespan of the strengthened structure.

Source:

- Lane, Fashole-Luke, and Skwarski [136]

7.4.10 Further on Bridge Strengthening with Fibre-Reinforced Polymers

The strengthening of bridges may be done using fibre-reinforced polymers (FRPs) as mentioned. The schemes for strengthening must be effective, economical, non-disruptive, and aesthetically acceptable for the structure being strengthened.

The performance of FRPs for strengthening bridges can be considered good.

The materials are made using high-strength fibres like carbon, aramid, and glass in a resin matrix, which is adhered to structural members such as girders/beams, columns, or the deck slab of the bridge using selective adhesives.

The application method may involve the wet-lay-up technique, where the open fabric is impregnated with the resin on-site before being applied to the structural members. Alternatively, preformed FRPs may be used by bonding with adhesives. The thickness of the laminates may be 1–2 mm.

Tests and trials have been conducted by researchers on FRPs to assess their performance. For long-term economic performance, the application of FRP materials may be considered suitable instead of traditional materials.

This is particularly applicable for structural members subjected to de-icing salts.

The applications will include bridge supports, especially in seismically active areas.

Tests were conducted in laboratories to assess the enhancement of strength across different types of FRPs.

The FRP strengthening method has been preferred over steel plate bonding for soffit strengthening of bridges in the UK.

FRPs require special care regarding correct specifications and applications.

Source:

- Sadka [124]

7.4.11 Fibre-Reinforced Polymers (FRP) in Construction – Strengthening Concrete Structures

The applications of FRP in strengthening structural members have been successfully implemented in many projects, as observed. FRP materials consist of fibrous reinforcement held in a polymer matrix to form various items. The main purpose of the matrix (resin) is to hold the fibres in the desired orientation to provide proper performance.

FRP materials exhibit good resistance to fatigue and creep and are very durable.

To enhance fire resistance, additives may be mixed with the resin, or intumescent paints can be used as a protective surface coating.

For beams, CFRP plates may be applied to the sides or the bottom flange of the beam.

FRP materials are easy to handle due to their lightweight compared to conventional steel plate bonding. This approach is less labour-intensive and needs minimal equipment on-site. Non-destructive tests may be organised to assess voids related to bond failure.

It has been reported that the FRP system can significantly increase the stiffness and load-carrying capacity of concrete structures beyond the levels achievable with conventional strengthening methods.

Source:

- Halliwell [122]

7.4.12 More on Strengthening of Concrete with Fibre Composites

7.4.12.1 Use of CFRP Composites and Wraps

To increase the load-carrying capacity of structures, CFRPs are found to have special advantages in increasing flexural strength, redistributing loads around openings, and improving shear, punching shear, and impact resistance of the member.

The following are some comments on salient items:

- **CFRP Composites and Wraps** – Very useful for strengthening, especially for carrying additional loads.
- **Cost Comparisons** – The high cost of composites can be offset by reduced labour and plant costs.
- **Quality Control Procedures** – Strict quality control procedures must be followed to develop confidence in the product.

- **System Design** – Various factors are involved, such as the flatness and quality of the substrate, potential corrosion, and the current stress conditions in the element.
- **Preparation and Application** – Cleanliness, surface preparation, product mixing methods, and application at the correct temperature are important.
- **Initial Project** – The initial attempt with a pultruded composite plate was carried out in the UK in 1996.
- **Further Comments** – The applications of fibre composites in projects represent a major advancement, especially for strengthening concrete structures. Success will largely depend on the quality of workmanship, and it is important that the repair work is executed by an experienced contractor familiar with the products and the method of application.

Source:

- Hackett [130]

7.4.13 Strengthening of Highway Structures with Fibre-Reinforced Composites

7.4.13.1 General Comments

There was a need for strengthening highway structures in the mid-1990s in the UK to accommodate additional loads that may arise from accidental impacts. Initial attempts involved concrete collars and infill between supports.

In this context, proposals were made to use fibre-reinforced composites (FRP) to strengthen existing structures. The composite consists of high-strength fibres, typically carbon, aramid, or glass, embedded in a resin matrix. These materials do not corrode and offer a durable, low-maintenance option for strengthening structures. However, long-term durability must be assessed based on the materials' performance over an extended period

The practical application of column wrapping was conducted in 1997 on a bridge in Cornwall, UK. The work was associated with using materials with glass, carbon, and aramid fibres following the wet lay-up methods. It was easy to use the material; however, adequate care and awareness regarding environmental issues would be necessary during execution.

For another strengthening project for the columns, pre-fabricated shells made of glass fibre-reinforced composites were used around the columns. The use of two half-shells was considered, which were connected on-site with a special fixing arrangement. Ordinary grouting was used to fill the gaps between the shell and the column. The application was found to be suitable and satisfactory.

Similar applications had been done for other bridges to improve bending and shear capacities.

Source:

- Loudon [154]

7.4.14 Strengthening Concrete Bridges with Fibre Composites

7.4.14.1 Review of Basic Items

Fibre-reinforced polymers (FRP) composite materials have been used for strengthening structural elements as stated.

The different types that are used have been mentioned in item 7.3.

- Pre-formed fibre composite plates.
- Sheets of fibres.
- Pre-formed fibre composite shells
- Fibre composite rods or strips

- Pre-formed shear straps
- The suitable type of fibre-reinforced polymer will depend on the requirements and applications.
- The composite material will be efficient and useful for applications compared to other conventional materials.
- The main intention of using FRP is to improve the structural capacity of the bridge, which may otherwise be inadequate. The concrete must be sound, with nominal reinforcement corrosion or other deteriorations.
- Apart from being lightweight, limited preparation of the concrete surface will be required for the application.
- The overall performance will include a cost-effective solution, fast implementation, and less possession time.

The Applications may cover different areas of the bridges to address various problems. A few may be mentioned below:

- Soffit
- Top surface of the deck
- Column strengthening
- Shear strengthening
- **Continuity** – by bonding glass fibre composite plates across the joints
- **Repair after Damage** – the repair work may be needed for a portion (i.e., edge beam of a bridge) after being hit by a vehicle.

The items that will be important and need careful assessment for the rectification work are:

- Cost covering both short and long-term.
- Cost of access and possession time will have a large impact.
- Strengthening with FRP material may be 30% cheaper than equivalent strengthening using steel plates.
- The quality of execution will be very important specially related to concrete surface preparation and the installation of the composite material as per the specifications for success.
- Proper inspection and monitoring must be done regularly to assess the strengthening work executed.

The critical items for strengthening bridges include fast execution/installation and short possession time, as already mentioned.

Source:

- Clarke [123]

7.4.15 Further on Strengthening a Bridge Using CFRP Plates

This example is for a single-span steel bridge of 26.06 m and a clear width of 8.31 m. There were two main girders with deep cross beams at 2.25 m centres.

For the bridge deck, CFRP was selected as the strengthening material considering its lightness, easy installation, and minimal disruption to rail traffic.

Ultra-high-modulus (UHM) CFRP plates were used for the strengthening work.

CFRP plates were bonded to the outer faces of the main girders, maintaining symmetrical structural steel sections.

Further details of the strengthening work have been included in item 7.7.
Source:

- Dodds [138]

7.4.16 Additional Comments on Designing with Carbon Fibre Composites

The comments on the design guidelines may be mentioned as follows:

7.4.16.1 Joints

- The joints are the weak links of the composites and must be designed properly.
- Open holes should be avoided for compression loading, and filled holes should be avoided for tension loads.
- There is a risk of damage to the material while making holes, which can lead to degradation. Machine-made holes may be preferable.
- Galvanic corrosion may present problems, so the materials used in the joints should be selected accordingly to control this risk. Steel and aluminium may be sensitive to galvanic corrosion.
- The selection of adhesive is critical; it must ensure that the loads are properly transferred to all the fibres in the composite.
- The thickness and orientation of the laminate are important for the joints.
- Adhesive bonding may be suitable for thin composites, while mechanical fasteners may be used for thicker composites.
- Efficient load transfer in an adhesive joint will depend on the joint's design.

7.4.16.2 To Minimise the Reasons for Peeling Stresses

- Sudden laminate closures or structural parts with large differences in flexural stiffness should be avoided.
- Peeling stresses are likely to occur in the weakest direction of the laminates.

7.4.16.3 Thickness

- The thickness tolerances may vary; thicker parts may require higher tolerances.
- Thicker laminates will have better resistance to impact damage and fatigue.

7.4.16.4 Design May Be Done as Large as Possible

- The number of parts will be reduced.
- The number of joints may be minimised.
- Costs may be reduced.

7.4.16.5 Treatment on the Surface Area

- The risk of moisture absorption in the matrix may be controlled by selecting an effective surface finish material as a coating.
- The epoxy material may be damaged by ultraviolet rays; therefore, the surface need not be treated with paint, which may not be suitable.
- Rain erosion coating may be applied to control erosion.
- Suitable surface coating may also protect the composites from damage.

The following points may be noted for the composites:

- Issues such as de-lamination, galvanic corrosion, jointing, and fatigue should be considered the most relevant and important.
- These factors will have significant effects on the mechanical properties of the composites.
- It is therefore relevant to study further and control the effects to minimise any adverse impacts on the performance of the composites.
- Additional tests may be carried out for a better understanding of the behaviour of CFRP under different loads and environmental conditions.

The interactions amongst the concerned parties will be important for further knowledge and understanding of materials for the successful application of the composites.

Source:

- Uusitalo [140]

7.4.17 FRP Shells for Strengthening Bridge Columns

Bridges may need protection against impacts from heavy vehicles travelling at high speeds.

For providing protection against vehicular impact, fibre-reinforced polymer materials have often been used in different countries.

High-strength aramid fibre-reinforced polymers (FRP) may be preferred considering economy and meeting technical requirements.

Kevlar aramid, along with traditional materials, has been used for bridges in the UK. In two bridges, 24 tapered columns were strengthened using the FRP system after considering the following:

- Difficult access for execution
- Minimising traffic disruption during execution
- Amount of strengthening work
- Aesthetically acceptable after completion
- Time for execution
- Cost of execution

Conventional alternative methods, such as tying the columns to walls, were not considered after reviewing time for execution, acceptable aesthetics, cost, and access.

7.4.17.1 FRP Shells

Kevlar fibre was used for its high energy absorption and impact resistance characteristics. Other fibres, like carbon or glass, were not taken into account as they may not have sufficient ductility for dynamic loads.

To meet higher requirements for strength upgrades, conventional in-situ wrapping was deemed technically and economically unacceptable. The decision was made to use pre-fabricated FRP jackets. The tapered columns were changed to circular columns with permanent encasement and fibre shell cylindrical moulding.

Two semi-cylindrical composite shells with an annulus between the pre-assembled mantles and the existing columns were used. The gap was filled with non-shrink, high-strength grout.

The FRP shell will act as permanent formwork, providing a shell that is impermeable, aesthetically pleasing, and maintenance-free during the service life of the structure.

The initial cost was equal to that of the reinforced concrete wall method, and tying the columns would have been connected with extended lane closures.

Furthermore, over the remaining service life of the structure, additional savings may be realised due to the impact resistance properties and low maintenance costs of fibre shells.

Source:

- Pinzelli [155]

7.4.18 Wrapping the Harbour City Car Park

In this case, aramid fibre structural reinforcement was applied to strengthen 164 columns at the Harbour City car park in the UK.

Kevlar FRP was used to accommodate additional stories as per the requirements. It was possible to complete the work within the specified and limited time schedule. The fibre option was beneficial, as the selection provided flexibility in application, requiring only small areas to be cordoned off, leaving most of the parking spaces open and unaffected.

The preparation and work could also be undertaken at night, further limiting disturbances to users. As mentioned, FRP is a lightweight material that contributed little additional loading to the structure, and its sheet form did not appreciably alter the width of the columns.

Source:

- Russell [156]

7.4.19 Confinement of FRP Jacketed Reinforced Concrete Columns

The studies focused on various parameters associated with the effectiveness of FRP jackets around rectangular columns.

The important items are listed below:

- Cross-sectional aspect ratio (ratio of long side to short side).
- The sharpness of the corners of columns.
- The amount and type of fibre in the FRP jacket.
- The amount of longitudinal and transverse steel reinforcement in the column.

The tests were conducted under uniaxial loading on FRP-jacketed reinforced concrete square/rectangular and circular columns. The grade of concrete was 21 MPa, and high tensile fibres of carbon, glass, and aramid were used for FRP jackets.

Further comments on the tests are presented in item 7.6.

Source:

- Belarbi [157]

7.4.20 Ribbed Floor Slab Strengthening, UK

The strengthening of the ribbed floor slab was carried out with carbon fibre laminates to control slab deflections and to strengthen the structure to meet the requirements of additional loads on the floor. Carbon fibre laminates were selected for the design and bonded to the bottom of the ribs. The width of the ribs was 150 mm. The advantages of carbon fibres, which have greater stiffness compared to glass and aramid, were utilised. This treatment helped in controlling the slab deflections.

Further comments are provided in item 7.7.

Source:

- Farmer [158]

7.4.21 Review of Preliminary Evaluation of CFRP Plates for Strengthening

7.4.21.1 Reinforced Concrete Members

An experimental study covering different parameters related to strengthening reinforced concrete beams by bonding CFRP composite plates was carried out.

A few comments are as follows:

- The bonded plates were able to carry additional moments beyond the yield of the reinforcement.
- The ultimate capacity decreases with the reduction of plate aspect ratio and the beam shear-to-span ratio.
- It was observed that, in most cases, failure occurred with the development of high longitudinal shear stresses.
- Concrete failed in many cases; however, the de-bonding of the adhesive from the concrete was not observed.
- Plate end anchorage was developed by resisting plate separation.

The tests were conducted with CFRP plates for the study as stated.

The comments related to the tests are included in item 7.6.

Source:

- Garden, Holloway, and Thorne [159]

7.4.22 Further on Wrapping It Up

The applications for strengthening concrete members may be performed with carbon fibre fabric and epoxy resin, as mentioned already. In the process, the structural members will not impose much additional load on the elements or the foundations. The strengthening with FRP composites has been covered in item 7.7.

The uses would include concrete tanks, pre-stressed water pipes, culverts, etc. Apart from strengthening the structures, FRP will provide an impermeable layer.

7.4.22.1 Installation

The installation of FRP should not be considered difficult, but it will require experience from the applicator. The success of the performance will depend largely on workmanship.

Surface preparation will be critical, and the concrete surface must be cleaned to remove any dirt or chemicals.

Water blasting or shot blasting may be performed to achieve a roughened surface.

There may be two types of applications: bond critical and contact critical.

Bond Critical Applications – This will be associated with the bond of the material to the concrete surface to transfer the stresses.

7.4.22.2 Contact Applications

In this case, FRP is bonded directly and provides confinement to the structural members, wrapping completely around surfaces such as columns to be strengthened.

Source:

- Palmer [134]

7.4.23 Repair and Strengthening of the East Pier, Pre-cast Slabs, Glasgow International Airport

7.4.23.1 Introduction

The information relates to Glasgow International Airport. The project consists of a passenger walkway at the first floor level and stairs leading to the passenger lounges and departure gates located on the ground floor. The structure was constructed in 1966 and was designed and built with lightweight steel frames. The columns were spaced in a 6×6 m grid with main and transverse beams supporting lightweight aerated concrete Siporex slabs spanning between the transverse beams.

A few slabs needed total replacement, while others required repair or strengthening with carbon fibre plates or with cementitious or epoxy mortar.

7.4.23.2 Damage

A detailed survey was conducted for the assessment.

The comments were as follows:

- Local physical damage to Siporex slabs at the first floor level close to the main terminal building.
- Deterioration may be due to previous physical actions and not due to any corrosion of the reinforcing steel in the pre-cast slab.
- No evidence of corrosion of reinforcement was found within the Siporex.
- The Siporex slabs performed well during their service life.

7.4.23.3 Options for Repair

The following options were considered:

- Concrete repair of pre-cast panels/slabs.
- Total replacement of pre-cast panels.
- Strengthening of pre-cast slabs/panels.

The damaged slabs were first repaired using either cementitious concrete repair or epoxy concrete repair materials. This was followed by the application of carbon fibre plates using epoxy adhesive, which were installed at the soffit of the slabs.

The Siporex slabs were replaced with a lightweight ready-mix concrete using Lytag aggregate to maintain the density of the material within acceptable limits.

Source:

- Couper [160]

7.4.24 Application of NSM Reinforcement for a Multi-storey Car Park

In this case, NSM reinforcement was used for the multi-storey car park after observing cracking and spalling of concrete. The design for the repair and strengthening involved using fibre-reinforced polymer (FRP) bars that were placed into grooves and bonded to the top surface of the car park decks. The bars were made of CFRP and are known as NSM reinforcing bars.

Increased use of NSM bars in future designs may be expected for strengthening and repair work.

Source:

- Farmer [161]

7.4.25 Design Codes/Guidance [162]

The design codes, guidance, and specifications will be necessary for the knowledge, acceptance, and confidence of those involved with the materials.

The American Concrete Institute (ACI) developed design and construction specifications related to composite re-bars and composite wraps that may be used to strengthen different structures, including infrastructure related to development work.

A pre-standard for Load and Resistance Factor Design (LRFD) related to pultruded fibre-reinforced polymer (FRP) structures was developed by the American Composites Manufacturers' Association (ACMA) and the American Society of Civil Engineers (ASCE). Following the publication, these guidelines will help structural engineers and architects design and construct safer, stronger, and more durable buildings with FRP composite materials. This will also help FRP composites compete with other conventional materials such as steel, concrete, timber, and aluminium.

It should be noted that the use of composite materials, as per the recommended guidance/code and specifications, will not only give confidence to structural engineers and architects but also to clients.

For design assistance, comments from different authors, supported by test results, have been also included in also been included in various sections. These may also be referenced for design and comparisons.

The design and construction of structures with FRP composite materials would be efficient in producing stronger and more durable structures, and their performance is expected to be very good during the service life.

Source:

- Hota and Liang [162]

7.4.25.1 Guidance from TR 55 Document [163,164]

For design guidance related to the strengthening of concrete structures and for applying and bonding fibre composite materials, the TR 55 document may be consulted.

This document refers to strengthening by placing externally bonded fibre composite materials [163]. Following the design based on the limit state method, it covers flexural strengthening of beams and slabs, shear strengthening of beams and columns, and also flexural and compression strengthening of columns.

Other issues, such as material types and properties, workmanship and installation, field applications, and long-term monitoring requirements, along with the design approach for flexural strengthening, are also addressed.

Furthermore, it deals with column design [164], extreme loading, NSM reinforcement, and the factors for FRP strains instead of stresses, as well as the design of the elements in shear.

Source:

- Arya, Clarke, Kay, and O'Regan [163]
- Concrete Society [164]

7.5 BONDING OF FRP COMPOSITE MATERIALS

7.5.1 Introduction

The applications of fibre-reinforced polymer (FRP) composites have been carried out for concrete structures requiring retrofitting or strengthening work for load-carrying members. The advantages of gaining flexural strength and load-carrying capacity of the structural elements will be significant. However, there may be problems associated with this system.

Failures may be brittle and related to the de-bonding of concrete layers, de-lamination of FRP, and shear failures. This may occur at a load lower than the computed theoretical strength value of the retrofit method.

It may be due to the de-bonding or peeling of the FRP from the concrete surface, which is connected with mechanical and environmental effects as well as epoxy de-cohesion.

Slow de-bonding of FRP composites under service load may also occur, leading to premature failures of the retrofitted technique.

In such situations, relevant non-destructive evaluation (NDE) methods are necessary, especially for inspection activities related to de-bonding, to ensure the satisfactory use of FRP composites in strengthening concrete members. The effects of moisture and temperature are also very important for de-bonding failures of the FRP composite method.

7.5.1.1 De-bonding Mechanisms

These may include

- Shear failure with concrete cover de-bonding.
- FRP de-bonding from flexure–shear crack.
- FRP de-bonding from plate end.
- FRP de-bonding from flexure crack.
- Plate end shear failure.
- Shear failure and de-bonding.

Source:

- Büyüköztürk and Yu [165]

7.5.2 Fatigue in Composite Materials

Due to the anisotropic characteristics of the composite materials, failure may be complicated under static and fatigue loading. The composites may have good fatigue resistance when the fibres are unidirectional, and it may follow a linear relationship to failure. When the composites are made of off-axis plies, different damage mechanisms may take place with non-linear stress–strain performance.

The following basic failure mechanisms may occur:

- De-lamination.
- Matrix cracking.
- Fibre breakage.
- Interface de-bonding.

The type and extent of damage may vary and depend on

- Material properties.
- Type of fatigue
- Stacking sequence

7.5.2.1 Brief Details

Some comments on the items will be provided:

7.5.2.1.1 De-lamination

De-lamination in composite materials may occur between the laminates due to the presence of interlaminar stresses, which may be present near the free edge of the material and under in-plane loading. These stresses may arise from transverse and shear stresses and may be compressive or tensile, depending on the stacking sequence. The amount and distribution of stresses may vary and depend on the laminate used, as well as the properties of the component materials, stacking sequence, and type of loading.

De-lamination may start at the free edge of the composite and proceed inward. Other factors, such as transverse cracking and interlaminar shearing, could contribute to the process of de-lamination.

7.5.3.2 Matrix Cracking

This may be the first damage mechanism to occur. With the increase of loads, the density of cracks will increase. When the laminate is subjected to a tensile load, cracks may occur in the 90° plies, followed by the ±45° plies. The first crack to occur in the composite may be computed, but it may not be easy to do so analytically for the ±45° plies, as it may be due to original cracks in the 90° plies. The density of cracks increases with the application of loads and may stabilise for a particular laminate. This condition may be known as the characteristic damage state.

Axial cracking in the 0° direction plies may not occur under transverse stress; however, under fatigue loading, cracking may take place in cross-laminates.

7.5.3.2.1 Fibre Breakage and Interface De-bonding

The critical factor for reinforced polymers will be the properties of the component materials and reliance on the fibre interface strengths.

Under tensile loading, a unidirectional composite using carbon fibre with a polymer matrix may experience defects in the fibres that cause failure before the matrix. With the increase of loads, the cracks may extend into the matrix, and the path will be influenced by the properties of the interface bond between the fibre and the matrix. When the bond is strong, the cracks may extend into the matrix. The cracks may cause interfacial de-bonding and fibre pull-out when the interface bond is weak.

The sensitivity of the matrix, interface, and fibre to fatigue loading must be considered in the design. Under cyclic loading, numerous changes may occur in the mechanical response of the specimen.

The changes may be described as follows:

- The strength and modulus of elasticity will be reduced with an increase in damage.
- The residual strength of the composite will be reduced with an increase in fatigue cycling.
- With an increase in the number of load cycles, hysteresis losses will increase.

The consequence of damage to the composite would depend on the understanding of the type of laminate and the kind of loading.

7.5.3.3 Fatigue Behaviour

The fatigue performance of polymer fibre composites is normally represented as a stress number of cycles relationship. This is connected to the properties of the component materials. Fibre orientation is important, and unidirectional materials are likely to show better fatigue resistance. The study indicates that there are not significant differences amongst glass fibre-reinforced polymers, Kevlar fibre-reinforced polymers, and CFRPs regarding fatigue performance. Furthermore, higher static strengths may be obtained by using the same fibre with different epoxy resins. The damage accumulation associated with fatigue failure in GFRP cross-ply laminates may indicate rapid stiffness reduction at the initial stage. This may lead to an increase in the density of transverse ply cracks. Some fibre breakage may occur in the 0° direction. This may be followed by a linear region of stiffness reduction and a slow increase in matrix crack density in the transverse direction.

7.5.3.4 Summary

These points may be noted below:

- The reduction of stiffness of the composite materials due to the failure mechanism may be significant.
- The reduction of stiffness may be small in relation to the limited cracking occurring until near failure in homogeneous materials.
- The design of the composites may be based on stiffness instead of strength according to the failure criterion.

Source:

- Hollaway [1]

7.5.4 Further on De-lamination of the Composites

De-lamination can be described as a mode of failure associated with the fracture of composite materials into layers. Composite laminated materials and concrete may fail due to de-lamination. This is a layer separation that may occur during the processing method. This situation can also be observed in steel formed by rolling, as well as in plastics and metals, which may de-laminate due to layer separation. A similar situation can occur with paints and films that may de-laminate from the treated substrate.

Processing may create layer separation as mentioned. Material may be removed from the concrete surface during finishing work. When finishing is done by trowelling, underlying concrete may bleed water or air. In such conditions, the top layer may separate later.

7.5.4.1 Inspection Technique

There are different non-destructive testing methods that can be used to detect de-lamination in structures. Visual inspection, tap testing (i.e., sounding), radiography, ultrasound, and infrared imaging can be considered. Visual inspection may be effective for detecting de-lamination at the surface and edges of materials. However, it may not be possible to detect internal issues without cutting or opening the material.

7.5.4.2 De-lamination Resistance Testing Technique

The following may be relevant:

- **Coating De-lamination Tests –** ASTM standards may be followed. This includes the cross-cut test, scrape adhesion, and pull-off test.
- Inter-laminar fracture toughness testing.

Source:

- Wikipedia [166]

7.5.5 More on De-bonding Failures

7.5.5.1 De-bonding Problems

The situation is akin to the crushing of compressive concrete or to the de-bonding of FRP plates from concrete beams.

Premature failure of FRP-strengthened beams with or without plate end anchorage may occur due to different reasons as mentioned.

These may be connected with high interfacial stresses, low concrete strength, and/or extensive shear cracking.

7.5.5.2 Modelling of De-bonding

- Strength approach
- Fracture approach
- Semi-empirical and empirical methods
- Moisture effect on de-bonding

7.5.5.3 Inspection for De-bonding with NDE Techniques

7.5.5.3.1 Review of NDE Techniques for De-bonding

7.5.5.3.1.1 Far-Field Radar NDE Method

- Physical radar measurement
- Measurement schemes

7.5.5.4 Further Comments

Experimental and model studies indicate that FRP de-bonding is a complex phenomenon. It may indicate failure within the concrete substrate, within FRP laminates, within the adhesive, and along the interfaces of the constituent materials. The contributions of moisture and temperature may play an important role in de-bonding, creating physical changes in the interface region that may lead to failure.

The interface fracture toughness of the joint will be very relevant for the assessment.

Source:

- Büyüköztürk and Yu [165]

7.5.6 Review of De-bonding of FRP Composites

The de-bonding problem is a critical issue for proper assessment of performance related to FRP composites in structural strengthening and repair applications. Some comments are made on FRP to steel and FRP to concrete based on experimental and theoretical studies [167] as given below:

7.5.6.1 FRP Repair to Fatigue-Damaged Steel Members [167]

The comments from tests on the repair of fatigue-damaged steel members using FRP systems are given below:

- Notched steel specimens with FRP patches indicated large increases in remaining fatigue lives of the specimens.
- Symmetric bonding of FRP patches may be recommended where possible.
- The length of the patch (perpendicular to crack orientation) is found to be more effective than its width (parallel to crack orientation).
- Surface preparation is very important for performance. Sand blasting may be the most effective method of surface preparation as it yields a uniform rough surface.
- Fracture mechanics is commonly used for modelling fatigue crack growth in metals. Considering steel members in civil engineering, the thickness of FRP composites indicates that gradual de-bonding of the patch is likely to be a common design consideration.
- The durability of bond repair when CFRP is used may be a concern due to galvanic corrosion.
- Further experimental and model studies will be required to develop fatigue life predictions for design guidelines in civil engineering projects with confidence.

Source:

- Büyüköztürk, Gunes, and Karaca [167]

7.5.6.2 FRP Strengthening of RC Beams [167]

The following are comments from studies and research work on de-bonding failures in FRP-strengthened RC beams:

- Assessment of proper de-bonding failures is the main part of RC beam strengthening using FRP composites.
- De-bonding of the cover takes place along with the re-bar level and differs largely from FRP de-bonding mechanisms in its causes and behaviour.
- Despite the brittle nature of de-bonding failures, test results indicated that the initiation of FRP de-bonding occurred at laminate ends and at crack locations under early load conditions.
- Under cyclic loading conditions, the stiffness and load capacity of FRP-strengthened beams underwent progressive reduction compared to the monotonic load case.
- The developed fracture model indicates a better prediction of de-bonding failure load compared to ACI-440F.
- Experimental studies concluded that a proper design of FRP-strengthened beams against de-bonding failures may increase their performance level compared to that determined by classical ultimate strength analysis.

Source:

- Büyüköztürk, Gunes, and Karaca [167]

7.5.6.3 The Effect of Surface Profile on Flexural Strengthening of RC Sections with FRP

It has been mentioned that flexural strengthening with externally bonded FRP composites has been used for RC structures.

Experimental studies have been carried out to assess the effect of surface profile on the flexural strength of RC sections with FRP. The failure of concrete sections strengthened by externally bonded FRP can occur due to

- Failure where composite action is maintained
- Failure where composite action is lost.

The failure plane can be at the surface of the concrete or at the level of the steel reinforcement.

The four important initiators of peeling failure are

• Shear cracks	• Surface profile flexural cracks
• Flexural cracks	• Inadequate anchorage.

It is noted that the presence of a concave surface profile at the soffit of a beam will reduce the degree of flexural strengthening provided by externally bonded FRP composites. From experimental studies, it may be observed that a curvature of 5 mm/m may indicate a loss in the degree of strengthening provided by an externally bonded FRP laminate.

Source:

- Porter [168]

7.5.7 De-bonding Failures in CFRP – Strengthened Steel Structures [169]

Studies, including experimental work on the performance of CFRP laminates/plates, have been conducted to assess the performance of the joint, particularly the role of adhesive in steel structures.

De-bonding failures between steel members and CFRP are critical and may occur in one of the following modes:

- Within the adhesive (cohesion failure)
- Between the adhesive and the adherents (adhesion failure)
- A combination of adhesion and cohesion failure

The following items are reported to be important from the study:

- Adhesion failure of CFRP-to-steel bonded joints may be avoided by adopting suitable treatment of the steel and FRP surfaces.
- With linear adhesive and mixed-mode cohesive law, de-bonding failures in CFRP-strengthened steel members may be accurately predicted.
- The performance of the joint will primarily depend on the properties of the adhesive selected/used.
- Further work is recommended to properly predict the performance of CFRP–steel interfaces with linear and non-linear adhesives.

Source:

- Teng, Fernando, Yu, and Zhao [169]

7.5.8 De-lamination Mechanism of GFRP Sheet Bonded to Reinforced Concrete Beams

Fibre-reinforced polymer (FRP) has been used for strengthening deteriorated structures. This may be used as a sheet that can be bonded to the concrete structure to be strengthened using suitable adhesive. Glass composite FRP sheets may be used for such strengthening work, considering the cost compared to other FRP materials.

This strengthening technique would be an alternative to using conventional steel reinforcement.

There are different types of FRP strengthening, such as glass fibre-reinforced plastics (GFRP), CFRP plastics, and aramid fibre-reinforced plastics (AFRP), which may be used for strengthening structural members. Here, the study of GFRP sheets bonded to the concrete surface for strengthening will be discussed.

The bonding/interaction of GFRP and concrete is very important and critical for achieving full mechanical flexural performance. Failure may occur when the bond stress in GFRP exceeds its bond capacity. The bonding capacity is associated with the bond strength of GFRP to the concrete surface where the sheet has been applied. Hence, the assessment mechanism is important.

The bonding capacity of the FRP sheet and its behaviour under flexural loading may not be the same as when the bonding capacity is subjected to direct axial loads.

For flexurally strengthened beams, de-lamination may begin with the opening of flexural cracks.

This study has been carried out to assess the GFRP sheet bonded to the concrete beam under flexural loads.

The experimental work was carried out using e-glass and epoxy resin.

7.5.8.1 Some Details

- Span of the beam = 3,000 mm subjected to two point loads located centrally.

- The lengths of GFRP were different such as 3,000, 2,250, and 1,500 mm and placed centrally.
- Beam size = 150 × 200 mm (depth) and reinforced with three bars at the bottom.
- Concrete compressive strength = 35 MPa
- Tensile strength of GFRP = 575 MPa
- Bonding strength of epoxy resin to concrete = 2.12 MPa
- GFRP sheets were applied to the concrete surface after curing for 28 days.
- The beams were tested under simply supported conditions with two point loads using a universal testing machine.

Basic assumptions are as follows:

- Sections perpendicular to the axis of bending will remain plane before and after bending.
- The strain in the reinforcement will remain the same as in the concrete at the same level.
- The stresses in the concrete and reinforcement can be computed for the changes using stress–strain curves for concrete and steel.
- Concrete and steel will act together to carry the loads.
- A perfect bond between the concrete, steel, and GFRP will develop.

7.5.8.2 Comments from the Tests

- At the initial stage, all the beams were uncracked, and the concrete resisted both compressive and tensile forces.
- The applied load was increased slowly, and when it reached the rupture strength of concrete, cracking started.
- The beam's flexural stiffness was reduced with the cracking.
- For the beams with GFRP sheet reinforcement, both steel and GFRP sheet resisted tensile forces.
- The increase in deflection occurred up to the yielding of the steel reinforcement with the added loads.
- Due to the presence of GFRP sheet reinforcement, the yielding of tensile reinforcement occurred at higher loads.
- The increase in deflection happened without an increase in load capacity.
- The post-yielding behaviour of beams without GFRP was similar to that of steel reinforced concrete beams, as expected.
- For beams with GFRP sheets, extra tensile force was carried after the yielding of the steel reinforcement.
- With proper bonding of the GFRP sheet to concrete, the load–deflection curve was smooth, and the applied load could be increased.
- With local de-lamination, a decrease in load capacity occurred. The deflection was about 40 mm.
- The specimen failed due to final de-lamination when the deflection was approximately 60 mm.
- The length of bonding was critical and influenced the flexural behaviour during post-yielding stages.
- The de-bonding mechanism can be identified by studying load–deflection behaviour.
- The decrease in load associated with local de-bonding may be identified by some 'ting' sounds.

7.5.8.3 Comments on Failure Modes

- The ultimate capacity of beams associated with de-lamination of GFRP reinforcement was studied.

- Some local de-lamination occurred before the final de-lamination.
- The final de-lamination was connected with the crushing of compressive concrete.
- The local de-lamination may be influenced by the development of cracks in the member.
- The increase in shear in the epoxy layer in another location may be responsible for maintaining the balance.
- The longitudinal shear stress in the epoxy layer between the GFRP sheet and the concrete surface was greater at the edge of the cracked concrete sections.
- The stress balancing cannot continue for a long time; the final de-lamination may take place.

Source:

- Djamaluddin, Hijriah, Irmawati, Faharuddin, and Wahyuningsih [170]

7.5.9 Detecting De-lamination in FRP Retrofitted Concrete Structures [171]

For strengthening applications, FRP sheets may be used. The sheets are impregnated with a resin and bonded to the concrete structures after proper surface treatment to ensure a good bond between the concrete and the FRP, as mentioned. Concerns exist regarding the weakening of the bond due to environmental factors such as wet and dry cycles, thermal variation, and freeze–thaw cycles, as applicable. If failureoccurs, it will be of a brittle type and will have catastrophic consequences. Hence, a reliable method of assessment on a regular basis will be essential.

Traditional ultrasonic methods (linear) have long been used for the condition assessment of the materials. In the study for assessment, non-linear elastic wave spectroscopy (NEWS) was used. In NEWS, the frequency characteristics of the received signal are analysed to understand the condition of the material.

Following the laboratory experiments, the comments are as follows:

- Total harmonic distortion (THD) computed from the amplitudes of the harmonics generally increases with applied load. It shows a degree of similarity across various tests, indicating a certain level of correlation with de-bonding.
- NEWS cannot be used as a baseline-free technique for condition assessment. The technique can be used where prior measurements and data are compared against new measurements for assessing the condition of the FRP retrofit.

Source:

- Kripakaran and Armitage [171]

7.5.10 Shear Peeling of Steel Plates Bonded to Reinforced Concrete Beams Using Adhesives

The steel plates may be bonded to the structural elements to strengthen the members using adhesives as mentioned.

The important items are as follows [172]:

- The plates may be attached to the tension face of the structural members.
- They will behave in a similar way to tension reinforcing bars in the members.
- They may be considered more efficient as the plates are located at the farthest distance from the neutral axis.
- Care should, however, be taken to ensure that the beams are not over-reinforced to meet durability issues.
- The plates may be placed on the sides of the members or even in the compression zones.
- The plates may increase both shear and flexural capacities of the members.

- It has been reported that the ends of the plates may be de-bonded prematurely due to shear and flexure.
- The bonding due to flexure may be controlled or prevented by terminating the plate at or adjacent to the point of zero moment.
- There may be shear peeling near the end of the plate.

7.5.10.1 Tests for Shear Peeling

The tests were conducted on beams simply supported over a span of 4,700 mm. The cross-sections of the beams were 200 × 370 mm and were subjected to point loads during the tests.

7.5.10.2 Some Details

These are as follows:

- Mean compressive strength (cylinder) = 48.9 N/mm^2
- Indirect tensile strength = 4.47 N/mm^2
- The beams were reinforced with three 20 mm bars (tension) and two 10 mm bars (compression), along with 10 mm stirrups.

7.5.10.3 Plated Beams for Tests

- Plated in the tension zone
- Plated in the compression zone
- Plated at mid-depth

The plates were placed on both sides of the beams, with dimensions of 130 × 10 mm thick.

7.5.10.4 Comments from the Tests

The main comments are

- It would be possible to increase both shear and flexural capacities by bonding steel plates to concrete members for retrofitting/strengthening work.
- The position of the plates is likely to influence to a large extent the shear load for developing diagonal shear cracks.
- Due to development of diagonal shear cracks near the steel plate end, bonded steel plates by using adhesives may be subject to de-bonding.
- From the tests, it can be concluded that side-bonded steel plates can control the formation of diagonal shear cracks and can restrict shear peeling.

Source:

- Mohamed Ali, Oehlers, and Bradford [172]

7.5.11 Review on the De-bonding Failures of FRP Laminates Externally Bonded to Concrete Surfaces [173]

The application of externally bonded fibre-reinforced polymers (FRP) to members has been carried out to repair/strengthen structures and extend their service life, as previously stated. From the experimental study, it appears that the theoretical high ultimate capacity of FRP-strengthened reinforced concrete members may not be achievable due to FRP plate de-bonding. Alternatively, horizontal cracking followed by loss of concrete cover below the reinforcing bars may be responsible. This will reduce the increased strength provided by FRP and can lead to brittle types of failures.

Consequently, this has raised doubts about the safety and usefulness of the externally bonded FRP strengthening technique when applied to reinforced concrete members.

- Generally, de-bonding failures in reinforced concrete members may occur in areas of high-stress concentrations. This may be related to the termination of FRP materials and the development of cracks in the concrete substrate. De-bonding propagation is associated with the properties of the base concrete, interface adhesive, and FRP, and may tend to follow the route of least resistance. It has been observed that extending FRP reinforcement closer to the support may reduce the chances of de-bonding; however, it cannot be completely avoided. Furthermore, bond transfer strength is related to concrete compressive strength and the bond development length, which must be provided adequately.

The items mentioned below may be relevant from the study/review work carried out

- The ultimate strength or bond transfer strength is proportional to f'_C or $\sqrt{f}\, f'_C$ (i.e., square root of f'_C).
- The application of more ductile adhesives will result in a weaker bond and extended failure strains.
- Proper surface preparation is important for ultimate bond strength, and cleaning/roughening of the composite plates will also be relevant.
- The distribution of strain in the composites reduces linearly along the bond length and, at the ultimate stage, is generally uniform along the length.
- Stiff bond layers may not be very efficient. The use of thinner adhesive layers may lower the risk of concrete–adhesive interface failures.
- When the FRP plate is narrower than the concrete element, there may be a non-uniform stress condition at the surface of the concrete. There may be higher shear stresses in the adhesive at failure.
- The deflection under service load may be related to FRP thickness and associated with stiffness.
- The load at failure cannot be increased beyond the bond development length.
- The bonded length of FRP should be maximised with the adhesive.
- It has been observed that by placing transverse straps, the de-bonding of FRP may be prevented. It is more effective to place transverse straps in high moment areas and within the shear span than near the end of the span.
- The key issues are the interface and the bond developing between the FRP laminates and the concrete substrate.
- The lack of proper understanding of the bond between the concrete and FRP and the composite action may lead to premature failures of FRP-strengthened concrete elements/members.

Further study and research may be necessary to develop reliable, accurate, and simple design methods/models for strengthening and retrofitting work. This will instil confidence in designers, applicators, and clients/owners to carry out work with FRP composites.

Source:

- Kang, Howell, Kim, and Lee [173]

7.6 TESTS CARRIED OUT ON THE MATERIAL

7.6.1 Explosive Testing of Fibre-Reinforced Cement Composites

Fibre-reinforced cement concrete can withstand shock waves from high-velocity impacts. Different tests have been carried out for explosive testing on fibre-reinforced cement composites. Comments will be summarised from tests conducted for explosive testing in this connection.

Experimental Programme – Nineteen square section bars measuring 25×25×210 mm have been tested. Ordinary Portland cement was used with a water-to-cement ratio of 0.3. The explosive charges were from ICI detonators.

Comments on Fibres – The fibres in the test are listed as follows:

- E-glass fibre
- Polypropylene fibre
- Carbon fibre
- Steel fibre

7.6.1.1 Discussion of Test Results

Brief comments from the tests are presented below:

A quantitative analysis has been attempted. Based on the performance, the specimens may be divided into four groups:

- **Multiple Cracking and Spalling** – The specimens were broken into three or four pieces. Each spalled off contained both unreinforced and OPC with 0.79% carbon fibres.
- **Multiple Cracking But No Spalling** – The specimens contained 1.57% carbon fibres. The specimens failed but remained in one piece.
- **Hair Line Cracking and No Spalling** – This group contained 1.71% polypropylene fibres and 1.39% steel fibres, with some containing 2.73% steel fibres.
- **No Visible Cracks** – This group contained E-glass fibres along with 3.47% polypropylene fibres.

  The fibre content was more than the critical volume.
- Greater resistance was observed for E-glass fibre composites, primarily due to uniform fibre distribution throughout the mix.
- E-glass fibres have the highest tensile strength among the four fibre materials used for the tests.
- The drawback of E-glass fibres is the deterioration of strength due to alkaline attack.
- Carbon fibre cement composites showed low impact resistance.

Source:

- Raouf, Al-Hassani, and Simpson [52]

7.6.2 A Preliminary Evaluation of CFRP Plates for Strengthening Reinforced Concrete Members

Concrete structures and beams deteriorate, in most cases, due to internal reinforcement corrosion, additional loading, freeze–thaw action, and poor initial design. To meet present requirements, old structures, including bridges, must be repaired and upgraded. It is preferable and more economical to upgrade them instead of rebuild.

An experimental work related to different parameters was conducted on strengthening reinforced concrete beams using bonded composite plates. The bonded plates effectively increase the ultimate capacity of the beams while potentially reducing durability. CFRP plates will increase beam stiffness beyond the yield of internal reinforcement.

7.6.2.1 Some Details of the Tests

7.6.2.1.1 Concrete

- Ordinary Portland cement was used.
- Free water:cement = 0.4:1.

- Cement:coarse aggregate:fine aggregate = 1:1.1:1.9
- Coarse aggregate was 10 mm in size and the largest fine aggregate was 5 mm
- 28-day compressive strength of concrete = 59.1 MPa

7.6.2.1.2 CFRP Plates

The plates contained 58% high-strength fibres and 42% resin. They were cut into strips 25 mm wide and 300 mm long, with electrical resistance strain gauges used for instrumentation. The plates were bonded to the beams at the bottom. The sizes of the plates were 5×90, 0.7×65, and 1.0×45 mm.

7.6.2.1.3 Adhesive

Adhesive from SIKA, SIKADUR-31, was used for plate bonding.

7.6.2.1.4 Loading on the Beams

The beams were tested for point loads. They were under-reinforced with three bars, and the steel area ratio was 1.0% and planned for flexural failure.

7.6.2.1.5 Test Beams

The beams were 100×100 mm in depth and were simply supported over a span of 900 mm. Point loads were applied 150 mm from the central line of the beam on either side. The bars were placed at the bottom of all the beams and instrumented with electrical resistance strain gauges.

GFRP angle sections were used near the support.

7.6.2.2 Comments from the Tests

These are given below:

- CFRP plates are suitable for upgrading or strengthening steel and concrete structures. The advantages include easy handling and a high strength-to-weight ratio. Joints can be avoided, as in the case of steel plate bonding.
- CFRP plates can increase beam stiffness beyond the yield of internal reinforcement.
- When bonded to the soffits of beams, CFRP plates help reduce tensile concrete strain during loading, which may lead to fine cracks, thus providing better serviceability performance.
- Anchorage at the ends of the plates may reduce the chances of premature failure due to high peel and shear stresses.

Failures are associated with relatively high longitudinal shear stresses at the adhesive–concrete interface.

Source:

- Garden, Holloway and Thorne [159]

7.6.3 Externally Bonded Steel Plated RC Beams after Long Period of Exposure

The externally bonded steel plated RC beams exposed to an industrially polluted area were tested in the laboratory after a period of 11–12 years. Twenty-one such beams were tested. The concrete beams measured 102×102×508 mm, strengthened with a mild steel plate measuring 38×3×508 mm, bonded centrally to one face using epoxy adhesive. This technique for strengthening beams will be suitable as an efficient, cost-effective, and technically sound method for increasing both serviceability and ultimate load-carrying capacity of damaged or inadequately reinforced or pre-stressed concrete load-bearing elements.

Experimental Work – Twenty-one beams were tested in the laboratory after being exposed directly and continuously to natural environments for a period of 11–12 years. A few beams were kept under sustained loading, while the others were left unloaded.

7.6.3.1 Variables Studied

The following four variables were studied:

- Number of steel plates.
- Thickness of adhesive layer.
- Configuration of plate (i.e., lapped plates or continuous)
- Loading condition (i.e., kept under load or stacked unloaded)

7.6.3.2 Main Comments from Tests

- The overall performance of the plated beams was satisfactory for those that were not maintained properly.
- The failure loads of the plated beams were 1% and 29% higher than the failure load of similar short-term controlled plated beams.
- Deflection and crack widths under service load conditions were found to be similar to or less than those of control plated beams.
- The effect of corrosion from exposure and its effect on structural performance varied and depended on the plating system.
- Great care is required for plate bonding, particularly in the application and maintenance of the protective system in critical areas, i.e., interfaces between plate layers, corners, and ends of the plates.
- Beams with a single continuous plate carried about 10% higher load than theoretical calculations.
- 'U' shaped anchor plates are suitable for anchor zones.
- For multiple plates, a general recommendation for acceptable plate breadth-to-thickness ratio for total plate thickness should be followed.
- With proper detailing and maintenance, the plate bonding technique may be used for strengthening and rehabilitating structures. This method may be considered durable and reliable, and the work may be done with confidence.
- All the beams failed at loads in excess of 90% of the theoretical failure load for plated beams. In some cases, the failure loads were higher than the theoretical loads by 20%.

Source:

- Swamy, Hobbs, and Roberts [174]

7.6.4 Steel Fibre-Reinforced High-Strength Concrete Corbels: Test Results and Analysis

Sixty corbels reinforced with conventional reinforcement along with steel fibres were tested. Parallel samples were tested without fibre reinforcement. The compressive strength of concrete ranged between 43 and 84 N/mm^2.

7.6.4.1 Experimental Programme

The items may be summarised as follows:

- 60 reinforced corbels were tested with shear span-to-depth ratios varying from 0.11 to 1.1.
- 29 corbel specimens were reinforced with steel fibres, and one specimen had polypropylene fibres.
- The corbels were divided into eight series for testing.
- Each corbel specimen had a 160×300-mm column, and two corbels were arranged symmetrically.

- The maximum depth of the corbel specimens was 465 mm with a width of 160 mm.
- The concrete mix proportion was 1:2.1:1.2 (by weight) with a water-to-cement ratio of 0.3.
- Melt-extract fibres were used for most of the specimens, with a length of 35 mm and an equivalent diameter of 0.47 mm.

7.6.4.2 Test Procedure

- Demec gauges were used on the concrete faces to measure strains and were located at selected positions.
- Corbels were tested as simply supported in a 2,500 kN capacity load frame.

7.6.4.3 Test Results and Discussion

These are as follows:

- The first crack that appeared in all the corbels was a flexural crack, irrespective of span-to-depth ratio and quantity of reinforcement.
- Diagonal splitting cracks were more prominent in corbels with tension reinforcement and without stirrup reinforcement.
- Corbels with steel fibre reinforcement exhibited greater post-ultimate ductility compared to plain concrete corbels in all span ratios.
- Sudden failure with concrete spalling could be mitigated by using steel fibre reinforcement.
- Good performance was observed for the corbels with main tension reinforcement and steel fibres compared to corbels without steel fibres.

7.6.4.4 Further Comments

The following may be noted:

- High strength concrete corbels reinforced with high tensile steel reinforcement demonstrated sufficient inelastic de-formation and ductility at failure when steel fibres were incorporated for both under-reinforced and over-reinforced samples.
- Tension and steel stirrup bars developed significantly higher tensile strains when steel fibres were added to the corbels.
- Steel fibre reinforcement resulted in a significant increase in the load-carrying capacity of the corbels, with an increase of up to 35% observed.
- The mode of failure was by tensile splitting across the diagonal plane.
- The load-carrying capacity of normal and high strength concrete corbels can be expressed as given.

Source:

- Mangat and Halabi [63]

7.6.5 Flexural Properties of Steel FRC

Tests were conducted on fibre-reinforced elements to study the increase in flexural strength by using steel fibres.

7.6.5.1 Observations from Tests [71]

The tests were performed on 100×100×500 mm prisms at 28 days and 2 years. The beams were constructed using a mix of 1:2.5:2.5 with a water-to-cement ratio of 0.55 and contained round or shaped steel fibres. Ordinary Portland cement, crushed graded gravel with a maximum size of 19 mm, and natural sand were used.

The following comments from the tests may be noted:

- The increase in flexural modulus observed due to the inclusion of fibres was in the order of 5%–10%.
- The flexural strength was doubled. The tensile strain capacity at the maximum load increased approximately tenfold compared to unreinforced concrete.
- The ability to sustain deflection and control cracks was noticeable, including improved ductile properties.
- The shaped fibres indicated improved strength, strain, and elastic properties compared to the straight round fibres.
- The brittle failure mode of plain concrete was transformed into ductile failure by the presence of steel fibres.

Source:

- Swamy and AL-Noori [71]

7.6.6 Strengthening for Shear of RC Beams by External Plate Bonding

External plate bonding to strengthen the shear resistance of RC beams without any internal shear reinforcement has been undertaken. A considerable amount of work has been done on the strengthening of beams in flexure. Both steel plates and fibre-reinforced polymers (FRP) have been applied as plate materials. RC beam theory can be used to satisfactorily predict the ultimate failure load of such strengthened beams that fail in flexure.

7.6.6.1 Test Programme

A total of 12 beams (150×250×3,000mm) were tested with a shear span to effective depth ratio of 3. The effective depth of the beams was 215mm. The plates, either steel or GFRP, were bonded externally to the beams. The beams had no or inadequate shear reinforcement. The intention was to convert shear failures to flexural failures of RC beams.

7.6.6.2 Comments on Placement of Bonded Plates

These may be mentioned as follows:

- Plates bonded on the tension face and on the web will increase shear capacity.
- U-shaped stirrups will control diagonal shear cracking.
- Bonding at the ends will provide adequate anchorage to the tension plate and will stop the initiation and propagation of shear cracks.
- Plates below the load points will prevent crushing of concrete in the compression zone.

7.6.6.3 Brief Comments from the Study

The following may be mentioned:

- When steel plates are bonded externally to the tension face of the RC beams, it will be possible to convert the brittle shear failure of the beams, which have no internal stirrups, into ductile flexural failure.
- RC beams can be strengthened with externally bonded plates that are properly designed and placed on beams with inadequate shear and flexural reinforcement, allowing them to fail in a ductile flexural manner.
- The increase in load required for the development of the first visible diagonal crack in the beam will be more than three times.

- Further study/tests will be required to better understand the mechanism of failure of RC beams bonded with plates for shear strengthening.

Source:

- Swamy, Mukhopadhyay, and Lynsdale [175]

7.6.7 Strengthening Approach with Fibre-Reinforced Polymers

The relevant items are as follows:

- The scheme for strengthening should be economical, with minimum disruption to traffic during execution.
- Concrete strength and strain capacity are enhanced due to confinement from hoop fibres.
- Additional compressive reinforcement will be provided by axial fibres.
- The member will be prevented from buckling by the use of overlying hoop fibres.
- FRP wrapping can increase bending and shear strength.
- The resin material is versatile and may be preferred over steel plate bonding, especially for soffit strengthening of bridges.
- The resin properties must be appropriately assessed to meet conflicting requirements of wetting ability, viscosity, gel, curing method, and time.

More comments on this are given in item 7.3 separately.

Source:

- Sadka [124]

7.6.8 Large-Diameter Columns Reinforced with Wrapped Glass Fibre-Reinforced Plastic Tubing

Large-diameter (greater than 300mm) GFRP tubes filled with concrete have been tested at Southampton University, UK, for both axially and eccentrically loaded columns, with aspect ratios (height to diameter) varying from 2:1 to 10:1.

Brief comments from the observations from tests are summarised:

Axially Loaded Columns	Eccentrically Loaded Columns
• Axially loaded columns failed at mid-height with fracture of GFRP	• Eccentrically loaded columns failed by breaking in half at or near mid-height. The failure mechanism was brittle
• The failure occurred suddenly and was explosive	• It developed large deflections before failure
• At failure, the fibres tend to unwrap themselves from the specimen	• The ultimate strengths were comparable with conventional concrete columns with 4% longitudinal steel reinforcement (as per BS-8110)
• Axially loaded columns failed at much higher loads than the eccentrically loaded specimens of the same diameter and strength	• The method of predicting failure loads had reasonable agreement with measured values
• The enhancement of compressive strength was found to be 3	• Full benefit from confinement will be available for accidental loads under seismic conditions

Source: Moss [176]

7.6.9 Confinement of FRP-Jacketed Reinforced Concrete Columns

The tests were conducted to study different aspects related to the confinement of FRP-jacketed reinforcement in columns. Various items associated with the effective confinement of FRP-jacketed reinforcement surrounding rectangular columns were studied. The issues included aspect ratio (ratio of long to short side), amount and type of fibre for the FRP jacket, sharpness of corners, and both longitudinal and transverse reinforcement in columns. FRP-strengthened columns performed better. The failure of the columns may be sudden. Columns with a larger aspect ratio exhibited decreased strength and axial deformation. The largest increase in axial strength was possible with carbon fibre reinforcement.

The comments from the tests are given below:

- FRP-strengthened columns performed better, but the failure may be sudden.
- Decreasing corner sharpness increased both ultimate strength and ultimate axial deformation in square cross-sections with carbon fibre-reinforced jackets.
- Circular reinforced columns experienced uniform confined pressure around the circumference with a significant increase in strength compared to columns with steel or only carbon fibre jacketing. However, dual confinement for square cross-section provided strength similar to carbon fibre confinement alone.
- Larger aspect ratios decreased strength and axial deformation.
- Increasing the thickness of jackets reinforced with any type of fibre increased both strength and axial deformation.
- For square columns, glass fibre-reinforced jackets increased the ultimate axial strain more effectively than either carbon or aramid fibre-reinforced jackets. For multiple-ply jackets on square columns, glass fibre was most effective in increasing the ultimate axial stress.
- Carbon fibre-reinforced columns provided the largest axial strength increase for single-ply jackets.
- Jackets with large fibre strains at ultimate load stages normally correspond to the most ductile columns.

Further comments are summarised in item 7.4 later.

Source:

- Belarbi [157]

7.6.10 Flexural Behaviour of Steel and GFRP-Reinforced Concrete Beams

The flexural behaviours of GFRP and steel-reinforced concrete beams have been investigated. The tests were conducted on beams. GFRP-reinforced beams exhibited higher failure loads, reinforcement strains, deflections, and cracking compared to similar steel-reinforced beams. The deflections and service loads in GFRP-reinforced beams were higher than the limiting value. The steel-reinforced beams exhibited less deflection. The theoretical deflections of GFRP-reinforced beams were 10% of the experimental results covering the service load range.

Further comments are given separately in item 7.2.

Source:

- Rooney and Taylor [92]

7.6.11 SFRC Suspended Slab in Flexural Behaviour

The applications of steel FRC (SFRC) include ground bearing slabs and spread concrete tunnel linings. The intention of using steel fibres in the concrete mix is to improve the performance of the serviceability limit state by providing better control over cracks and limited deflections. Steel fibres increase the ductility and tensile strength of the concrete mix.

Experimental Programme – Different FRC slabs were tested with dimensions of 1,055×1,055×60mm and 1,750×450×85mm, incorporating fibres in the mix at 0.2%, 0.5%, and 0.8% by volume. The slabs were tested under simply supported conditions after 28days from the date of concreting. Central patch loading was applied, and the corners were restrained for the square slabs.

Results – Good agreement was observed between elastic theory and actual results for the deflection curve. Reasonable agreement was achieved between plastic theory lines and the actual test results for both rectangular and square slabs.

Comments – The tests indicate that close agreement between the theoretical and actual results was achieved. This indicates that established methods can be used as a design basis within the serviceability limit state consideration. Furthermore, the average stress coefficient suggests that the results reasonably represent the stress distribution over the tensile zone in the plastic condition. Further applications would be required for full-scale structures. Furthermore, there is potential for the development of design methods based on yield line theory and a simplified plastic model.

Source:

- Ward [61]

7.6.12 Advanced Polymer Composite and Concrete Slab and Test

A concrete slab was designed with a CFRP corrugated sheet as permanent formwork and tension reinforcement, as stated. The CFRP corrugated sheet serves as permanent formwork. The slab was tested simply supported with an effective span of 900mm, supported in the direction of the corrugation [143]. The slab was tested for a point load at the centre. The CFRP sheet was 0.5mm thick with high-strength carbon fibre fabric in epoxy resin. The thickness of the concrete slab was 35mm over the top of corrugation. Proper bond between the CFRP and concrete was developed by using two-component epoxy adhesive, which was applied on the CFRP sheet before the concrete.

7.6.12.1 Some Details

- The concrete was designed with a characteristic strength of concrete of 40 N/mm^2 at 28days.
- 10mm aggregate was used in the mix
- Water-to-cement ratio=0.42
- Target slump=30–60mm
- The failure strain of concrete was assumed to be 0.0035
- The modulus of elasticity of CFRP was considered to be 164 N/mm^2, and tensile strength was 860 N/mm^2.

7.6.12.2 Comments from the Test

The following may be mentioned:

- The slab was tested under simply supported conditions with the corrugation along the span.
- Strain gauges were placed at the top and bottom at salient locations.

- The point load was applied at the centre of the span using an area of 150 mm^2.
- During the loading, a peak load was recorded and the failure was observed in the transverse direction of bending.
- The CFRP layer straightened locally at this stage, and de-bonding from the concrete occurred over one corrugation.
- Cracking occurred along the whole width of the slab.
- Even after cracking, it was possible to place additional loads on the slab to induce failure in the major direction of bending.
- The recorded strain showed good agreement with the expected strain capacity of CFRP.
- After unloading, a permanent deflection of 5 mm was recorded.
- The slab was reloaded, and it was possible to achieve a maximum deflection of 25 mm with the formation of a plastic hinge across the total width of the slab.
- After reaching a concrete strain of about 0.0035, crushing of the concrete surface was noticed. The slab failed suddenly with the de-bonding of CFRP from the concrete, covering one quarter of the slab.
- At the serviceability limit state, the performance of the composite slab may be considered good.
- It was observed that at the ultimate limit state, the expected failure strains of CFRP and concrete were reached.
- Reasonable agreement was found between experimental observations and the elastic analytical approach.
- The performance of the slab at the serviceability limit state was also satisfactory.
- Following the test, it was noted that the slab was capable of sustaining large plastic deformations.

Source:

- Szczerbicki [141]

7.6.13 CFRP Plates as Shear Strengthening for Beams

The work was carried out to assess the efficiency of various CFRP configurations in increasing shear resistance.

Test Programme – A total of 38 beams were considered, designed to fail in shear. Strengthening was done using CFRP methods along with other factors. The direction of CFRP fibres was horizontal, vertical, and at a 45° incline.

Variables included

- Percentage of main (bottom) reinforcement.
- Different configurations of CFRP strengthening.
- Link spacing in the shear span.

SIKADUR-30 adhesive bonding material was used for the concrete, irrespective of curing method and surface treatment.

The application of two horizontal CFRP strips over the shear span and tension zone proved to be an efficient method for shear strengthening. A better increase in shear strength could be achieved by bonding plates over the total depth and shear span.

The use of Sika Wrap-103 C on the soffit and sides of the shear span could increase the shear strength by 80%–90%.

Source:

- Walker [137]

7.6.14 Fracture of Latex Modified Steel FRC in Plane Shear

The application of fibres in concrete as shear reinforcement to eliminate brittle fracture has already been established. It may be possible to make brittle concrete more homogeneous and isotropic by using discontinuous fibres.

7.6.14.1 Test Method and Details

The concrete specimens were prepared with ordinary Portland cement, natural river sand, crushed granite aggregates with a maximum size of 19 mm, along with steel fibres of 0.53 mm and an aspect ratio of 50, with ultimate strength of 790 MPa. Styrene butadiene rubber (SBR) latex polymer and super-plasticisers were used for the SBR concrete mix.

7.6.14.2 Test Specimens

Two types were considered for the test.

These were

- Plain steel FRC (SFRC)
- Latest modified SFRC (LMSFRC)

7.6.14.3 Items

- 150×150×500 mm prism with two 15 mm deep all-a notches and 2 mm wide spacing in the middle for shear tests.
- 100 mm diameter and 200 mm long cylinders used for compressive and tensile split strengths.

The specimens were supported on knife edges and subjected to concentrated loads causing shear failures.

The cylinders were tested for compressive (direct, uniaxial) and split tensile strengths.

The addition of fibres marginally increased the compressive strengths.

The maximum increases of 31.3% and 30.8% were observed for the splitting tensile strengths of concrete with fibres for SFRC and LMSFRC types, respectively.

7.6.14.4 Comments/Observations from Tests

Steel fibres were found to effectively improve the shear strength and shear toughness in SFRC and LMSFRC specimens.

The inclusion of latex in SFRC increased the fracture properties.

The average increase of LMSFRC over SFRC in the following items was

- **Ultimate shear strength** – 17.52%
- **Fracture energy (Mode-II)** – 8.18%
- **Fracture toughness (Mode-II)** – 10.56%

By including fibres and latex polymers in the mix, the type of failure changed from brittle to ductile.

The ultimate shear strains were observed as

- **Plain** – 0.5%
- **LMSFRC** – 8.8%
- **SFRC** – 8.0%
- A 10% increase in ductility may be possible due to the addition of latex to SFRC.

Source:

- Sreenivas Rao and Appa Rao [177]

7.6.15 Test Methods of Flexural Strength and Toughness Parameters of FRC

7.6.15.1 Recommendations of ICI Technical Committee

The guidelines are provided for assessing the flexural strength and toughness of FRC by conducting tests. Only a few items will be referred to.

The assessment may be carried out by increased post-cracking capacity expressed in terms of set toughness parameters taken from the load–deflection curve against the loading test of a non-notched moulded specimen.

7.6.15.2 Test Specimens

The test specimens may have a cross-section of 150×150mm, and the length may vary from 550 to 700mm. The maximum size of the aggregates in the concrete would be 25mm, and fibres up to 60mm in length may be used.

The tolerances on the cross-section of the test specimens would be ±2%.

7.6.15.3 Preparation and Curing of Test Specimens

Guidance may be obtained from ICI recommendations.

7.6.15.4 Test Procedure

Reference may be made to ICI recommendations. A few items are mentioned below:

- The span length shall be 450mm and should be accurate to within 1mm.
- The specimen should be properly placed for testing as per the guidelines.
- Before applying the load, checks should be done to ensure the specimen is resting against the rollers properly.
- The rate and range of loading shall follow the guidelines.
- The measurements of deflections must be properly recorded during the test.
- Cracks outside the middle third region may be ignored.
- Generally, six specimens must be tested to obtain an average value.

7.6.15.5 Flexural Strength

The load–deflection curve must be prepared as per the testing.

The modulus of rupture may be calculated from the first peak load, P_{max}

$$f_{ct} = P_{max} \times L/(bd^2)$$

where

f_{ct} = the flexural strength of the specimen in MPa.
P_{max} = the peak load, N.
L = span of the specimen in mm.
b = the average width of the test specimen.
d = the average depth of the test specimen.

The mean value may be noted as f_{ctm} and the equivalent flexural strength as $f_{e,n}$. The equivalent flexural strength may be computed for specified deflections provided for L/n for values n = 150 and 300 or other intermediate values as per the requirements.

The equivalent flexural strength of the specimen may be computed as per the guidelines of ICI.

The documentation/report should include various items as per the guidelines.

Some of these are

- Identification of the test specimen.
- Date of testing.
- Information on the mixed proportion, fibre type, dosage, slump, curing, date of casting, storing, etc.
- The average dimensions of specimens to the nearest 0.1 mm.
- Rate of loading.
- Number of specimens tested.
- Load–deflection curve of the specimen.
- The number of fibres crossing the fractured section.
- Flexural strengths of each specimen.
- Location of failure.
- Any deviations from the recommendations.
- Others from the guidelines of ICI.

Source:

- Recommendation of ICI Technical Committee [72]

7.6.16 Beam Tests, Tests on Concrete Beams in Flexure with Basalt Reinforcement

For the study, 12 reinforced concrete beams were tested for performance in flexure. The beams were 100 (width)×200 (height)×1,200 mm (length). Six beams were constructed with basalt reinforcement, and the other six were made with conventional steel reinforcement for comparison. 10 mm diameter bars were used for both basalt and steel reinforcement. High-yield steel reinforcement with $f_y = 450$ MPa was used for six beams. The bars were placed at the bottom, with nominal top and shear reinforcement.

7.6.16.1 Tests

The beams were tested using four-point loads to create a uniformly distributed load for simulation. The testing of beams was conducted under simply supported conditions with a span of 1,100 mm. The concrete compressive strength was 30 MPa at the time of testing.

7.6.16.2 Comments from the Tests

The following may be noted:

- For poor ductility, brittle types of failure were expected.
- Basalt strands were wrapped around the bars in spiral form to increase mechanical bonding. This was not adequate to prevent slippage.
- The beams did not fail abruptly and gave the impression of ductile failure due to bar slippage. The slippage can be easily observed by the formation of gaps or cavities behind the ends of the bars.
- Poisson's effect may also contribute to bar slippage.
- The beams with basalt reinforcement showed greater deflections, approximately twice those of the steel-reinforced beams.
- The construction with basalt reinforcement was challenging as only straight bars were available.
- The polymeric material used to bind basalt fibres may not be very suitable and exhibited large Poisson's ratio effects, leading to bar slippage within the concrete.

- The favourable properties of basalt reinforcement include its lightness, high tensile strength, and resistance to heat, as previously highlighted.
- Basalt and steel may be considered a hybrid system for retrofitting work; basalt may be used for strength, and steel for ductility.
- The potential of basalt reinforcement looks promising, and further development work will be necessary.

Source:

- Gohnert, Gool and Benjamin [49]

7.6.17 Pull-Out Behaviour of Axially Loaded Basalt Fibre-Reinforced Polymer (BFRP) Rods Bonded Parallel to the Grain of Glulam Elements

An experimental program was carried out to study the behaviour of bonded BFRP bars loaded parallel to the grain of glulam members. Tensile pull-out tests were carried out to study the effect of bond length and bond stress–slip on the structural capacity of the connection.

Glued-in rods (GIR) can be effectively used for connections in timber structures. The connecting roads may be steel and fibre-reinforced polymers (FRPs) bonded into resin-filled holes in the timber. The work had been carried out using GFRP and CFRP rods.

The tensile pull-out tests had been conducted with 12 mm BFRP rods bonded parallel into glulam timber members to study the effects of some factors on the capacity of the joints. The variables taken into account were bonded length and the joint–slip behaviour measured from the loaded end of the specimen.

The following types may be involved in pull-out tests:

Pull–pull, pull–compression, pull–beam, pull–pile foundation types.

Pull–pull and pull–conditions will be suitable for both parallel and perpendicular to the grain tests. Pull–pile foundation and pull–beam situations will be suitable for perpendicular to grain tests.

7.6.17.1 Test Program

7.6.17.1.1 Timber

Glulam was used as its mechanical properties are more uniform, and in spite of the presence of knots, the member will be more homogeneous.

7.6.17.1.2 Basalt Fibre-Reinforced Polymer (BFRP) Rods

Basalt fibre- reinforced polymer (BFRP) rods are inorganic fibre materials. These are extruded from volcanic basalt rock deposits and from an inert rock formation. The pultrusion process is followed to form the FRP rods.

7.6.17.1.3 Adhesives

The adhesive selected was a two-part epoxy thixotropic gap filling adhesive. The thickness may vary from 2 to 12 mm. The product performs better as a gap filler than others like polyurethane (PUR) and phenol-resorcinol formaldehyde (PRF). A suitable and matching hardener was also used.

7.6.17.1.4 Testing

The pull-out specimens were tested using a direct testing machine fitted with a 100 kN load cell.

7.6.17.2 Comments

The items mentioned below should be noted:

- It was observed that the pull-out capacity increased almost linearly with the bonded length.
- The maximum value was for a bonded length of 15 times the hole diameter.
- This value did not increase further with additional bonded length.
- The main failure modes were at the timber-adhesive interface and pull-out of the BFRP rods.
- The ultimate bonded length using 12 mm BFRP rods would be 250 mm.
- The failure mode was mainly shear failure of timber close to the joint with pull-out of the rods.
- A very good bond between the rods and the adhesive was noticed.
- The shear failure of the grain samples resulted in failure of the timber at the joint with variable thicknesses surrounding the rod.
- With the increase in bonded length, the shear failure of parallel-to-grain samples changed.
- It shifted from a fracture greater than 5 mm thick to a layer less than 5 mm in thickness.
- The bond stress–slip behaviour of the bonded-in BFRP rod joints indicated a brittle type of response.
- The distribution of stress at the timber–adhesive interface of the samples was non-linear. The increase in bonded length provided an increase in joint slip.

Source:

- Yeboah, Taylor, McPolin, and Gilfillan [50]

7.6.18 Basalt Fibre-Reinforced Polymer for Bridges

By using durable, light, and high-strength basalt fibre-reinforced polymer along with SCC, it is possible to produce durable concrete structures with better service life performance.

BFRP bars have been used for the single-span Thompson skew bridge in the UK. The mid-span was constructed with 12 mm diameter basalt fibre-reinforced polymer (BFRP), and the balanced portions were completed with 12 mm diameter steel reinforcing bars.

7.6.18.1 Details of the Tests

Six test panel areas were identified and used for the study of the bridge. The main objective of the test programme was to examine the performance of different reinforcement types on the service behaviour of the bridge deck slab.

The central portion of the bridge had 0.6% BFRP reinforcement, while the remaining slab contained 0.7% steel bars. Concrete cubes were prepared from each batch to assess the strength during testing.

A simulated wheel load was applied to each test panel using a self-sustaining test rig. A 300 mm diameter plate was used to apply a circular concentrated load at the mid-span of each test panel. Strain in the reinforcement and an electronic displacement transducer were used along the central line of the panels to assess the deflected shape.

7.6.18.2 Comments from the Tests/Observations

The following items are relevant:

- BFRP-reinforced concrete bridge deck slabs showed lower deflections than similar bridge decks with steel reinforcement.
- The bridge deck could support a wheel load of 40 tonnes with no harmful effects. The strain values remained within service limits.
- Upon removal of loads, all the test regions showed good recovery in deflection and strain.
- The maximum deflection in the BFRP slab was 0.78 mm under the applied load of 40 tonnes. This is approximately span/2054, which is well within acceptable limits for deflection.

- The maximum deflection occurred at the mid-span.
- The strain values were low with an applied load of 40 tonnes.
- BFRP can provide an alternative corrosion-resistant design for bridges to improve the service life performance of the structure.

Source:

- Taylor, Robinson, and Sonebi [178]

7.6.19 MSFs, Long-Term Behaviour and the Importance of CE Marking

The study commented on the free-cracked-reinforced panels using round fibres subjected to different amounts of sustained loading, comparing the creep coefficients of steel and Strux 90/40 reinforced sprayed concrete.

Further work was carried out to study the differences in creep behaviour between steel fibre and Strux 90/40 reinforced beam samples.

The cylinder compressive strength of concrete varied between 35 and 40 MPa. The dosage of Strux 90/40 ranged from 2 to 20 kg/m^3, with 60 mm long hooked-end steel fibres used. The beams measured 150 × 150 × 550 mm and were cured for 28 days in lime-saturated water. The beams were tested using a four-point bending method. The test was stopped at a deflection of 1.75 mm, before reaching the final deflection of 3 mm, to assess the amount of elastic energy stored in the crack FRC elements and the response due to load relaxation.

The deflections were measured with a mechanical dial gauge for the two fibre-reinforced materials under sustained loading.

Both materials exhibited similar long-term creep behaviour. However, the deflection of Strux 90/40 reinforced concrete beams was greater at the initial stage compared to steel FRC beams. Later, the slopes of the curves were almost similar.

Following the low coefficient of creep, it can be reasonably assumed that the steel FRC will not experience significant creep effects. It was found from the tests that steel FRC beams exhibited creep effects. This may be due to the slow pulling out of the steel fibres from the concrete matrix and the interface, which also creeps. For MSFs, it will be a combination of the creep of the fibre material along with the creep of the fibre/matrix interface.

Strux 90/40 will have a high elastic modulus. The creep coefficient will be smaller compared to MSFs that have a lower elastic modulus.

It is important for the manufacturer to provide information on fibres such as description of the product, diameter, length, shape, tensile strength, modulus of elasticity, dosage, and effect on strength, etc.

7.6.19.1 CE Marking

Without CE marking, the products/items cannot be used in the European Economic Area.

The fibres, as per BS code, may be manufactured from polymeric materials such as polypropylene, polyethylene, polyester, PVA, polyacrylic, nylon, aramids, and blends of these.

These may be

- **Class I** – Micro-fibres
- **Class Ia** – Micro-fibres < 0.3 mm in diameter, mono-filament.
- **Class Ib** – Micro-fibres < 0.3 mm in diameter, fibrillated.
- **Class II** – Micro-fibres > 0.3 mm in diameter.

The important property of MSFs will be the tensile strength. It may vary to a large extent due to the polymer used and the conditions and profiles of the fibres. The tensile strength of fibres may vary from 100 to 650 MPa, and this will have an important impact on the performance and dosage of fibres.

There is an increasing interest in using MSFs for ground slab applications and to replace both steel fabric reinforcement and steel fibre reinforcement. From different applications, it can be mentioned that MSFs can provide concrete with the required level of post-cracking performance.

Source:

- Attree [179]

7.6.20 UHPFRC Panel Testing

Explosive testing was conducted on full-size ultra-high performance FRC (UHPFRC) panels. Each panel measured 3.5 m × 1.3 m and had a thickness of 100 mm. The panels were located within a large concrete area and placed between 7 m and 12 m from a 100 kg.

TNT equivalent explosive charge. Each panel was placed carefully at a suitable distance from the explosive charge to cause permanent damage but not total collapse of the structure.

UHPFRC panels were prepared with fine powders of cement, glass-grade sand, silica fume, silica flour, super-plasticiser, water, and fibres. No other aggregates were used. The panels were fabricated with different quantities of steel fibre doses and made into panels with steel fibres along with additional steel reinforcement. High tensile steel fibres, with lengths and thicknesses of 13 and 0.2 mm, respectively, were used to provide mechanical strength and can be compared with architectural applications. High binder cement over 1,000 kg/m^3 with a low water binder ratio of 0.15–0.18 was possible for the mix by applying a high dosage of super-plasticiser and well-graded materials in the mix.

By adopting a high curing temperature of 90°C and 100% humidity, it was possible to achieve early compressive strength between 150 and 200 MPa and flexural strength of 25–50 MPa by using steel fibres. Furthermore, UHPFRC also provided high fracture energy of about 20,000–40,000 J/m^2 compared to around 100 J/m^2 for normal concrete.

In the layout area, the targets were positioned at different stand-off distances ranging from 7 to 12 m around the central explosive charge. The blast waves produced by detonating the explosive charge radiated in all directions and could be used to load a certain number of targets simultaneously.

Simplified modelling using a single degree of freedom (SDOF) procedure was adopted for comparison and assessment.

7.6.20.1 Comments from Tests

The following may be noted:

- Two of the panels had both steel bars and steel fibre reinforcement. The other panels had only steel fibre reinforcement.
- One panel contained 4% steel fibres by volume, while the other panels had 2% steel fibres by volume.
- The advantages of using steel fibres included anti-spall properties and additional energy dissipation once the steel bars began to yield.
- The blast loading caused different permanent damages and formed yield lines at or near the middle of the span, but did not result in total structural collapse due to the stand-off distance from the charge.
- The application procedure for one-way spanning members indicated good agreement for the maximum deflection and practical use of the simplified SDOF method.

Source:

- Schleyer, Barnett, Millard, Wight, and Rebentrost [180]

7.6.21 Explosion and Impact Resistance of UHPFRC

The study of explosion and impact resistance of ultra-high-performance FRC (UHPFRC) has been undertaken for anti-terrorism applications.

UHPFRC is a cementitious material that can provide compressive strength in the range of 150–200 MPa and tensile strength in the range of 30–60 MPa, which is significantly higher than that of normal concrete. Furthermore, UHPFRC exhibits high fracture resistance and high ductility. The fracture energies may range between 20,000 and 40,000 J/m^2. The material has a low water-to-binder ratio, high cement content, and a high dosage of super-plasticiser. Micro-silica may be used, and the only aggregate is fine silica sand less than 400 μm to ensure homogeneity. Post-set heat treatment at 90°C is applied for further improvement of the micro-structure, and high dosages of short, small-diameter steel fibres are used to improve ductility.

UHPFRC can be mixed and cast like normal concrete with no special requirementsfor handling.

When an explosion occurs close to a panel, the loading applied will be localised. A high-speed projectile will also provide localised loading. Alternatively, an explosion with a large stand-off distance will provide loading that is more uniformly distributed over the whole panel.

High-speed rates of localised loading were studied using drop hammer facilities.

The main intention was to assess the following initially:

- Size and edge effects due to preferential alignment of steel fibres along moulded surfaces.
- Loading rate
- Panel thickness, 25–50 mm
- Fibre content
- Secondary reinforcing bars.

This material, with further developments, will be suitable for applications aimed at protecting people and buildings from terrorism.

Source:

- Barnett and Tyas [64]

7.6.22 Blast Loading on Concrete Frame Structures

Explosive devices can produce blast loads for a short duration, typically lasting milliseconds. This time period may be short compared to the natural period of typical reinforced concrete elements, and hence it produces an impulsive load

A huge amount of energy is released from a detonation. Gas will be produced at high pressure and temperature, forming a shock wave front that travels radially at high speed into the atmosphere, compressing the air molecules in its path.

Reinforced concrete columns in a frame structure may be vulnerable to blast loads. The progressive effects or collapse may be avoided by providing a proper stand-off distance between the vehicles and the buildings. When this is not possible according to the layout, the dimensions of the columns and the amount of shear reinforcement should be significantly increased. The provision of emergency or additional bracing may be introduced, if possible, to redistribute the loads away from the damaged columns. This situation may be addressed by introducing internal shear walls or strong facades, a system that may be absent in buildings designed for accommodating glazed facades and open-plan architecture.

Source:

- Byfield and Paramasivam [181]

7.6.23 Ultrasonic Pulse Technology for Checking the Performance of FRC

Ultrasonic pulse technology may be used to detect voids in FRC slabs.

7.6.23.1 Use and Limitation of GPR

GPR operates by emitting radio waves into the material for study. The radio waves are reflected by metal objects in the concrete. By using GPS, in FRC slabs, only limited information can be obtained. The results may be erroneous in the case of densely packed steel fibres in the concrete slabs.

7.6.23.2 Ultrasonic Pulse Technology

Using wireless inspection with the multichannel ultrasonic pulse echo method is appropriate. From studies applying this technique in different projects, it has been noted that the limitations of GPR may be overcome by this method.

Ultrasonic technology (tomography), when applied, emits ultrasonic pulses into the concrete, with 40% to 50% of the pulse being reflected by the metal objects.

The remaining 50% of the pulse penetrates deep into the concrete material. These pulses, when encountering an air gap, will indicate and produce a target on the screen.

Tomography can detect voids in concrete; however, the actual volume of voids and their connectivity cannot be determined accurately. More experience and engineering judgement will be required to use this technology properly and for the evaluation of non-destructive techniques.

Source:

- Thomas [119]

7.6.24 Creep Tests for FRC

7.6.24.1 Experimental Study of Creep Properties

A uniaxial tensile load may be applied to a shaped test specimen, and the creep deformations of the polymer are measured as a function of time. A typical creep curve for the polymers under uniaxial stress and at constant temperature may be divided into five parts, as mentioned below:

7.6.25 Initial Elastic Strain, Initial Plastic Strain, Primary Creep, Secondary Creep and Tertiary Creep

This is shown in Figure 7.6

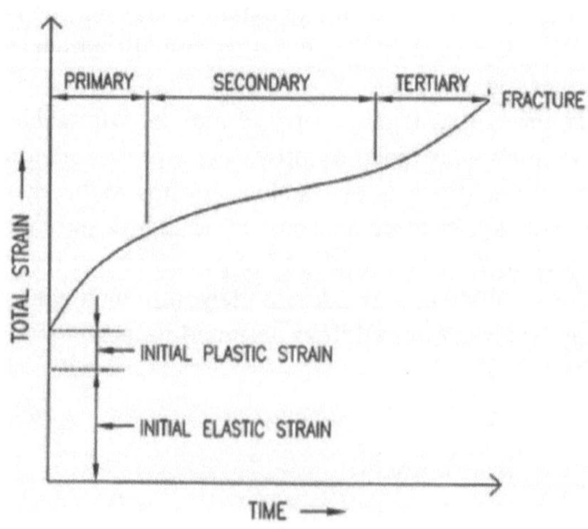

FIGURE 7.6 Highly idealised creep–time graph. (Source: Hollaway [1]).

7.6.26 Further on FRC Creep Tests

7.6.26.1 Introduction

In this study, six specimens of plain and fibrous concrete beams measuring 100×100×500 mm were considered for the test.

These are listed below:

- **Class C** – These beams contained no polypropylene fibres
- **Class F** – These beams were constructed with 4 kg/m^3 and 50 mm long ADFIL Durus synthetic fibres. The beams were cured at 75% relative humidity for 3 months.

These were wrapped with hessian/plastic and transported to Imperial College, UK, where they were kept in laboratory air for at least 1 year before testing. The 28-day compressive strength was approximately

Two types of tests were considered:

- Flexural strength test done before and after the creep test.
- Creep test for 12 months was carried out under a load of 30% of the ultimate flexural strength assessed prior to creep loading. The loading was maintained, and deformations were recorded at the central point during the 12-month period.

The following measurements were taken for the creep test:

- Initial elastic deformation.
- After this, creep with two components, i.e., delayed elastic strain and irrecoverable flow, and finally, after unloading, were measured.
- Final elastic recovery, which is the opposite of the elastic component.

7.6.26.2 Flexural Strength Results

The average flexural strength of three plain concrete and three fibrous specimens was 5.6 and 6.03 MPa, respectively. These were 12% and 13.4% the 28-day compressive strength of plain and fibrous concrete. It may be mentioned that fibrous concrete exhibited a slightly higher percentage of compressive strength.

7.6.26.3 Flexural Strength after 12 Months Creep

After 12 months of creep under a 30% load, the specimens were unloaded.

The average flexural strengths of plain concrete and fibrous concrete specimens were 4.98 and 5.56 MPa respectively.

It appears that loading both plain and fibrous concrete beams for 12 months has reduced the flexural strengths of both by about 10%.

No increase in cracking during the 12-month loading period was observed compared to the initial state of the specimens.

The flexural strengths were 10.7% and 12.4% of the 28-day compressive strength for plain and fibrous concrete. It seems that fibrous concrete achieved a marginally higher percentage of the compressive strength.

7.6.26.4 Flexural Strength Comparisons

Fibrous concrete attained higher flexural strength before and after 12 months of loading by approximately 10%, and with loading for 12 months, it reduced the flexural strength by about 10%.

Fibrous concrete specimens did not totally collapse after the flexural tests. It was observed that some fibres remained intact and supported the beam across the main crack.

7.6.26.5 Flexural Creep Results

Larger initial elastic deflection was observed for the plain concrete compared to fibrous concrete beams.

The deformation of the fibrous concrete specimens indicated higher values than plain concrete beams. The reverse was true for flexural strength tests.

The initial elastic deformation was 17% higher for fibrous concrete beams compared to plain concrete beams.

For creep only, the deformation was 41% higher, and the elastic recovery was 100% (i.e., 2 times).

Source:

- Khoury and Algar [182]

7.7 DIFFERENT APPLICATIONS

7.7.1 Introduction

The applications of FRC and FRP polymer composite materials are more prevalent in advanced countries over a long period of time. It was interesting to search for applications in European countries, especially in the UK. In this connection, help from the Concrete Society's journal, Concrete (UK), was very interesting and valuable. Furthermore, journals from other institutions/universities, along with different publications, have been consulted. Many applications from the UK are presented to highlight the extensive use of these materials in different projects over time.

Hope the information will be useful to young engineers.

A large number of applications with FRC have been implemented in different countries around the world. Only a few applications with brief details will be presented here.

7.7.1.1 Applications of FRC

Some common applications included here will cover projects such as

Industrial floors, pavements, runways and aircraft parking, canal and tunnel lining, curtain walls, thin shells, slope stabilisation, different rehabilitation projects, man holes, pipes, pre-cast applications, dams and hydraulic structures, geotechnical applications, roof tiles, composite deck, impact-resisting structures, sprayed concrete work, etc.

Some of the applications will be described in more detail later.

7.7.1.2 Special Applications

Examples of a few well-known structures with fibre concrete (FRC) are referenced. A few of these structures are mentioned below:

- Roman Colosseum built in 80 AD. Horsehair was used as secondary reinforcement.
- Pueblo house built in 1540 using straw-reinforced adobe bricks. It appears to be the oldest house in the USA.
- Tipu Sultan's palace at Srirangapatnam, which was built using sheep's wool.

The use of horsehair in plaster has been applied in many historical structures.

Source:

- Kaur and Talwar [33]

7.7.2 Common Applications

Some common applications with different types of fibres will be mentioned below:

7.7.2.1 With Steel Fibres

Steel FRC was selected for tunnel lining [57].

A higher dosage rate would be required with a lower aspect ratio to meet performance requirements.

Source:

- Greenhalgh [57]

Similar applications with steel fibre-reinforced sprayed concrete were done for rock support [58].

It was observed that with a higher aspect ratio and volume concentration, better performance could be achieved for hardened concrete, but it is difficult to mix, pump, and spray.

Source:

- Swannell [58]

Steel fibres were also selected for a suspended two-way spanning concrete slab for a car park in Belfast supported on piles [62].

Source:

- Cameron [62]

Applications have been made to use fibres in ground floor slab construction with the intention of eliminating the commonly applied reinforcement with steel. The design may be done for slabs supported on piles using steel fibre reinforcement. The effects of using fibres will depend on fibre type, dimensions, properties, and the dosage of fibres to be used in concrete, etc.

For the extension of the Barcelona (Spain) metro [65], construction has been done with pre-cast tunnel lining segments using steel reinforcement fibres. It was noted after studying the site trials that the use of steel fibres in concrete could satisfy the design requirements without using conventional steel bar reinforcement.

Source:

- Berge [65]

7.7.2.2 With Glass Fibres

GRC cladding was used at Hays Wharf Galleria [85]. A steel sub-frame supporting the GRC cladding was given two coats of treatment for protection against corrosion.

The thickness of GRC members varied. After thorough cleaning, a smooth surface finish was provided on GRC elements. Further information on this has been included in item 7. 2.

Source:

- Spence [85]

Other Applications of e-GRC will be [87]

Different applications have been made with e-GRC. These are mentioned separately.

The main benefits [6] of using e-GRC for construction are tabulated here:

- Much better appearance, specially the brightness of colour in exposed concrete structures, along with the freedom of shapes and sizes
- Lightweight
- Better air quality, especially in congested urban areas
- Reduced loads on elevated structures

Source: Bartos [87].

7.7.2.3 With Synthetic Fibres

STRUX 90/40 synthetic macro-fibres have been [183] successfully used over steel decking in various projects. It was found that the combination of synthetic macro-fibres with steel decking could reduce construction time and was cheaper to use.

Source:

- Perry [183]

Strux 90/40, synthetic macro-FRC [97], has been used in Blackpool's promenade. Such applications of synthetic fibres will be potentially very beneficial in aggressive marine/coastal areas.

Source:

- Perry [97]

Applications of MSFs [184] in Sweden have been carried out for sprayed concrete, underground car parks, and tunnel linings in Norway, Sweden, and Japan.

Source:

- Ridout [184]

It has been commented [32] that the combination of macro- and micro-synthetic fibres will provide significant advantages in both design and performance for external applications. The dosage rate has been mentioned in item

Source:

- Atkinson [32]

The use of macro- and micro-synthetic fibres in concrete, along with three-dimensional reinforcement, will improve anti-seismic characteristics and inhibit the development of plastic and settlement cracks. Furthermore, it will enhance resistance to impact, shattering, abrasion, and freeze/thaw cycles.

Further information has been given before.

Source:

- Atkinson [32]
- Ridout [98]

FRC and steel FRC containing [106] an adequate amount of fibres can provide better or equal crack control compared to normal fabric reinforcement and can improve the flexural capacity of ground-bearing concrete.

FRC concrete containing MSFs with high-end properties can deliver similar flexural toughness and creep resistance performance when compared to FRC containing steel fibres [106].

Source:

- Hathaway [106]

7.7.2.4 With Natural Fibres

- The large applications of natural FRC for structural purposes are rather limited.
- It seems that only a few special applications have been made where natural FRC has formed an outer layer over some other materials.
- The outer layer can serve as a permanent, strong, and tough covering over a weak core and can help provide architectural and ornamental features to the design.
- The greatest advantage of natural fibre cement composite is that such fibres are available in a variety of forms in abundance, freely, in many countries, making them suitable for housing, especially for low-cost construction as an alternative building material.

Source:

- Swamy [21]

7.7.3 GRC at Height – Creating a Cathedral Dome

The case study is related to a dome in Ukraine. Glass FRC was used for the construction.

The cathedral must accommodate 1,500 persons for a congregation. The construction of the lower large dome was challenging.

It had a diameter of 22 m and the top of the dome was 26 m above ground level. As per the specification, the dome must to be watertight, insulated, and the surface should be smooth enough for elaborate painting.

The spraying of GRC was done at heights of 16–26 m, in the air, with workers who were not familiar with the material. It was important to check that alkali-resistant glass fibres flow freely through the spray gun without blocking and are easy to compact.

High zirconia material with 19% zirconia content was selected for the application. It was critical to ensure that GRC produced in these difficult site conditions with a well-trained workforce is comparable to the products prepared under factory conditions

A small 8 m diameter dome was constructed at a lower level and hoisted to 47 m high, passing through the opening in the top of the larger dome and fixed to the corbels properly.

The smooth internal surface of the GRC as constructed was suitable for decorative painting.

Source:

- Peter and Crocker [185]

7.7.3.1 GRC as Cladding Material

GRC as a cladding material is becoming popular in recent years. The material has been successfully used in many countries. A few comments are made below:

- GRC may be available with a range of architectural finishes, textures, and colours.
- Cladding panels will be lighter than traditional counterparts (i.e., stone) without loss of visual appearance.
- GRC may be manufactured in a variety of shapes and sizes compared to either sun-dried or wet-cast stone.
- GRC will provide more durable and versatile methods of replicating natural stone for larger cladding panels.
- From small rain screens to cladding panels up to 30 m², complex building envelopes for modern projects.

Source:

- Faulding [186]

7.7.4 Further on Applications

Concrete bridges are being widely strengthened by suitable fibre composite materials. Fibre composites [124] may be considered an efficient and cost-effective means of strengthening concrete bridges. The major advantages include the speed of installation leading to short possession time. Tests have been conducted at the Transport Research Laboratory on half-scale column specimens with varying amounts of axial and hoop fibre wrapping. An enhancement of strength has been observed.

The workmanship will be important at all different stages of the installation process.
Source:

- Sadka [124]

CFRP plates were applied to strengthen a 65-m-tall fertiliser prill tower [139] in Qatar. A specially formulated adhesive suitable for the Middle East was selected. Other options were also considered. However, the CFRP option was finally selected as it would have the smallest effect on the operations of the tower.
Source:

- Luke and Godman [139]

7.7.5 Special Comments on FRP Strengthening

Confinement of FRP Jacketed Reinforced Concrete Columns.

Following the tests on FRP jacketed reinforced concrete square/rectangular and circular columns with 21 MPa concrete and using high-tensile fibres of carbon, glass, and aramid, the comments are [157]

- FRP-strengthening columns performed better.
- Increasing the thickness of jackets reinforced with any type of fibres increased both strength and axial deformation.

Further information may be obtained from item 7.4.
Source:

- Belarbi [157]

7.7.5.1 More Comments

- **Inspection** – The presence of moisture may be the main cause of deterioration of FRP strengthening [126]. Hence, gutters, drains, etc., should be kept clear of debris so that rainwater does not accumulate. Cleaning the surface with water jetting or grit blasting will not be appropriate for FRP, as these methods are likely to cause damage. Steam cleaning will not be appropriate either, as it is likely to soften the adhesive.

Inspection should be conducted at all stages of the strengthening process. All strengthened structures should be subjected to regular routine visual inspections.
Source:

- Clarke [126]

7.7.6 Other Applications of Fibres Including Repair Works

A few additional comments will be made regarding the use of fibres and fibre composites related to repair/retrofitting work in different projects. Some of these items may have already been briefly covered in this publication. A few typical applications will be highlighted below:

7.7.7 Concrete Bridges [154]

Fibre composite materials using carbon, glass, or aramid fibres with epoxy resin or adhesives are being utilised for the repair and strengthening of concrete bridges in the UK. Fibre composite polymers (FRP) are available in the forms mentioned in item 7.4.

Source:

- Loudon [154]

Various applications of fibre composites have already been made in advanced countries, including the UK. Some brief details are mentioned in item 7.4.

The installation time will be less with FRP materials. However, workmanship and inspection at all stages of operation will be very important, as already mentioned. Strengthening with FRP materials may be cheaper [123] than equivalent strengthening with steel plates.

Source:

- Clarke [123]

7.7.8 Steel Bridges

CFRP plates were selected to strengthen a single-span steel bridge in the UK [138].

The details of strengthening are as follows:

Main Girders in Bending	–	Ultra-high modulus (UHM) CFRP plates of equal section were placed to outer faces of both top and bottom flanges of both girders for bending
Cross Girders in Bending	–	UHM CFRP plates were put to outer face of bottom flange only on all cross beams for bending
Main Girders in Shear	–	Addition of high modulus CFRP plates were provided to outer face of web end panels for shear requirement

SIKADUR-30 (a two pack epoxy) was used as adhesive. After studying a lot of alternatives CFRP was preferred considering ease of installation and the ability to carry out the strengthening work without interrupting rail traffic.

(CFRP) composites materials was selected for strengthening Jack-arch Structures for London underground limited (LUL) [187].

Source:

- Dodds [138]
- Research Focus [187]

7.7.9 Strengthening Structures with Carbon Fibres

7.7.9.1 General Comments

The strengthening of buildings and bridges by applying CFRP plates had been done as already stated [136].

The use of CFRP plates will be an attractive solution for the strengthening work.

The advantages of carbon composite material will be high strength, low weight, easy to handle, having a range of elastic moduli, good fatigue, fire and creep resistance and higher durability performance.

CFRP plates may be bonded to the structural elements using adhesive. The adhesive may be applied to the plate and substrate.

CFRP plates had been used to strengthen different structures like timber structures, concrete floor slabs and rectifying deflection problems of structural members and shoring up of masonry walls.

7.7.9.2 A Completed Project with CFRP

In a Bridge Project – A single CFRP plate of 120×1.2 mm thickness was attached to the underside of the bridge to bring back the lost flexural capacity. Specially prepare adhesive was used for bonding. The bridge was kept open during the period of repair work.

7.7.9.3 Tests Carried Out

The tests in the laboratory and on the site were conducted to study the flexural strength, creep, fatigue, and durability of beams.

The results indicated an increase in strength in the range of 24%–80% over unplated beams.

7.7.9.4 Applications Technique

The applications were mostly carried out on concrete structures. Following a similar technique, it may also be used for timber, wrought iron, cast iron, and steel structures. However, the use of proper materials, correct adhesives, and efficient applications will have significant implications for the life of the strengthened structure.

Source:

- Lane, Fashole-Luke, and Skwarski [136]

7.7.9.5 More on Concrete Bridges

Three simply supported RC slabs with longitudinal reinforcement only were treated in the USA. The bridges were repaired using an FRP system. The use of NSM tapes/FRP bars was adopted as a complement to externally bonded CFRP laminates [127]. FRP composites are anisotropic and have excellent tensile strength in the direction of the fibres. They are also more resistant to corrosion than other construction materials. The use of such material may be considered ideal for deficient RC members of bridges needing strengthening and for unreinforced masonry walls.

Source:

- Stone, Tumialan, and Nanni [127]

7.7.10 Strengthening Structures Highway Structures with Fibre-Reinforced Composites

7.7.10.1 General Comments

Proposals were made to use fibre-reinforced composites (FRP) to strengthen existing structures, especially bridges. The composite consists of high-strength fibres, usually carbon, aramid, or glass, embedded in a resin matrix as stated. These materials will not corrode and will be durable. However, long-term durability must be assessed by observing the performance of the materials over an extended period.

For strengthening bridge columns, pre-fabricated shells made of glass-fibre-reinforced composites around the columns were also used. The use of two half-shells was considered, which were connected at the site using a special fixing arrangement. Ordinary grouting was carried out to fill the gaps between the shell and the column. The application was found to be suitable and satisfactory.

Further details are presented separately in item 7.4.

Source:

- Loudon [154]

7.7.11 Widening a Bridge and Stitching the New Section to the Existing One

It was necessary to widen the existing bridge and to provide a new section for the widening. The new portion of the bridge had to be connected to the existing one with minimal disruption.

The relevant items are mentioned below:

- Demolition was required for a cantilever side section of the original bridge.
  Hydraulic shears were used.
- The existing reinforcing bars were straightened, and these bars were used to stitch the new section to the old structure along with dowel bar connections.
- A composite structure was conceived for the new section.
- Fabricated steel beams along with an M 40/50 reinforced concrete deck were used.
- The steel beams were placed conveniently, and concrete was poured over glass FRC permanent formwork between the steel beams.
- Pre-cast units were placed to create the outer parapet of the new bridge.
- Curing was done, and time was allowed for the new section to settle/mature.
- The stitching and final pouring of concrete were completed.

The items mentioned below may be considered important:

- Weathering steel was used to reduce maintenance work.
- The use of pre-cast concrete edge elements avoided working at heights, and the traffic disruption was reduced.
- Extending and retaining the life of the existing bridge deck.

Source:

- Chong [148]

7.7.12 FRP Shells for Strengthening Bridge Columns

Twenty-four tapered columns on two bridges at the A19/A174 interchange (in the UK) on the [155] outskirts of Middlesbrough were strengthened using aramid FRP treatment. Carbon or glass fibres were not ductile enough for the dynamic load applications. Two semi-cylindrical composite shells with an annulus between the pre-assembled mantles and existing columns were injected with high-strength, non-shrink grout. The schedule time was cut by this application.

The low maintenance cost of the fibre shells with impact resistance properties will provide further savings for the remaining part of the service life.

Source:

- Pinzelli [155]

7.7.13 Strengthening Harbour City Car Park

Existing structures (i.e., 164 columns) at Harbour City Car Park, Salford Quays (UK), had to be strengthened to accommodate additional loads for extra storeys. Aramid fibre sheets [156] were selected for the strengthening work instead of steel plates. Two additional storeys were added after assessment. The car park had to remain open throughout the work, which was completed within a tight time schedule.

Source:

- Russell [156]

7.7.14 Car Park for Merseyside Police

This is a multi-storey car park in Liverpool, UK.

During use, the structure developed cracks and spalling of concrete, indicating that major repair work would be necessary for strengthening the structure. The presence of chloride and structural cracks in the same location was responsible for the corrosion and deterioration of the steel reinforcement.

Fibre-reinforced polymer (FRP) bars bonded into grooves on the top surfaces of the car park decks were finalised as part of the solution. This treatment is referred to as execution with NSM reinforcement.

Grooves 20 mm wide and 20 mm deep were made in the slab above the level of the retained steel reinforcement. The grooves were primed and filled with epoxy adhesive to a depth of 10 mm. CFRP bars were placed in the centre of the groove and finally finished with epoxy adhesive and levelled. The application of NSM CFRP was found to be suitable for strengthening car park decks in flexure.

Source:

- Farmer [161]

7.7.15 International Airport, Belfast

MSFs in concrete were used [188] to replace a section of the drainage channel at Belfast International Airport. 60 MPa MSF (Barchip Shogun Fibres) at 5 kg/m^3 of concrete was selected, with the addition of 18 μm, 12 mm polypropylene fibres to improve plastic shrinkage cracking and freeze/thaw resistance. Barchip "Shogun" is a structural synthetic fibre reinforcement added to concrete to replace welded reinforcing mesh.

Source:

- Martin and Donnelly [188]

7.7.16 Glasgow International Airport

A few Siporex slabs needed total replacement. Others required repair/strengthening with carbon fibre plates or repair with cementitious or epoxy mortar.

7.7.16.1 Options for the Repair work

Different options were considered.

The damaged slabs were repaired first. A few slabs had to be replaced.

The removal of the existing slabs was not easy because of the presence of different services.

Further information will be available in item 7.4.

Source:

- Couper [160]

7.7.17 Sprayed Fibrous Concrete Tunnel Support

Stainless fibrous concrete in an Aqueduct project in Derbyshire may be referred to.

Fibrous concrete is a dry mix; water is added at the nozzle spray gun, and 25 mm stainless steel fibres were added at the mixing stage.

Initially, 6 mm aggregate was tried in the mix, but it bounced off the rock surface.

Finally, it was a dry mix with sulphate-resistant cement, zone 2 sand with 3% stainless steel fibres (by dry weight), and 5% powder accelerator by cement weight. The use of steel mesh was also tried. The fibrous concrete was sprayed using a Swiss Meyco Piccola 020 machine.

Source:

- Barfoot [78]

7.7.18 Tunnel Linings, Channel Tunnel, UK

SCC, along with steel fibres, had been considered the best option for pre-cast tunnel lining work [57]. In the second phase of the CTRL, about 40km in length had been conceived with fibre-reinforced pre-cast concrete segmental lining containing no conventional reinforcing bar cages. These tunnel linings also contained polypropylene fibres for fire resistance and to avoid explosive spalling.

7.7.18.1 Segmental Lining – Future of Steel FRC

The comments are [57]

- It had been used in the Channel Tunnel, railway link, CTRL, and the new works at Heathrow Airport, Terminal-5.
- The tunnel linings also included polypropylene fibres.
- Normally, for lower dosage rates, fibres with a high aspect ratio are found to be more effective in the long run. With a lower aspect ratio, higher dosage will be required to meet performance criteria.
- Self-compacting concrete was used.

Source:

- Greenhalgh [57]

MSFs have been applied [99] successfully in tunnel lining works during recent years. In such applications, MSFs not only performed technically as well as steel fibres but also performed better. When a long service life is required, MSFs may be a very good alternative.

Source:

- Ridout and Martin [99]

7.7.19 Channel Tunnel Rail Link

The CTRL is the first new mainline railway constructed in Britain.

The tunnels were lined with 7,150mm internal diameter×350mm thick pre-cast segmental linings. Each was made of ten segments forming a ring 1.5m wide and 14mm taper. 2.5m long segments were joined with 24mm bolts screwed into cast-in plastic sockets. The segments were made watertight.

The concrete was specified as 60MPa compressive strength with limestone or granite aggregates, and 30kg/m^3 Novocon HE 1050 steel fibre reinforcement to increase tensile strength. Furthermore, 1kg/m^3 Fibermesh Stealth polypropylene fibres were used to prevent explosive spalling in fire. Different trials and tests were conducted for the finalisation of the mix.

Source:

- Court [189]

7.7.20 The Important Factors/Items for Fire Safety in Tunnels

The items may be mentioned as follows:

7.7.20.1 Smoke Extraction, Ventilation, Detection, Alarms

Concrete is an inert, non-flammable, and stable material with high fire safety factor, which will not contribute to fire load.

Source:

- Chana [190]

7.7.21 Further on Tunnel Fires

7.7.21.1 General Comments

In tunnel fires, the temperature may rise quickly and drastically. The immediate danger will be the evacuation of the persons.

The heat produced by the fire will vaporise the water in the concrete, leading to explosive spalling.

The thickness of the concrete will be reduced, affecting its load-bearing capacity. This will be the spalling effect.

7.7.21.2 Fibres Will Melt

Polymeric fibres have been used for tunnels in Germany.

When the temperature of 160°C is reached, the fibres in the concrete will melt. Small voids will be created in the concrete to release the pressure.

7.7.21.3 Experimental Study

700×700×300mm slabs were tested. The concrete was heated up to 1,300°C quickly. Initially, the micro-cracking process related to inner deterioration was studied.

Ultrasonic sensors were used to monitor changes in ultrasound waveforms. A micro-bolometric infrared thermography camera was applied to control the thermal heat flow of the specimens.

7.7.21.4 Numerical Study

It has been noted that the explosive spalling of concrete is a local phenomenon. The local material, geometrical, and loading properties, as well as their variations, will be relevant.

7.7.21.5 Comments

Following the study, it has been noted that non-destructive testing methods, such as acoustic emission and ultrasound methods, may be used to investigate concrete spalling under fire conditions.

Source:

- Richter, Grosse, Juknat, and Dehn [191]

7.7.22 SFRC Shaft Lining Replaced Traditional Reinforcement, Sewer Project

SFRC shaft lining replaced traditional reinforcement with steel fibres in a sewer project.

It was possible to remove the majority of the traditional reinforcement and replace it with a steel fibre solution.

Four main shafts have been completed. The maximum dimensions of the pumping shaft were 38 m×86.5 m deep. Following this method, it was possible to save time, manpower, and reduce risks during construction and material handling activities.

The operation had to be conducted in a confined space.

Extensive testing in the laboratory and on-site has been carried out.

The steel fibres showed strain-hardening properties at 40kg/m^3. The design was adequate to carry surcharge loading from the surrounding ground and other underground structures.

Source:

- Greenhalgh [192]

7.7.23 Ribbed Floor Slab Strengthening, Poole, Dorset (UK)

The strengthening of the ribbed floor slab was carried out with carbon fibre laminates to control deflection and strengthen the structural elements to meet the requirements of additional loads on the floor. There were design restrictions related to space. The carbon fibre laminates were selected and bonded to the bottom of the ribs.

Source:

- Farmer [158]

7.7.24 Suspended Floors for Industrial Use: An Overview

Suspended floor slabs may be used for the ground floor where the ground conditions are poor, and the floor may not be able to support the required loads. To avoid the risk of settlement, an elevated floor slab may be required above the ground.

7.7.24.1 Types

- Monolithic with downstand beams or ribs.
- Flat slabs with piles; the slab will not be tied to the piles.
- Flat slabs on piles; the slab will be tied to the piles.

The relevant performance requirements for the design will include

the load-carrying capacity, including the dynamic loads as applicable; settlement within the limits; abrasion resistance; durability; flatness; curling, joints; drying shrinkage; random cracking; concrete grading M30/40 will be (preferably without high cement content); structural performance; floor finish; uniform colour; waterproofing; insulation; no seepage water from below; and any special items that have to be considered.

Source:

- Simpson [193]
- Ganguly [7]

7.7.25 Biggest SFRC Flooring, Scotland, UK

A steel FRC floor slab was constructed, covering an area of about 250,000 m^2. It featured joint-less warehouse flooring and had to accommodate a uniformly distributed load of 50 kN/m^2, along with each racking leg loading of 85 kN. The design included a 150 mm thick slab of 40 MPa concrete reinforced with 40 kg/m^3 steel fibres.

In chilled and freezer zones, 40 MPa concrete was used for construction with a 240 mm thick SFRC slab for the ambient area, reinforced with 45 kg/m^3 of AFT+1/60 undulated steel fibres. A super-plasticiser and curing agent were used in the construction. The surface of the floor was nearly fibre-free. This method may be suitable for both ground and suspended ground floor slabs.

Source:

- Eddy [194]

7.7.26 SFRC Floor Slab for a Warehouse, UK

The project involved an area of 60,000 m^2 for internal and external floor slabs.

The slab had to be designed for individual racking leg loads of 90 kN and a uniformly distributed load of 50 kN/m^2.

A 175 mm thick slab and M40 concrete with steel fibres at 40 kg/m^3 were specified in the design.

The whole slab was cast in 12 panels.

The concrete was directly placed on the ground, and laser screed machines were used to place, compact, and level the concrete.

Source:

- Eddy [195]

7.7.27 Joint-less Floors

Joint-less floors are often constructed with steel fibre reinforcement and may include construction joints in panels to accommodate concrete shrinkage. Jointed floors may be constructed with steel mesh fabric and include construction and sawn joints in the slab to control shrinkage.

Joint-less ground-supported slabs may be more costly due to the steel fibre reinforcement required for the same slab thickness. The cost is likely to exceed that of an equivalent slab with steel mesh fabric.

Furthermore, steel fibre-reinforced floors may require dry shake topping to control the number of fibres appearing at the surface. The dry shake topping may not provide additional durability to the concrete floors; on the contrary, it may create problems associated with de-lamination.

Joint-less floors may be made thinner than jointed floors. Care should be taken to ensure performance at the joints, especially concerning load transfer issues, before deciding on the use of thinner slabs. The calculated capacity of sawn joints is often about 30% of the applied load.

It should be noted that joint-less floors, particularly when heavily loaded with racking or block stacking, may develop cracks.

For joint-less floors, it is recommended to provide construction joints at closer intervals.

The normal treatment for cracks in the floor slab would be to fill them with low viscosity resin.

For this treatment, the cracks must have a minimum width at the surface. The minimum crack width for treatment may be recommended as between 0.5 and 0.8 mm.

A crack width of 0.4 mm on the floor slab surface may be subject to damage. This width may be considered narrow for carrying out proper repair work.

Not all cracks may extend to the full depth of the floor slab, and they may not create structural problems.

It has been noted that, in open areas with pallet truck activity, there has been more wear on sawn joints.

Joint-less floors may be a good choice where there are heavy static loads, especially if applied immediately after construction.

Joint-less floors may also be a suitable option for open areas designated for light block stacking, such as fresh food distribution or marshalling areas. Joint-less floors may also be considered in food processing facilities where a resin type of finish can be applied on the top.

It has been suggested that both types of construction can be used in a project to satisfy different requirements.

Source:

- Hulett [196]

7.7.28 GRC for Slabs on Ground

Alkali-resistant (AR) glass has been used as reinforcement for high-performance cementitious concrete for durable construction. The fibres may be used for load-bearing slabs on the ground.

7.7.28.1 Material Properties

Structural GRC will require short integral alkali-resistant macro-glass fibres with a length of 36 mm and equivalent diameters of 0.54 mm. Glass macro-fibres will not corrode and are flexible, supporting the concrete well during setting and controlling the risk of cracking in the initial stage.

7.7.28.2 Production

The bonding of AR glass fibres and the cementitious matrix will be good, and no deformation or embossment will be required to increase the bond of fibres to the matrix.

7.7.28.3 Design of FRC Floors

The following items are relevant:

- FRC is a composite material made of fibres and a cementitious matrix.
- The fibres may be made from different materials such as glass, steel, and polymers.
- FRC provides post-cracking residual strength through fibre reinforcement.
- Minimum mechanical performance must be ensured by FRC.
- Fibres may be useful for improving performance under the serviceability limit state (SLS). By using fibres, it may be possible to reduce crack spacing, crack widths, and improve durability.
- Important factors for structural design will include the performance of post-cracking tensile strength of fibre composites.

7.7.28.4 Case Studies

Two case studies will be mentioned

7.7.28.5 Brazil

As a case study, reference may be made to the Industrial Floor.

- Area = 10,500 m^2 floor and M30/37 concrete with 7 kg/m^3 of glass macro-fibre Anti-CRACK HP 67/36.
- The slab thickness was 200 mm.
- The joints were cut every 5 m in both directions.

7.7.28.6 Germany

A Roundabout in Road -

- Area = 1,300 m^2 with M30/37 concrete.
- Reinforced with 10 kg/m^3 of anti-CRACK HP 67/36 glass macro-fibres.
- Thickness of the slab = 260 mm.
- The joints were made in radial directions every 5–6 m.

Source:

- Heiden, Holschemacher, Lober and Barragan [197]

7.7.29 Fibres in Floor Screed

The main advantages of using fibres in screeds are as follows:

- Reduction of plastic shrinkage cracking.
- Increased impact resistance of screed surfaces.

7.7.29.1 Testing In-Situ Screed

The testing of screed made over a concrete floor base may be done by dropping a load.

A 4 kg weight is dropped from 1 m to impact using a steel test anvil for an area of 500 mm^2.

Four blows are applied in the same area for the indentation, which is then measured for assessment. This may provide an indication of the screed's crushing resistance.

Source:

- Kitchen [108]

7.7.30 Composite Reinforcement for Coastal Defence Project, Blackpool, UK

Com BAR – Glass fibre-reinforced polymer (GFRP) with tensile strengths of 435 and 500 MPa for steel was used [198]. This has been found suitable for highly corrosive environments and for structures exposed to electro-magnetic fields. It is corrosion-resistant, non-conductive, and highly durable. This type of treated bar is expected to have a design life of 100 years.

The material is used in pre-cast and in-situ construction and has been performing well. By using these bars, it will be possible to save costs, meet safety requirements, and achieve environmental benefits.

Source:

- Cunningham, Burgess, and Spitz [198]

7.7.31 Fibre Reinforcement in Underground Civil Construction

The applications for sprayed concrete often involve steel fibres and welded wire fabric. MSFs have also been used for sprayed concrete construction. The advantages of using macro-synthetic fibres include being non-rusting, easy to handle, lightweight, and potentially causing less wear on the pumping equipment.

It should be noted that the replacement of steel fibres with synthetic fibres has been implemented in some projects in Norway.

7.7.31.1 Test Program

The tests were conducted using steel and MSFs. It was reported that the energy absorption requirements were satisfied by using Barchip MSFs of 5 and 7 kg/m^3. The results were compared with steel dosage rates of 25 and 35 kg/m^3 for the same energy absorption class.

7.7.31.2 Halsnoy Tunnel

This is a tunnel in western Norway. The entire tunnel was sprayed with MSFs.

7.7.31.3 Sweden

MSFs were used for sprayed concrete reinforcement. Barchip Shogun at a rate of 10 kg/m^3, 48 mm structural synthetic fibres were used for a project.

7.7.31.4 Tunnels in Japan

MSFs have been used to replace steel fibres and welded fabric in various projects.

Various tests have also been conducted.

7.7.31.5 Comments

It can be summarised that MSFs have already been used in many projects. The stated advantages include resistance to corrosion and cost-effective application. Furthermore, maintenance costs on concrete equipment will be reduced.

Source:

- Ridout [184]

7.7.32 Scrabster Harbour Sea Wall, Scotland

It should be noted that a 70% ground granulated blast furnace slag cement replacement mix was used for the construction of the new quay and wave wall in this project. The deck was supported on steel tubular piles. The deck was 1 m thick, and the widths varied from 16 to 30 m. The wave wall was 0.6 m deep and 3 m high.

Pavix was used as a one-application treatment for the protection of concrete. The solution penetrates into the concrete surface and sticks tightly to the concrete pores. The crystals are likely to attract moisture and swell to provide a water-repellent barrier, protecting the concrete against harmful substances like de-icing salts, oils, jet fuels, and other corrosive materials. The applications/treatments may be considered suitable for marine structures.

Source:

- Jenkins [199]

7.7.33 Further on Strengthening with Polymer Composites

Two case studies are presented below:

7.7.33.1 Boots Building, Nottingham

This was a skeletal steel-framed building structure. Following the investigation, it was found that the two main beams had a 30% loss of flange and web sections due to electrochemical corrosion. The options were to replace them or to strengthen the beams.

It was decided to use pre-impregnated carbon fibre-based composite for strengthening. The execution was carried out using vacuum bag technology and applying a localised heating method to cure the resin.

The thickness of the composite material fixed to the original steel section was 5 mm and did not pose any problems with other treatments of the structure.

7.7.33.1.1 Nuclear Power Station

Some structural items needed strengthening to withstand hazard loading, including seismic effects.

The problems were associated with mechanical and thermal loading and corrosion of reinforcing bars. It was decided to strengthen/retrofit concrete walls with CFRP strips, which could be easily located at the required places.

The work had to be done considering different factors, such as shutdown periods, economic solutions, and the structure's location in a coastal environment.

It was proposed to use CFRP strips that could be bonded to the concrete directly over the embedded reinforcing bars.

Roped access was utilised for execution, which was faster, easier to install CFRP strips, and cheaper.

CFRP strips were lightweight and could be placed directly over the embedded bars at the required locations.

7.7.33.1.2 Impact of Applications

The following will be important:

Financial, social, engineering, environmental, robustness

Source:

- Polymer Composites as Construction Materials [200]

7.7.34 Non-metallic Reinforcement for the Third Bosphorus Bridge, Turkey

For the Bosphorus Bridge in Istanbul, Turkey, curtain walls with solid reinforcement were applied for the construction of the pylons. The bridge structure is a combination of a suspension and cable-stayed bridge and may be subjected to a wind load of 300 kg/m^2. The facade panels are for the protection of the bridge and should support a maximum weight of 110 kg/m^2. Initially, the panel dimensions of 3×4.5 m and 30 mm thick panels were assumed. Hybrid reinforcement of alkali-resistant glass (AR glass) and carbon-fibre fabric was finalised.

7.7.34.1 Performance

Fibre-reinforced plastic concrete utilises carbon and AR glass fibres in place of steel and can provide corrosion-free, lightweight, and durable construction. Concrete structures in poor condition may be repaired or renovated using textile-reinforced concrete to improve durability.

7.7.34.2 Production

The production of panels started in February 2016. The material was expected to provide 100-year durability.

The production of the panels at a Turkish pre-cast factory was carried out under the supervision of German engineers.

The Bosphorus Bridge was inaugurated in the summer of 2016.

Source:

- Shams [201]

7.7.35 Strengthening/Retrofitting of Chimneys

7.7.35.1 Introduction

Chimneys may be subjected to extreme external environmental conditions, such as chemical attacks, as well as extreme heat and cold. Furthermore, they may face operating conditions that change over time. Repair and strengthening of concrete chimneys were carried out using concrete enlargement or externally connected steel items to extend the service life of damaged or deteriorated chimneys. Later on, an effective FRP strengthening technique was developed and used as a better alternative for the repair of tall chimneys.

Source:

- Alkhrdaji and Galati, SCRIBD [202]

7.7.35.2 Applications

- A 400-ft-tall concrete chimney at Portland Station was retrofitted. The chimney was constructed in 1955 and was in need of structural retrofitting/strengthening treatment. The use of fibre-reinforced polymer composites (FRP) or fibre wrap was recommended for strengthening/retrofitting work to extend the service life of the chimney. It will be a lightweight application with a waterproofing system, and the design will be based on strength and durability.

Source:

- Gedalia and Sanchez [203]

Concrete chimneys associated with blast furnaces were studied. Tensile tests were carried out on reinforcing bars to assess the material properties of the steel bars. Compressive tests of concrete were done on concrete core samples.

After a detailed study, it was finalised that the concrete jacketing technique would be adopted to strengthen the 59.2 m long chimney.

Source:

- Kaçın and Gürsoy [204]

7.7.35.3 FRP as a Retrofitting Solution

The strength of the member may be compromised due to the application of substandard materials in construction, due to load changes in functional, or due to seismic effects for which the structure was not originally designed. FRP composites are being used increasingly for strengthening/retrofitting work.

The advantages of FRP composites have been mentioned previously.

As a repair material, fibre-reinforced plastic (FRP) composites have many advantages compared to traditional methods. The cross-sectional dimensions of the member will not be increased, which will satisfy architectural restraints.

The sheets of FRP, with a thickness of 0.05 inch (1.3 mm) approximately, may be applied similarly to wallpapers to strengthen chimneys.

Source:

- AFZIR Retrofitting [205]

7.7.36 Rigid Pavements

Steel fibres have been used for the design of rigid pavements. The fibres are distributed during mixing in the concrete, which is desirable for the pavements. The important contribution of steel fibres in the concrete slab may be to delay and control the tensile cracking of the composite material. Furthermore, they can improve the properties of the material, such as ductility, fatigue resistance, wear and impact resistance, and will control the development of temperature and shrinkage cracks.

The studies were conducted using fibre doses of 0%, 0.5%, 1%, and 1.5%.

Steel FRC has been used for airfield pavements, road overlays, and bridge decks.

7.7.36.1 Comments

The comments from the study are as follows:

The optimum dosage was found to be 0.5% of steel fibres.
The compressive strength was about the same with the optimum dosage.
Compared to conventional concrete, the flexural strength of the material increased.
The slab thickness may be reduced with SFRC compared to IRC guidelines.

Source:

- Gawande, Pawar, Puja, and Mohod [206]

With proper dosage, polypropylene fibres may be used for rigid pavements.

Polypropylene fibres can enhance the strength of concrete. Polypropylene fibres in the mix will reduce water permeability, plastic shrinkage, settlement, and carbonation depth.

Concrete reinforced with polypropylene fibres may be used for the construction of concrete rigid pavements.

Source:

- Ali and Mohod [207]

7.7.36.2 Experimental Study on Polypropylene-Reinforced Concrete Pavement

Polypropylene fibres have been used at low volume fractions (i.e., <0.3%) as secondary temperature shrinkage reinforcement related to payments, overlays, flooring systems, canals, tunnel linings, etc.

Normally, steel and polypropylene fibres are considered for these applications.

Polypropylene is a by-product of petroleum.

Polypropylene fibres have hydrophobic characteristics and have no effect on the amount of water used for concreting.

In this study, the effects of polypropylene fibre content on concrete properties have been examined for compressive strength, split tensile strength, flexural strength, and thermal effects. The fibre amounts in the mix varied from 0%, 0.5%, 1%, 1.5%, and 2%.

7.7.36.3 Comments

The following comments may be made:

- The polypropylene-reinforced concrete exhibited better performance for all strengths.
- The compressive strength and split tensile strength increased proportionately with the increase in volume ratio of polypropylene fibres up to 1.5%. This was compared with concrete with no fibres.
- Slump reduction occurred with the increase of fibre content. Beyond a 1.5% dosage, the mix became fibrous.
- The maximum increases in compressive and splitting strengths were 34% and 40%, respectively, when compared to the mix without fibres.
- Polypropylene-reinforced concrete may be used for strong and durable applications, including rigid pavements.

Source:

- Iswarya, Tharunkumar, Vignesh, Vignesh, and Suriya Kumar [208].

7.7.37 Geo-Polymer Resin and Subsoil Injection for Soil Stabilisation

The main idea is to improve the load-bearing capacity of the soil structure below the foundation. The soil may be associated with geotechnical problems and may need strengthening to carry additional loads from the structure. Strengthening may be achieved by deep injection methods.

The problems may be connected with poor compaction or may be affected by external forces like flooding, movement, or washout. Signs may include the development of cracks in internal/external walls and problems with door and window frames, resulting in misalignment with the frames.

7.7.37.1 Technical Issue

For consolidation, when the material is injected under the foundation load, it may provide upward thrust of up to 2,500 kN/m^2, potentially lifting the building. This remedy may be achieved through a single application of a deep injection technique. Furthermore, the method may eliminate the need for excessive ground excavation or structural disruption during execution.

7.7.37.2 Deep Injection Method

The injection may be done through points located at 1 m c/c in the affected area. Liquid polymer in two components, i.e., resin and catalyst, may be injected until the required lift has occurred. By using this method, reliable ground consolidation can be achieved.

7.7.37.3 Case Studies

A few projects may be briefly mentioned:

Increased load on Hoxton property.

The additional load-carrying capacity of the foundation was achieved. The soil was densified to provide this additional load-carrying capacity.

7.7.37.4 Gothic Ruin, Durham County Council

The deep injection technique was adopted for consolidation, restoration, and increasing the bearing capacity.

7.7.37.5 Eel Pie Island Innovation

Geo-polymer injection was performed for each stanchion. The soft soil required additional stabilisation to carry extra loads. Tests were conducted for verification.

7.7.37.6 Comments

The foundation weaknesses may be rectified, in many cases, by adopting the geo-polymer injection method to avoid demolition and reconstruction.

Source:

- Brierley [209]

7.7.38 Synthetic Fibre-Reinforced Marine Concrete

FRC will have good durability provided by the fibres and protected by the cement paste. The deterioration of concrete structures starts at the surfaces and the skin of the concrete, which will be the main issue in providing satisfactory performance of the material with fibres.

The use of synthetic fibres may slow down the deterioration process at the surface of concrete.

The fibres considered for concrete may include acrylic, aramid, carbon, nylon, polypropylene, and polyethylene, as already mentioned.

MSFs are capable of improving the post-cracking properties of hardened concrete.

Micro-synthetic fibres are often used on-site for ground-supported slabs to reduce plastic shrinkage and settlement cracking. The long-term effects will be important.

A study has been carried out that is particularly relevant for the effects of algae, predominantly Ulva, which may create problems for slip hazards in FRC.

From site observations, it can be noted that many fibres of varying orientations were observed at the surface of the new units, and many were also protruding from the surface.

The hairy substratum will offer more surface area to algae and provide more entry points to the interior of the concrete.

When there is an adequate presence of algae in concrete structures, they may start to absorb calcium, silica, and magnesium. The growth may be observed on both the underside and the fibre side.

Furthermore, algae can attract other bio-fouling organisms like fungi and bacteria, and the deterioration process accelerates with suitable structural conditions for the survival of one or more such organisms.

The presence of such growth at the fibre–cement interface may be 0.15–0.25 cm/day and can weaken the bond at a faster rate. Consequently, the fibres will be more exposed to environmental conditions and may allow the fibres to pop out from the surface of the cement.

The comments from the study are associated with microbial assemblages, fouling, and degradation of synthetic fibres in marine concrete.

The degradation may occur through algae attack or other mechanical means (e.g., power washing), which may not be the main cause of degradation but can activate the condition. The exposed fibres on the new concrete areas and the eroded surfaces will facilitate further algal attachment.

Further work and study on the contribution of algal collection/growth on the material and the mechanical bio-degradation of fibre-reinforced marine concrete will enhance knowledge and confidence.

Source:

- Hughes [210]

7.7.39 Further Applications of MSFs in Marine Structures

The use of MSFs has been implemented in the following marine structures:

- **Port of Tyne**

  The concrete pavement was subjected to heavy dynamic loads from freight handling equipment, high impacts from container movement, and abrasion from HGV traffic tyres. The concrete pavements deteriorated in the process. Synthetic macro-fibres have been used to replace steel fabric to mitigate this type of damage from traffic and site handling equipment tyres.
- **Port of Boston**

  The concrete pavement was damaged from heavy traffic use and corrosion of the reinforcing bars.

  MSFs were used in the concrete to repair the damaged areas. The application of MSFs provided minimal disruption to port operations during execution.

Source:

- Boggust [116]

7.7.40 Earth Reinforcing

The reinforcing system for stabilisation and improved performance of earth may be carried out by

Anchored earth	Soil nailing
Lateral earth support system	Polymer impregnation

7.7.40.1 Reinforcement Types

The types of reinforcing materials are

Strips	Planks
Grids and geo-grids	Sheet reinforcement (geo-textiles)
Composite reinforcement	Anchors

The important properties will be the coefficient of friction, durability, and properties of polymeric reinforcing materials. The proprietary reinforcing materials include glass fibre-reinforced plastic, geogrid-tensar, paragrid, paralink, and engtex.

Source:

- Jones [211]

7.7.40.1.1 Further Review and Comments on Geotechnical Applications

The composites and polymers may be used for geotechnical requirements.

These may include

7.7.40.1.2 Soil Anchoring Method

- Soil nailing
- Ground anchors
- Reinforced soil
- Friction anchoring

7.7.40.1.3 Soil Reinforcement Technique

- Embankment slope reinforcement
- Reinforced soil rafts

Polymers may be used to provide reinforcement to the soil.

Considering lightweight, foamed polystyrene polymer blocks have been used for sub-bases in roads constructed on soft sub-soils to control settlements.

Source:

- Hollaway [1]

7.7.41 Summary of Different Applications

The applications of composite materials have been successfully implemented in various projects across different countries. Only a few are selected as presented in Table 7.8, with some brief details.

TABLE 7.8
Different Applications

No.	Item and Brief Details	Use/Strengthening Materials	Country	Remarks
01.	Well-known Structures [33]	Horsehair/straw and sheep's wool	Different countries	Performed well
02.	Tunnel Lining [57,78,99,189]	Steel fibres	Netherlands, Germany	SFRC exhibited reduced crack development
03.	Rock Support, Tunnel [58]	Steel fibres, sprayed concrete	U.K.	50–100 aspect ratio was used
04.	Suspended floor slab supported on piles [11,32,62]	Steel fibres	U.K.	Crack width may be critical than deflection
05.	Pile supported Car Park [62]	Steel fibres	Belfast, U.K.	To withstand 100 kN/m^2
06.	External Cladding [85]	GRC, Single skin panels	Wharf Galleria, U.K.	Stone-like finish was required
07.	The Atlantis Hotel, External Cladding [212]	GRC façade Units	Dubai	Ribbed façade panels, lighter units
08.	Re-cladding West Chester County Courthouse [213]	GRC panels	N.Y.	External cladding
09.	Dome Structure, Catholic-Church [185]	GRC	Ukraine	22 m dia dome, 26 m above ground level
10.	Black pool's promenade [97]	Macro-synthetic fibres	U.K.	Strux 90/40 fibres used

(Continued)

TABLE 7.8 (*Continued*)
Different Applications

No.	Item and Brief Details	Use/Strengthening Materials	Country	Remarks
11.	Scrabster Harbour Sea Wall [199]	Ground granulated blast furnace slag (ggbs)	Scotland	Largest marine engineering application
12.	Underground Car Park, Tunnel Lining [184]	Macro-synthetic fibres	Sweden, Norway and Japan	Sprayed concrete
13.	Belfast International Airport, drainage channel [188]	Macro-synthetic fibres	Belfast, U.K.	Synthetic fibres replaced welded reinforcement mesh
14.	Tunnel lining, Channel Tunnel Rail Link [214]	Steel and polypropylene fibres	U.K.	Pre-cast segments, use of Pfa
15.	Tunnel linings [101]	Polypropylene fibres	Netherlands	Less sensitive to spalling
16.	Marine concrete [210]	Monofilament polypropylene fibres	U.K.	Better durability and cost benefits
17.	Industrial floors [68,196,197]	Steel fibres (SFRC)	U.K.	Joint-less, different projects
18.	Hertfordshire Warehouse [195]	Steel fibres (50 mm long 1 mm diameter)	U.K.	Large area of 60,000 m^2 excellent flatness achieved
19.	Biggest SF RC Flooring [97,195]	Steel fibres	U.K.	250,000 m^2 (approx.) area, joint-less ground bearing slab
20.	Fibres in floor screed [108]	Polypropylene fibres 60 mm long and 60 μm in diameter	U.K.	4 kg drop tests for screed crushing resistance
21.	Rigid pavement [206,208]	Steel fibres	India	As reinforcement
22.	Road Tunnel [99]	Steel/macro-synthetic fibres, sprayed concrete	Hind-Head tunnel, U.K.	Executed quickly
23.	Pre-stressed concrete Bridge strengthening [136]	CFRP plates	Skipton, U.K.	Pre-stressed concrete beams strengthened
24.	Steel bridge strengthening [138,187]	Ultra-high-modulus (UHV) CFRP plates	U.K.	Job executed without interrupting rail traffic
25.	Concrete Bridges [123,154]	CFRP plates, NSM reinforcement	Park House Bridge (U.K.), Tay Road Bridge (U.K.)	Used for different parts of the bridge
26.	Strengthening of Concrete bridges [127,154]	Carbon fibre composite plates, NSM reinforcement	Japan, USA, U.K.	Strengthening soffit, shear, top deck
27.	Concrete bridge strengthening [127,148]	NSM tapes and FRP bars	USA	Very effective in increasing flexural/shear strengths
28.	Strengthening bridge Columns [155]	Glass FRP shells	Middles-brough, U.K.	24 tapered columns strengthened on two bridges
29.	Jack-arch beams strengthening [215]	CFRP composite materials	London UG Ltd. (LUL)	Cost benefits from initial assessment
30.	Harbour City Car Park [156]	Aramid fibre sheets	U.K.	164 columns strengthened, addition of extra floors
31.	Car park Multi-storied, Merseyside Police [161]	NSM reinf. bars CFRP strips	U.K., Liverpool	Car park decks strengthened against flexure
32.	Fertiliser prill tower [139]	CFRP plates	Qatar	Specially formulated adhesive used

(*Continued*)

TABLE 7.8 (*Continued*)
Different Applications

No.	Item and Brief Details	Use/Strengthening Materials	Country	Remarks
33.	Ribbed floor slab strengthening [158]	Carbon fibre laminates	U.K.	Carbon fibre laminates bonded at the bottom
34.	Composite Metal deck Applications [28]	Fibre reinforcement	U.K.	Time saving up to 20% may be achieved
35.	Coastal Defence Project, Blackpool [198]	Glass fibre-reinforced polymer (GFRP)	U.K.	Safety and cost savings
36.	Concrete Chimney strength retrofitting [202,203,204,205,216]	CFRP composites FRP composites	USA, Switzerland, Japan USA, Europe	Lower installation and maintenance cost. To control rapid deterioration and corrosion to reinforcement
37.	Soil stabilisation [209]	Geo-polymer injection, Hoxton Property Ecl/Pie Island Innovation	U.K. U.K.	To increase foundation load bearing capacity Additional soil stabilisation
38.	Earth Reinforcement [1,211]	Strips, sheet reinf. (geotextile), composite reinf., anchors	U.K.	Bridge abutments, vaults, land reclamation, slope stability

7.8 CASE STUDY, USING CARBON FIBRE COMPOSITE WRAPPING MATERIAL

7.8.1 Introduction

The case study for the repair of an RC column using a carbon fibre composite wrapping system has been carried out in a building. This will be briefly discussed below:

7.8.2 Case Study

The case study related to the strengthening of a column in a residential building will be presented. The building is approximately 55 years old. One column of the building was damaged mainly due to the effects of carbonation, followed by corrosion of the reinforcing bars. Problems with proper cover and the quality of construction in this column may also be mentioned. A detailed visual survey was conducted for the assessment.

The strengthening of a reinforced concrete column in a residential building with FRP wrapping, along with photographs, will be presented. The building is located in Kolkata and is a two-storied residential building. The building has been reasonably maintained during its present use. However, one column at the first-floor level developed cracks just above the floor level. The building was designed by a well-known consultant and may be assumed to be adequately designed.

There were no problems at the lower level of the affected column. Other columns of the building appeared to be in good condition at the time of inspection.

The case study will focus on the repair/strengthening of the affected RC column using the carbon fibre wrapping technique, along with some repair and injection work.

The steps that were followed for the repair work will be presented.

A few photographs have been included to describe the method adopted for the repair of the column.

The supporting beams along with the slab (locally) have also been repaired. Polymer-modified repair mortar was used for the repair of the beams and slab.

7.8.2.1 Steps Followed for the Execution

The steps are described below (Figure 7.7):

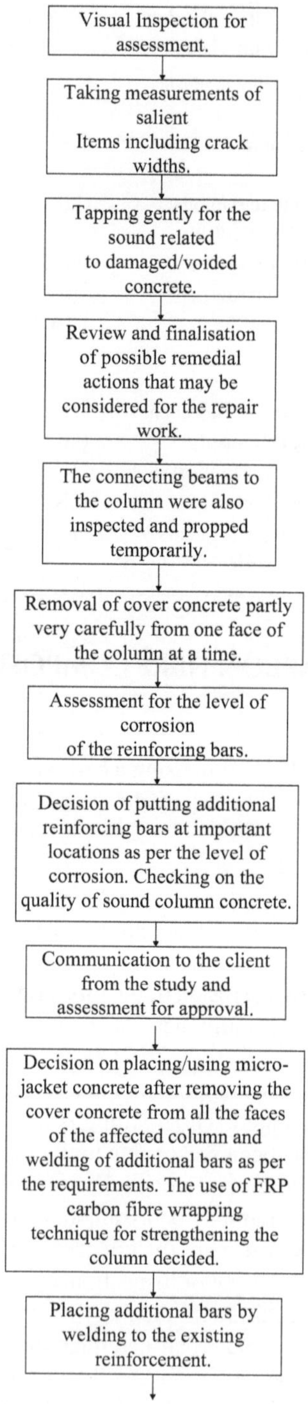

FIGURE 7.7 Steps followed for the execution.

(*Continued*)

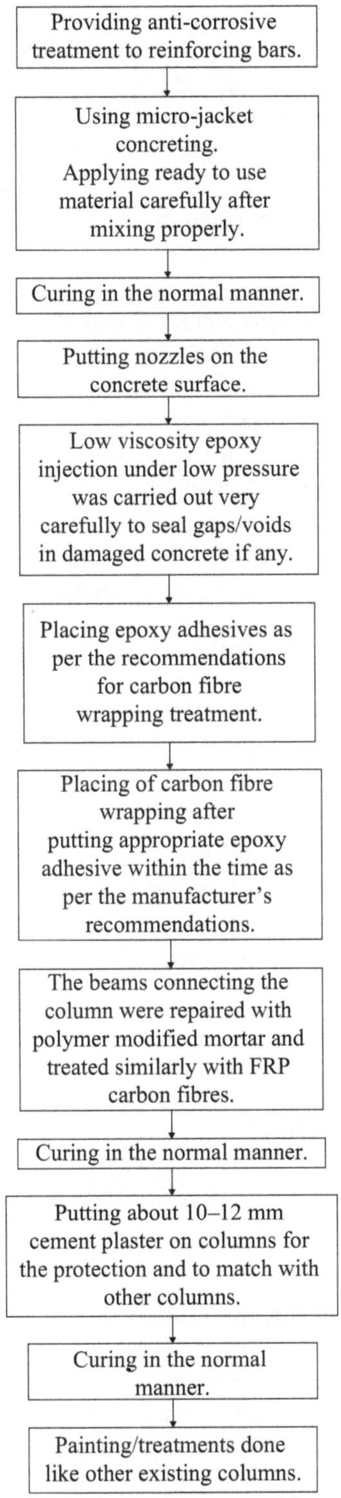

FIGURE 7.7 (*Continued*) Steps followed for the execution.

7.8.2.1.1 Repairing Materials (Major Items)

A few main repairing materials will be briefly described below:

7.8.2.1.1.1 Sika Rustoff 100 It is a single-component liquid that removes rust and prevents further rusting. It is easy to apply and may be applied with a brush.

7.8.2.1.1.2 Friazinc R This is a two-component, zinc-rich epoxy resin-based primer for steel. It may be used as a protective coating or as a primer.

7.8.2.1.1.3 Sika Top 77 It is an acrylic emulsion used to improve the adhesion of cementitious mortar and concrete.

7.8.2.1.1.4 Sika Top 122 HS This is suitable for structural repairs and is a polymer-modified, pre-batched, one-component repair mortar for hand and machine applications. One day's compressive strength at an ambient temperature of about 30°C exceed 15 N/mm^2.

7.8.2.1.1.5 Sika Rep Micro-crete-4 This is a ready-to-use, non-shrink cementitious micro-concrete. It is a pourable repair concrete material with selected cement, aggregate, and other chemicals. It is easy to mix and apply.

The strength is generated quickly, it develops a good bond with the old concrete, and it does not contain any additional chloride.

7.8.2.1.1.6 Carbon Fibre Wrapping (Sika Wrap) The wrapping material is a unidirectional woven carbon fibre-reinforced fabric that may be used for dry or wet application processes.

7.8.2.1.1.7 Sikadur-330 (Adhesive) It is a two-component epoxy impregnation resin system that may be used for Sika Wrap fabric reinforcement for the dry application method.

7.8.2.1.1.8 Sikadur-53 UF It is a low-viscosity, epoxy resin-based injection grout.

Source:

- Sika Literature [1–3,7,8,21,74]

7.8.3 Photographs from Execution

A few photographs are presented to explain the steps followed for the repair work (Figures 7.8–7.47).

7.8.4 Comments

The execution was carried out smoothly. As it was a residential house, specific timing had to be followed during the execution. There were some restrictions on making noise during the process. SIKA materials were selected for the strengthening/retrofitting work.

The applicator was experienced and had previously worked with similar materials for repair/retrofitting. The work was completed within the agreed timeframe, and the final finish/look was acceptable.

FIGURE 7.8 Cracks in the column.

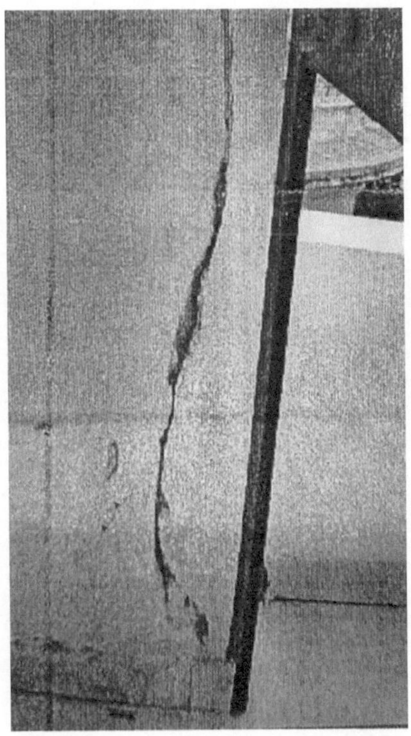

FIGURE 7.9 Cracks in the column.

FIGURE 7.10 Cracks in the column.

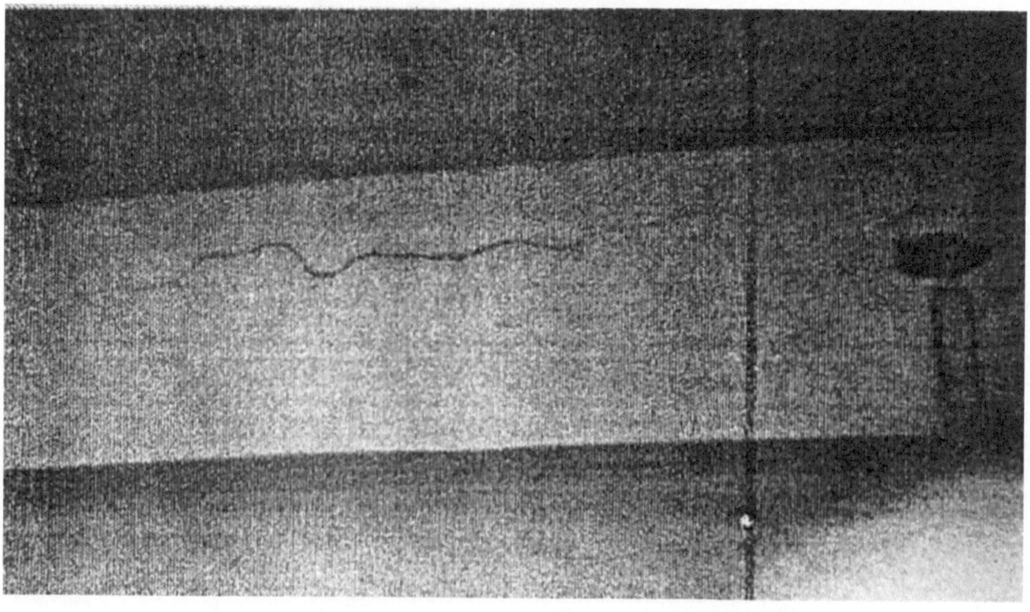

FIGURE 7.11 Cracks in the beam.

FIGURE 7.12 View after removing cover concrete.

FIGURE 7.13 View after removing cover concrete.

FIGURE 7.14 Welding reinforcing bars.

FIGURE 7.15 Welding reinforcing bars.

FIGURE 7.16 Welding reinforcing bars.

FIGURE 7.17 Welding reinforcing bars.

FIGURE 7.18 Welding reinforcing bars.

FIGURE 7.19 Welding reinforcing bars.

FIGURE 7.20 Welding reinforcing bars.

FIGURE 7.21 Mixing the material.

FIGURE 7.22 First part of concreting.

FIGURE 7.23 Carbon fibre (wrapping) strip.

FIGURE 7.24 Carbon fibre (wrapping) strip.

FIGURE 7.25 Pouring micro-concrete material.

FIGURE 7.26 Surface after concreting.

FIGURE 7.27 Nozzles for injection.

FIGURE 7.28 Nozzles for injection.

FIGURE 7.29 Preparation for carbon fibre wrapping treatment.

FIGURE 7.30 Carbon fibre wrapping treatment.

FIGURE 7.31 Carbon fibre wrapping treatment.

FIGURE 7.32 Carbon fibre wrapping treatment.

FIGURE 7.33 Exposing bars from the beam.

FIGURE 7.34 Exposing bars from the beam.

FIGURE 7.35 Beam repair/strengthening work, propping.

FIGURE 7.36 Beam repair work, propping.

FIGURE 7.37 Beam repair work.

FIGURE 7.38 Beam repair work.

FIGURE 7.39 Beam repair work in progress.

FIGURE 7.40 Beam repair/strengthening work.

FIGURE 7.41 Beam repair/strengthening work.

FIGURE 7.42 Beam repair/strengthening work.

FIGURE 7.43 Beam repair/strengthening work.

FIGURE 7.44 Beam repair/strengthening work.

FIGURE 7.45 Repair work for slab.

FIGURE 7.46 Repair work for slab.

FIGURE 7.47 Repair work for slab.

Source:

- Hollaway [1]
- Neville [2]
- Mehta and Monteiro [3]
- Ganguly [7]
- Ganguly [8]
- Swamy [21]
- Perkins [74]

REFERENCES

1. Hollaway L, *Polymer Composites for Civil and Structural Engineering*, Springer Science & Business Media, 2012.
2. Neville A M, *Properties of Concrete*, Longman, 1995.
3. Mehta P K and Monteiro P J M, Concrete: Microstructure, *Properties, and Materials*, 4th Edition, McGraw-Hill Education, 2014.
4. Sinha P K, Chapter 1, in *Composite Materials and Structures*, Composite Centre of Excellence, AR & DB, Department of Aerospace Engineering, I.I.T. Kharagpur, pp. 1–16.
5. Humphreys M F, The use of polymer composites in construction, in *Proceedings of the CIB 2003 Int'l Conference on Smart and Sustainable Built Environment*, 2003, pp. 1–9.
6. Technology Information, Forecasting and Assessment Council, Department of Science and Technology, Government of India, TIFAC, pp. 1–9.
7. Ganguly K K, Composite materials with particular reference to concrete, in *Seventh International Conference - Geomate 2017, Geotechnique, Construction Materials and Environment, Tsu Mie Japan, 21–24 Nov. 2017*, 2017, pp. 511–523.
8. Ganguly K K, *Reinforced Concrete Design: Repair/Restoration, Strengthening of R.C. Structures and Concrete Technology*, Medtech, 2020.
9. Qiao P, *Composite Materials in Civil Infrastructure (Structural Composites)*, Washington State University, Pullman.
10. Ganguly K K, Fibre reinforced concrete, in *Workshop on Advances in Special Concrete*, January 1995.
11. Vollum R, Design of steel-fibre-reinforced pile-supported slabs, *Concrete*, March 2007, Vol. 41(2), pp. 12–14.
12. Materials Technology Divisional Committee, *Concrete*, March 1971, pp. 97–98.
13. Dixon J and Mayfield B, Concrete reinforced with fibrous wire, *Concrete*, March 1971, Vol. 5(3), pp. 73–76.
14. ACI 544.2R: 2017 Report on the Measurement of Fresh State Properties and Fiber Dispersion of Fiber-Reinforced Concrete, 2017.
15. Armfield S and Birkinshaw A, Fibre potential, *Concrete*, October 2014, pp. 51–52.
16. Beddar M, Fibre-reinforced concrete: past, present and future, *Concrete*, 2004, Vol. 38(4), pp. 47–49.
17. US ARMY, Technical Report 1757-TR.
18. Kirby A, Not all fibres are the same, *Concrete*, Vol. 44(8), pp. 23–24, 2010.
19. Crowther D, Fibre-reinforced concrete - origin of the species, *Concrete*, April 2009, Vol. 43(3), pp. 22–23.
20. Gray M, Steel-fibre-reinforced concrete: breaking new ground, *Concrete*, September 2010, Vol. 43(3), pp. 54–56.
21. Swamy R N (Editor), *New Reinforced Concretes*, Surrey University Press.
22. Atkinson T, Fibres: the traditional method of reinforcement? *Concrete*, September 2010, Vol. 44(8), pp. 45–46.
23. Ganguly K K, Fibre reinforced concrete, in ACE-2011, October 2011.
24. Bussell M N, The era of the proprietary reinforcing systems, *Proceedings of the Institution of Civil Engineers - Structures and Buildings*, 1996, Vol. 116(3), pp. 295–316.
25. Bureau of Indian Standards, IS 456 (2000): Plain and Reinforced Concrete - Code of Practice.
26. Vanheuverzwijn B, Hybrid reinforcement is gaining in popularity, *Concrete*, April 2020, pp. 38–39.
27. Camiola P, The perfect union, *Concrete*, February 2019, pp. 43–44.

28. Evans S, The use of fibre-reinforced concrete in composite metal deck application, *Concrete*, October 2005, Vol. 39(10, pp. 14–15.
29. Burgoyne C, Department of Engineering, University of Cambridge, Concrete, June 2001, pp. 14–15.
30. Jowett N, Fibre reinforcement the non-metallic way, *Concrete*, May 2004, Vol. 38(5), pp. 43–44.
31. Clarke J L, Fibre-reinforced plastic reinforcement for concrete, current practice sheet number-114, *Concrete*, January 1999, pp. 15–16.
32. Atkinson T, The next generation of fibre reinforcement for external applications, *Concrete*, September 2004, Vol. 38(8), pp. 37–38.
33. Kaur P and Talwar M, Different types of fibres used in FRC, *International Journal of Advanced Research in Computer Science*, May 2017, Vol. 8(4), pp. 380–383.
34. Narayanan R and Kareem-Palanjian A S, Factors influencing the workability of steel-fibre reinforced concrete - Part I, *Concrete*, October 1982, Vol. 16(10), pp. 45–48.
35. Narayanan R and Kareem-Palanjian A S, Factors influencing the workability of steel-fibre reinforced concrete - Part II, *Concrete*, February 1983, Vol. 17(2), pp. 42–44.
36. Structural Strengthening Materials, ARROWTEX, August 2019, pp. 1–3. https://www.linkedin.com/company/arrowtex-structural-strengthening-systems-408658206/?originalSubdomain=in. Accessed Oct. 13, 2024.
37. Wikipedia, Silica Fiber, p.1.
38. Wikipedia, Kevlar, pp. 1–9. https://en.wikipedia.org/wiki/Kevlar. Accessed Oct. 13, 2024.
39. CSR Braids, p. 1.
40. Quartz Fiber, CAMEO, pp. 1–2.
41. ADFIL Construction Fibres, Fibrin XT, Product Information.
42. Frazão C, Camões A, Barros J and Gonçalves D, Durability of steel fibre reinforced self-compacting concrete, *Construction and Building Materials*, April 2015, Vol. 80, pp. 155–166.
43. Guildhall East, The Structural Engineer, December 1998, p. 461.
44. Gedge G, The use of stainless steel reinforcement in concrete, *Concrete*, May 2000, Vol. 34(5), pp. 12–14.
45. Baddoo N and Cutler P, Stainless steel in indoor swimming pool buildings, *The Structural Engineer*, 4 May 2004, Vol. 82(9), pp. 24–26.
46. James E and Skovsgaard T, Sustainable high performance concrete structures, *Concrete*, March 2004, Vol. 38, pp. 36–37.
47. Wells T, Stainless steel - the durable solution in corrosive environments, *Concrete*, April 2015, pp. 46–47.
48. Buitelaar P and Muller S, Fibres in ultra-high-performance concrete, *Concrete*, April 2020, pp. 32–34.
49. Gohnert M, Gool R V and Benjamin M, Basalt reinforcement in concrete beams in flexure, *The Structural Engineer*, February 2014, Vol. 92, pp. 38–43.
50. Yeboah D, Taylor S, McPolin D and Gilfillan R, Pullout behaviour of axially loaded basalt fibre reinforced polymer (BFRP) rods bonded parallel to the grain of glulam elements, *The Structural Engineer*, May 2012, Vol. 90(5), pp. 42–50.
51. Evans D, Fibre solution for aggressive environments, *Concrete*, October 2014, pp. 49–50.
52. Raouf Z A, Al-Hassain S T S and Simpson J W, Explosive testing of fibre-reinforced cement composites, *Concrete*, April 1976, Vol. 10(4), pp. 28–30.
53. Taurah A, Time to call in reinforcements, *Concrete*, April 2020, pp. 30–31.
54. Greenhalg J, Ten years of steel fibre concrete: a manufacturer's view, *Concrete*, September 2001, Vol. 35(8), pp. 51–52.
55. Steel Fibre Reinforced Concrete Industrial Ground Floors: An Introductory Guide, Concrete, ACIFC.
56. Destree X, Structural steel-fibre-reinforced concrete construction, *Concrete*, September 2007, pp. 23–24.
57. Greenhalgh J, Segmental linings, *Concrete*, October 2003, pp. 19–20.
58. Swannell N G, Fibre-reinforced sprayed concrete for rock support, *Concrete*, September 2007, Vol. 41(8), pp. 35–36.
59. Better concrete with fibre mesh polypropylene fibres, *Concrete*, June 1997, pp. 28–29.
60. Beckett D, Ground-supported slabs made from steel-fibre-reinforced concrete: an appraisal of punching shear, *Concrete*, September 2004, Vol. 38(8), pp. 30–31.
61. Ward P, *The Structural Engineer*, 5 January 2010, pp. 16–19.
62. Cameron G, Steel-fibre- reinforced pile-supported car park in Belfast, *Concrete*, June 2002, Vol. 36(6), p. 26.
63. Mangat P S and Halabi W, Steel-fibre-reinforced high strength concrete corbels: test results and analysis, *The Structural Engineers*, December 1996, Vol. 74(23), pp. 412–421.

64. Barnett S and Tyas A, Explosion-and impact-resistance of ultra-high-performance fibre-reinforced concrete, *Concrete*, February 2007, Vol. 41, pp. 32–33.
65. Berge B, FRC - exclusive to continental Europe or will the UK join the mix?, *The Structural Engineer*, 18 January 2011, Vol. 89(2), pp. 21–23.
66. Viney T, Steel fibres for steel decks, *Concrete*, September 2007, Vol. 41(8), pp. 38–39.
67. Destree X, Steel fibre reinforcement for suspended slabs, *Concrete*, September 2001, Vol. 35(8), pp. 58–59.
68. Williamson N, Steel fibre-reinforced concrete floors, *Concrete*, April 2011, pp. 20–22.
69. Steel Fibres - Advantages and Disadvantages, CANZAC Knowledge Centre. https://www.canzac.com/knowledge-center/steel-fibres-advantages-and-disadvantages. Accessed Oct. 13, 2024.
70. Robbins P J and Austin S A, Sprayed steel fibre concrete, *Concrete*, April 1985, pp.18–20.
71. Swamy R N and Al-Noori K, Flexural properties of steel fibre reinforced concrete, *Concrete*, June 1975, Vol. 9(6), pp. 30–31.
72. ICI Technical Committee Recommendations, Reference no. ICI-TC/1.1, *ICI Journal*, September 2014, pp. 39–43.
73. Rao A S and Rao G A, Fracture of latex modified steel fibre reinforced concrete under mode-II loading, *ICI Journal*, July–September 2017, pp. 31–37.
74. Perkins P H, Steel fibre reinforced concrete for sewer pipes, *Concrete*, March 1977, pp.16–17.
75. Mangat P S and Swami R N, Compactibility of steel fibre reinforced concrete, *Concrete*, May 1974, Vol. 8(5), pp. 34–35.
76. Martin D, Micro-silica steel-fibre reinforced concrete flooring, *Concrete*, April 2000, pp. 14–16.
77. Thooft H, Structural behaviour of steel-fibre reinforced pile-supported concrete floors, *Concrete*, September 2000, Vol. 34(8), pp. 50–54.
78. Barfoot J, Sprayed fibrous concrete tunnel support, *Concrete*, November 1983, Vol. 17(11), pp. 9–11.
79. Stainless steel fibre, fibre for concrete, *Fibrerzone*, April 2016, pp. 1–2.
80. 12th GRCA Congress in Dublin, 14–16 May 2001, in *Concrete*, April 2001, pp. 19–22.
81. Muller D, Recent developments in glass fibre reinforced concrete, *Concrete*, May 2000, Vol. 34(5), pp. 47–49.
82. Ferry R, where is glass fibre reinforced concrete going? *Concrete*, September 2000, Vol. 34(8), pp. 62–63.
83. Morrison I, Specifying GRC, Concrete, June 2010, Vol. 44, pp. 14–15.
84. Morrison I, GRC Standards and testing, *Concrete*, June 2008, Vol. 42, pp. 12–13.
85. Spence F E, GRC cladding at hays wharf galleria, *Concrete*, January 1987, Vol. 21(1), pp. 5–7.
86. Domingos J and Finzi F, GLASSFIBRE REINFORCED CONCRETE-Limit-state behaviour of GRC structural elements, *Concrete-Crowthorne*, 2002, Vol. 36(2), p. 58.
87. Bartos P J M, E-GRC: improving the appearance of concrete buildings and quality of urban environment, *Concrete*, April 2009, Vol. 43(3), pp. 25–27.
88. Faulding B, GRC in house building, *Concrete*, October 2009, Vol. 43(9), pp. 11–12.
89. The Glass-fibre Reinforced Concrete Association GRCA, *Concrete*, April 2001, pp. 19–22.
90. Ferry R, Thirty years of GRC permanent formwork, *Concrete*, March 2000, Vol. 34(3), p. 32.
91. McClelland R, Permanent formwork for composite bridge decks, *Concrete*, April 2007, Vol. 41, pp. 10–11.
92. Rooney J and Taylor S, Flexure behaviour of steel and GFRP reinforced concrete beams, *Concrete*, May 2004, Vol. 38, pp. 38–40.
93. Sadegzadeh M, Kettle R and Vassou V, The influence of glass, polypropylene and steel fibres on the physical properties of concrete, *Concrete*, April 2001, Vol. 35(4), pp. 12–13.
94. Chen P-W and Chung D D L, Concrete reinforced with up-to 0.2 vol% of short carbon fibres, *Composites*, 1993, Vol. 24(1), pp. 33–52.
95. Kitchen A, Improving concrete performance in fires, *Concrete*, September 2004, Vol. 38(8), p. 40.
96. Kitchen A, Polypropylene fibres reduce explosive spalling in fire, *Concrete*, April 2001, Vol. 35(4), pp. 40–41.
97. Perry B, Synthetic macro fibres storm to the front of coastal defence innovation, *Concrete*, November 2006, Vol. 40(10), pp. 72–73.
98. Ridout A, Macro synthetic fibres replace steel fabric in Chilean housing, *Concrete*, September 2007, Vol. 41(8), pp. 32–34.
99. Ridout A and Martin G, Macro-synthetic fibres for macro-sized projects, *Concrete*, April 2009, Vol. 43(3), pp. 15–17.

100. Kitchen A, Micro polypropylene fibres for external concrete slabs and hard-standings, *Concrete*, September 2007, Vol. 41(8), pp. 28–31.
101. Varley N, Fire protection of concrete linings in tunnels, *Concrete*, May 1999, Vol. 33(5), pp. 27–30.
102. Kitchen A, New regulations will lead to high-fibre diet, *Concrete*, September 2006, Vol. 40(8), pp. 29–30.
103. Atkinson T, Polypropylene fibres control explosive spalling in high-performance concrete, *Concrete*, November/December 2004, Vol. 38(10), pp. 69–70.
104. Wikipedia, Basalt Fibre, pp. 1–9. https://en.wikipedia.org/wiki/Basalt_fiber. Accessed Oct. 14, 2024.
105. Kirthika S K and Singh S K, Experimental investigations on basalt fibre-reinforced concrete, *Journal of the Institution of Engineers (India): Series A*, December 2018, Vol. 99(4), pp. 661–670.
106. Hathaway B, A comparison of macro-synthetic and steel-fibre-reinforced concretes, *Concrete*, October 2007, Vol. 41, pp. 43–46.
107. Carr K, Polypropylene and steel fibres combination, *Concrete*, September 2000, pp. 60–61.
108. Kitchen A, Fibres in floor screeds, *Concrete*, September 2001, pp. 54–55.
109. CementConcrete.org, Properties of Fibre Reinforced Concrete (FRC) - Types, Uses and Advantages, pp. 1–27. https://cementconcrete.org/concrete/fiber-reinforced-concrete-frc/2524/. Accessed Oct. 14, 2024.
110. The Constructor, Fibre Reinforced Concrete - Types, Properties and Advantages of Fibre Reinforced Concrete, pp. 1–13, 38–39. https://theconstructor.org/concrete/fiber-reinforced-concrete/150/. Accessed Oct. 14, 2024.
111. Witerberg R, Durability of Fibre Reinforced Concrete, Concrete Performance, Corrosion & Durability, Part 1 - Corrosion/EPC, August 2016, pp, 1–5.
112. Kitchen A, Fibres reduce marine concrete permeability, *Concrete*, April 2007, Vol. 41(3), pp. 43–44.
113. Wikipedia, Traslucent Concrete, pp. 1–3. https://en.wikipedia.org/wiki/Translucent_concrete. Accessed Sep. 28, 2024.
114. Gahrana S, Raj V, Chouhan S and Krishnia S, Application of transparent concrete in reen construction, *International Journal of Scientific & Technology Research*, April 2018, Vol. 7(4), pp. 119–122.
115. Prasad A, New Bendable Concrete that is Stronger and More Durable, Linkedin, February 15, 2018, pp. 1–5. https://www.linkedin.com/pulse/new-bendable-concrete-stronger-more-durable-ajin-prasad/. Accessed Oct. 16, 2024.
116. Boggust M, Synthetic fibres in concrete for port and dock applications, *Concrete*, July 2018, pp. 55–56.
117. Brookfield T, The use of macro polypropylene fibres in ground-bearing concrete floors, *Concrete*, April 2012, pp. 34–35.
118. Brady S, Failure investigations and the forensic process (part-I), *The Structural Engineer*, December 2013, Vol. 91(12), pp. 24–25.
119. Thomas W, Ultrasonic pulse technology, April 2020, pp. 35–36.
120. Viney T, Codes and standards for steel-fibre concrete, *Concrete*, August 2006, Vol. 40(7), pp. 18–19.
121. Clarke J L and Waldron P, The reinforcement of concrete structures with advanced composites, *The Institution of Structural Engineers*, 7 November 1996, Vol. 74(17), pp. 283–288.
122. Halliwell S, Fibre-reinforced polymers in construction – strengthening concrete structures, *Concrete*, May 2004, Vol. 38(5), pp. 41–42.
123. Clarke J, Strengthening concrete bridges with fibre composites, Concrete, October 2004, Vol. 38(9), pp. 20–22.
124. Sadka B, Strengthening bridges with fibre-reinforced polymers, *Concrete*, February 2000, Vol. 34(2), pp. 42–43.
125. Clarke J, Installing, inspecting and monitoring FRP-strengthened concrete structures, *Concrete*, February 2002, Vol. 36(2), pp. 34–36.
126. Stone D, Tumialan G, Nanni A and Parretti R, Near-surface mounted FRP reinforcement: application of an emerging technology, *Concrete*, May 2002, Vol. 36(5), pp. 42–44.
127. *Polymer Composites as Construction Materials*, A Patterns in Innovation project sponsored by DTI.
128. Matthys S and Taerwe L, FRP reinforcement: developments in Europe, *Concrete*, June 2001, Vol. 35(6), pp. 20–21.
129. Jeanes C, The FRP advantage, *Concrete*, May 2000, p. 20.
130. Hackett M, Concrete strengthening with fibre composites, *Concrete*, March 2005, Vol. 39(3), pp. 20–21.
131. Parvin A and Brighton D, FRP Composites strengthening of concrete columns under various loading conditions, Polymers, April 2014, Vol. 6(4), pp. 1040–1056.
132. Abdelrahman K and El-Hacha R, Cost and ductility effectiveness of concrete columns strengthened with CFRP and SFRP sheets, Polymers, May 2014, Vol. 6(5), pp. 1381–1402.

133. Bakis C E, Bank L C, Brown V L, Cosenza E, Davalos J F, Lesco J J, Machida A, Rizkalla S H and Triantafillou T C, Fibre reinforced polymer composites for construction –state of the art review, *Journal of Composites for Construction*, May 2002, Vol. 6(2), pp. 73–87.
134. Palmer B, Wrapping It Up: Strengthening Concrete Members with Carbon Fiber Fabric and Epoxy Composites, Concrete Construction, February 19, 2010. https://www.concreteconstruction.net/how-to/materials/wrapping-it-up_o (accessed Sep. 28, 2024).
135. Sika Information, Carbon Fibre Wrap.
136. Lane J, Fashole-Luke S and Skwarski A, Strengthening with carbon fibre, *Concrete*, January 1998, Vol. 32 No. 1, pp. 14–15.
137. Walker P, Carbon fibre-reinforced polymer plates as shear strengthening for beams, *Concrete*, May 2004, Vol. 38(5), pp. 34–36.
138. Dodds N, Strengthening a bridge using carbon fibre reinforced plates, *The Structural Engineer*, 4 March 2003, Vol. 81(5), pp. 17–19.
139. Luke S, Godman S and Hamnevoll H, Strengthening the Qafco prill tower, Qatar, with FRP plates, *Concrete*, February 2002, Vol. 36(2), pp. 32–33.
140. Uusitalo K, *Department of Product and Production Development*, Chalmers University of Technology, Gothenburg, Sweden 2013.
141. Szczerbicki J R and Bloodworth A, An advanced polymer composite and concrete slab system, *The Structural Engineer*, 17 August 2004, Vol. 82(16), pp. 12–15.
142. Plum D R, The behaviour of polymer materials in concrete repair and factors influencing selection, *The Structural Engineer*, September 1990, Vol. 68(17), pp. 337–345.
143. Carr K, Designing with high-performance polymer fibres, *Concrete*, September 2001, Vol. 35(8), pp. 48–50.
144. NPTEL: Textile Engineering - High Performance and Specially Fibres. NPTEL. https://archive.nptel.ac.in/courses/116/102/116102006/. Accessed Sep. 28, 2024.
145. Ahmed S F U and Mihashi H, A review on durability properties of strain hardening fibre reinforced cementitious composites (SHFRCC), *Cement and Concrete Composites*, May 2007, Vol. 29(5), pp. 365–376.
146. FRP in Building Construction, Civil Engineering and Construction Review, February 1991.
147. Fy-Composites, Structural Composite.
148. Chong P, A novel bridge stitching scheme over the M20 J4 in Kent, *Concrete*, July 2007, Vol. 41(6), pp. 30–31.
149. Marsh B K, Creep of fibre-reinforced corrugated concrete roofing, *Concrete*, June 1982, pp. 13–15.
150. Donnelly E, Keeping a wrap on it, *Concrete*, May 2006, Vol. 40, pp. 33–34.
151. Smale K and Smith I, *Plastic Fantastic*, New Civil Engineer, ICE, February 2016, p. 35.
152. Clarke J, Design guidance for strengthening concrete structures using fibre composite materials, *Concrete*, January 2001, Vol. 35(1), pp. 16–18.
153. Mays G and Calder A, External plates extend reinforcement's reach, *Concrete*, November 1988, Vol. 22(11), pp. 25–28.
154. Loudon N, Strengthening highway structures with fibre-reinforced concrete, *Concrete*, June 2001, Vol. 35(6), pp. 16–17.
155. Pinzelli R, FRP shells for strengthening bridge columns, *Concrete*, June 2001, Vol. 35(6), pp. 22–23.
156. Russell P and Modi A, Wrapping the Harbour City car park, *Concrete*, June 2001, Vol. 35(6), p. 24.
157. Belarbi A, Confinement of FRP-jacketed reinforced concrete columns, *Concrete*, June 2001, Vol. 35(6), pp. 25–26.
158. Farmer N, Using FRP composites to strengthen concrete structures externally, *Concrete*, February 2000, Vol. 34(2), pp. 47–49.
159. Garden H N, Holloway L C and Thorne A M, A preliminary evaluation of carbon fibre reinforced polymer plates for strengthening reinforced concrete members, *Proceedings of the Institution of Civil Engineers - Structures and Buildings*, May 1977, Vol. 123(2), pp. 127–142.
160. Couper A, BAA Glasgow International Airport - Repair and strengthening of the East Pier, *Concrete*, March 2011, Vol. 45, pp. 52–54.
161. Farmer N, Near-surface-mounted reinforcement preserves car park for Merseyside Police, *Concrete*, October 2003, Vol. 37(9), pp. 36–38.
162. Hota G and Liang R, Advanced fibre reinforced polymer composites for sustainable civil infrastructures, in *Proceedings of the International Symposium on Innovation & Sustainability of Structures in Civil Engineering*, 2011.

163. Arya C, Clarke J L, Kay E A and O'Regan P D, TR 55: design guidance for strengthening concrete structures using fibre composite materials: a review, *Engineering Structures*, July 2002, Vol. 24(7), pp. 889–900.
164. *TR55 Design Guidance for Strengthening Concrete Structures Using Fibre Composite Materials*, 2nd edition, Concrete Society.
165. Büyüköztürk O and Yu T-Y, Understanding and assessment of de-bonding failures in FRP-concrete systems, in *Seventh International Congress on Advances in Civil Engineering,* October 11–13, 2006, Yildiz Technical University, Istanbul, Turkey, 2006, pp. 11–13.
166. Wikipedia, De-lamination, pp. 1–4. https://en.wikipedia.org/wiki/Delamination. Accessed Sep. 29, 2024.
167. Büyüköztürk O, Gunes O and Karaca E, *A Fracture Based Approach to Understanding De-Bonding in FRP Bonded Structural Members*, National Science Foundation, IST Group 2004, pp. 1–6.
168. Porter A, The effect of surface profile on flexural strengthening of R.C. Sections with FRP, *The Structural Engineer*, March 2003, Vol. 81(6), pp. 17–19.
169. Teng J G, Fernando D, Yu T and Zhao X L, Debonding failures in CFRP-strengthened steel structures, in *3rd Asia-Pacific Conference on FRP in Structures, APFIS 2012*, 2012.
170. Djamaluddin R, Hijriah, Irmawati R, Faharuddin and Wahyuningsih R T, De-lamination mechanism of GFRP sheet bonded on the reinforced concrete beams, *MATEC Web of Conferences*, January 2019, Vol. 258, pp. 1–6.
171. Kripakaran P and Armitage P, Detecting de-lamination in FRP retrofitted concrete structures using nonlinear ultrasonics, *The Structural Engineer*: Journal of the Institution of Structural Engineer, June 2014, Vol. 92(6), pp. 46–50.
172. Mohamed Ali M S, Oehlers D J and Bradford M A, Shear peeling of steel plates adhesively bonded to the sides of reinforced concrete beams, *Proceedings of the Institution of Civil Engineers - Structures and Buildings*, August 2000, Vol. 140(3), pp. 249–259.
173. Kang T H K, Howell J, Kim S and Lee D J, A state-of-the-art review on de-bonding failures of FRP laminates externally adhered to concrete, *International Journal of Concrete Structures and Materials*, July 2012, Vol. 6(2), pp. 123–134.
174. Swamy R N, Hobbs B and Roberts M, Structural behaviour of externally bonded steel plated RC beams after long term exposure, *The Structural Engineer*, August 1995, Vol. 73(16), pp. 255–261.
175. Swamy R N, Mukhopadhyay P and Lynsdale C J, *The Structural Engineers*, January 1999, pp. 19–30.
176. Moss R M, Behaviour of large diameter columns reinforced with wrapped glass fibre reinforced plastic tubing, *The Structural Engineer*, 17 July 2001, Vol. 79(14), pp. 23–31.
177. Sreenivas Rao A and Appa Rao G, Fracture of latex modified steel fibre reinforced concrete, *ICI Journal*, July–September 2017, pp. 31–37.
178. Taylor S, Robinson D and Sonebi M, Basalt-fibre reinforced polymer reinforcement, *Concrete*, April 2011, Vol. 45(4), pp. 48–50.
179. Attree G, Macro-synthetic fibres, long-term behaviour and importance of CE marking, *Concrete*, September 2011, Vol. 45(9), pp. 55–56.
180. Schleyer G K, Barnett S J, Millard S G, Wight G and Rebentrost M, UHPFRC panel testing, *The Structural Engineer*, 6 December 2011, Vol. 89(23–24), pp. 34–40.
181. Byfield M and Paramasivam S, Blast loading on concrete frame structures, *Concrete*, February 2007, Vol. 41(1), pp. 34–35.
182. Khoury G A and Algar S, Fibre-reinforced concrete creep tests, *Concrete*, October 2008, Vol. 42(9), pp. 40–42.
183. Perry B, Synthetic macro-fibre concrete in composite steel deck floor construction, *Concrete*, September 2006, Vol. 40(8), pp. 26–29.
184. Ridout A, The use of macro synthetic fibre reinforcement in underground civil construction, *Concrete*, April 2007, Vol. 41(3), pp. 39–41.
185. Peter I D and Crocker I A, GRC at height-creating a cathedral dome, *Concrete*, May 2014, pp. 47–48.
186. Faulding B, Glass-fibre-reinforced concrete-coming of age, GRC facade cladding, *Concrete*, October 2012, Vol. 46(10), pp. 38–40.
187. Research Focus, August 2001.
188. Martin G and Donnelly C, Macro-synthetic fibres take off at Belfast International Airport, *Concrete*, September 2010, Vol. 46(10), pp. 42–43.
189. Court D, Channel tunnel rail link, *Concrete*, October 2003, pp. 9–10.
190. Chana P, Improving fire safety in tunnels: the concrete pavement solution, *Concrete*, November/December 2004, Vol. 38(10), pp. 65–66.

191. Richter R, Grosse C U, Juknat M and Dehn F, Tunnel fires: preventing explosive spalling of concrete, *Concrete*, July 2015, Vol. 49(6), pp. 26–28.
192. Greenhalg J, Green home, *Concrete*, April 2013, pp. 37–38.
193. Simson D, Suspended floors for industrial use: an overview, *Concrete*, February 2004, Vol. 38(2), pp. 16–18.
194. Eddy D, UK'S biggest ever SFRC flooring project, *Concrete*, March 2008, Vol. 42(2), pp. 24–25.
195. Eddy D, CONCRETE ON SITE-SFRC floor slab for Hertfordshire ware house, *Concrete-Crowthorne*, March 2006, Vol. 40(2), p. 18.
196. Hulett T, Joint-less floors, *Concrete*, May 2013, pp. 22–23.
197. Heiden B, Holschemacher K, Lober P and Barragan B, GRC for slabs on ground, *Concrete*, July 2015, Vol. 49, pp. 48–50.
198. Cunnigham L, Burgess A and Spitz C, Composite reinforcement for coastal defence project, *Concrete*, May 2008, Vol. 42(4), pp. 22–24.
199. Jenkins T, CONCRETE AND WATER-Durable solution for Scrabster Harbour sea wall, *Concrete-Crowthorne*, November 2006, Vol. 40(10), p. 84.
200. Polymer Composites as Construction Materials, Strengthening, Application Summary Sheet-7, pp. 1–8.
201. Shams A, Non-metallic reinforcement for the third Bosphorus Bridge, *Concrete*, August 2018, pp. 15–17.
202. Alkhrdaji T and Galati N, Strengthening of RC chimneys with FRP composites, in *The 7th International Conference on FRP Composites in Civil Engineering*, International Institute for FRP in Construction, pp. 1–6.
203. Gedalia B and Sanchez J, Chimney structural strengthening and retrofit, *Power Engineering*, June 2010, Vol. 114, p. 24.
204. Kaçın S and Gürsoy G, Structural retrofitting of a reinforced concrete chimney, *International Journal of Advanced Engineering Research and Science (IJAERS)*, December 2017, Vol. 4(12), pp. 108–113.
205. Chimney, Afzir Retrofitting Company, pp. 9–12.
206. Gawande D S, Pawar A A, Puja S and Mohod M V, Innovative use of steel fibres in rigid pavements, in *International Conference on Science and Technology for Sustainable Development (ICSTSD)*, 2016, pp. 73–78.
207. Ali A A and Mohod M V, A review on effect of fibre reinforced concrete on rigid pavement, The International Journal on Engineering, Science and Technology, 2015, Vol. 1, pp. 222–227.
208. Iswarya N, Tharunkumar K, Vignesh T P, Vignesh D, Suriya Kumar M, Experimental study on polypropylene reinforced concrete pavement, *International Journal of Latest Engineering Research and Applications (IJLERA)*, March 2017, Vol. 2, pp. 98–102.
209. Brierly M, Geo-polymer resin and sub-soil injection for soil stabilisation, *Concrete*, February 2004, Vol. 38(2), pp. 20–21.
210. Hughes P, Fibres-a new mechanism for accelerated degradation of synthetic-fibre-reinforced marine concrete, *Concrete-Camberley*, September 2012, Vol. 46(9), pp. 18–20.
211. Jones C J F P, *Earth Reinforcement & Soil Structures*, Thomas Telford, ASCE Press, 2010.
212. Hanley M, Designing the GRC façade for the Atlantis Hotel, *Concrete*, July 2010, Vol. 44(6), pp. 24–26.
213. McDougle Ed, Re-cladding the Westchester County Courthouse, New York with GRC panels, *Concrete*, July/August 2001, Vol. 35(7), pp. 10–12.
214. Craig R, Pre-cast tunnel linings in the Channel Tunnel Rail Link, *Concrete*, October 2004, Vol. 39(8), pp. 39–42.
215. RESEARCH FOCUS, Carbon fibre Composites, Issue No. 46, August 2001, pp. 1–8.
216. Ballinger C, Meada T and Hoshijima T, Strengthening of reinforced concrete chimneys, columns and beams with carbon fibre reinforced plastics. *Special Publication*, 1993, Vol. 138, pp. 233–248.

8 Brief Comments on Building Materials

8.1 TO DESIGN STRUCTURES WITH A BUILDING MATERIAL

Designing with a building material in a project is associated with various factors. This may be particularly critical when the material is an alternative to traditional building materials. One must examine the issues to ascertain the details of the alternative material required for the design. The alternative material must meet the design requirements. Familiarity with traditional materials is especially important when the alternative material is used alongside traditional materials in the same project.

Guidance from standard codes will be necessary for the designer to carry out the design work. Additionally, checks are necessary to determine whether the same material has been used before and what information is available from past performance. A few trial results from the past may be desirable to build confidence. Without knowledge of past performance, it will be difficult to convince the client to accept the alternative material.

There are different aspects to satisfy in the design, as mentioned. It may be possible that one particular alternative material cannot satisfy all the requirements of the design. In such cases, composite actions may need to be considered by applying other suitable materials to fill the gaps and meet the design issues. Such combinations have been employed even with traditional materials like concrete, where steel reinforcing bars are used to address the weaknesses of concrete under tensile stresses when developed in a concrete element subjected to applied or self-loading. Steel reinforcing bars are normally placed in the tensile zones of concrete elements. The bars may also be required in the compression zone to meet design requirements or to address changes in stresses due to applied loading, particularly from wind or seismic effects.

It is recommended that two traditional materials should be selected for comparison with the alternative materials and to explain the applications, in different areas, of projects where these materials have been used successfully. Photographs have been included for clarity.

Some basic and brief details are provided for the two traditional materials to demonstrate the fundamental issues of the design. The limit state method of design is applied for the two traditional materials selected for comparison.

The items for the design checks may vary for a particular material. The checks for the limit states will also be applicable to the alternative materials. The construction methods may differ with the materials. The present guidance from BIS may not be adequate for the alternative materials.

The alternative materials must be examined thoroughly in the context of the design requirements, as highlighted, to replace the traditional materials and to ensure that the selected alternative materials can be suitably used in their respective applications, employing composite actions if required.

It is interesting to note that some alternative materials, along with other materials (i.e., composite action), have already been used successfully in the past.

It is hoped that the design steps/guidance from concrete and steel materials will be of great help in using the alternative materials efficiently for a structure when sufficient guidance is not available from the code for a particular composite material. The design checks for assessing the stresses under different load conditions will be the most critical factors in ensuring the stability and durability of the structure during its service life.

DOI: 10.1201/9781003319108-8

A review will still be necessary in light of modern design developments, particularly to meet durability and sustainability criteria with nominal maintenance during the service life of the structure. If we are successful in applying the alternative materials (with composite actions) and utilising other developments and our experience with the products, it will significantly help to reduce the use of popular traditional materials that would otherwise have been employed in the projects. The depletion of the constituent materials along with the saving of energy will be possible in that case. We must remain optimistic and continually improve the use of alternative materials based on our experience from different applications.

Regular inspection, supported by maintenance repairs, may be required from time to time. Concrete repair methods are also mentioned briefly as a reference.

We must strive to use alternative materials wherever possible in the future.

There are other items associated with design and construction that are important for success. These may be mentioned as follows:

- Computer-aided design and calculations
- Structural detailing
- Education and training of engineers
- Façade engineering
- Hybrid construction
- Construction materials
- Construction work
- Diverse developments in concrete
- Site supervision
- Site inspection of structural steel work
- Project management
- Risks in construction, etc.
- Safety/security
- Repair and restoration
- Communications

The above items are briefly mentioned in the remaining chapters to cover the challenges that will be faced by design and construction engineers, along with others, particularly in meeting the expectations of the public and developments.

8.2 PHOTOGRAPHS

A few photographs are presented as examples to demonstrate the aesthetics, simplicity, and flexibility achieved with traditional concrete materials, which have been used successfully in projects.

Figures 8.1–8.5 may be referred to.

FIGURE 8.1 Concrete structure.

FIGURE 8.2 Concrete structures.

FIGURE 8.3 Concrete building.

FIGURE 8.4 Water tower in concrete.

FIGURE 8.5 Concrete buildings.

9 Concrete as a Building Material

9.1 INTRODUCTION

Engineers and architects have extensively used concrete to construct innovative structures over the last hundred years. Understanding the weaknesses of concrete in tension, pre-stressing techniques have been developed to construct large-span structures, including bridges.

Concrete has become the most indispensable material for construction and is used in greater quantities than any other man-made materials. Our current consumption rate of cement is significantly lower than that of advanced countries. Furthermore, with the increase in population, the requirements for various developmental projects are likely to increase in the future. Hence, concrete may need to be used much more extensively in the years to come. This demand will be further increased by the repair and rehabilitation work required to maintain existing structures for a longer period.

The weaknesses of reinforced concrete, as observed over the years, have been briefly mentioned. Suggestions will be made to address some of these weaknesses during the lifespan of the structure.

Regular inspection and maintenance will be closely associated with the service life of concrete structures.

Source:

- Ganguly [1]

9.2 AIMS OF STRUCTURAL DESIGN

A good design will be connected with simplicity, unity, and meeting needs. Simplicity can be considered a hallmark of excellence. Simple does not necessarily mean elementary. A structure that is properly designed and constructed may still present many complications in complex analysis. Unity refers to a single entity, and when the whole and its parts are well designed, unity can be realised. Necessity is a key factor; the required elements in the design must be duly taken care of. It is beneficial when necessary needs have been properly considered in the design. Otherwise, the result may be wasteful, superfluous, and artificial, leading to a missed main issue. The characteristics mentioned above can be realised in a good design developed to meet its function, safety, economy, and environmental protection. These are expectations the minimum from the designer. However, efforts to achieve excellence and improvement must continue.

9.2.1 Design and Analysis

The structural design will ensure that the structure satisfies functional requirements, covering support of loads, watertightness, fire resistance, etc. After construction, the structure must have adequate resistance to weathering, ageing, misuse (to some extent), and protection against other causes of deterioration.

Once the design items are finalised, proper structural analysis can be carried out. Structural analysis will confirm integrity and stability and identify items required for construction. The support conditions, especially under accidental loading, will also need to be checked.

DOI: 10.1201/9781003319108-9

9.2.2 Safety

The following may be mentioned:

- The structure will be able to support its own weight and other applicable loads to meet requirements.
- The structure will be stable in all directions against applied loads.
- The sub-structure, including the foundation, will support loads based on the type of sub-soil at depth.
- The structure will be able to support loads within acceptable limits of deformation and movement.

9.2.3 Responsibilities

The designer has responsibilities towards both his profession and society.

The former is associated with his skills, integrity, and ability for innovative work.

The responsibility to society involves minimising construction risks and ensuring minimal waste of society's resources.

Source:

- Cambell [2]

9.3 DESIGN ISSUES FOR THE STRUCTURES

The structural design must be carried out for the elements within a structure to accommodate the loads to which the structural elements will be subjected. These loads will include dead, imposed, wind, seismic, and any other applicable loads. The load combinations will be important for assessing the critical cases for the design. The analyses may be performed using software packages for different load cases. The material properties and permissible stresses are to be applied as per the guidance of IS codes. The design items for the two materials, i.e., concrete and steel, have been mentioned for clarity. A few types of structures are provided as examples.

9.3.1 Limit State Design

Limit state design is typically followed as per the guidance of IS codes. Essentially, the limit states to be followed are

- Collapse
- Limit states of serviceability
  Deflection
  Cracking and others – vibration, fatigue, durability, fire, etc.

The purpose of design is to ensure that the structures being designed will not become unfit for their intended use, i.e., that they will not reach a Limit State. The essential basis of the design method, therefore, is to consider each limit state and provide a suitable margin of safety. To obtain values for this margin of safety, it is proposed that probability considerations should be applied, and the design process should aim at providing acceptable probabilities to prevent the structure from becoming unfit for use throughout its specified life.

Accepting the fact that the strengths of construction materials may vary, along with the loads on the structure, two partial safety factors must be applied. One will be for the material and is designated γ_m, while the other for loading is termed γ_f.

9.3.2 Characteristic Strength of Materials

The term characteristic strength refers to the strength of material below which not more than 5% of the test results are expected to fall.

9.3.2.1 Materials

$$f_d = f/\gamma_m$$

where

f_d = design strength of the material
f = characteristic strength of the material.
γ_m = partial safety factor appropriate to the material and the limit state being considered for limit state of collapse.

γ_m = 1.5 (for concrete) and 1.15 (for steel)

9.3.2.2 Loads

$$F_d = F \cdot \gamma_f$$

where

F_d = design load
F = characteristic Load
γ_f = partial safety factor appropriate to the nature of the loading and the limit state being considered.

Partial safety factor γ_f for loads under different load combinations is given in IS codes.

Design charts may be used for the design of the structural elements.

Source:

- Ganguly [3]

9.3.2.3 Resilient Building Design

The issues related to resilient building design have been discussed in Chapter 12.

9.4 ADDITIONAL COMMENTS ON CONCRETE DESIGN AND CONSTRUCTION

Some basic and related items are listed below:

9.4.1 Composition of Reinforced Concrete

Cement, aggregate, water, reinforcement, admixtures, concrete mix design, mixing, placing and compacting concrete, curing, etc.

9.4.2 Properties of Reinforced Concrete

Weight, durability, strength and variation with age, elasticity, shrinkage, creep, fire resistance, temperature stresses, movements, joints, etc.

Source

- Faber and Mead [4]

9.5 REINFORCED CONCRETE DESIGN

A few items may be summarised as follows:

9.5.1 Design and Analysis

Loads – dead load, imposed load, wind load, seismic, snow, dust, any special loads, method of analysis, 2D-3D analyses may be required. A suitable software package is often used, limit state design – serviceability limit state, ultimate limit state, design for shear, design for torsion, bond, anchorage, stiffness of members, checks for cracks and widths, checks for deflection, long-term effects.

The common items for the structure to be considered in the design are

Slab design, beam design, column design, spread foundations, retaining walls, piling, bunkers and silos, concrete shell roofs, tanks, water towers and reservoirs, chimneys, pre-cast concrete in different applications, design of roads – various categories, pre-stressed concrete – this item will not be discussed further.

9.5.2 Design Guidance as Per IS Codes

Guidance will be available from IS Codes for design issues.

Source:

- Murashev, Sigalov, and Baikov [5]
- Faber and Mead [4]

9.6 FURTHER ON REINFORCED CONCRETE

Design of members for buildings

Reinforced concrete foundation

9.6.1 General

These may be related to

Isolated footing, combined footing, strap footing, raft foundation, pile foundation – different types.

Reinforced concrete columns, beams, walls, columns of different shapes, different types of beams, walls, shear walls with or without openings, concrete core structures, staircases with different types.

9.6.2 Slabs

The following will be relevant:

9.6.2.1 Classification of Floors

Ribbed site cast floors – one way, ribbed site cast floors – two-way, composite floors with pre-cast or steel beams, flat slab pre-cast floors, beam–slab in-situ construction, flat–slab in situ construction, flat–slab composite floors, selection of floor type considering different options including available time and cost.

Roof Beams, Frames Arusses, and Arches

Roof beams – different types, frame structures, trusses, arches, cable stayed slab elements, other non-concrete treatments with flexible sheets/membranes.

Single-storey Framed Buildings, Industrial Structures

Layout of buildings, lateral frames to satisfy different requirements, crane girders, roof design – especially satisfying different requirements of commissioning.

Other items for projects may include

Retaining walls, more on piling, bunkers and silos, water tanks, towers and reservoirs, industrial chimneys, roads – different types, pre-cast concrete, pre-stressed concrete.

9.6.3 Industrial Projects

A few items may be mentioned as follows:

Track hopper – application of concrete and steel, wagon tippler, concrete buildings in chemical plants, concrete buildings in water treatment plants, large storage structures, large water storage tanks, workshop buildings, pump houses, concrete ducts, water towers, bunkers, silos, other structures for containing materials, industrial chimneys, etc.

9.6.4 Other Projects

Academic buildings, hospitals, hotels, residential units for staff, school buildings, sports complex, swimming pools, etc.

9.6.5 Concrete Bridges

Reinforced concrete/pre-stressed concrete bridges, associated items like roads and pavings as mentioned.

9.6.6 Special Applications

More on space structures, lighting masts.

Source:

- Murashev, Sigalov, and Baikov [5]
- Faber and Mead [4]

9.7 INDUSTRIAL AND OTHER PROJECTS

In non-industrial projects such as university complexes, hotels/resorts, hospitals, housing projects, commercial complexes, etc., architects normally play a major role in the development of layout, building form, aesthetics, etc., of different buildings. The cost of architectural finishes along with services may also be very high in such projects. The challenges are greater here for the structural designers, and there will be pressure to reduce the dimensions of structural elements for visual and other requirements, as well as to produce economical designs for different structural schemes. The services of structural engineers are more demanding and appreciated compared to those in industrial projects.

In some cases, higher grades of concrete have to be used to keep the structural dimensions within the limits set by other requirements.

Engineers and architects depend on each other to design and construct innovative modern buildings.

Structural engineers must learn to work with the architects in harmony for creative and innovative design.

A few auxiliary and secondary structures are presented in Figures 9.1–9.3 as examples.

Source:

- Ganguly [6]

FIGURE 9.1 Auxiliary structure.

FIGURE 9.2 Secondary structures.

9.8 HYBRID CONSTRUCTION

Hybrid construction using different materials, especially the mixing of pre-cast and in-situ concrete, has been in use for some time. The structural engineers generally will need more experience in such techniques, especially for the performance of joints and their detailing. A few items, such as aesthetics, speed, function, safety, integration, buildability, confidence, and responsiveness,

FIGURE 9.3 Secondary structures.

will require very special care. Hybrid concrete construction may, in some cases, be more expensive than conventional in-situ concrete or structural steel frames with concrete decks.

A high standard of finish is possible from pre-cast concrete elements, along with high-quality, fast, and long-lasting exposed frame elements and soffits, which may be accommodated properly to the structural forms. Composite action should be considered for economy wherever possible.

Hybrid concrete construction has high potential and benefits, which have been recognised for applications by design engineers, architects, and contractors.

Source:

- Downing [7]
- Ganguly [6]

9.9 FAÇADE ENGINEERING

A few items related to facade engineering covering design detailing and retention will be briefly discussed below:

9.9.1 Masonry Facade Retention

The items mentioned below are relevant:

9.9.1.1 Responsibilities

- The appointment of a properly qualified and experienced co-ordinator related to temporary works for the total facade retention scheme is important.
- In the design many different parties may get involved during the progress of the project and they may offer comments.

- It is preferable that there is ownership of the scheme covering different stages from inception to completion of the total work.
- The outline solution and method statement will be required, covering the items by the permanent works designer.
- The scheme may be changed to a large extent by the contactor's temporary work designers.
- It is important that the responsibilities of temporary works are well maintained and identified at all times. Unplanned delays for implementation must be avoided.

9.9.2 Design for Wind

The guidance given in the IS code may be followed:

9.9.2.1 Health and Safety Items

The façade retention may present different hazards, and specific recommendations are required covering

- The risk assessments should address the issues of inspection requirements for the items.
- An experienced and competent person should be involved in the inspection, following the site-specific schedule.
- Residual hazards have to be identified as far as possible, considering the health and safety plan.

CIRIA guidance may be referred to.

9.9.2.2 Other Design Parameters for the Temporary Facade Retention System

The façade offsets and out-of-plumb conditions will generate lateral loading effects, which must be considered along with wind and other loads.

It is recommended that the temporary façade retention structures may have lateral deflection up to the limit of height/750 under loads.

9.9.3 Fixings

The proper fixing of the facade will be very important between the façade and the permanent and temporary station structures. In temporary cases, clamping of the two structures may be done using the existing openings to a large extent to eliminate damage. For masonry construction, the permanent connections may be made with bonded anchors involving an injection system, and the method of application will be suitable for masonry construction. The method may also be used for old masonry structures.

There will be concerns regarding the performance of resin, especially under fire and long-term creep effects of the material.

9.9.4 Monitoring

The monitoring techniques and the methods of recording by the team must be established at the early stage.

The retention structures may be connected considering the movements due to wind. The movements may be due to thermal affects, which may vary to a large extent on a daily basis, and also due to ground movement.

Source:

- Lazarus [8]

9.9.5 More on Facade

The façade is the envelope of the building, and it is generally connected with all exposed walls, roofs, and floors. The engineering relates to various aspects and parts of the envelope design to accommodate and provide an innovative look in the end. Façade engineering presents more challenges to structural engineers to maintain the aesthetic appearance of the buildings alongside the safety of the cladding system adopted. The design problems are complex, and the design detailing may fail during or shortly after construction. Often, the façade design problems are associated with leakage. The initial watertightness tests may suggest that the treatment will require either remedial work or design changes in wall areas.

The interaction between the cladding and the main structure is very important and is primarily related to

- Load transfer; apart from vertical loads, horizontal loads from wind and seismic effects, including settlement, will also be applicable.
- Accommodation of floor movement.
- Consideration of other movements during the service life.

The design of facade engineering will need study and knowledge of architectural elements, structural detailing, material performance, building physics, material engineering, construction processes, and special knowledge of facade covering joining and sealing technology.

Structural engineers must invest time in learning the techniques of hybrid construction in order to ensure good performance.

The use of stones and glass curtain walls with different types of fixing arrangements is quite common. The complexity of façade design must be understood, especially when glass and thin stone panels are adopted. In recent years, the use of coloured aluminium sheets in façade design has also gained popularity. Other materials include metals, plastics, timber, ceramics, polymers, and fabrics. The external envelope endures most of the wear; thus, the cladding or roof will be subjected to risks of premature failure well before the service life of the structure, as mentioned. One must be careful about leakage problems through the joints. The use of glass and stone panels in façade systems may be regarded as a concern. The provision for cleaning and maintenance of the façade materials should be given due importance in the design.

The guidance provided by the Institution of Structural Engineers may be followed.

Facade material is evolving to meet design requirements. Working with new materials can be considered a challenging, innovative, and satisfying activity for designers.

Higher emphasis has been placed on tests for water and wind resistance.

The following issues will also be important:

- The role of workmanship in quality
- The effects of poor project design and improper application of the system components.

Proper construction will become more complex in the future. It will require competent and experienced personnel for applications involving new materials, methods, and systems

The building envelope may cost about 15%–35% of the total building cost, excluding land. The façade primarily needs to control elements such as the internal environment of the building. This includes thermal performance, comfort, lighting, acoustics, and ultimately, the appearance.

Facade engineering is progressing rapidly, as mentioned, adding more value to the building and its design and construction, which is becoming more complicated in the process.

A façade with a glass curtain wall is shown in Figure 9.4 as an example.

Source:

- Ledbetter [9]
- Ganguly [6]

FIGURE 9.4 Facade with glass curtain wall.

9.10 REVIEW OF WHAT'S BEHIND THE FACADE

Facade engineering has become a very important issue in our design activities. Initially, building envelopes served merely as protection from the climate. However, the requirements for envelopes have changed in recent years. Apart from the basic requirements, building envelopes must also control the internal environment. The modified requirements may include acoustic performance, blast resistance, lighting, and glare to accommodate new activities within the building. The function of the building can be described as a skin acting as a filter.

The required performance of the facade may be summarised as follows:

- **To Keep Out** – Rain, wind, snow (as applicable), noise, and intruders
- **To Keep In** – Heat, temperature, occupants, privacy, and data
- **Let In** – Ventilation, daylight, views, and occupants
- **Let Out** – Extra heat and moisture

The façade can create an impression of the building and its occupants and may create a conflict between the art and science of the building.

Buildings in modern times are generally dependent on energy to condition the internal space and may not always provide the primary function, i.e., delivering a comfortable internal environment.

To control the internal environment mechanically, building services engineers have taken on greater responsibility. Their main tasks are related to electrical and mechanical equipment. Experts are suggesting the requirement of building physicists in the team to give advice during the initial stages of complex buildings. It seems that this practice is already being adopted in some European countries.

In a small space of 200 mm in thickness, many requirements, as mentioned above, must be accommodated while providing a competent structure and presenting an attractive appearance.

9.10.1 Building Physics

The physics of the building is closely related to its structural form. Acoustic performance is well associated with the form and mass of the walls. Thermal performance, heat transfer, and condensation are affected by the structural form and can be dominated by it. The structural penetrations of the building envelope will need special consideration and proper detailing.

9.10.2 Structural Aspects

The increasing use of glass, plastics, and thin stone is included in the design, as stated. The application of double-skin facades, ventilated rain screens, and large shading devices has changed the loading patterns for buildings. Existing design practices must accommodate larger panels, fewer joints, more slender frames, and frameless solutions.

9.10.3 Weathering and Durability

Whole life costing and environmental impact are the main issues for building design.

Refurbishment of buildings will be necessary when the building envelope is no longer watertight, and some aspects of its performance have deteriorated to an unacceptable level compared to advancing standards. The whole life cost can be reduced if the wall is designed to last longer. Environmental impact may increase if premature failure occurs, such as the failure of a seal or gasket, which can shorten the service life of the wall.

Maintenance and repair costs will depend on the wall design, materials used, type of construction, component sizes, and access arrangements. Adequate and proper assessment of all these factors at the design stage can allow for durable and robust types of construction, which may also simplify and reduce maintenance and repair requirements.

Facade engineering is a complex field of activities, covering many traditional disciplines. The façade design requires detailed knowledge of materials, building physics, structure, manufacturing, and construction techniques to be adopted. There is a need for individuals with general knowledge of facade design.

Efficient facade design and construction are important for making buildings energy efficient. Furthermore, it will extend the service life of the building and help control/reduce the whole life impact of the buildings. It will be helpful to have more persons who understand façade design, performance, and construction. Additionally, it will require knowledge of items such as glazing, finishes, and sealing.

The facade design team should consist of an architect, structural engineer, building physicist, and facade engineer. The constructor and manufacturer may be associated with the design team from the very beginning.

Source:

- Ledbetter [10]

9.11 IMPROVING FENESTRATION AND CLADDING PERFORMANCE

Guidance from the Fenestration and Cladding Engineering Technology Scheme should be considered, as it deals with the design, manufacture, erection, and maintenance of building envelopes.

The following items are connected:

Glass and glazing, advanced glazing, rain-screens, envelope sealing, durability, corrosion and finishes, buildability, structural systems, stone cladding, structural glazing, thermal properties, procurement, building comfort, lighting.

Source:

- Research Focus [11]

9.12 WEAKNESS OF CONCRETE

This topic has been discussed in Chapter 14.

Source:

- Ganguly [1]

9.13 COMMON PROBLEMS IN CONCRETE STRUCTURES

This item has been covered in Chapter 14.

Source:

- Ganguly [3]

9.14 MAJOR PARAMETERS FOR DURABLE CONCRETE

Some important items for the production of durable concrete structures are summarised below:

- **Cement** – Good quality cement must be selected for construction. The chemical composition of the cement is important for alkali–aggregate reaction, corrosion of reinforcement, setting time, heat of hydration, and finally, the strength of the concrete.
- **Other Raw Materials** – The other necessary raw materials for concrete, such as sand, coarse aggregate, and water, should be checked before starting the construction work.
- **Formwork** – Formwork is a very important item for successful and economic concrete construction with respect to durability. It has a large influence on the appearance and accuracy of the finished concrete structures.
- **Water-to-Cement Ratio** – In order to satisfy durability requirements, codes generally give an upper limit for water-to-cement ratio based on exposure. Better durability will be achieved if the water-to-cement ratio can be reduced further.
- **Admixtures** – When any admixture is used in concrete construction to reduce the water-to-cement ratio for strength and/or to increase the workability of the mix, it should not contain chloride and sulphate ions outside the specified limits given in the specification/standard codes of practice.
- **Fibres in Concrete** – Fibres may be used in concrete for different advantages.
- **Detailing of Reinforcement** – Proper and simple detailing should be aimed for in the design. Care should be taken to ensure that concrete can flow easily between the bars using acceptable gaps, especially when detailing beam/column joint reinforcement and splicing areas.
- **Construction Method** – Mixing, handling, placing, compaction of concrete, and curing shall be done as per the specification and good engineering practice for best results. Controlling cover and avoiding vibrations will need special care.
- **Cover to Reinforcing Bars and Embedment** – Cover to reinforcing bars shall be provided as per the construction drawings. Proper concrete cover is required to protect re-bars against corrosion, to provide adequate resistance against fire as per the design requirements, and to ensure sufficient embedding of bars to be stressed without slipping.

- **Computer Aided Design** – The idealisation, analysis, and design of the structure must be reasonable and practical, and the results from computer output must be cross-checked before acceptance.
- **Selection of Concrete Grade for Design** – The grade of concrete may be finalised after checking the availability of raw materials to be used for concreting. High strength concrete should be aimed for in the design.
- **Joints in the Structure** – Different joints, as per the design assumptions, must be properly detailed and protected. The treatment of construction joints, particularly the preparation of the joint surface before starting the next pour, is often neglected, resulting in leakage through the joints.
- **Proper Curing** – This is very important to control/reduce premature cracking in concrete.
- **Shape and Proper Drainage** – Since the deterioration of concrete may occur due to the ingress of moisture/water through the cracks, the structure should be designed, wherever possible, to minimise ponding of water or exposure to moisture.
- **External Factors, Environmental Conditions** – Most of the items mentioned above will, to a large extent, be able to withstand external factors, including environmental conditions.
- **Fire and Durability** – The minimum requirements for adequate protection against fire related to the durability of concrete are given in different Codes of Practice.

Source:

- Ganguly [3]
- Ganguly [12]
- Ganguly[1]

9.15 USE OF PRE-CAST CONCRETE

Over the years, different types of pre-cast concrete units have been designed and constructed for industrial and other projects. A few important items that have already been successfully tried are listed below:

Piles for plant and non-plant buildings and equipment foundations, roof/floor slab units of many types, bridge beams/girders of various shapes, concrete blocks of different types, lintels and shelf units, retaining walls, building columns and beams, staircases, paving units of different shapes and designs, concrete pipes of different diameters and strengths, pre-cast concrete sleepers, pre-cast poles, road-kerbs and bridge furniture, small retaining wall/box culverts/tanks, louvers/chajjas, jallis of different designs and fascia elements, road signs, wall/floor tiles, manhole/pit/trench covers, stair units, etc.

Considering better quality concrete under controlled conditions, the trend of using more pre-cast units is expected to increase in the future.

It may not always be possible to obtain innovative and new designs easily from the available units for use. To achieve this, there will need to be more interactions between designers and manufacturers regarding their knowledge, experience of mix proportions, colour, mould design, control and handling of units, and joint detailing. This way, the confidence of the whole team will grow, and better pre-cast and durable structural elements will be available for use.

It has been observed that pre-cast elements can be very effective in rehabilitation work, especially when a few structural elements are in distressed condition. The uniformity in texture, finish, and colour of pre-cast concrete allows for an exposed finish, thus saving considerable time and money, especially on plastering, which is often adopted in many concrete structures.

Source:

- Ganguly [1]

9.16 READY MIXED CONCRETE

To use ready-mix concrete to maintain the quality of construction will be important. Its use on different construction sites is increasing rapidly. The use of ready-mix concrete will ensure good quality concrete from all contractors, including the smaller ones who are unable to arrange concrete from batching plants for their projects. The suitable use of admixtures will also help in transporting concrete over long distances. The advantages of using ready-mix concrete are as follows:

- **Economy** – The cost/per cubic metre will be economical, as locally available raw materials are likely to be used for production.
- **Ensures Strength Requirement** – The design requirements will be met with confidence. Higher grades of concrete may be used by engineers in the design for slender and stronger structures.
- **Easy and Safe Use** – Using ready-mix concrete will facilitate speed in construction, workability and better hygienic conditions. .
- **Versatility** – It adapts to the specific needs of different worksites.
- **Services** – The services will be regular and uniform to maintain the schedule as planned for the construction. Pumps, conveyors and deliveries for difficult worksites may be arranged.
- **Quality** – Ready-mix concrete will guarantee quality. The constituent materials are subjected to quality checks initially, and rigorous quality control checks are carried out throughout the manufacturing and delivery process.
- **Comfort** – Deliveries of ready-mix concrete can be picked up from ready-mix plants, or the concrete can be delivered to worksites by concrete mixer trucks. The speed of delivery will be as per the customer's needs. The use of ready-mix concrete will help keep the workplace clean, while dirt, congestion, and noise will be minimised.

Ready-mix concrete will offer different varieties of mix designs. It can utilise different types of fillers, cement and binders to produce special concrete as per the customer's requirements.

To optimise costs, it will be necessary to have uniform and high-quality concrete for durable and better structures as per specifications.

Source:

- ganguly [6]

9.17 FURTHER ON READY MIXED CONCRETE

For ready-mix concrete, four-axle truck mixers are likely to be required. The capacity of a truck mixer may range from 5 to 7½ m^3. With the introduction of ready-mix concrete, it will be possible to consistently maintain good quality concrete at construction sites, as mentioned. The demand for ready-mix concrete is expected to be very high in most parts of the country where large-scale construction programmes will be undertaken. Similar trends have been observed in other developing countries. It is strongly felt that this improvement must be introduced immediately in our country to keep the minimum standard of concrete mix at a higher level and to regularly construct better quality and more durable structures. It is needless to say that, in large projects, the main contractors have been using batching plants to produce quality concrete for construction.

The use of ready-mix concrete will ensure good quality concrete from all contractors, including small ones who are unable to arrange concrete from batching plants for their projects. The suitable use of building chemicals will also help in transporting concrete over longer distances.

It is expected that the ready-mix concrete industry will not only be efficient in the production and delivery of large volumes of ready-mixed concrete but will also be able to handle building chemicals efficiently, which might be necessary to meet specific requirements.

It is understood that a good network of ready-mix concrete plants, across all parts of the country, will take some time to develop. However, it may not be too long before we witness ready-mixed concrete plants supported by pumping concrete facilities at many important sites in this country.

Source:

- Ganguly [1]

9.18 FURTHER ADVANTAGES OF USING CONCRETE

9.18.1 Fire Engineering and Performance of Concrete under Fire

Following recent fires in various parts of the world, the performance of structures during fire has become a very important design issue for engineers. Concrete has performed very well during fires over the years, and the same has been observed during tests carried out.

It is important to note that the stability of the members and overall structural performance will depend upon

- Maximum temperature and temperature distribution (including penetration)
- Material properties
- Applied loads
- The level of composite action with the slab
- The restraints provided by the surrounding structure
- The continuity of interaction with other cold/heated members
- Factors related to engineering.

The change in properties of concrete due to high temperature will depend upon the type of concrete aggregates used.

The systems that are often considered for fire engineering design are

- Prescriptive code of practice
- Performance-based approach
- Fire testing

It may be noted that the mechanical properties of concrete, structural steel and reinforcing steel bars will decrease as the temperature rises.

A proper performance-based approach, apart from high temperature, must consider ventilation conditions and active protective systems, e.g., sprinklers or fire brigades, on the growth of fire.

For an analytical model and the analysis to tackle fire conditions in structural components, the following will be important, apart from external service loads:

- Internal forces due to restraints that prevent free expansion.
- Internal forces due to changes in pore pressure.
- Re-distribution of internal forces due to changes in mechanical properties of concrete and steel.
- Secondary effects on the internal forces due to interaction of external loads and deformation from the above effects.

It has been reported that concrete with a strength higher than 55 MP is more susceptible to spalling and may show lower fire resistance. Carbonate aggregates are preferable in concrete mix for enhancing fire performance. In order to reduce spalling in high strength concrete under fire, fibres may be used in the mix to protect the concrete surfaces at salient places or to provide thermal barriers.

It has been observed that the use of polypropylene fibres in concrete will increase the permeability of heated concrete under fire to release the built-up pressure and prevent explosive spalling. Hence, the performance of concrete in fires can be improved by using these fibres. Furthermore, it has been found that the greater the quantity of individual fibres present in the concrete, the higher the number of channels and voids that will be created during the fire.

Source:

- Ganguly [3]
- Ganguly [13]

9.18.2 Design for Blasts

A blast will result in producing high-amplitude impulse loading, which will last for a very short time period and produce high pressure loads. The forces that will be experienced by the structural elements will depend on the size, geometry and proximity of the explosion.

Concrete has performed well against blast loads. The presence of continuous reinforcement with good detailing will provide reserve capacity to develop alternate load paths and resistance mechanisms under extreme conditions.

Steel fibre-reinforced concrete can make an important contribution to the integrity and resistance of blast-resistant structures. Further comments have been made in Chapter 12.

Source:

- Ganguly [3]
- Ganguly [13]

9.19 UHPFRC PANEL TESTING FOR EXPLOSION AND IMPACT

Ultra-high-performance fibre-reinforced concrete (UHPFRC) has been developed with significantly improved material properties compared to conventional concrete. UHPFRC can be produced using fine powders (cement, glass-grade sand, silica fume, and silica flour), super-plasticiser, water, and fibres. No other aggregates were used. High tensile steel fibres with lengths and thicknesses of 13 and 0.2 mm, respectively, were used to provide improved mechanical strength. A higher binder content (i.e., in excess of 1,000 kg/m^3) and a low water-to-binder ratio (i.e., 0.15–0.18) were achieved using a super-plasticiser with a high dosage and well-graded materials.

Studies, supported by experimental work, were carried out to investigate the properties of UHPFRC under impact and explosive loading from a 100 kg TNT equivalent charge. Some panels contained both steel bars and steel fibre reinforcement. The benefits of steel fibres include anti-spall properties and additional energy dissipation once the re-bars begin to yield. Blast loading caused various degrees of permanent damage in the form of yield lines, but not total structural collapse.

The material has a low water-to-binder ratio, high cement content, and high super-plasticiser dosage. Micro-silica and fine silica sand (<400/μm) were used in the mix for enhancing homogeneity. Post set heat treatment is done at 90°C for further improvement. High dosages of short small diameter steel fibres are used to improve ductility.

UHPFRC will be suitable for protecting people and buildings from acts of terrorism.

Source:

- Ganguly [3]
- Ganguly [13]

9.20 INFLUENCE OF CONCRETE ON THE ENVIRONMENT

Concrete is an environmentally friendly and sustainable building material, as mentioned. In this context, some related items are

Resource Efficiency – Limestone is used for the production of cement and is available in sufficient quantities in many countries. Waste products from different industries, such as fly ash, GGBS, and silica fume, are also used efficiently with cement to produce the right type of concrete for different applications.

Durability – With proper design and construction, concrete structures will be durable, as already stated, and will last for a long time. Concrete is a sustainable material, and its performance against fire is better than that of other common building materials.

Thermal Mass – Concrete has the ability to absorb and retain heat, making it energy efficient. Hence, expenditure on heating or cooling the building as per requirements will be significantly reduced.

Reflectivity – Light-coloured concrete will absorb less heat and reflect more solar radiation than dark-coloured materials. The demand for air conditioning will be reduced in summer.

Applicable to Retaining Storm Water – Pervious concrete can help retain storm water to some extent when used for driveways, parking areas, side walls, and later replenish local water supplies.

Minimum Waste – Concrete can be produced as per the requirements of the project with minimal waste.

Source:

- Ganguly [3]

9.21 GREENHOUSE GAS EMISSION DUE TO CONCRETE

The emission of CO_2 is considered to be the basis for a rating technique to compare the environmental impact of different construction materials in environmentally sustainable design.

CO_2 emissions from concrete will include all activities related to manufacture and placement of the concrete material. CO_2 emissions from cement production have been controlled to a large extent by adopting efficient production systems and using alternative renewable energy.

As per data from Australia, the equivalent CO_2 emissions from normal strength concrete using Portland cement binder is 290–320 kg CO_2-e/m^3.

CO_2 emissions per ton of cementitious binder, according to South Africa = 765 kg.

Source:

- Ganguly [3]

9.22 RECYCLING CONCRETE MATERIALS

Concrete material can be recycled for further use after prolonged use and demolition. (However, proper checking and testing will be essential for making such a decision.) Common applications include road work as coarse sub-grade material. As the aggregates are sourced from demolished concrete, they can be used for normal concrete after proper checking and testing.

The tests are related to

- Determining chloride content
- Occurrence of alkali–silica reaction

- Risks of using high-alumina cement concrete as aggregates
- Effects of hydrated gypsum powder on concrete properties

Recycled Aggregates – Up to 30% recycled concrete aggregates as a replacement for natural aggregates will have no adverse effects on the engineering and durability properties of concrete. With higher RCA content, minor alterations to the mix may be necessary to ensure equivalent performance to natural aggregate concrete is achieved. The density of RCA is generally 5%–15% lower than that of natural aggregates, and water absorption may be higher. Checks should be made on chloride, sulphate, and alkali content.

The potential of glass waste in concrete has been examined. Glass aggregates (0.2–12 mm) and glass powder (˂0.2 mm), along with cement, sand, aggregates, and admixtures, have been considered for full-scale trials. The following comments are important from the study:

- Alkali content has a significant influence on the rate of ASR of glass in concrete.
- Glass, both as pozzolan and aggregates, can be used with no detrimental effects for up to 1 year.
- Zero ASR expansion was observed in a series of tests for up to 1 year of age.

Recycled Glass (RG) – RG is a graded material that may be obtained from glass cullet with particle sizes ranging from powder to 8 mm. It can be used in concrete as glass powder filler or fine aggregate as RG aggregate.

9.22.1 Reuse of Structural Components and Materials

A lot of emphasis has been placed in recent years on reusing and recycling construction materials. It is preferred to recycle materials rather than reuse components.

The construction materials that will be available are

Hot rolled and cold-formed steel sections, structural timber, timber sheets, masonry items – bricks, stone, concrete, pre-cast concrete units, sheet piling, portal frame buildings, foundation items, etc.

The following items will be important in this connection:

Prevention	–	To avoid the use of excessive materials through efficient design and structure repetition.
Reuse	–	Recovery and direct use of a component, e.g., a brick reused as a brick.
Recycle	–	Recovery and manufacture of a material into a component of equal quality.
Safe Design	–	Disposal of non-recoverable materials with a negative impact on human health or the environment.

Source: Ganguly [3]

9.23 DEVELOPMENT OF SPECIAL CONCRETE

The summary of the items may be mentioned as

Ferro-cement, fibre-reinforced concrete, high-strength concrete, micro-silica concrete, benefits of using silica fume in concrete, high-performance concrete, further on silica fumes for high-performance concrete, UHPFRC, self-compacting concrete, self-compacting premix GRC, polymer concrete, perlite–polymer concrete, magnesia–phosphate cement (MPC) concrete, self-curing concrete, smart concrete, sprayed concrete, high-volume fly ash concrete, lightweight aggregate concrete, high-performance lightweight aggregate concrete, autoclaved aerated concrete, concreting with expanded clay aggregates, expanded clay aggregate blocks, foamed concrete, air-entrained concrete, vacuum concrete, saw dust concrete, mass concrete, roller compacted concrete, dry lean concrete, no fines concrete, concrete without coarse aggregates, water-resistant concrete, waterproofing concrete, waterproofing concrete with sprayed acrylics, high-density concrete, heat-resisting concrete, coloured concrete, special nano-concrete, geo-polymer concrete, reactive powder concrete, bacterial and self-healing concrete.

Source:

- Ganguly [3]

9.24 TESTING OF MATERIALS

Only concrete materials will be discussed briefly. Testing of other materials has been covered, in a limited manner, separately.

A few items will be discussed.

9.24.1 Testing Concrete

Apart from testing each item of concrete ingredients, the following tests are common for the assessment of concrete:

- Cube/cylinder test
- Core from the element
- NDT tests

Guidance may be obtained from the IS code for testing concrete.

9.24.2 Petrography Examination of Concrete

Petrography of concrete will entail a detailed examination of all types of hardened cementitious mixtures, applying a combination of visual, microscopical, and often scanning electron microscopical methods.

Important information may be obtained about mix proportions, workmanship, and the original quality of concrete.

Source:

- Smart and Sims [14]

9.24.3 Temperature Monitoring

It may be relevant to assess the long-term strength and stability of concrete through such monitoring. This may contribute to understanding the temperature difference between the surface of the concrete and the core. It will be important to keep the temperature within safe limits.

9.24.4 Data Loggers

Thermocouple probes may be provided at salient places before pouring the concrete. These may be tied using plastic cable ties on the steel reinforcement and connected to data loggers for monitoring. Readings from the central core will be important.

9.24.5 Measurements/Information

The information will be useful for the engineers to take corrective actions if required. The information for curing concrete and the necessary corrections may be very relevant and is associated with long-term structural performance and integrity. The curing of concrete must be carried out properly, as this will provide important information.

Source:

- Petford [15]

9.25 CASE STUDY

9.25.1 From Research to High-Profile Application

Recommendations from the research on the in-situ concrete frame test building, as applied to construction in the UK, are briefly mentioned below as guidance:

- Calculations for slab deflections with reduced costs of internal finishes.
- Early removal of formwork for more efficient use and to avoid delays.
- Recommendations for using pre-fabricated reinforcement.
- Application and fixing of cladding with the CRC joint system.
- Use of pre-cast units with the CRC joint cast method for rapid construction.

Source:

- Research Focus [11]

9.26 DIFFERENT APPLICATIONS OF REINFORCED CONCRETE

To demonstrate a few applications of reinforced concrete over the years, associated with the completed industrial and other projects experienced by the author, some photographs are included as shown in Chapter 15. These are mostly related to different power plants, fertiliser projects, material handling plants, and other different projects.

It can be said from the above photographs that various shapes and structures have already been constructed to meet the project requirements. These applications have been successfully carried out with both in-situ and pre-cast concrete. With the advancement of concrete technology, the demands on engineers will be much greater in the future for producing creative and attractive structures with the material.

Source:

- Ganguly [1]

REFERENCES

1. Ganguly K K, A brief review of Different Applications with Reinforced Concrete, in *Industrial Projects, with Present and Future Trend, International Symposium on Innovative World of Concrete (ICI-IWC-93), Proceedings*, Expressions in Concrete, pp. 1–12.
2. Campbell P, Aims of structural design, view point, *The Structural Engineer*, January 1983, Vol. 61, pp. 3.
3. Ganguly K K, *Reinforced Concrete Design, Repair/Restoration, Strengthening of R C Structures and Concrete Technology*, Scientific International.
4. Faber J and Mead F H, *Reinforced Concrete* ELBS.

5. Murashev V I, Sigalov E E and Baikov V N, *Design of Reinforced Concrete Structures*, MIR Publishers.
6. Ganguly K K, Applications of structural engineering in projects and other associated issues, *Indian Concrete Journal*, January 2008, pp. 41–56.
7. Downing M, *The Structural Engineer*, June 2001, pp. 18–19.
8. Lazarus D, Masonry facade retention, *The Structural Engineer*, 19 November 2002, pp. 17–18.
9. Ledbetter S, Façade Engineering: the challenge for structural engineers, *The Structural Engineers*, June 2001, Vol. 79, pp. 13–14.
10. Ledbetter S, Façade engineering – what's behind the façade? *The Structural Engineers*, November 2002, Vol. 80, pp. 21–24.
11. Research Focus, Issue No. 44, February 2001.
12. Ganguly K K, Durability requirements of concrete structures, in *National Seminar of ICI with Institution of Engineers (I)*, Nagpur Chapter, January 1990.
13. Ganguly K K, Composite materials with particular reference to concrete, in *Geomate-2017*, 21–24 November 2017, TSU City, Mie, Japan.
14. Smart S and Sims I, Concrete petrography, *Concrete*, April 1997, Vol. 31, pp. 31–34.
15. Petford R, Guarding against temperature issues during concrete curing, *Concrete*, November 2018, pp. 36–37.

10 Steel Structures

10.1 INTRODUCTION

The selection of steel as a construction material in buildings and industrial projects has been successfully made by engineers and architects in the past. The preference for using steel in such projects has been increasing in recent years. With proper planning, faster construction is possible with steel material. An attempt has been made to review some aspects of steel design associated with buildings and industrial projects. Different stages of design activities have been reviewed briefly. The construction of steel structures is quite common in industrial applications, and this has been effectively applied by clients, engineers, and architects for a long time, as mentioned. The flexibility to accommodate changes later is a significant advantage of steel structures over similar structural elements in concrete. The discussion will be restricted to frame structures, which are normally adopted for industries and buildings. Engineers must understand structural mechanics and the behaviour of actual and idealised structures, and they must appreciate the requirements of safety, feasibility, performance during the expected lifetime of the structure, and economy. Idealisation of the structural system is required for structural analyses to compute internal forces in the members, which are expected to be close to the actual ones. Finally, the results must be assessed considering experience and judgement, which are very important and should be used effectively in structural design and detailing.

It is noted that by adopting steel material in construction, it is possible to produce fast and economical structures for industrial and other projects.

It is important to examine some basic issues related to steel structural design and construction as applicable in a project.

Source:

- Ganguly [1]

10.2 DESIGN

10.2.1 Basic Design/Conceptual Design

At this stage, different alternatives are considered. The following factors are important:

- First cost
- Ease of maintenance
- Aesthetics
- Flexibility in the design for future extensions

It also includes

- Preparation of layout showing the structural arrangement, i.e., beams, columns, bracings, and supporting systems
- Determining the type of frames

DOI: 10.1201/9781003319108-10

In most cases, a lower first cost will dictate the choice. However, in some public projects, aesthetics may be more important than first cost or maintenance issues.

Source:

- Ganguly [1]

10.3 DESIGN WITH STRUCTURAL STEEL

At the initial stage, it will be necessary to work out the cost on the basic design using a structural system covering the following items:

- Detailed design with some connection details
- Fabrication cost
- Transport cost
- Erection
- Painting
- Cladding and finishing

Typical trusses will be presented later.

Source:

- Ganguly [1]

10.4 GUIDANCE AS PER IS CODE

As per IS-800 [2], options are available for structural steel design based on two methods :

- Limit state design
- Working stress design

10.4.1 Limit State Method of Design

The following limit states are to be considered:

- Limit state of strength
- Limit state of serviceability

The reliability of design has to be satisfied by ensuring.

Design action ≤ Design Strength

The load factors for different limit states are given in IS-800 [2]. Accidental loads from explosions and the impact of vehicles, etc., have to be considered for the limit state of strength calculation. The design for accidental loads is generally not required for buildings unless specifically requested by the client or approving authority.

To avoid disproportionate collapse, the columns are to be effectively tied in both directions. The steel ties and their end connections should be capable of resisting the factored tensile force as per the guidance in the code [2].

10.4.2 Working Stress Method

Where the limit state method cannot be conveniently adopted, the working stress design may be used. The elastic analysis method shall be used in working stress design. The stresses may be calculated using the respective partial load factors for service/working loads as given in the code [1].

The limit state of serviceability will be associated with

- Deformation or deflection that will affect the appearance or use of the structure.
- Vibrations in the structure causing discomfort to people and damage to the structure.
- Repairable damage or cracks due to fatigue
- Corrosion and durability
- Fire resistance

Source:

- IS-800 [2]
- Ganguly [1]

10.5 METHOD OF STRUCTURAL ANALYSIS

Following the assumptions in the IS code, the structural analysis may be carried out by

- Elastic analysis as per IS-800
- Advanced analysis as per Annex - B, IS-800
- Plastic analysis as per IS-800
- Dynamic analysis as per IS 1893 (Part 1)

The information on the analysis and design for resisting earthquake effects shall be carried out as per IS 1893 (Part 1)

Source:

- IS 800 [2]

10.6 LOAD AND LOAD COMBINATIONS

IS-800 may be referred to. For earthquake loads, IS 1893 will be applicable.

Source:

- IS-800 [2]

10.7 IMPORTANT FACTORS/ITEMS FOR DESIGN

The items mentioned below will be relevant:

- Dead loads
- Imposed loads (i.e., live load, crane load, and snow load, as per IS-800)
- Effective length factors
- Eccentricity of beam reactions
- Permissible axial stresses
- Distribution of bending moments produced by eccentricity of beam reactions
- Permissible bending stresses
- Connection details

The permissible stresses may be obtained from IS-800.

Source:

- Steel Designers' Manual [3]

TABLE 10.1
Response Reduction Factor (R)

Serial Number	Lateral Load Resisting System	R
1	• Braced frame systems	
	Ordinary concentrically braced frames	4
	Special concentrically braced frames	4.5
	Eccentrically braced frames	5
2	• Moment frame system	
	Ordinary moment frame	4
	Special moment frame	5

Source: IS-800 [2]
The revised values have to be used in the design as per the changes.

10.8 RESPONSE REDUCTION FACTOR

The Response Reduction Factor (R) for the building system shall be in conjunction with IS 1893, as mentioned in Table 10.1:

10.9 DESIGN AND DETAILING FOR EARTHQUAKE LOADS

The steel frames must be designed and detailed as per the guidance given in IS 1893.
Source:

- IS-800 [2]

10.10 CONNECTIONS, JOINTS, AND FASTENERS

IS-800 may be referred to.
Source:

- IS 800 [2]

10.11 EFFECTIVE LENGTH OF COMPRESSION MEMBERS

The theoretical values for effective lengths related to idealised cases are mentioned below in Table 10.2:

The code [2] recommends the following for computing the effective length of a column under different boundary conditions:

10.12 REVIEW OF OTHER ASSOCIATED ITEMS FOR DESIGN

10.12.1 Type of Structures

- The types of structures that are common include trusses, portals, column/beam arrangements, space structures, etc.
- A simple column–beam arrangement with bracing appears to be economical.
- Portal actions in both directions may be adopted to eliminate bracing.

TABLE 10.2
Effective Length

Boundary Conditions	Recommended Value of Effective Length KL
Effectively held in position and restrained against rotation at both ends	0.65 L
Effectively held in position at both ends and restrained against rotation at one end	0.80 L
Effectively held in position at both ends, but not restrained against rotation	1.00 L
Effectively held in position and restrained against rotation at one end and, at the other end, restrained against rotation but not held in position	1.20 L
Effectively held in position and restrained against rotation at one end but not held in position nor restrained against rotation at the other end	2.00 L

Source: IS-800 [2]
where, KL = effective length of the compression member
L = unsupported length of the compression member

10.12.2 Common Applications

In many projects, the arrangement and applications of steel structures may be standardised. The common types of structures that are often adopted for industrial applications, along with a common truss, are shown in Figures 10.1 and 10.2.

10.12.3 Flooring System

These may be described as

- Composite flooring with metal decking
- Composite flooring without metal decking

Hence, steel column–beam arrangements with metal decking, as permanent formwork, become the most suitable solution for the project. Profile sheets may be in the order of 0.8–1.2 mm thick and can cover a span of 2.5 m -3.0 m.

10.12.4 Detailed Design

The design of load-bearing elements for steel structures is primarily related to the design of tension members, compression members, or members under bending with or without axial forces. Some of the basic expressions related to the design are mentioned below:

- **Tension Members** – For tension members, the factored design tension T shall satisfy the following:
  $T < T_d$ [2]
  where
  T_d = design strength of a member under axial tension; T_d shall be the lowest of the design strength due to yielding of the gross section, T_{dg}, rupture strength of the critical section, T_{dn}, and block shear, T_{db}.
- **Compression Members** – The design compressive strength P_d shall be $P < P_d$ [2].
  $P_d = A_e f_{cd}$
  where
  A_e = effective sectional area

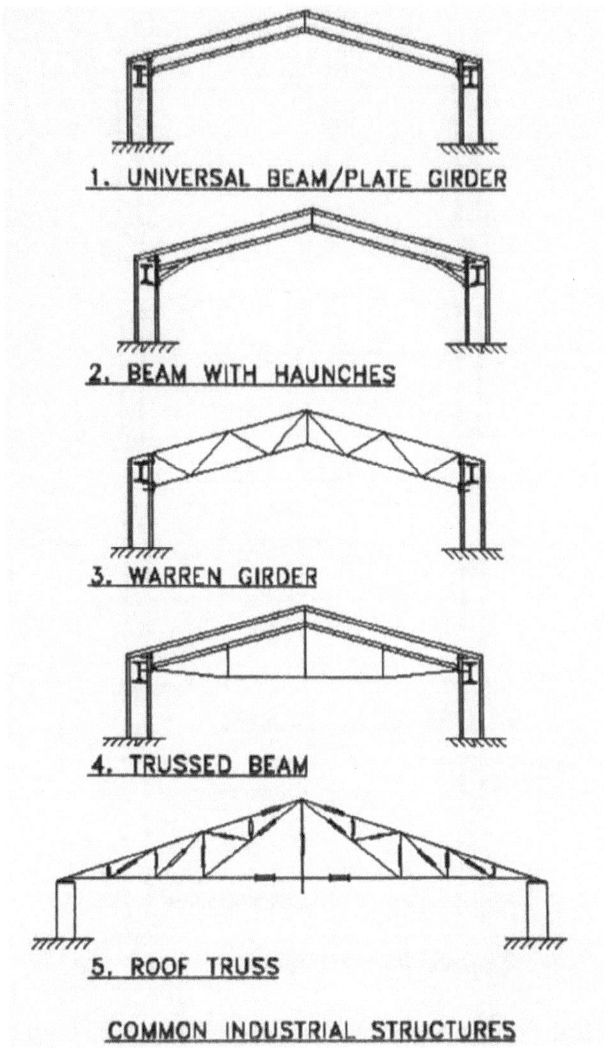

FIGURE 10.1 (Source: IS 800 [2]; Bresler and Lin [4]; Leech [5]; Subramanian [6]; Steel Designers' Manual [3]; Ganguly [1]; Ganguly, Pal, and Ghosh [7]).

f_{cd} = design compressive strength as per IS 800 [2]
The values of different buckling classes are given in the code.

- **Members Subjected to Bending** – The factored design moment 'M' shall satisfy $M \leq M_d$, where M_d = design bending strength of the section.
- **Combined Axial Force and Bending Moment** – Under this situation, checks are to be made for material failure and member strength as governed by buckling failure as per the guidelines given in the Code [2].

Similar guidance will be available for the design based on working stress. Furthermore, the items that will be important for detailed design are

- Following the structural arrangement, the support conditions are finalised and the accommodated in the analysis for the structure.
- Care will be required to develop details of roller/pin connections accordingly, as per the assumptions.

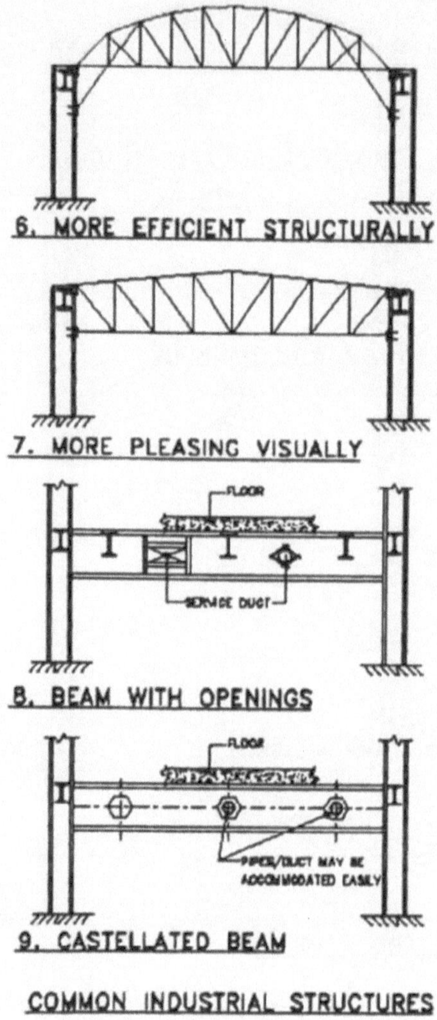

FIGURE 10.2 (Source: IS 800 [2]; Bresler and Lin [4]; Leech [5]; Subramanian [6]; Steel Designers' Manual [3]; Ganguly [1]; Ganguly, Pal, and Ghosh [7]).

- Detailed drawings must be developed in a manner that satisfies design assumptions and facilitates easy erection.
- The design assumptions should be simple and relevant to assess the behaviour of the structure correctly.

Source:

- IS-800 [2]
- Ganguly [1]

10.12.5 Bracing against Lateral Loads

- The function of bracing is to transfer horizontal forces from the frame to the foundation.
- Provisions for adequate bracing must be provided if the joints are assumed to be hinged.
- Different systems of bracing include eaves bracing, knee bracing, column bracing, rafter bracing, gable end bracing, horizontal bracing, etc.

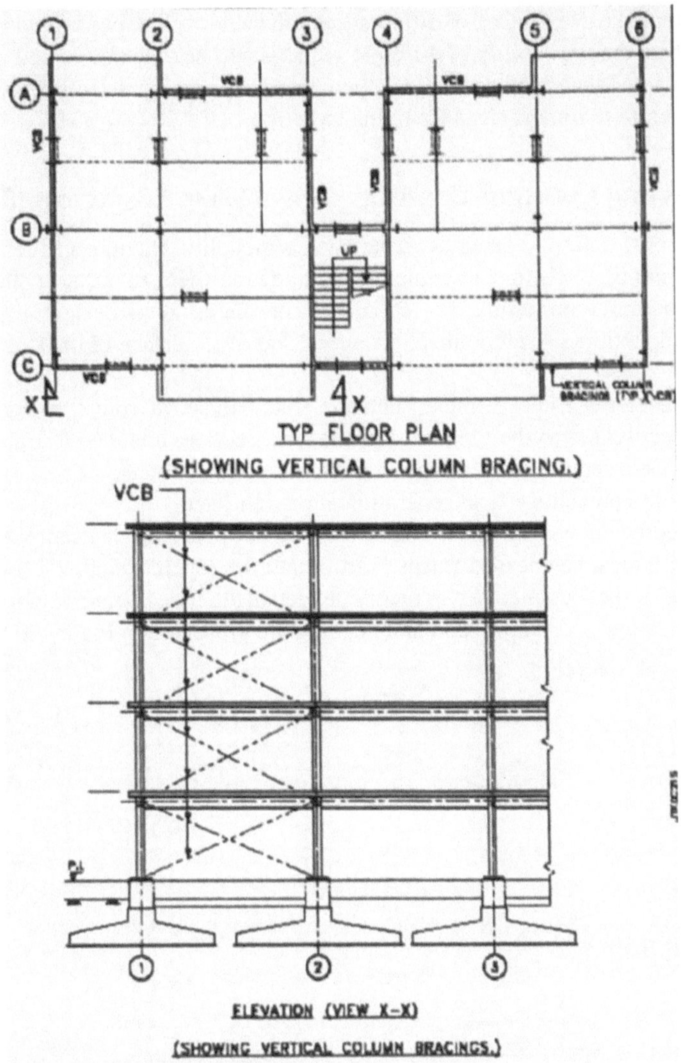

FIGURE 10.3 Bracing arrangement. (Source: IS 800 [2]; Bresler and Lin [4]; Leech [5]; Subramanian [6]; Steel Designers' Manual [3]; Ganguly [1]; Ganguly, Pal, and Ghosh [7]).

- In buildings where light-speed travelling cranes are supported by the structure or where a building or structure is otherwise subjected to vibration or sway, triangulated bracing may be provided to reduce the vibration or sway to a suitable minimum.
- Temporary bracing must be provided for all applicable parts of the structure during erection or to support the erection loads.
- A typical bracing arrangement applicable for a building project is shown in Figure 10.3.

10.12.6 Member Selection

- Member selection is an important exercise in the design process.
- Apart from minimising weight, it is relevant to check the suitability for easy connection on-site.

- Information on the availability of different sections will need to be verified in this context.
- Information on the availability of different sections will need to be verified in this context. It should be noted that good aesthetic judgement is required for selecting the proper member type and size to impart elegance to the structure.

10.12.7 System with Composite Construction and Shear Connectors [8]

- Composite construction (steel-to-concrete) means monolithic action between the structural steel beam and cast-in-situ slab components, allowing them to act as a single structural unit. For economy, composite action should be considered in the design.
- To make it monolithic, shear connectors are used at the interface of the steel and concrete surfaces.
- A shear connector is a steel element, such as a stud, bar, spiral, rolled section, or any other similar device welded to the top flange of the steel section, intended to transmit the horizontal shear between the steel beam and the cast-in-situ concrete.
- It is also used to prevent vertical separation at the interface.
- In an encased beam system, shear connectors are not provided; instead, the top flange of the steel beam must be encased at least 50 mm into the concrete slab.
- Several types of shear connectors, as recommended in the code [8], are shown in Figure 10.4.
- Composite sections in compression and bending are presented in Figure 10.5.

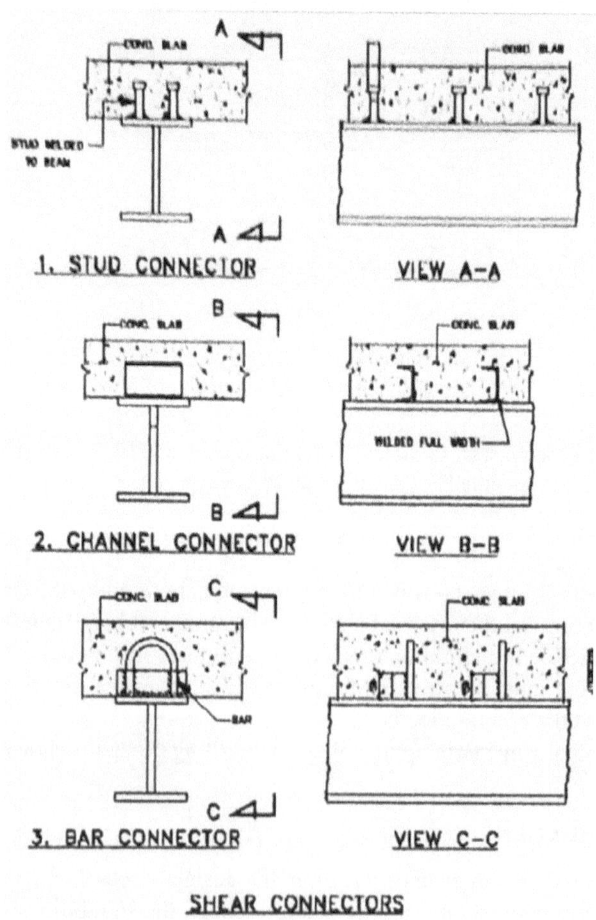

FIGURE 10.4 Shear connectors.

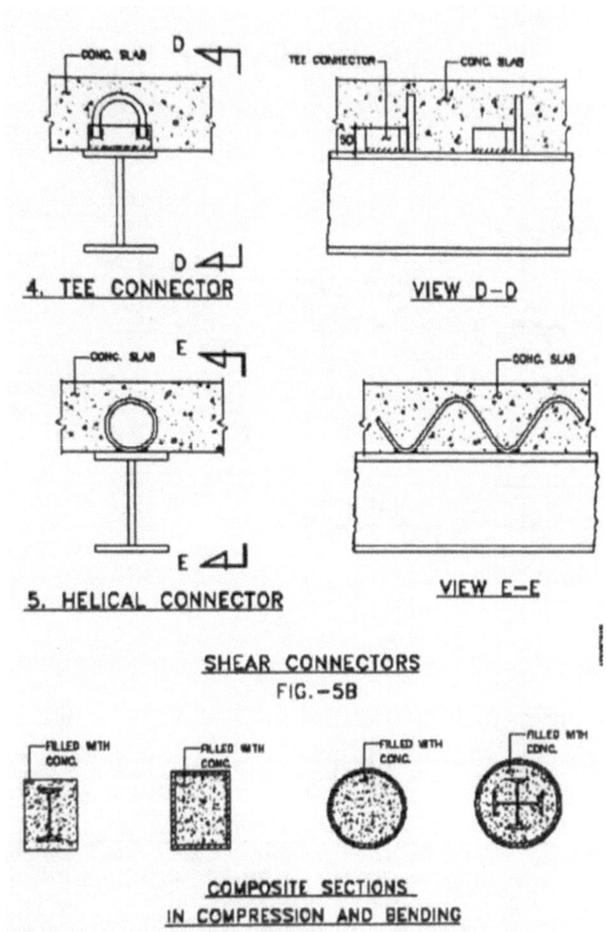

FIGURE 10.5 Composite sections. (Source: IS 800 [2]; Bresler and Lin [4]; Leech [5]; Subramanian [6]; Steel Designers' Manual [3]; Ganguly [1]; Ganguly, Pal, and Ghosh [7]).

Source:

- IS-11384: 1985 [8]
- IS-800 [2]

10.12.8 Welded Beams and Types of Weld

In some projects, welded beams are required, especially for plate girder applications. The common types of beams include

- Equal flanges
- Wide top flange
- Narrow top flange
- Jumbo column section.

These are presented in Figure 10.6.

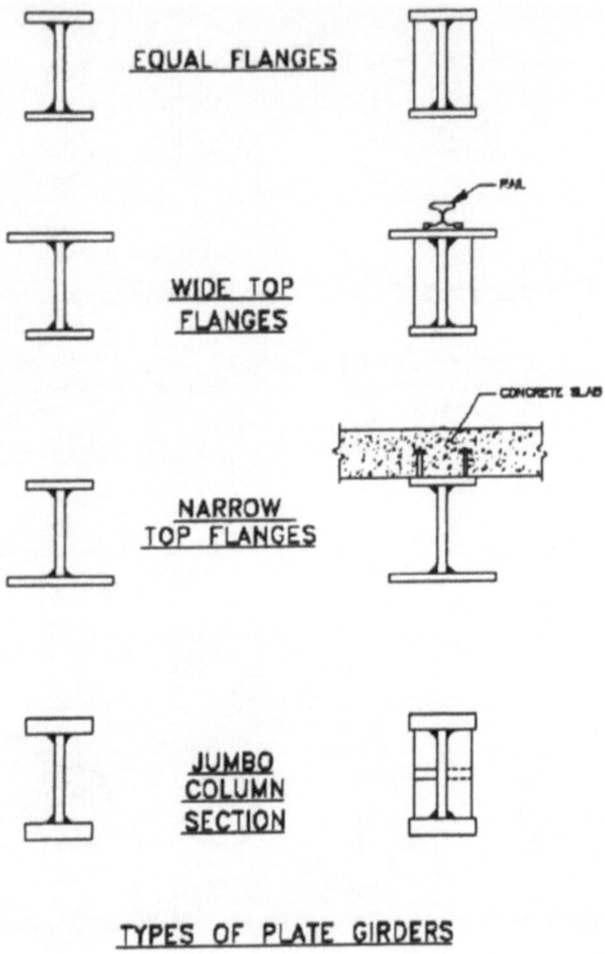

FIGURE 10.6 Types of plates girders. (Source: IS 800 [2]; Bresler and Lin [4]; Leech [5]; Subramanian [6]; Steel Designers' Manual [3]; Ganguly [1]; Ganguly, Pal, and Ghosh [7]).

10.12.9 Typical Joint Details

For connecting columns and beams, different types of joint details may be followed depending on the assumptions made in the analysis/design. Typical moment connections between a column and beam, as well as a beam-to-beam connection, along with a typical framing in an industrial building, are shown in Figure 10.7.

10.12.10 Presentation of the Design

- The structural arrangement must be idealised for the software package that will be used for the analysis.
- Data preparation, input, output, and the detailed design, along with preliminary design drawings, are often presented to the client for checking/approval/information as applicable.
- Following the detailed design, drawings are prepared for tendering and later for construction.

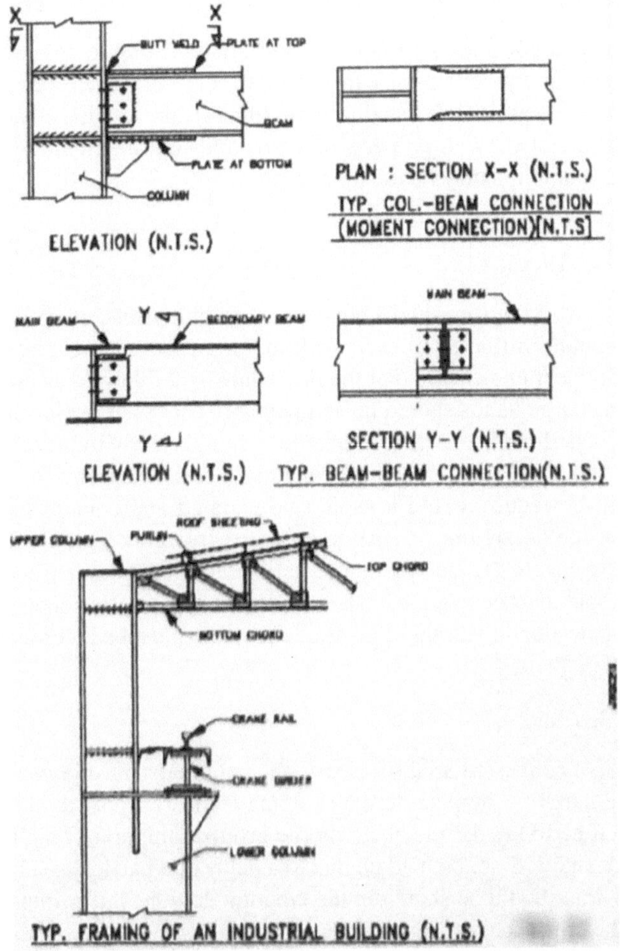

FIGURE 10.7 Typ. framing of an industrial building. (Source: IS 800 [2]; Bresler and Lin [4]; Leech [5]; Subramanian [6]; Steel Designers' Manual [3]; Ganguly [1]; Ganguly, Pal, and Ghosh [7]).

- Detailed fabrication drawings are normally prepared by the fabricator based on the design and scope drawings.
- For steel structures, it is important to check different temporary conditions that may arise during the execution period. This is important and must not be overlooked in the design work.

10.13 DURABILITY

IS-800 may be referred to.

10.14 DESIGN ASSISTED BY TESTING

IS-800 may be referred to.

10.15 FABRICATION AND ERECTION

The tolerances for fabrication and erection shall be carried out as per the recommendations of IS Codes.

10.16 FURTHER ON SITE INSPECTION OF STRUCTURAL STEELWORK

Site supervision is very important, as mentioned, and it is essential to ensure that the steelwork is erected as per the design and specifications, following the correct dimensions and using safe procedures as quickly as possible as per the schedule. The supervision activities from both the contractor and the client must be effective, with proper interaction and coordination.

The following relevant items may be mentioned [9]:

10.16.1 Preliminary Work

A rational sequence of working steps and a good standard of housekeeping will be very important.

The supervisor should familiarise himself with the proposed method and sequence of erection as approved by the engineer and ensure that the procedures are followed as per contract terms. The erection foreman and charge hands should be competent to carry out the job safely.

The relevant drawings should be available, and they should clearly indicate the erection sequence and the necessity of using any temporary materials that will be required for stability during erection. A bar chart will be required, and it should be updated and coloured every day to indicate the progress made on site at any time. Insisting on proper record-keeping will be very helpful for smooth progress and to ensure that any corrective action required to maintain the schedule is taken.

A site bolt list should also be available, specifying the types of fasteners, diameters, and grip lengths. Types of fasteners and the total number of each type must be recorded.

10.16.2 Laying-Out Space

If the site is located in a congested area (i.e., city), the contractor must plan and arrange the deliveries to avoid double handling by placing major items from transport directly into position. The normal procedure will be to lay the members on the ground temporarily so that the erection gang will have buffer stock and will not be held up due to a delay of a particular lorry load. A position is provided on site, and this should be shown in the erection drawings. It is important to visit the site initially to identify the area specifically for access.

The contractor must ensure that the materials are not damaged during handling or delivery (i.e., projecting cleats and fittings, etc.) and separate the materials properly for later handling.

10.16.3 Delivery Schedule

The sequence of deliveries should be determined during the early stages to match the erection schedule. Lower lengths of columns should be delivered first. It is desirable to avoid accepting the delivery of items that will be required later, as per the programme. This may occur for items like purlins, which may have been outsourced to other parties who are eager to settle their accounts. If an early delivery is accepted, care should be taken to avoid the development of corrosion on items if they are left lying for a long time.

10.16.4 Cranage

A plan should be made available for tower and other mobile cranes showing the track layout to reach each crane at its planned positions. If this is not available, the feasibility of the layout should be studied, and discussions should be held with the contractor at an early stage.

The cranes are to be examined before assembly. In case of doubt, damages must be repaired before assembly. One should look for previous repair records to ensure that cranes are in sound condition.

10.16.5 Setting Out

The position, time, and level should be checked with reference to other adjacent permanent items. Accurate theodolite work may be required for verification. The levels for the bases should be checked. Generally, a 25–50 mm allowance is given for grouting between the top of the foundation and the underside of column bases. Additional dimensions will not be needed to counteract errors. The location of holding-down bolts will be very important. The foundation bolts may be installed by the contractor, but in some cases, the location may be incorrect. The responsibility for the correct positioning of bolts will lie with the foundation contractor.

It will be helpful if the two teams can check each other's setting out with the progress of work. The outstanding lengths of bolts will also be relevant. It is better to handle bolts that are too proud rather than too short. If the bolts are too short, the extensions must be done properly, which will be expensive and time-consuming.

10.16.6 Loose Packs

Loose plates may be placed on the bases, accurately levelled, and grouted. For heavy columns, the base plates may be supplied loose. In such case, the base plates should be fixed in position first, and the grouting work can then be completed beneath them. The contractor may normally carry out the work to a tolerance of ±3 mm. The steelwork may be erected on them and may become self-aligning, requiring little or no subsequent adjustment to line and level. After this, the erected columns should not be disturbed, and mortar fillets may be used for this.

For lighter columns, the base plates may be supplied with the columns, and grouting may be carried out only after adjustments to line and level and the final erection of permanent members and connections.

10.16.7 Inspection of Steel Work on Delivery

Each delivery of steelwork should be inspected carefully with respect to the delivery schedule.

The members must be straight and free from accidental damage.

The loose pieces/fittings should be delivered along with the steelwork to avoid delays.

10.16.8 Erection of Steel Work

Proper supervision during the erection and inspection on completion will be critical. The erection of steelwork may be divided into three phases, although there will be some overlap.

These are

10.16.8.1 Preliminary Erection

Tack-bolted state, lining, levelling, etc., related to permanent connections.

10.16.8.2 Second Stage

Related to bracings covering both permanent and temporary.

10.16.8.3 Third Stage

The stresses imposed on the structure during erection may exceed the assumed values in the design. In some cases, reversal of stresses may occur, leading to buckling conditions in members. This condition must be avoided.

The erection should be considered as a whole.

10.16.9 Lining and Levelling

Once the erection of a portion is finished, lining and levelling should proceed. After checking, the final connections may be made.

When a good amount of erection work is completed, it may be very difficult to make adjustments.

Conversely, if the erection work is allowed without making permanent connections, unwanted movements may occur.

10.16.10 Permanent Connections

The permanent connections may be made in the form of bolts in clearance holes, high-strength friction grip bolts, or site welding.

10.16.11 Painting

Previous paintwork may be damaged during steel erection. Final touch-up of base coats and the finishing coat may be done as per the specifications after erection and completion of connections.

Special attention will be required for bolts unless they are supplied with metallic protection.

10.16.12 Handing Over

When all work at any stage has been finished, the engineer will issue a completion certificate.

This should not be delayed unnecessarily.

10.16.13 Comments

The contractor or sub-contractor will be legally responsible for the satisfactory completion of the work.

However, the client's engineers can help keep the work on schedule and ensure completion without delay and at no extra cost to the contract.

Source:

- Needham [9]
- IS-800 [2]

10.17 FOUNDATIONS

The design for the foundation of the building and other structures shall be done as per the analysis and design assumptions to provide rigidity and strength to the superstructure.

Special care will be required for the design of base plates and holding-down bolts as per the guidelines provided in the codes.

Source:

- IS-800 [2]

10.18 OTHER IMPORTANT ITEMS IN THE PROJECT

10.18.1 Cost Aspect

As mentioned earlier, the initial cost of construction may be slightly high for structural steel structures. This will be offset by the advantages of saving time. Further reduction in cost will be possible if life cycle cost (LCC) cost is considered in the total evaluation.

10.18.2 Aesthetics

With proper selection of steel members supported by good detailing an attractive steel structure can be constructed.

10.18.3 Durability

Construction with structural steel will be highly durable. The use of steel sections manufactured under strict quality control will eventually produce the most durable construction. This has been referred to separately.

10.18.4 Completion Time

It should be noted that with proper planning and efficient control, the construction time can be reduced significantly.

10.18.5 Fire Protection

Essentially, steel building elements must exhibit the required fire resistance level (FRL) as per the specifications. The thickness of the protection material (hi) provided shall be greater than or equal to the period of structural adequacy (PSA) and greater than or equal to the required FRL [2].

The required FRL [10–12] may be obtained from the IS code. The specified FRL value will indicate the requirement of standard fire load on affected structural elements without collapse. This is related to [2]

- The purpose for which this structure is used.
- The time that will be required to evacuate in case of fire.

The mechanical properties of steel, covering both yield stress and modulus of elasticity, will be affected by an increase in temperature. The expressions for evaluating values with the rise in temperature are given in the code. It is needless to mention that with higher temperatures, the properties will be severely affected. For temperature less than 215°C [2], no reductions need to be considered in yield stress.

The following comments may be made to summarise fire design for steel structures and affected elements [4]:

- Fire protection for structural elements is required for the safe exit of occupants during a fire, the safety of firefighters combating the fire, the prevention of spread, the safety of adjoining structures, and to minimise damage to the property involved in the fire.
- Steel structures are generally considered incombustible and provide reasonable safety for light occupancy structures.
- The fire resistance of steel structures can be increased by applying fire protective covers such as concrete, vermiculite, gypsum, special lightweight plaster, or by providing fire retardants on the steel surface.
- The use of intumescent coating/paint may be considered an efficient way to protect load-bearing structures made of structural steel.
- During the design, a protection period of 2–4 hours is often considered to be adequate for the structures.

Source:

- Bresler and Lin [4]
- IS-1641: 1988 [10]
- IS-1642: 1989 [11]
- IS-1643: 1988 [12]
- IS-800 [2]

10.19 FURTHER ON FIRE RESISTANCE

Some basic items are

- **Requirements –** Steel building elements must provide the required FRL of performance as per the relevant specifications.
- Fire resistance level.
- Limiting steel temperature.
- PSA.
- Variations of mechanical properties of steel with temperature.
- Temperature increase over time in protected members.
- Temperature increase over time in unprotected members.
- Three-sided fire exposure condition.
- Determination of PSA from a single test.
- Special considerations.
- **Fire Resistance Rating –** The fire resistance ratings of different building elements are given in IS-800.

Source:

- IS-800 [2]

10.20 FATIGUE

The different considerations related to fatigue failures that must be addressed in the design are outlined in IS-800.

For the purpose of design against fatigue, different details of the members and connections are classified into different fatigue classes.

However, the following items will not be covered:

- Corrosion fatigue
- Thermal fatigue
- Low cycle (high stress) fatigue
- Stress corrosion cracking
- Effects of high temperature (>150°C)
- Effects of low temperature.

Source:

- IS-800 [2]

10.21 LIFE CYCLE ASSESSMENT FOR STEEL CONSTRUCTION

The Steel Construction Institute (SCI) is involved in European Union-funded research to study and assess the life cycle environmental impact of all applications of steel products. This includes cladding, decking, and light gauge steel products. SCI will conduct audits involving site visits, interviews, and the completion of detailed questionnaires. The study on life cycle phases will include the following:

- Steel production
- Transportation
- Fabrication and manufacture of steel construction components
- Product assembly and erection methods
- Monitoring, in-house maintenance, and repair strategies to be followed
- At the end of life, items should be considered for demolition, reuse, and recycling.
- Scrap processing
- Preparation of the report after the detailed study.

Source:

- Research Focus [13]

10.22 DRAWINGS/SKETCHES FOR STRUCTURAL ARRANGEMENT/SYSTEM

These are enclosed below: There are a number of sheets, as shown in Figure 10.8–10.12.

Source:

- Steel Designers' Manual, Ref. No. 3
- B Bresler and T Y Lin, Ref. No. 4
- L V Leech, Ref. No. 5
- MERO Information Document, Ref. No. 14

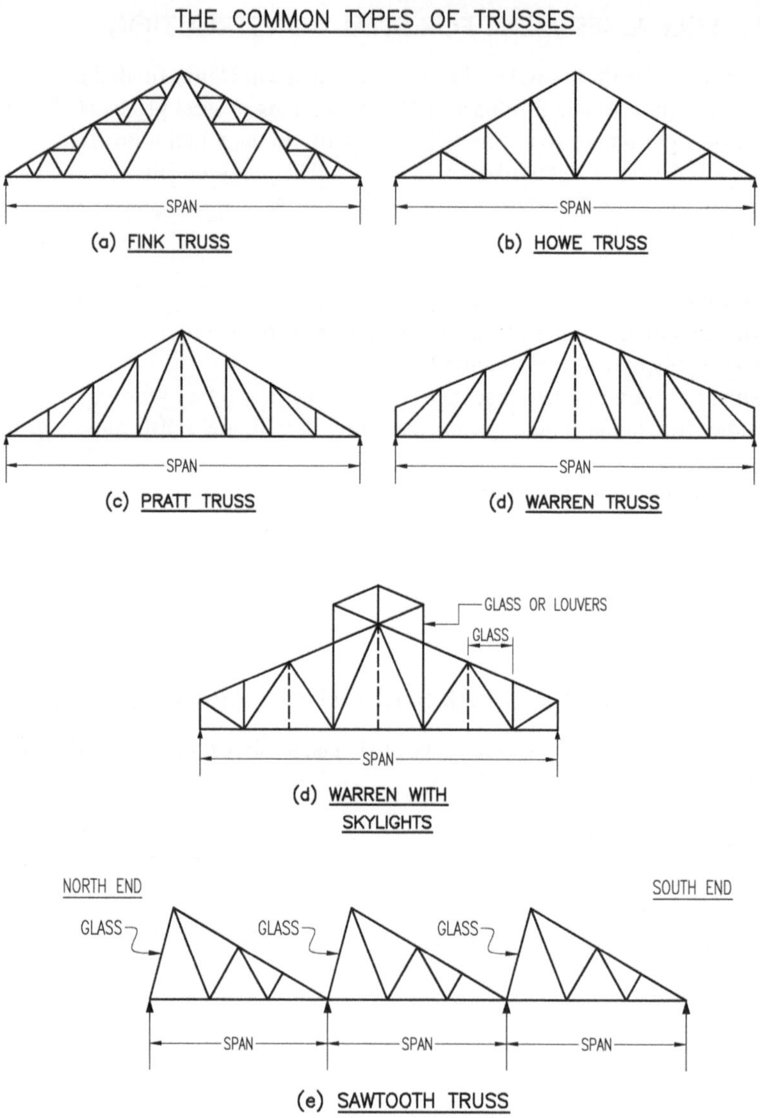

FIGURE 10.8 The common types of trusses. (Source: Steel Designers' Manual [3]; Bresler and Lin [4]; Leech [5]; MERO Information Document [14]).

(Continued)

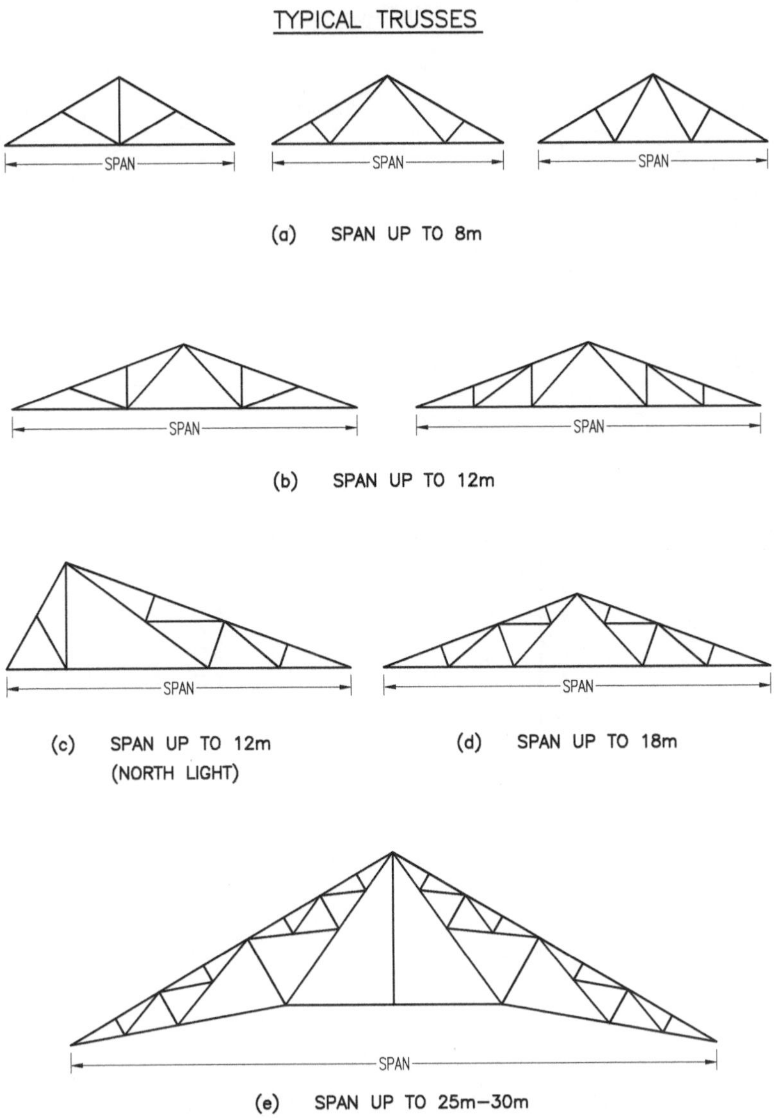

FIGURE 10.8 (*Continued*) The common types of trusses. (Source: Steel Designers' Manual [3]; Bresler and Lin [4]; Leech [5]; MERO Information Document [14]).

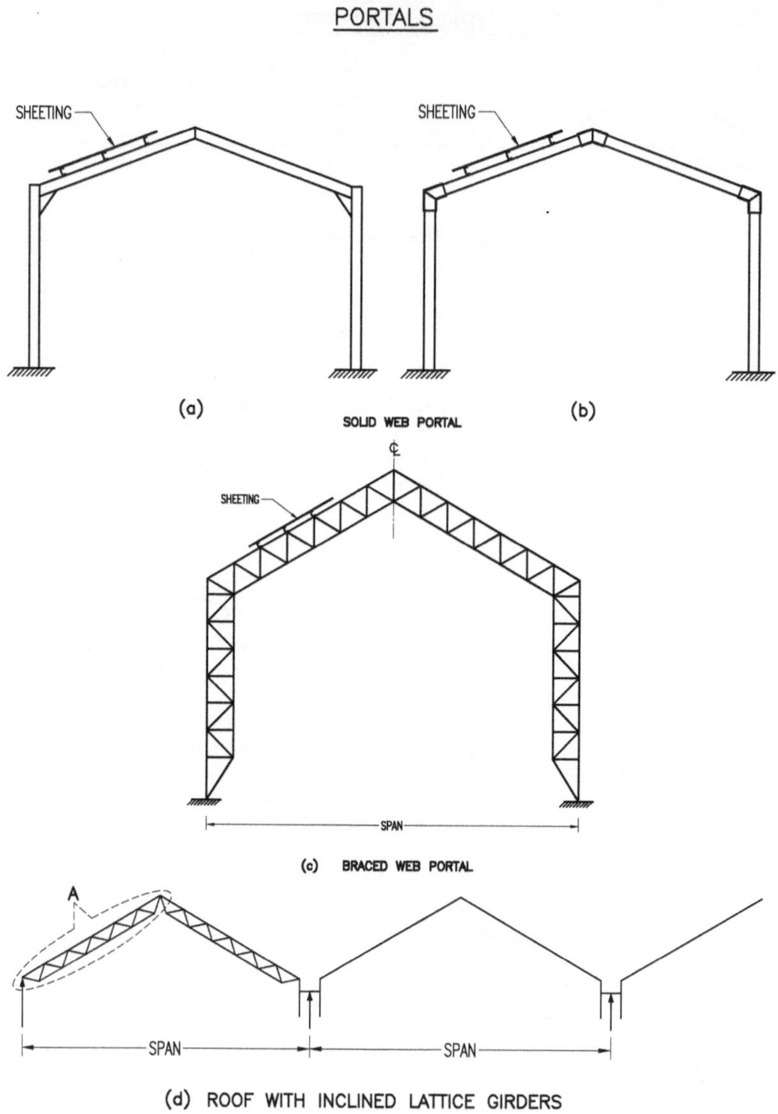

FIGURE 10.9 Portals & inclined lattice girders. (Source: Steel Designers' Manual [3]; Bresler and Lin [4]; Leech [5]; MERO Information Document [14]).

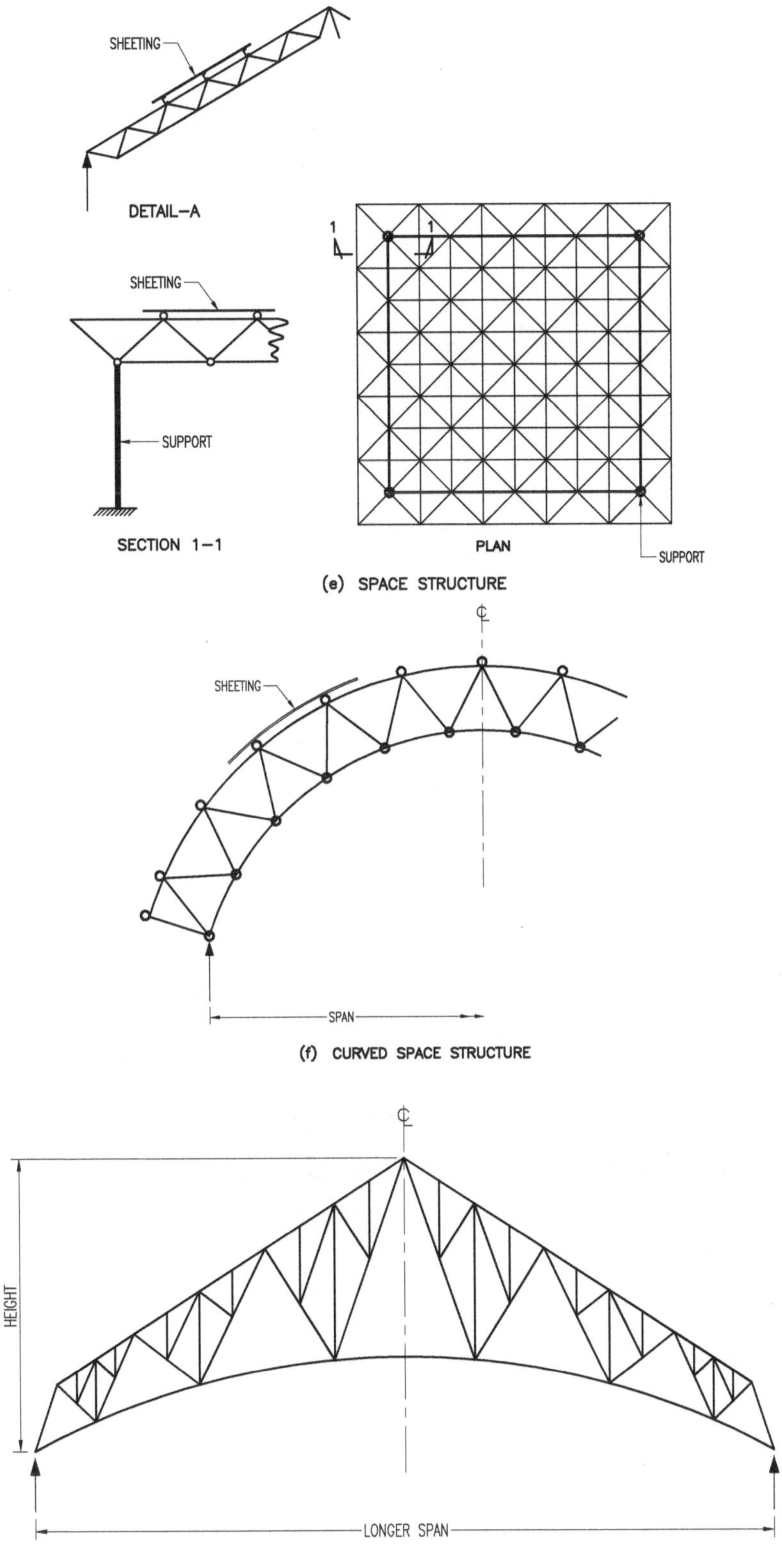

FIGURE 10.10 Space structure & curved space structure. (Source: Steel Designers' Manual [3]; Bresler and Lin [4]; Leech [5]; MERO Information Document [14]).

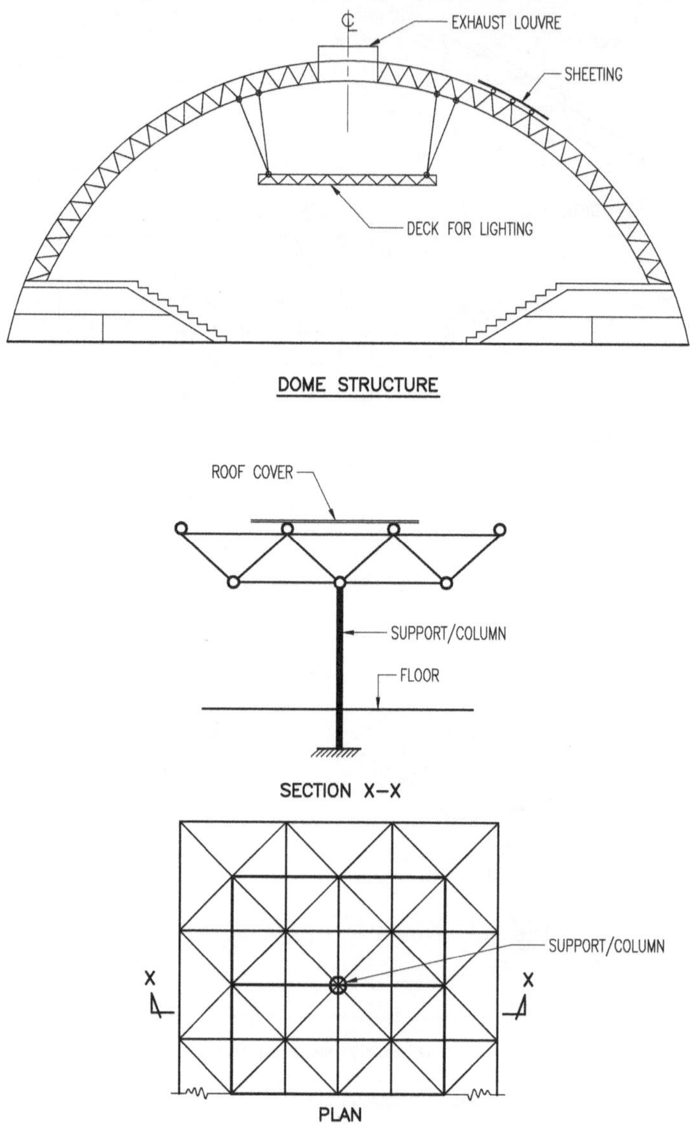

FIGURE 10.11 Dome structure & space structure. (Source: Steel Designers' Manual [3]; Bresler and Lin [4]; Leech [5]; MERO Information Document [14]).

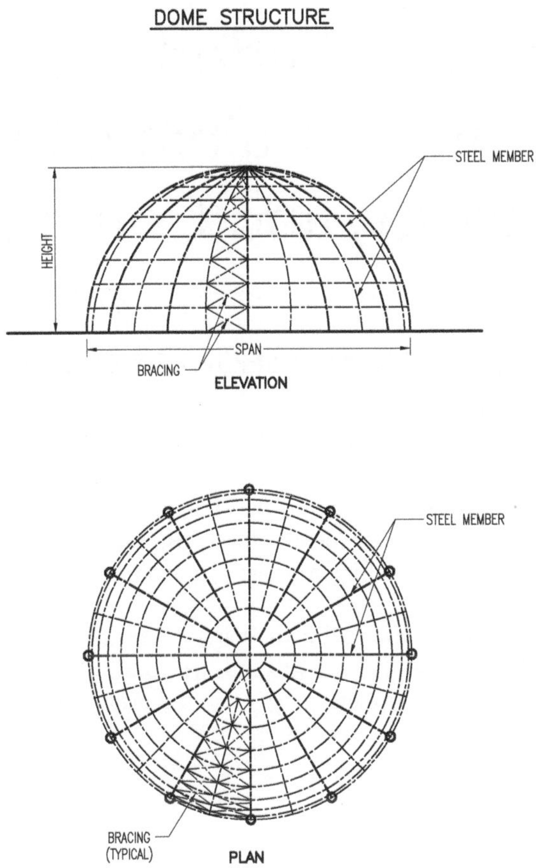

FIGURE 10.12 Dome structure. (Source: Steel Designers' Manual [3]; Bresler and Lin [4]; Leech [5]; MERO Information Document [14]).

REFERENCES

1. Ganguly, K K, Applications of structural steel in buildings and industrial projects, *INSDAG's Steel in Construction*, July 2008, Vol. 9, No. 2.
2. IS-800: 2007, General Construction in Steel - Code of Practice.
3. Steel Designers' Manual, Crossby Lockwood and Son Ltd.
4. Bresler B, Lin T Y and Scalzi J B, *Design of Steel Structures*, John Wiley & Sons Inc.
5. Leech L V, *Structural Steel Work for Students*, Butterworths, London.
6. Subramanian N, *Design of Steel Structures*, Oxford University Press, 2008.
7. Ganguly K K, Pal D C and Ghosh S, Steel Structures in Buildings and Industrial Projects, May 2002, JPC Auditorium, Kolkata.
8. Bureau of Indian Standards, IS-11384: 1985, Code of Practice for Composite Construction in Structural Steel and Concrete.
9. Needham F H, Site inspection of structural steel work, in *Proceedings of the Institution of Civil Engineers,* Part-1, August 1981, Vol. 70(3), pp. 393–402.
10. Bureau of Indian Standards, IS-1641: 1988, Code of Practice for Fire Safety of Buildings (General): General Principles of Fire Grading and Classification (First Revision).
11. Bureau of Indian Standards, IS-1642: 1989, Code of Practice for Fire Safety of Buildings (General): Details of Construction (First Revision) (Superseding IS-1645).
12. Bureau of Indian Standards, IS-1643: 1988, Code of Practice for Fire Safety of Buildings (General): Exposure Hazard (First Revision).
13. Research Focus, Issue No. 44, February 2001.
14. MERO Information Document, Applications of Steel in Buildings and Industrial Projects.

11 Computer-Aided Work, Education, Training, Quality, Reliability, and Communication

11.1 USE OF COMPUTERS FOR ENGINEERING CALCULATIONS

In all structural engineering design activities, available software packages are predominantly used. It has been observed that there has been a shift in emphasis from design calculations to structural modelling [1]. This shift may lead, in some cases, to inappropriate modelling and incorrect results. The use of computers has significantly reduced the time required for design computation work, which was a major time-consuming factor for structural designers in the past. Engineering calculations may be treated as a series of modelling exercises [1] and may be based on

- An engineering model representing the actual physical existence of the structure
- A conceptual model that allows for material and structural behaviour
- A computational model (a description of the structural arrangement in a form suitable for computer analysis)

The assessment of loads in the analyses must be performed properly. The persons involved should have proper training and experience to carry out their part of the work. However, the person with overall responsibility should be a senior and experienced chartered engineer.

Each stage should be checked against acceptance criteria, and the final outcome should be assessed against an overall modelling review.

Untrained people should not be involved in work beyond their abilities or use programmes that are unsuitable for the work at hand without carrying out fundamental checks. This could result in an unsafe structure. A proper and systematic approach is essential.

The following basic items must be satisfied:

- The model must be verified to ensure it is correct.
- The selected programme should be suitable for the work and utilised correctly.
- Checks must be made to ensure that the results are accurate.

The items associated with modelling an engineering structure include

- Setting up a system related to the purpose of the work and how best to carry it out.
- Preparing a number of models that represent the structure and the relevant surrounding environment.
- Selecting suitable hardware, software, and qualified persons to prepare the models.
- Carrying out the calculations.
- Checking and validating the models.
- Checking/reviewing the modelling exercise.

Apart from the computer, hardware, and software resources, critical requirements will include experienced personnel, time, and other facilities.

DOI: 10.1201/9781003319108-11

The analyst may consult with a software specialist to ensure that the software is suitable and is being used to its best advantage.

The responsibility for checking the modelling exercise should not fall solely on the analyst. This review should be done in-house by a senior engineer. In large and important projects, such responsibility may also be given to an independent consultant.

The engineering model will consider the overall geometry of the structure, all load-bearing components with dimensions, supports, materials used, and loads, including lateral loads. Non-structural components are not considered, but their effects are assessed and allowed for in the calculations.

The detailed analysis of the model will need to be carried out using the theory of structural analysis, material behaviour, boundary conditions, loads, and the geometry of the structure.

The guidelines published by the Institution of Structural Engineers [1], London, related to the use of computers for engineering calculations may be referenced.

Source:

- Farmer [1]
- Ganguly [2]

11.2 STRATEGY FOR USING COMPUTERS IN STRUCTURAL ENGINEERING

The items may include

11.2.1 Steps for the Modelling Process [3]

- The model should be created as per the requirements.
- Evaluation of the software for use with the model.
- Use of the software for obtaining results.
- Validation of the model.
- Verification of results.

11.2.2 Creating the Model

The following may be relevant:

- The support conditions – pinned or fixed.
- Main objectives of the analysis.
- Are the joints to be treated as pin-jointed or moment-resistant?
- What will be the loading types – i.e., dead, live, wind, seismic, snow, dynamic, etc.?
- Any secondary effects need to be considered.

11.2.3 Other Information

Other items may include

- Type of elements and the list of associated items
- Dimensions of the system and member sizes
- Member/node numbers
- Support conditions
- Material properties
- Loading cases.

Information has to be checked from computer output.

Evaluation of the Software – The software applied must be capable of handling the model. A verification exercise may be conducted to check at least a portion using a standard software model.

11.2.4 Verification of the Results

These are

- The data to be checked.
- Local equilibrium should be checked, if possible.
- Checking overall equilibrium, i.e., sum of the reactions = sum of the applied loads.
- Checking that support restraints are properly used by verifying deformations of the restrained nodes.
- Checking the overall form of the results.
- Check for symmetry as applicable.
- Checking the deflected shape of the structure and whether this matches the expected results.
- Creating a checking model for comparing the results.

The important part will be to create the model and to check/ensure that the results are correct.

Based on assumptions, the results from the output should match the expected behaviour of the real structure.

Source:

- MacLeod [3]

11.3 PRODUCTIVITY IN 3D MODELLING AND STRUCTURAL DESIGN

The analysis, design, and drafting activities may be carried out using a software package to increase the productivity of the design office.

3D modelling for analysis and design and the use of software packages have produced a tremendous impact in increasing performance in the design office, which until recently was unimaginable. This, in turn, has given engineers the opportunity to provide better service.

11.3.1 3D Modelling and Design

A model may be generated by inputting information from the architectural drawings, including different structural elements defining slab types, other structural elements, and applying all loads as applicable to complete the model. The design of the structural elements can be carried out very quickly.

The advantage of such modelling is that it will reduce the time required to accommodate and implement repetitive and daily design changes to the structure. The calculations for contributory areas for live loads, the reduction of live loads as per the code on the supporting members, etc., can be done quickly and correctly. The whole operation will save a significant amount of time. The structural members can be checked quickly for finalisation.

For composite action, as applicable, the computation can be done by calculating the effective flange widths and composite section properties prior to performing the member design. The work will be completed quickly, and different requirements such as width or depth reductions of structural members can be checked quickly for finalisations.

3D design and modelling can facilitate faster building design than manual and general analysis systems.

The same 3D model of a structure used for gravity-resisting loads can also be used for the lateral resisting system.

For the lateral systems of the building, the following items need to be finalised:

- Where to place the braced frames.
- Whether the walls may be considered to reduce lateral forces.
- The need for moment frames and the extent of such frames.
- The building may be sway, non-sway, or partially sway frame.
- The amount of loads that can be supported by each frame or wall.
- The generation of the appropriate analysis for the structure and the design checks for each member.
- 3D modelling related to analysis and design, with the help of software, is a significant step toward improving productivity and providing better services to the client. Proper training and skill of the user will be very important.

Many software packages are available for the job. Proper assessment will be important for selecting a particular package for use. A poor selection may be very costly.

Source:

- Miller [4]

11.4 COMPUTER-AIDED DESIGN (CAD)

This is a design process carried out using a computer system, as mentioned.

11.4.1 Advantages of Using CAD

- Time will be saved.
- Accuracy will be improved.
- The error percentage should be reduced.
- Design effort will be reduced.
- Editing will be easy.
- Reuse of code.
- Sharing information will be easier.

11.4.2 Disadvantages

- Breakdown of the computer.
- Regular updating of the software operating system may be necessary.
- The work may be prone to viruses.
- The work may be lost or may be hacked.
- High production or purchasing costs for new systems.
- Time will be required to learn the details and operate the software.
- Time and cost for training the staff must be considered.
- It will require less employment for the CAD/CAM system.

Source:

- ARCVERTEX [5]

11.5 FURTHER ON ANALYSIS AND CAD

In recent years, with the availability of different software packages, engineers have been able to analyse complicated structures with confidence. Rigorous analyses are possible and are being conducted in many areas, including simple cases. Young engineers are keen to pursue 3D analysis even for concept design. Working out preliminary sizes through hand calculations may be considered old-fashioned. The hand calculations technique is not practiced by the younger generation, and in some cases, they may take longer to carry out simple design calculations due to a lack of practice.

The use of software packages, on the contrary, is convenient and easy to use, provided one is familiar with the package. Input data preparation will need some guidance at the beginning. After some use, it becomes easier and can be used by persons who are not necessarily very experienced in structural engineering. Even a non-technical person with little familiarity can be trained to use sophisticated software packages to analyse and design structural elements. The output results will be correct and reliable for a conventional structure, as long as the input data is prepared correctly. The idealisation of the structural system is important, and proper experience is required for the assumptions to achieve satisfactory results. Unusual and complicated structures will definitely need greater care from the designers. It may be dangerous if such packages are allowed to be used by inexperienced or non-technical persons in the design office.

One must understand the behaviour of the structure well and should be able to compare it with the computer output. The person should be able to carry out simple hand calculations or use a different software package, if required, before accepting the results. Accepting computer output blindly is not advisable.

It will be essential for the engineers to understand the following:

- How the structure will work
- How to model the structure to obtain a sufficiently accurate solution
- The supporting system of the structure
- How to use appropriate methods of analysis
- How to consider seismic loads and carry out dynamic analysis

In the design output package, the detailing part may need thorough checking against the analysis output. In some cases, poor or incorrect detailing may need to be corrected prior to issuing it for construction. Easy and construction-friendly detailing should be aimed for as far as possible, as mentioned.

Source:

- Ganguly [2]

11.6 STRUCTURAL DETAILING

Simple and practical detailing should always be aimed for during the design. Aesthetically, detailing should be presentable and must avoid clumsy joints. Special emphasis must be placed on seismic detailing to provide ductile behaviour. Column–beam junctions will need additional care to ensure the flow of concrete between bars, especially when ductile detailing has been done for the confinement of concrete. Often, the details for joints are not fully shown in the drawings to reduce time. Proper seismic detailing will be essential to control damage to the structure. Providing the correct cover to reinforcement during construction is very important for structural performance, especially related to strength, durability, and fire resistance. Some concrete structures may not achieve their intended design lives due to deficiencies in the cover actually provided during construction.

Due to inadequate cover, structures may have to be demolished prematurely or may require expensive repair and protection work. It is reported that for an external concrete wall sheltered from rain and having a 30 mm cover, the reinforcement will be protected for 135 years, while a 10 mm cover will only provide 10 years of life for the same wall.

Spacers and chairs, as per BS-7973, may be considered for construction. Plastic and cementitious spacers and wire chairs are recommended for use. It will be important to specify the use of spacers and chairs in the specifications and to show the correct type of spacers and chairs in the drawings, i.e., at least for standard details for easy understanding and use in construction.

The major comments made after the Chi-Chi (Taiwan) earthquake in 1999 are improper lap splices of rebar, column tie hooks without a 135° bend, and poor concrete quality control at job sites. These are typical observations made after examining the earthquake damage to structures. The above comments are quite common and may also be applicable in our country based on an earthquake disaster site. Site engineers should take care to ensure that construction is done correctly as per the drawings.

Source:

- Ganguly [2]

11.7 DRAFTING STANDARD BY CAD

It seems that the quality of drafting may have declined to some extent when using CAD facilities, but the overall presentation has improved, and the output is quick. There is now a more mechanical approach to the production of drawings from CAD. The thinking aspect of drafting and good detailing is generally left to others. Users are often unable to view the complete drawings properly on the monitor. Blow-up details in important areas are often left to the contractors to develop and execute on site. In the process, problems may arise, and alternative detailing may be required to resolve these problems. Input from contractors may be important while developing details in critical areas, and this should be encouraged. Communicating structural details in the form of tables will need improvement, especially for important structures and elements. The scales selected for some important details may be erratic. Here again, training for CAD operators will be necessary to ensure that drafting is not done mechanically. There must be a thinking process to improve and present information in a better form for others to understand. Competent operators who can think critically and produce drawings of good standards that satisfy basic requirements will be required.

Source:

- Ganguly [2]

11.8 EDUCATION AND TRAINING OF ENGINEERS

There are very few organised places for civil engineers to receive proper training during and after graduation. Many industries and consulting engineers are not very keen to provide training to fresh graduates. This may be considered a waste of time, as the trainee engineers cannot contribute much at the initial stages of their services. This perspective is misguided, as it only satisfies short-term objectives. More organised training programmes will be essential for fresh engineering graduates to start their careers [2]. The curriculum may need special consideration to meet training requirements. It may not be wise to leave this matter entirely to the placement officers of the universities to arrange such training. Job opportunities are also very limited at this stage. Institutions, industries, consulting engineers, and contracting firms should collaborate on this issue and strive to develop proper training programmes for engineers to cope with the situation. Young engineers will expect assurance that during the course and after graduation, they will receive formal training from the respective industries and employers. A package may need to be developed by the group to address the financial burden on a particular sector. Help may be sought for developing education and training programmes for engineers, along with procedures followed in other advanced countries. Universities may need to play an important role in arranging proper coordination with industries and consulting/contracting organisations.

A brief review will be made.

Universities may not produce fully trained designers.

It may take 5–8 years of practice with competent/senior designers [6], in the team to become a trained and competent designer.

A complex mix of skills is necessary to generate the product or design, identify the requirements of users/clients, and understand the role of the engineering designer.

Some skills, such as visualising and conceiving the end product, which will be the practical output, may be difficult to teach.

Universities may seek more help from industries by involving practicing engineers to teach students, especially in the areas of design and analysis of structural systems. It seems that bright students are not very keen to pursue civil/structural engineering courses. Young engineers are always looking for a change after graduating from a particular discipline, presumably for better salaries.

Graduate engineers should be encouraged to learn and develop the following:

- Knowledge of the civil/structural engineering industries.
- Types of major projects that will need civil/structural engineers.
- Different innovative structures already constructed.
- Special design skills and proper planning, materials, and construction techniques required to produce such structures.
- Power of judgement.
- Ability to define objectives.
- Skill in the use of resources.
- Decision-making abilities.

The need for management training will be important for engineers to take on further responsibilities in the organisation. This will also help an engineer perform their current duties better. The items that may be covered are

- Report writing
- Effective speaking and communication
- Interactions with engineers of other disciplines in the project team
- General and special conditions of contract and law as applicable
- Industrial relations
- Sanctioning procedures and some basic rules of local authorities as applicable
- Health and safety at work
- Interaction with other engineering disciplines in the project team
- Checking the availability and cost of construction materials
- Awareness of financial control

11.8.1 Further Comments

It will be very important to motivate students for design awareness, which will help them develop a total range of product skills [6].

11.8.2 Advantages [6]

- Increased understanding and enjoyment of engineering.
- Appreciation of the combination of design and analysis.
- Increased technological repertoire.
- Development of management, leadership, and communication skills.
- Learning how to work in a multi-disciplinary team
- Motivation for engineering studies.

11.8.3 Problem Based Learning [6]

Students may experience difficulties working on design problems, which may include

- Developing their own technical model
- Appreciating the iterative nature of design
- Realising that there is no single correct solution.
- Communicating with a limited engineering vocabulary
- Paying insufficient attention to details
- Copying incomplete information.
- Coping with a shortage of technical repertoire and knowledge of existing solutions.

11.8.4 Further Benefits [6]

- Encouragement of changes at senior levels in universities.
- reviewing design courses across disciplines, and fostering new insights that actively involve staff and students.
- Resolution of issues arising between disciplines.
- To emphasise the importance and value of engineering design and education.
- To set up multi-disciplinary committees and industrial liaison teams.
- To bring universities and industries closer together.

Source:

- Stansfield [6]
- Ganguly [2]

11.8.5 Additional Comments

The structural and architectural designs and the construction of buildings are evolving with the advancement of knowledge regarding concrete as a material [7].

We are aware that industries are constantly changing with the introduction of new techniques and procedures [8]. It is important that engineers in the current degree programme are able to cope with these changes and meet the requirements of industries and employers.

Some key areas in this regard are

11.8.6 Physical Modelling [7]

Physical modelling of concrete and exposure to research work will be relevant.

This will prepare students to understand how the material performs.

It appears that some universities in the UK, with postgraduate programmes, are closely associated with research groups in concrete technology with the research group of concrete technology.

11.8.7 Industry Interaction [7]

It is vitally important to maintain a close relationship with the construction industry. This will ensure that the topics covered are relevant, and that graduate engineers are well-suited for the industry in terms of knowledge and skills.

11.8.8 Future Development

There will be an ambitious drive to utilise concrete materials.

The introduction of digital modelling and design, 3D additive manufacturing techniques, and the use of smart materials will be significant. This may necessitate changes to the traditional design and construction processes.

Concrete will continue to present challenges and will face pressure from other materials such as composites, lightweight steel, and timber.

Understanding the fundamentals of concrete will remain important.

Futurecrete may play a vital role in the education of concrete designers and producers.

Source:

- Newlands and Jones [7]
- Richardson [8]

11.9 REVIEW AND SUMMARY

We need to establish a rigorous training programme for young and fresh graduated engineers to prepare them adequately for the requirements of design, construction, and different aspects of project management. This is not being taken seriously at present. Much more is required to train our engineers.

Employers are always looking for experienced engineers. Who will provide training for our engineers? It has been observed that university education may not fully prepare engineers immediately after graduation. They often need to learn initially from senior professionals about different aspects of analysis, design, and detailing of the structures under consideration. There is a tendency among young structural engineers to be more interested in learning data preparation using software packages. However, understanding the behaviour of structures and validating the acceptability of analysis results may not initially engage these young engineers. It is important to learn how to verify computer output from software packages before accepting the results of analysis or design. Engineers should be exposed to practical aspects of design and detailing to support construction activities. Carrying out analysis using software is only a small (but very important) part of the overall work. To run a software package blindly without understanding the behaviour of the idealised structure will not be useful to the engineer. He has to visualise the behaviour largely and will be able to check some salient parts of the computer output with simple calculations or by using other software packages.

We have to re-examine the nature of education and training we are currently offering to our students so that they are given proper knowledge and can develop skills to cope with the challenges of modern times.

Engineers should be encouraged to develop their own analytical models. This will increase their understanding and enjoyment of engineering. Undertaking design projects in a multi-disciplinary team may be the best way to develop engineering design abilities in students [6].

Industries, universities, potential employers, including consulting and contracting houses, must spend time and money to formulate procedures to train our young engineers. Students must be encouraged to stay in the profession and carry out their work efficiently, as they will face a competitive market in the future. Professional institutions may also play an important role in this respect.

Source:

- Stansfield [6]
- Ganguly [2]

11.10 COMMUNICATION

The term communication may be expressed as sharing or exchanging information to provide achieve results.

For effective communication, the information must flow in both directions. The receiver may not be listening to or reading the information that has been communicated. It is important to ensure that the receiver has received the message properly and understood it.

Civil engineering works are carried out at construction sites involving different disciplines at various stages of design and construction. The work is carried out as per the customer's requirements/specifications and may be unique for each project.

The owner/customer communicates the requirements to the consulting engineer, who will design and accommodate them. The contractor, after understanding the requirements, will execute the design on-site by using subcontractors as needed.

Hence effective communication between the groups is important for the success of the project.

Effective communication can help deliver the project on time, within budget, and to an agreed quality.

Improper communication or breakdowns in communication will yield the opposite results.

Source:

- UK essays/Construction [9]

11.11 FURTHER ON COMMUNICATION

11.11.1 Design Work

The effective communication among different engineering disciplines is very important for design activities. In a multi-disciplinary project, proper and timely communication will play a critical role. This is also relevant for reducing abortive work. The engineer must understand the requirements of other disciplines and incorporate those requirements into the design. These may include cutouts, their locations, and the application of loads to the structural elements.

Communication must be conducted in a simple manner so that the issues are understandable to others associated with the project.

11.11.2 Comments on Construction Work

The following steps may be relevant:

- **To Avoid Jargon –** Do not use special industry-specific terms
- **General Communication Skills to Be Encouraged –** Some communication training may be helpful.
  Some relevant items may be
  - To write clearly
  - Drawing clearly
  - Taking correct measurements
  - To avoid jargon
  - Listening attentively

- To utilise a clear chain of command
- **To Encourage Communication Skills –** Communication training provided on-site.
- Attempts should be made to become better listeners
- Quality checks on communication
- To encourage open communication
- Updating/reviewing new information
- Educate and try to create a bilingual workforce

Source:

- Big Rentz, Inc. [10]

11.12 MORE ON COMMUNICATION

Effective communication is crucial for carrying out any job properly.

Effective Communication – The items mentioned below will be relevant:

- To decide on the means of communication
- To understand the audience
- Appropriate language and method will be important
- To assess potential areas of misunderstanding
- Feedback will be critical for assessing whether communication has been understood correctly as intended.
- Regular interaction/communication will be useful. Long gaps should be avoided.

11.12.1 Practical Steps for Effective Communication

The steps mentioned below may be followed:

- To be polite and use simple common language. We should think before we speak. Communication may be spoken or written.
- To decide on the best/most effective method of communication. Better methods should always be attempted for improvement.
- To determine the message or desired action one is trying to communicate.
- To assess that the message has been received and understood properly by observing positive responses.
- **To Show Empathy and Listen to the Responses –** These responses may give rise to other issues that may need consideration.
- Observe and check on interpersonal communications and note the methods used. Inconsistencies or distortions in the messages should be noted. The context of communication will also be relevant.
- To summarise the information to help the readers at the beginning before going into detail.

11.12.2 Communicating Bad News

All the necessary facts must be known before speaking to others. Before communicating, one should be clear about all the facts and details. The items that will be useful are

- To know how the decision was made.
- To know what other possibilities were taken into account.
- To know who was consulted.
- To assess the fundamental reasons behind the final outcome.

11.12.3 Communicating with Management

The key points are

- To prepare how the bad news will be delivered after collecting and understanding the necessary details/facts.
- The scope for misinterpretation should be avoided, and the bad news should be communicated confidently.
- The audience should be warned in advance, and the bad news should not come as a surprise.
- The bad news should be fair, objective and complete as far as possible.
- The news should be communicated to managers quickly, and the truth should be told.
- Solutions should be worked out and presented to the managers instead of only communicating the problems.
- Maintaining regular communication with the managers will be important.

Source:

- Business Practice Notes [11]

11.12.4 Communication in the Event of Failure

In spite of using sophisticated technology, the quality of communication will be a key issue for the structural engineering profession. Being aware of safety, our profession is labour -intensive and extensively fragmented in its operations, so the scope for errors cannot be ruled out completely.

The professional institution can come forward and prepare the feasibility of developing

- a Code of Conduct to be incorporated for the members who may act as expert witnesses, providing information carefully that may have public benefit.
- The period of confidentiality may be limited to (say up to 6 years). The individuals involved might share the information with others after completing the limitation period. This would be a second choice due to the inherent delay.

For Criminal Investigation – The action may be

- Consideration of the current statutory provisions for health and safety and how communication can be used for better effect. How the principles could be particularly useful for the construction industry
- To study and examine foreign practices that may be useful and incorporated into our industry.

Source:

- Carpenter [12]

11.13 ACHIEVEMENT OF QUALITY

To achieve quality in engineering services related to design and construction, the following items appear to be important:

- **Setting out Quality Objectives** – These will be related to design basis reports, specifications, guidance on detailing, construction materials, and good construction and management practices.

- **Formulation of Quality System** – Procedures and methods should be identified to achieve the above objectives.
- **Implementation of the Quality System** – Procedures should be described.
- **Preparation of Quality Manual** – This will outline the quality system.
- **Human Resources**
- **Education and Training of Personnel** – Those who will be involved in the project.
- **Documentation of Methods and Procedures** – This will provide test reports, inspection reports, and audit reports.
- **Planning Product Realisation**
- **Maintenance of Records** – Covering all aspects of the projects.
- **Resource Management.**
- **MIS and Other Reports.**
- **Taking Corrective Actions** in all situations that may arise.
- **Checking/Auditing of the Quality** System procedure to ensure that it is working effectively.
- **Measurement, Analysis, and Improvement.**
- **Identifying/Monitoring** costs and financial benefits.
- **Good Communication** and regular interaction with the concerned persons will be important.

Source:

- Chakraborty [13]
- Ganguly [14]

11.14 ASSESSMENT FOR RELIABILITY

To carry out an assessment for reliability, the following items will need special consideration:

Scheme and Selection of Basic Materials	–	The adoption of a proper scheme along with selection of basic materials is applicable. It is well-known that some details will be more applicable for particular situations. Providing ductile detailing for the structure
Not to Allow over Stressing	–	Loadings and other details must be followed carefully so that overstressing is avoided, and an adequate safety margin is maintained for all possible conditions
Environment	–	Structural elements may be easily damaged in an unfriendly environment. In the design, proper and adequate actions should be taken so that the designed elements will be more reliable under the specific environment for which the structure is designed
Minimum Complexities	–	This should be the main target to achieve reliability in design and construction. The design should be construction-friendly. Initial interaction with the contractor regarding the scheme, assumptions, materials, and special comments on safety during execution will be helpful
Redundant Elements in the Design	–	Redundancy in design may indicate inadequate or standby elements that come into effect in unusual situations. This may call for additional load-bearing members or arrangements for energy dissipation, etc., to overcome such extreme conditions. To what extent this can be considered in the design of an important structure will need special thinking/discussion with others and for decision-making. This may mean some additional expenditure for the structure
Diversity	–	Diversity in design, especially with load-carrying members and detailing, will be relevant. Some elements may be identified, assumed, and designed carry additional loads. In case of a damaged member, which may occur due to carrying additional loads under a special circumstance, the identified member should be easily and quickly repaired
Protection to Structure	–	To minimise deterioration of structural materials over the years, it may be prudent to consider external protective treatment. This will ensure a longer life and will increase the reliability assessment with respect to design and construction. Discussion with the regular/competent manufacturer may be required

(*Continued*)

(*Continued*)

Assessment for System Reliability	–	Some exercises for system reliability may be carried out with repeated trials on the actual projects or structures. Changes or remedies, if required, should be addressed immediately

Source: Ganguly [14]

11.15 QUALITY CONTROL IN PRACTICE

Some relevant comments are made for quality control in practice, especially related to pre-cast industries.

The basic and important items may be

- Quality as a way of life.
- Uniform competition.
- Speed and Economy.
- Learning from history.
- Selection of cladding specialist – as applicable.
- Quality assurance – selection of quality assessors to visit and assess.
- Quality in management.

Source:

- Quinn [15]

REFERENCES

1. Farmer I, Using computers in engineering calculations, *The Structural Engineer*, 5 February 2002, pp. 8.
2. Ganguly K K, Applications of structural engineering in projects and other associated issues, *Indian Concrete Journal*, January 2008, Vol. 82(1), pp. 41–56.
3. Macleod I A, A strategy for the use of computers in structural engineering, *The Structural Engineer*, November 1995, Vol. 73, pp. 366–370.
4. Miller A, Productivity in 3D modelling and structural design, *The Structural Engineer*, 5 February 2002, pp. 14–16.
5. Advantages and Disadvantages of using Computer Aided Design (CAD), ARCVERTEX, pp. 1–5. https://www.arcvertex.com/article/advantages-and-disadvantages-of-using-computer-aided-design-cad/. Accessed Oct 07, 2024.
6. Stasfield K, The role of design in engineering education, *The Structural Engineer*, 5 April 2005, Vol. 83(7), pp. 16–17.
7. Newlands M and Jones R, A concrete education, *Concrete*, June 2018, pp. 38–40.
8. Richardson A, Meeting the demands of industry and employers, *Concrete*, June 2018, Vol. 52(5), pp. 41–42.
9. Importance of Effective Communication in Construction, UK Essays/Construction, August 2018. https://www.ukessays.com/essays/construction/the-importance-of-effective-communication-in-construction-construction-essay.php. Accessed Oct 07, 2024.
10. Big Rentz, Inc., August 2019, pp. 1–19.
11. Yusuf J, Business Practice Notes, Effective Communication, No. 18, The Structural Engineers, September 2018, Vol. 96(9), pp. 22–23. doi: 10.56330/MWIJ4982.
12. Carpenter J, Viewpoint: Communication in the event of failure, view point, *The Structural Engineer*, September 2014, Vol. 92(9), pp. 54–55.
13. Chakraborty G, Quality assurance in construction, *Joint Plant Committee*, September 2002, Kolkata.
14. Ganguly K K, Quality control and codes of practice, in *Conference on Infrastructure Development in India* organised by The Institution of Civil Engineers, UK, The Institution of Engineers (India) and CES Pvt. Ltd., November 2001, Kolkata.
15. Quinn K N, Quality control in practice, *Concrete*, April 1984, pp. 8–14.

12 Sustainability/Resilience, Hazards, Risks, and Structural and Site Safety

12.1 COMMENTS ON SUSTAINABILITY/RESILIENCE

The items sustainability and resilience, in design, may cover a wide range of areas. The comments/discussion on sustainability have been made previously.

However, some additional comments may be added on the following items:

12.1.1 Resilient Design

12.1.1.1 Introduction

In addition to considering the usual stress conditions of structural elements during normal use, challenges must take into account the effects from the environment that may pose dangers to the safety of the structure and its occupants.

The environmental effects will be important for making the building/structure resilient.

Hazards such as hurricanes, storm surges, flooding, blizzards, seismic effects, and heat loads may be important considerations for resilient design, alongside man-made hazards.

The relevant items may include

- Seismic design
- Resiliency and commercial building design
- Emergency resiliency and disaster preparedness.
- Extreme weather – storms, hurricanes, flooding, etc.
- Extreme heat and cold.
- Fire resistance/performance.
- Infrastructure failure and power outages.
- Everyday resiliency and normal wear and tear.
- Durable building envelopes.
- Interior finishes.

Comments: Resilient design may be quite complex, involving long-term consideration of worst-case disaster events alongside common ones.

Source:

- Fehrenbacher [1]

12.2 FURTHER ON RESILIENCE IN DESIGN

It is possible that a structure may be subjected to extreme effects. These may occur naturally or accidentally.

Earthquakes may be classified as natural events, while accidental effects may include blasts, extreme fires from explosions, or intentional events like bombing aimed at creating disruptions.

DOI: 10.1201/9781003319108-12

The term resilience can be defined as the capacity to recover quickly from difficulties and return to a previous good condition.

It is associated with the ability to absorb, adapt, and quickly recover from disruptive events.

Resilience may be achieved through a combination of different activities. The main and relevant items in this respect are resistance, redundancy, reliability, and response and recovery.

Resistance – Strength provisions are incorporated into the structure through traditional methods to resist hazards. The safety margin is based on acceptable failure probability.

Strength/protection alone may not be adequate.

Redundancy – This is associated with additional backup or spare capacity in the structural system. To control disproportionate collapse, it is possible to introduce redundancy into the structure.

The efforts may be made to develop alternative paths for transferring the loads (wherever possible) rather than depending on a single path.

Reliability – The components of the infrastructure must manage and mitigate the damage or loss resulting from events.

Various components may perform in different ways. All relevant components in the structure must be properly considered at the design stage.

Response and Recovery – This will be connected with proper planning, preparation, and practice well before the events. Pre-event planning, such as exploring opportunities to reduce risks or building resistance into the system during the recovery stage, will be helpful.

Passive provisions must be supported by active management to aim for effective solutions.

Resilient design may be achieved through combinations of the above items.

The client must be properly informed about expectations regarding the design and maintenance of the building/structures to achieve resilient design.

Source:

- Navaratnarajah [2]

12.3 USING CIRCULAR TECHNIQUE IN DESIGN

The items are as follows:

- **Maximum Reuse –** Reusing existing assets to recover materials/products on-site or from another site, and to share products or materials for further reuse.
- **Design to Aim Optimisation –** Longevity, flexibility, adaptability, assembly, disassembly, and recoverability.
- Standardisation should be followed.
- Products for use.
- **To Reduce Impact and Waste –** To use low-impact new materials.
    - To reduce waste in design.
    - To use recycled or secondary materials.
    - To minimise construction impacts.

The relevant items in this connection are as follows:

- Preparation and briefing the design.

Briefing on circular design techniques from the beginning will be important.

- Proper co-ordination

The client and the design team must be properly informed regarding the reuse of materials with minimum risks.

- Concept design

Reuse and consideration of disassembly in the future; the design approach must be sorted out.

- Technical design issues
- Construction
- Handover
- Use and
- Maintenance

Source:

- MacNamara [3]

12.4 FURTHER COMMENTS

12.4.1 Lean Design

The intention of lean design is to make the project more efficient to increase additional value with fewer resources. This may be achieved through a collaborative working approach instead of drawing from the spare capacity of the structure. The approach will mean working in an integrated manner for the design in order to provide various benefits.

12.4.2 More on Resilient Design

This type of design aims to include the capacity of the structure to absorb and accommodate the changing environment. The buildings should have the capacity to accommodate short-term hazards (i.e., shocks) and be adaptable for long-term changes (i.e., stresses). These may be related to climate change, changes in technologies, lack of maintenance, and degradation of the materials as mentioned.

Resilient buildings will be able to retain business function by protecting critical resources.

It will be important to assess the tolerable impact that a loss in function may cause.

The critical item will be the recovery time objective or return to service, and these should be discussed with the client and agreed upon.

12.4.3 Design Strategies for Resilience

- To include resilience issues in design.

It will depend on the risk and the performance needs of the building. It may not be costly or mean physical strengthening only.

- Consideration of risks and uncertainties for resilience in design

Some items for risks and uncertainties, in design, may include material properties, loads, and uncertainties in modelling.

The role of structural engineering for lean and resilient design.

The items involve

Design Approach – Mitigation and adaptation strategies are required to be handled throughout the life cycle of the building.

Scenario Analysis – Potential future scenarios that may occur and how to handle.

Performance Requirements – To assess the level of resilience to be considered in the design of the building.

Risk Assessment – To consider the type of structure, experience in working in geographical areas, and expected quality of construction. Furthermore, other items may include short-term hazards like impacts, fires, floods, or explosions. Additionally, long-term hazards may include climate change, lack of proper maintenance, and using the structure beyond its design life.

Design for Adaptability – Resilient buildings should be adaptable to future changes. These may cover

- The ease of upgrading the structure with the availability of new technology and time.
- Assessing and understanding the carbon content already present in the design, with attempts made to reduce it.
- Designing for flexibility of use
- Consideration of climate change in the design.

Stress Testing – Stress testing may be considered for assessment.

Collaboration – Structural engineers should work with all concerned persons, including owners and operators.

Time Monitoring – Situational awareness will be critical. Sensors may be used for monitoring purposes. Necessary actions should be taken based on the information/data received and recorded.

Repair and Maintenance – Repair and maintenance work should not be complicated and should be carried out easily without depending on specialists.

Source:

- Field [4]

12.5 MORE ON LEAN DESIGN

The items are

- Avoid rebuilding or construction.
- Utilisation must be maximised.
- Existing buildings should be upgraded wherever feasible.
- Serviceability criteria to be reviewed. These are
  - Facades
  - Long-span beams and slab deflections
  - Internal partitions
  - Dynamic criteria
  - Crack width requirements
- Loading criteria to be reviewed.
  - Minimum appropriate imposed loading.
- Requirements should be reviewed, and attempts should be made to reduce grids and floor slabs.
- Designs should be suitable for the present and may be strengthened if the use changes.
- The substructure should be minimised and designed/considered properly to
  - Eliminate basements and suspended floor slabs wherever possible.
  - The superstructure and the site should be studied thoroughly to reduce foundation work, sizes, etc. The superstructure should be lightweight.

    - Use low cement content concrete in the mix.
    - Use reclaimed steel tubing instead of steel piles.
    - The building design may be based on 56 days' strength.
    - Settlement criteria should be reviewed.
    - Adequate ground and site investigation will be helpful.
- The proper type of cement should be used.
    - A 20% cement replacement may be aimed for the building.
- Learning and sharing knowledge with others on the team.

Source:

- Watson [5]

12.6 REVIEW ON BUILDING-RESILIENCE CONSIDERATIONS

12.6.1 Comments on Hazards

12.6.1.1 Introduction [6]

Resilience in design will be associated with increasing the robustness of the structure and ensuring the guaranteed redundancies required to provide the assumed performance in response to hazards. The resourcefulness of the system recovery will be important.

To increase resilience, efforts will mainly focus on strengthening structures and minimising the impact from natural or man-made hazards.

12.6.1.2 Hazards [6]

Hazards can create threats to life, property, the environment, and personal integrity.

In a project, hazards may affect the success of project activities and the overall project.

Hazards may be potential or latent and can become active or effective, generating emergency situations. Hazardous conditions may cause an incident, accident, or disaster.

12.6.1.3 Comments on Uncertainty and Risk [6]

The meanings are as follows:

Uncertainty – This is connected with the occurrence of an incident that may not be fully known.
Risk – This may be the result of an event that can be assessed based on statistical probability.

Source:

- ICE Wiki [6]

12.7 FURTHER ON HAZARD-SPECIFIC BUILDING RESILIENCE CONSIDERATIONS

The following items will be relevant and will need careful consideration in design related to hazard-specific building resilience design:

- **Durability** – already commented
- **Climate change**
- **Long-Term Performance** – will be important and will reduce maintenance work
- **Technological obsolescence (i.e., system maintainability and ease of replacing/upgrading)**

- **Maintainability**
- **Safety**
- **Consideration of gravitational and lateral loads correctly**
- **Wind effects**
- **Seismic effects**
- **Shelter**
- **Flood**
- **Fire**

12.7.1 Security

A few comments on security are as follows:

A physical attack on a building may potentially disrupt its operations and functioning. The extent of the disruption will depend upon the magnitude and duration of the attack. The provision of security measures and hardening items associated with facility design will help provide robustness against such attacks. Design provisions must consider post-event security measures after the disruption.

Other Important Items may be

- **Design for blast resistance.**
- **Radiological and nuclear protection – may be necessary for some structures.**
- **Chemical and biological protection – may be necessary for some structures.**
- **Ballistic protection.**
- **Forced entry/physical attack resistance.**
- **Cyber security.**
- **Defending building control systems.**

Source:

- WBDG [7]

12.8 COMMENTS ON STRUCTURES FOR PROTECTION AGAINST HAZARDS

With the changing requirements, structural engineers must consider protections for structures under hazardous conditions. It is not possible to design structures against unknown or indefinable conditions.

Engineering judgement supported by experience will play a vital role in designing structures against hazards. The important aspect will be to protect the structure for the safety of the people and to avoid total collapse. Guidance may not be available to design or handle such hazards. Load tests may be required to obtain certain inputs. Partial collapse may be acceptable, but the emphasis should be on avoiding total collapse and saving human lives, as mentioned before. Full security against all hazards may not be possible, but structural robustness and redundancies can improve protection against hazards at a finite cost.

12.8.1 Types of Hazards

The following may be mentioned apart from hurricanes, floods, seismic effects etc.:

Missile Impact – Two types may be possible. One would be a hard type approaching at high speed and may cause damage/perforation to the structural elements. The other type may be a soft missile that crumples upon impact and produces a collapse mechanism similar to static loading, which can extend widely. The use of fibres or lightweight concrete sandwich elements may be used to increase impact resistance.

Explosion – Two cases may need to be considered. One may be unconfined, i.e., open air and its impact on the structure. The other will be a confined type, with the explosion occurring inside the structure. The type of building materials and the detailing may become relevant.

Temperature Effects – Both high and low temperatures may occur; hence, these must be considered in the design. For cryogenic storage conditions, pre-stressing may be preferred to ensure liquid tightness.

Fire – The effects may vary widely. Concrete may be considered a suitable material for fire resistance. Containment will be an important issue to aim for.

Cryogenic Conditions – Concrete would be suitable for cryogenic use. However, the performance of reinforcing bars at low temperatures may need to be examined very carefully.

Nuclear Radiation – The use of heavy concrete would improve shielding performance.

Concrete, along with other combinations of materials, would be suitable for hazards, and knowledge of the materials will be required to utilise their full potential for different applications against hazards.

Source:

- Long [8]
- Ganguly [9]

12.9 MORE ON HAZARDS, RISKS AND STRUCTURAL SAFETY [10]

The matter related to risks and safety may be reviewed to prevent problems. This can be categorised as follows:

12.9.1 Different Hazards in Structural Design

Problems may develop due to

- Inexperienced and incompetent persons/staff being involved in the design.
- Variations in ground conditions.
- Dead, imposed, temperature, and snow loads (as applicable) not being correctly considered.
- Degradation processes, e.g., corrosion of bars, chemical attack, etc., not being assessed properly in the design.
- Impact loads from vehicles, fire, explosion, flood, tsunami, earthquake, etc., may be applicable.
- Errors in the design and specification process such as
  - Non-compliance with codes.
  - Gross errors.
  - Current and updated information not being known/considered.
  - Misunderstandings and mistakes.
  - Future operational hazards not being considered.
  - Improper interactions/communication amongst concerned persons.
  - Quality design checks to arrest mistakes were missing.

12.9.2 Different Hazards in Construction

The items may include

- Involvement of inexperienced/incompetent staff in the construction work.
- Use of materials not per the specifications and non-compliance with construction drawings and documents.
- Inadequate design for temporary works.
- Local overloading.

- Unforeseen ground conditions; proper tests were avoided for assessment.
- Fire, explosion, impact, and hazardous substances.
- Power/plant failure.
- Extreme external events, i.e., flood, wind, temperature, snow (as applicable), etc.
- Supervision of construction activities was not carried out properly.

12.9.3 Different Hazards during Use

These may be summarised as follows:

- Exceptional natural events like floods, wind, temperature changes, earthquakes, and snow (as applicable), etc.
- Unforeseen movement and degradation.
- Inspection and maintenance were poor during use.
- Fire explosion impact.
- Change of use.
- Man-made accidents.
- Misuse – overloading, overcrowding, public demonstrations.
- Spillage of harmful substances.

Source:

- Menzies [10]
- Ganguly [9]

12.10 CARE AND ATTENTION

12.10.1 Introduction

Contact with wet concrete can be serious and may result in allergic reactions depending on the exposure. Wet cement is caustic and abrasive.

The use of protective clothing is important, including overalls, goggles, and gloves to protect the eyes and hands.

Waterproof knee pads may be used. Kneeling on concrete may cause the mix to seep through the protective overalls.

The most serious concrete injuries may take months to heal, and in some cases, skin grafting may be needed.

12.10.2 Further Comments

The employer must take the initiative to assess health risks and to regulate and prevent exposures.

Adequate welfare facilities must be arranged/provided for washing with both cold and hot running water, including changing and drying clothes. Employers must provide adequate personal protective equipment for employees.

12.10.3 Risk Assessments

The risk assessment may cover

- Assessment of hazards
- Determining whether the present arrangements are adequate or if improvements are necessary.

- Identifying who might be affected and how.
- Recording all assessment findings/details and discussing them with employees and other concerned persons.
- Regular review/assessment of the facilities. Improvements that may be required, especially after studying an accident/injury, must be arranged.

Source:

- Brooks [11]

12.11 DESIGN (FUTURE) TREND

The future trend, as mentioned before, will still be applicable now. However, some items are briefly reviewed in this connection.

12.11.1 Design for Security [12]

In recent years, security design has taken a prominent place, especially for buildings. Architects and engineers must understand the different aspects related to the safety of building components. Some basic issues associated with building design are mentioned briefly.

Blast – Blast pressures decrease significantly with distance. A blast within 5–10 m can cause damage to modern framed buildings and major damage to load-bearing masonry buildings. A canopy at the entrance can trap and enhance blast pressure.

Bomb Threats – This can cause significant local damage if detonated inside the building. Car and lorry bombs can also damage nearby structures. The use of chemicals is also dangerous, as they may be more harmful to people than to buildings. Much will depend on the size of the bomb, type of explosive, and how it is detonated.

Blast on Buildings – The ideal shape of a building against blast forces will be circular. Keeping the facade clean and smooth, i.e., without fins, balconies will be important; otherwise, blast forces may be amplified. The provision of a central core may provide space for shelter during an emergency. The use of stone cladding in buildings may not perform well. The consideration of replacement techniques for stone or glass when damaged, especially in multi-storey buildings, will be important. If glass can be replaced from the inside, it will save money.

Blast on Glass – Glass is often used in the façade of a building. This is a high-risk material and can cause a lot of damage when exposed to a bomb blast. Flying glass pieces can be responsible for most injuries. The type of glass used in the building will be important. A blast can destroy all the windows in a building while leaving the structure undamaged. Hence, care should be taken to ensure that the glass and the frame remain intact. Unprotected annealed plain glass will break from a blast up to 100 m away and cause potential injury up to 50 m from the place of explosion.

Toughened glass, on the contrary, will offer twice the protection of plain glass. Laminated glass can also offer good performance with proper frame design.

12.11.2 Structural Safety

Periodic inspection of the structure covering salient load-bearing elements will be very relevant in identifying weak areas and avoiding catastrophic collapse that may take place later. Based on the inspection, it will be necessary to provide warnings to engineers to carry out preventive measures quickly. Inadequacies in the structure may often manifest initially by the loss of serviceability and may finally lead to safety issues from corrosion of reinforcement, spalling of concrete, and large deflections. Errors in calculations may indicate that safety is much less than intended. During construction, situations may arise where safety will be in jeopardy. In all cases, detection by an alert engineer in advance will be required to avert catastrophic failure and ensure safety.

Source:

- Stansfield [12]
- Ganguly [9]

12.12 CAUSES OF ACCIDENTS IN CONSTRUCTION SITES

This item has been covered in Chapter 13.

Source:

- The top 3 causes of Accidents in Construction Site [13]

12.13 SITE SAFETY RULES

The following may be mentioned as site safety rules to prevent accidents:

Use PPE at all times – Keep the site tidy, do not start work without an induction, do not place oneself or others at risk, do not work in unsafe areas, follow safety sign and procedures, report defects and near-miss incidents, use proper/right equipment, do not tamper with equipment, and ask if in doubt.

Source:

- Simple Construction Site Safety Rules [14]

12.13.1 Further on Safety Measures

Theses may be mentioned as follows:

Safety training before starting, site security – restricted site access. Pedestrians are to be protected against construction hazards, manage and minimise risks – safety audits will be important, adopt safe work methods, clear signage on the site, entry and exit points – separate entry and exit points should be provided, chemical storage – to be stored carefully to avoid fires, environmental conditions – extreme weather conditions may create safety hazards, first aid facility – to appoint first aid officers, one person for every 25 workers, use PPE – as already stated, dropping of objects – the risk of falling objects must be reduced.

Source:

- Safety Measures Every Construction Site [15]

12.14 ADDITIONAL COMMENTS ON SITE SAFETY

The following comments may be summarised:

12.14.1 General Items and Emphasis on Safety

Designers can reduce the risks related to safety and health in the construction stage, during use, and subsequent maintenance. The emphasis is on health and safety to protect the lives of people, which are associated with the client/owner, designers, contractors, and site operations. The designer's role will be important.

12.14.2 General Methods of Precautions

These are

Unavoidable risks to be evaluated, risks to be avoided, risks at source to be combated, development of a prevention policy that may be adequate, distribution of work to the individual, particularly in the design of workspaces, adoption of technical progress in the workplace, avoidance of risky items, substances, or systems, and replacing them with non-dangerous or less dangerous items/substances. The emphasis should be on collective protective measures instead of individual measures. Appropriate training and instructions should be given to employees/workers.

Source:

- Ganguly [9]

More comments on minimising health hazards [9]

- Attempts should be made to use materials that are less hazardous, such as low-solvent adhesives and water-based paints.
- Avoid creating hazardous fumes, including dust, vapours, fumes, noise, vibration, and disturbance of existing asbestos, as well as making chases in the brickwork.
- Easy-to-handle materials should be specified.
- Access areas must be properly designed to accommodate work-at-height equipment.
- Block-paved areas should be adequately designed to be suitable for mechanical handling and laying of units

Source:

- Ganguly [9]

12.14.3 Specific Items [16]

The items mentioned below are relevant:

- **Fall protection**

  Workers – Should be aware of all hazards from falls and the protection systems available on-site. Workers may inspect items if required. They must be aware of the instructions to minimise hazards and prevent falls.

  Employers/Owners – Employers should provide all protective systems as recommended to protect workers before starting the work. Training should be arranged for workers who may be exposed to fall hazards.
- **Using Scaffolds**

  Workers – Hard hats should be worn when working on, under, or around scaffolds. Workers must be trained and authorised.

  Employers/Owners – A competent person should be involved in designing, erecting, and disassembling scaffolds. They must ensure that all workers are trained and can inspect/check the scaffolds before use.
- **Stairways and Ladders**

  Workers – Three points of contact must be maintained, i.e., using both feet and at least one hand while ascending and descending a ladder.

  Employers/Owners – Should arrange to inspect all ladders before use each day.
- **Personal Protective and Life-Saving Equipment**

Workers – Must use head protection while working on the construction site to protect against falling objects, tools, debris, etc.

Employers/Owners – Should provide all employees with necessary head protection as per the standards.

- **Toxic and Hazardous Substances**

  Workers – Must read and understand the Material Safety Data Sheet related to hazardous/chemical items being used on-site.

  Employers/Owners – Should implement a communication programme for all hazardous chemicals, including proper handling instructions.

- **General and Health Provisions**

  Workers – Must be aware that protection measures are in place while working on-site.

  Employers/Owners – Must implement safety programmes for the protection of workers. A competent person must be involved in inspecting and ensuring the safety of items on the construction site.

Source:

- Designing for safety, HS & A [16]
- 10 Construction Site Safety Tips [17]
- Ganguly [9]

12.15 REVIEW AND SUMMARY OF SAFETY RULES

These may be mentioned as -

Training, awareness, documentation, communication/reporting, supervision/checking, use of proper equipment, innovation, work should not be carried out in a hurry, use of PPE at all times, transparency.

12.15.1 Special Comments on COVID-19 Situation and Guidelines

Employees/workers may stay at home if they are sick. It must be ensured that the workers are using masks to cover their mouths and noses while working on-site. staggered timing, restrictions on visitors, thermal checking, hand sanitising, construction waste must be removed from the site in covered containers, stores should be disinfected daily, use of digital format, items that cannot be disinfected should be kept in containers for 72 hours, all vehicles entering and leaving the site should be disinfected, workers should be instructed to avoid physical/direct contact with others, a social distance of 6 ft between persons must be enforced, use of PPE as mentioned, workers must be trained to use PPE and clothing properly, wash stations with access to soap and water for handwashing must be maintained, additional areas for handwashing, toilets must be disinfected daily, Do's and Don'ts posters for safety in colony premises, using the language spoken by the labourers.

The workers must be aware of safety issues. In this respect proper communication will be very important. Regular supervision/checks will be necessary to reduce infections/accidents/injuries.

Source:

- 8 Best Practices to Improve Construction Site Safety [18]

12.16 EFFECTS OF SILICA DUST [19]

People working in the concrete repair industry for a long time may suffer from substances hazardous to health, including

- Hand–arm vibration syndrome
- Noise-induced hearing loss
- Skin and lung disease

Individuals must be educated and trained to understand and develop engineering controls, and designers must be involved with Construction Design Management regulations to develop solutions that will be safe.

The use of more grinders, breakers, and drills has increased considerably in the construction industry, generating a large amount of dust on-site. Control over the generation of silica dust and health hazard protection considerations have not been properly recommended and have not received enough attention. The primary important item will be to create safe systems of work and to reduce or eliminate exposure to dust.

It will be necessary to understand, in detail, the risk assessment along with a health surveillance programme for the workplace. The workplace must be made safe along with a personal air monitoring method.

Source:

- Roper [9]

12.17 PRESENCE OF CO_2 IN THE AIR [20]

The presence of CO_2 in the air can be hazardous, and its concentration level is important for the working environment. Its current level in the air is approximately 0. 038%. It may be higher in other places.

Considering health risks associated with CO_2, the average exposure of a health employee working for 8 hours should not exceed 0.5%.

12.17.1 Carbon Dioxide and Physiological Effects

Some values are indicated below in Table 12.1 related to the possible effects on people with the volume of carbon dioxide in the air:

Measurement of both oxygen and carbon dioxide in the air will be important for assessment.

Source:

- EIGA [20]
- Ganguly [9]

REFERENCES

1. Fehrenbacher J, Resilient Design: Is Resilience the New Sustainability? *Inhabitat,* December 2013, pp. 1–19. https://inhabitat.com/resilient-design-is-resilience-the-new-sustainability/. Accessed Oct 08, 2024.
2. Navaratnarajah K, Engineering towards a resilient world - managing client expectations, *The Structural Engineer: Journal of the Institution of Structural Engineer*, April 2016, Vol. 94(4), pp. 10–15.
3. Macnamara E, Applying circular principles to the design process, *The Structural* Engineer*: Journal of the Institution of Structural Engineer*, August 2020, Vol. 98(8), pp. 9–11.
4. Field C, Lean yet resilient - designing for the future, *The Structural Engineer: Journal of the Institution of Structural Engineer*, August 2020, Vol. 98(8), pp. 15–17.
5. Watson N, Lean design: 10 things to do now, *The Structural Engineer: Journal of the Institution of Structural Engineer*, August 2020, Vol. 98(8), pp. 12–14.
6. Risk in Building Design and Construction, ICE Wiki, November 2022. https://www.designingbuildings.co.uk/wiki/Risk_in_building_design_and_construction. Accessed Oct 08, 2024.

7. Hazard-Specific Building Resilience Considerations, WBDG, Whole Building Guide, August 2018, pp. 1–30. https://www.wbdg.org/resources/hazard-specific-building-resilience-considerations. Accessed Oct 08, 2024.
8. Long J E, Concrete in hazard protection, *Concrete (London)*, January 1986, Vol. 20(1), pp. 35–36.
9. Ganguly K K, *Reinforced Concrete Design, Repair/Restoration, Strengthening of RC Structures and Concrete Technology*, Medtech, 2020.
10. Menzies J B, Hazards, risks and structural safety, *The Structural Engineer*, November 1995, Vol. 73(21), pp. 357–363. Accessed: Oct 08, 2024. [Online].
11. Brooks T, Care and attention, Concrete, February 2019, pp. 42.
12. Stansfield K, Report on designing for security, *The Structural Engineer*, October 2001, pp. 16–18.
13. The Top Three Causes of Accidents in Construction Sites, Gauze Mat Brakes, December 2017, pp. 1–17.
14. 10 Simple Construction Site Safety Rules - HASpod, January 2023, pp. 1–11. https://www.haspod.com/blog/construction/10-simple-construction-site-safety-rules. Accessed Oct 08, 2024.
15. 11 Safety Measures Every Construction Site Should Have in Place, LATHAM, pp. 1–6. https://www.latham-australia.com/blog/safety-measures-construction-site. Accessed Oct 08, 2024.
16. Designing for Safety, HS & A, pp. 1–4. https://www.hsa.ie/eng/your_industry/construction/designing_for_safety/. Accessed Oct 08, 2024.
17. 10 Simple Construction Site Safety Rules – HASpod, January 2023, pp. 1–21. https://www.haspod.com/blog/construction/10-simple-construction-site-safety-rules. Accessed Oct 08, 2024.
18. 8 Best Practices to Improve Construction Site Safety, e SUB Project Management, December 2018, pp. 1–10. https://esub.com/blog/improve-construction-site-safety/. Accessed Oct 08, 2024.
19. Roper N, Industry must take action to control exposure to silica dust, Concrete, February 2013, pp. 18–19.
20. European Industrial Gases Association (EIGA), Safety Advisory Council, Safety Information 24/11/E, pp. 1–3.

13 Construction, Supervision, Site Accidents, Project, and Risk Management

13.1 CONSTRUCTION WORK

In industrial projects, it was possible to attract most of the resourceful contractors in the country. Since the total volume of civil work is typically high, it has been profitable for contractors with high turnover to participate in tendering. It has been observed that some of these contractors have attempted to use modern methods and improve the quality of construction. However, much more needs to be done quickly to ensure consistent quality using proper equipment and tools to produce high-strength and durable construction. With these improvements, better performance may be expected in future projects. The author feels that future improvements in construction will include the following areas:

- Better-quality formwork.
- The use of different building chemicals to ensure early high strength, workable, and durable concrete and other materials with better-finished surfaces.
- Increased use of equipment for handling materials (i.e., cranes and conveyors for weigh batching).
- More rigorous checks before accepting raw materials, ensuring good and consistent quality of cement for concrete work.

These may be

- Use of ready-mix concrete for construction.
- Pumping concrete to complete larger volumes of work.
- Greater use of pre-cast units in construction.
- Proper and effective curing of structural components.
- Better staging and scaffolding work.
- Rigorous checks and supervision during different construction activities.
- Better training for personnel to understand all important aspects of materials and construction.
- Increased use of building chemicals in the form of bonding aids, curing compounds, and proper concreting items to ensure more durable and impermeable construction.
- Strict adherence to quality control procedures.

Source:

- Ganguly [1]

13.2 CONSTRUCTION MATERIALS

Different materials for construction have been tested by engineers and architects over the years. Among these, concrete and structural steel have been the most widely used. Other materials include brick, stone, timber, glass, and different composites. The figure presented in Chapter 5

DOI: 10.1201/9781003319108-13

illustrates a combination of load-bearing rammed earth construction with timber roofing. This application was implemented by the architect in the north-eastern part of the country. Timber trusses were chosen to suit local conditions. Fly ash and lime-sand bricks have also been utilised in many recent projects. Additional photographs will demonstrate the use of compressed earth blocks (CEBs) as a replacement for conventional burnt bricks. By applying hand-operated machines, CEBs have been produced as mentioned in Chapter 3. Different types of arches, domes, etc. have been constructed using CEBs, as shown. Applications of ferro-cement roofing units have also been presented.

Source:

- Ganguly [2]

13.3 SITE SUPERVISION

In recent years, it has been observed that only periodic supervision is often given to the consulting engineers for some large projects to cut down on fees. The contract is prepared accordingly. Hence, detailed supervision is not carried out by the design team/consultant. The structural design engineer from the consulting house visits sites occasionally as per the contract and site requirements. Continuous supervision may be lacking if the structural engineer is not present from the design office of the consulting house. It would be beneficial if the design engineer were involved in the site supervision activities. However, for other projects, such as large industrial or power projects, proper supervision may be required from the consulting house with project management and full site supervision services in some cases. It is vitally important for the structural designer to have opportunities for continuous supervision, at least for a period, to observe the construction of the different structural elements he has already designed. This way, the designer will gain experience and understand the sequence of different aspects of construction as followed on site, ultimately becoming a better designer. Furthermore, the cost aspect, variation of quantities (due to incomplete details in some cases), non-scheduled items (due to changes made later), etc., are experienced, and these issues are transformed into meaningful realisations for the structural designer. Structural design does not simply mean carrying out an analysis mechanically by using software and providing reinforcement in the elements based on the computer output. The art of economical structural design must be developed and learned in stages through experience. This approach is relevant, including the codal guidance for design engineers. The design engineer should be given the opportunity to practice and learn. Experience from construction activities will be very important for the designer.

Source:

- Ganguly [2]

13.4 REVIEW OF SUSTAINABLE CONSTRUCTION

The sustainable construction rules may indicate a 4Rs cycle, i.e., reduced, refurbish, recycle, and reuse.

If this approach is adopted, it will benefit our planet. The following may be mentioned in this connection:

- Reducing the emission of greenhouse gases.
- Controlling and constructively reusing waste.
- Efficient use of resources.
- Reducing the negative effects of construction activities.

13.4.1 Further Comments on Sustainability

The items are

- Controlling/reducing disposal of soil offsite.
- Design loads to cover seismic, typhoon, and groundwater effects. Avoiding the need to over-design the structure.
- Standardising the design of formwork and temporary work.
- Assessing for the design life of the structure. The reality will be the service life.
- Selecting structural components that can be reused or recycled from other projects.
- Making attempts to reuse and recycled materials whenever possible.
- Incorporating used elements in the design of new buildings to reduce initial costs.
- The structural adaptability for future use and suitability for alternative uses.
- The calculations, along with relevant records, should be available to assist in reuse and adaptation.
- As-built documents/details may be important for ownership and demolition activities.
- Pre-fabrication items may be preferable to reduce waste and facilitate recycling later.
- The joints may be designed for dismantling and demolition, including the reuse of the same materials for future assembly.
- The contractors may be asked to prepare an environmental and waste management statement along with the tenders.
- Information on operational energy, embodied energy, and the maximum uptake of new materials for performance may be important.
- The building records must be updated by the users to cover the modifications undertaken.

It is needless to mention that the resulting product must be comfortable, inspiring, and pleasant.

Source:

- Dickson [3]

13.5 CAUSES OF ACCIDENTS IN CONSTRUCTION SITES

The main reasons for accidents on construction sites in India may be

Accidents can occur due to various reasons on a construction site. These may involve falls or slips, electrocution, ladder accidents, scaffolding accidents, the use of ineffective equipment, unsafe working conditions, and lack of adequate safety measures.

These will be briefly discussed below:

Falls – Accidents due to falls may be the leading cause of deaths on construction sites. The number of deaths may exceed 10,000 in a year. Construction safety plans must be properly executed to protect workers from falling from heights. A few items may be mentioned as

- Portable ladders not used correctly
- No protection for sides and large cut-outs
- Inadequate scaffolding construction

13.5.1 Fire

Fires have caused over 1,000 deaths in 2013 in India.

Incomplete electrical systems and unfinished piping may be responsible for fires/explosions on construction sites.

More common causes of industrial fires may be

- Hot work
- Equipment and machinery
- Combustible dust
- Flammable liquids and gases
- Electrical hazards

Electrocution – Deaths from electrocution are also significant on construction sites. The number of deaths may exceed 10,000 per year in India. Workers may be exposed to power lines, wires, and unfinished electrical systems. They may not be aware of the exposed lines/wirings.

Shocks or electrocution may cause different types of injuries, internal wounds, and cardiac arrest in some cases.

13.5.2 Consequences

Serious accidents may cause life-changing injuries to workers. Some accidents may require expensive treatments, leading to high medical bills that can keep workers away from work. This may lead to significant financial problems for the affected families.

Many accidents can be avoided by taking preventive measures and fostering a sense of responsibility among workers.

Prevention will include adequate awareness and maintaining a proper working environment that is safe and secure.

Source:

- The top three causes of accidents [4]

Further brief causes may include

Falls from heights or scaffolding as mentioned, falling debris, materials or objects, electrocutions, getting caught between objects or materials, overexertion, fires and explosions, being hit by a vehicle, machinery accidents, etc.

Consequences can affect health and livelihood, as stated.

Source:

- 10 Most Common Construction Site Accidents [5]

13.6 BASIC RISKS IN CONSTRUCTION PROJECTS

The general/common items in this context may include interface, risks, tax risks, and local site risks.

Other items may include

- Scope, performance, quality, and related technical items.
- Scope, cost, and schedule may not be clear.
- Environmental, safety, and health issues.
- Political interference/concern.

The risks may differ for each project.

A few common risk items may be mentioned as follows:

13.6.1 Construction

Changes in work, site access, geological and geotechnical issues, quality of detailed design provided by the client, delays in drawings/instructions, accidents, resource availability, damage to property or persons, faulty design, cost of testing and preparing samples, the actual quantities of the work and variations, equipment, and commissioning.

13.6.2 Performance

Productivity of labour, suitability of materials, equipment productivity, defective work, hindrance to the conduct and performance of work, labour disputes, etc.

13.6.3 Security

Terrorism, vandalism, corruption, negligence, assaults, intrusion.

13.6.4 Other Issues

Contractual and legal
Financial items
Physical
Political and societal

13.6.5 Technical Risks

The items may be mentioned as follows:

13.6.5.1 Design Work

- Owner involvement in design.
- Changes in seismic design.
- Incomplete and inadequate design work.
- Errors in completing structural/foundation work.
- Improper selection of materials.
- Take-off information (i.e., traffic demand/water consumption requirement)
- The need for design exceptions.

13.6.6 Construction Risks

- Incorrect contact time schedule/estimate.
- Work permissions.
- Construction procedures.
- Construction occupational safety.
- Surveys done late, incomplete, and incorrectly.
- Utilities.
- Late deliveries and disruptions.
- Innovative projects.
- Project site safety and worker.
- Unsuitable equipment and materials.
- Environmental factors (projects close to rivers, floodplains, coastal zones, etc.).

Environmental factors may include

Social factors
Environmental factors
Organisational risks

13.6.7 Project Management Risks

The following may be relevant:

- Project purpose definition, objectives, requirements, costs, etc., not properly defined or understood.
- Handling too many projects.
- Lack of control over staff priorities.
- Delays from the consultant or contractor.
- Inexperienced workforce and resource availability.
- Lack of coordination and communication.
- Communication breakdown with the project team.
- Errors found in estimation and/or scheduling.

Source:

- ICE, Wiki [6]

13.7 TRAINING AND PROJECT MANAGEMENT

Some comments have been made on this issue in Chapter 11.

Apart from design and construction methods, engineers must understand all aspects of planning and monitoring techniques, including coordination, scheduling, quality control, etc., to identify problems well in advance and prepare corrective actions on time. These are important items to complete a project within the allocated schedule of time and budget.

Control of projects during construction

Controlling projects is a very important requirement, especially related to cost control and the progress of construction. It has been observed that if the design is completed before starting a project, the project may be completed within the estimated time frame. The items mentioned below may be responsible for not achieving the ideal condition, i.e., completing the project within the budgeted time and cost:

- Changes may occur during construction activities, which will necessitate changes in design and the construction schedule.
- Design work depends on site conditions, availability of materials, different construction methods proposed by the contractor, special services, etc. The design may need to be modified after receiving input on these factors.
- The construction work may be delayed by slow progress, disputes, poor performance of suppliers, and other related parties.

The effects of risks in construction can be significant and may include

- Controlling/limiting initial expenditures to assess construction risks, and maintaining spare capacity to handle problems.
- Financing projects is costly, leading to the effects of delays.

- Increased specialisation in materials and methods may reduce the flexibility of construction sequences and techniques.

Delays from these reasons can disrupt the planning process and progress of work, potentially leading to disputes and contractual claims, which can also upset contractual and industrial relations.

13.7.1 Causes of Problems

These may be

- Problems arise in design, finance, construction, contractual relationships, testing, etc.
- Each concerned party should attempt to resolve their own problems.
- Problems that cannot be resolved by one party may need to be negotiated with others to avoid extra costs and delays.
- Each party should be involved in their own planning, costing, budgeting, and monitoring
- Problems should not be allowed to cause delays later.
- There have been repeated failures in the industry to demonstrate economic results.

13.7.2 Systems of Control

Control must be maintained at all stages of the construction period. This should be clearly communicated to all parties at the beginning of the project.

13.7.3 Project Team

The project team is employed by the client for execution and oversight of the project. This includes completing design, budgeting, planning, risk analysis, and preparing contract documents.

13.7.4 Project Director

The project director/manager is responsible for top-level management of the project. He will co-ordinate efforts through his command of information and strength of personality. If he must seek authority from others to make decisions, project control may slow down and become inaccurate.

13.7.5 Timing

The project director should be appointed at the beginning of the project. He should study the project brief, meet with key stakeholders, and review the assumptions in planning, design (as applicable), and budgeting to anticipate and assess potential problems.

13.7.6 Control of Information

The project director is expected to have the following information to carry out the work:

- Original project brief, budget and programme.
- Latest project brief, budget, and programming with approved changes.
- Possible potential problems that may impact construction.
- Reports on cost, quality, and progress of completed/committed work.
- Changes that may affect project performance.
- Estimations of the effects of changes.
- Reasons for changes related to decisions.

13.7.7 Committees and Contingencies

A team may be formed consisting of persons from all departments and sections who are responsible for the performance and success of the project.

13.7.8 Meetings

The meetings are arranged regularly to review the progress of construction work on the project and to make decisions/actions on problems.

13.7.9 Responsibility for Control Information

- The project team is responsible for processing information and informing/issuing it to the concerned persons.
- The available information should indicate trends and expected problems during construction in a timely manner.
- Time is required to arrive at decisions and make them effective.

Source:

- Ninos and Wearne [7]
- Ganguly [1]

13.8 REMOTE CONTROL TO IMPROVE EFFICIENCY IN CONSTRUCTION ACTIVITIES

With the development of construction activities, it should be noted that machine control techniques are taking a prominent position at construction sites. This has been particularly noted for a road widening job in the UK. Here, a specialised global positioning system with 3D technology has been adopted for execution.

In this method, sensors are installed on the plant equipment (which could be on a bucket of an excavator). The sensors send accurate data about the plant's location and combine it with 3D building information and survey models. This helps the plant operator work more accurately and efficiently.

Following this system will improve the accuracy of the execution work while it is being carried out and also improve checking accuracy.

Source:

- Smale [8]
- Ganguly [9]

13.9 FURTHER ON PROJECT MANAGEMENT

Apart from design, civil engineers must understand all aspects of planning and monitoring techniques, including co-ordination, scheduling, and quality control, so that problems are identified well in advance and corrective actions are prepared on time. The concept of total project activities must be understood by engineers.

More interactions between industries, universities, polytechnics, and technical colleges should occur to assess the details of the syllabus and the subjects taught in universities. Theoretically, there should be some relationship between the courses and their content as followed by universities and the requirements of industries for engineers. Engineering courses should ensure that computer-aided analysis, design, and codes of practice for a structure are not seen as substitutes for actual physical understanding.

In recent years, more colleges have been established for studying engineering courses, presumably in response to demand. The current job prospects are not particularly good, raising concerns about how students will cope with competition in the future, especially when job prospects have not improved for some time.

Young engineers will need time to develop professional skills, which cannot be cultivated solely in a classroom. The younger generation will need to be mentored by senior/experienced engineers to gain experience on real projects.

Further on Project Management Principles and Process

The following items may be mentioned:

- Initiation
- Execution
- Method
  Scope statement and document
  Breakdown of the work
  Communication system
  Risk management plan
- Performance and monitoring
- Project completion

Source:

- Ramos [10]

13.10 CONSTRUCTION MANAGEMENT

It is associated with different items.

The following may be mentioned:

Efficiently managing the team working on the project. To define and communicate each person's role and responsibilities

- Quality management
- Time management
- Contract administration
- Safety management
- Cost management
- Review, assessment, and planning of remedial work that may be necessary, especially when the project is running late.

Source:

- Ramos [10]

Contractor's Activities in Construction Management

These may be

- The design phases as applicable.
- Bidding process to interested contractors.
- The Contractor has to provide public safety.

- Engineering as per requirements.
- Decision-making ability.
- Drafting.
- Human resources.
- Quality management.
- Time and cost management.

The selection of the contractor based on low bid selection, best value selection, or qualifications is considered in a standard manner.

Source:

- Ramos [10]

13.11 CONSTRUCTION PROJECT MANAGEMENT, HOW TO WIN THE PROJECT

13.11.1 Project

There may be two types of bidding as given below:

- **Open Bid** – Normally advertised and associated with public projects. Any contractor can submit an offer in an open bid.
- **Closed Bid** – The owner may invite shortlisted contractors to submit their bids. Generally, this method is connected with private projects.

After receiving the bids, the selection procedure for the contractor will commence and will depend on the following items:

Low Bid Selection – The contractor may be selected based on the lowest price quotation.

Qualifications-based Selection – The contractor may be asked to submit, along with bids, a request for qualifications, detailing experience, management plans, past performance in completing a project within budget and on schedule. The project owner may select the contractor by examining their experience and qualifications

Best-Value Selection – The owner may select the contractor by considering both the bid price and the contractor's experience and qualifications.

Following this procedure, the owner discusses and negotiates with the contractor regarding the payment terms/methods as mentioned below:

Lump-Sum – A lump-sum contract may be agreed upon.

Cost-Plus-Fee – Cost plus fee will include the total cost of the project and a fixed percentage fee based on the overall cost, which will be paid to the contractor.

Guaranteed Maximum Price – This is agreed upon for the guaranteed maximum price contract. The total cost and fee cannot be changed.

Unit Price – The owner and the contractor may agree to execute the project based on the unit price method. The contractor will be paid according to the unit price method for the agreed items/quantities for the project.

Source:

- Ramos [10]

13.12 CONSTRUCTION RISK MANAGEMENT

The following items will need special attention:

- Analysis and forecasting considering the project deadlines to eliminate delays and cost overruns using effective strategies
- Managing the main contractor's relationship with the client to avoid contract disputes.
- Improving productivity management of the existing workforce while considering the shortage of skilled and unskilled workers.
- Efficiencies in project execution may require automation processes to reduce risk and improve efficiency.
- The procurement management system must be efficient to control price fluctuations and ensure the timely availability of construction materials.
- Developing and planning a mechanism to overcome uncertainties in construction.
- Proper communication with all concerned parties involved to avoid misunderstandings.
- Considering the current construction environment, strategies must be developed to overcome different risks during the construction period.
- Regular review and monitoring of progress and activities should be followed by interaction to organise any necessary measures required for correction or improvement.
- The safety regulation process must be enforced on-site to create a safe environment for staff and workers.

Source:

- Athmakuri [11]
- Ganguly [9]

REFERENCES

1. Ganguly K K, A Brief Review of Different Applications with Reinforced Concrete, in *Industrial Projects, with Present and Future Trend, International Symposium on Innovative World of Concrete (ICI-IWC-93), Proceedings*, Expressions in Concrete, pp. 1–12.
2. Ganguly K K, Applications of structural engineering in projects and other associated issues – A brief review, *Indian Concrete Journal*, January 2008, Vol. 82(1), pp. 41–56.
3. Dickson M, Engineering buildings for a small planet: towards construction without depletion, *The Structural Engineer*, 5 February 2002, Vol. 80(3), pp. 35–43. [Online]. Available: https://www.istructe.org/journal/volumes/volume-80-(published-in-2002)/issue-3/engineering-buildings-for-a-small-planet-towards-c/. Accessed: Oct 08, 2024.
4. The Top 3 Causes of Accidents in Construction Sites, gosmartbricks, December 2017, pp. 1–7. https://gosmartbricks.com/top-3-causes-accidents-construction-sites/. Accessed Oct 08, 2024.
5. 10 Most Common Construction Site Accidents, Law Offices of Pat Maloney, July 2015, pp. 1–6.
6. Designing Buildings Wiki *Institution of Civil Engineers (ICE)*, October 2020. https://www.designingbuildings.co.uk/wiki/ICE_articles_on_Designing_Buildings. Accessed Sep 30, 2024.
7. Ninos G E and Wearne S H, Control of Projects During Construction, in *Proceedings of the Institution of Civil Engineers,* Part-1, August 1986, Vol. 80(4), pp. 931–943.
8. Smale K, Remote control, *New Civil Engineer*, August 2016, pp. 32–33.
9. Ganguly K K, *Reinforced Concrete Design, Repair/Restoration, Strengthening of RC Structures and Concrete Technology*, Medtech, 2020.
10. Ramos D, Construction Project Management 101, Smart Sheet, 14 November 2015. https://www.smartsheet.com/construction-project-management-101. Accessed Oct 08, 2024.
11. Athmakuri S, Construction Risk Management. In *Seminar in Mumbai*, May 2017.

14 Repair Work to Structures

14.1 CONCRETE STRUCTURES

14.1.1 Common Problems

The problems may be described as follows [1,2]:

- Honeycombing, open texture, blow holes, and spalled concrete.
- Cracks and open construction joints.
- Carbonation and exposed reinforcement.
- Contamination of the concrete surface.

The types of cracks, their widths and locations will be important. The issues related to overloading and any changes that may have been made should be examined.

14.1.2 Other Items [1]

Dampness in a wall, leaking through the roof slab, expansion joints, spalling of concrete, corrosion problems, deflection problems, foundation settlements, leaking through joints, and efflorescence.

Tests may be carried out for investigation.

14.1.3 Non-destructive Tests [1]

The tests may be summarised as follows:

Visual inspection, rebound hammer test, core cutting and testing, ultrasonic pulse velocity test, pull-out/pull-off tests, cover meters and re-bar locators, radioactive methods, nuclear methods, electrical methods, maturity test method, microwave absorption, infrared thermography, tests for permeability and absorption, abrasion resistance and integrity testing, petrographic tests, etc.

The important factors of the test are listed here [1]:

Sampling and reporting, aggregate-to-cement ratio grading, cement content, cement type and cement replacement, aggregate type, chloride determination, sulphate and sulpho-aluminate determination, micrometric methods, admixtures and carbonation, etc.

Source:

- Ganguly [1]
- Ganguly [2]

14.1.4 Further Comments

When a concrete structure is constructed, the reinforcing bars in the structural elements are supposed to be well protected by good quality concrete as per the design. The pH value of the surrounding concrete to re-bars will be about 13. The problem starts when the pH value of the concrete protecting re-bars starts dropping due to the structure's exposure to the environment in which it was constructed. If the quality or thickness of the concrete cover is inadequate or if the concrete cover is cracked, the pH value will be progressively lowered by the ingress of oxygen, water, chloride, carbon dioxide, sulphur dioxide, etc., leading to corrosion of the reinforcing bars. Under these conditions, the problem worsens with atmospheric salinity in coastal areas or in areas with humidity, especially where there is inadequate or poor concrete cover.

DOI: 10.1201/9781003319108-14

Nature of problems often exhibited during service life

Following corrosion in reinforcing bars, the concrete surface is initially cracked. Later, spalling of concrete occurs in the affected areas. Voids in the structural elements may result from poor mix design and construction.

Other options to fight against corrosion of reinforcing bars

The following alternatives may be considered to avoid corrosion of reinforcing bars in concrete structures:

Stainless steel, FRP reinforcement, Tata Tiscon CRS, cement slurry-coated bars, galvanised reinforcing bars, fusion-bonded epoxy-coated bars, specially coated bars, non-metallic reinforcement.

14.1.5 Proper Construction

The various items listed below are generally involved in connection with concreting during construction. Proper construction as per design assumptions is very important.

The issues for construction are

Cement, aggregate, insufficient or unacceptable water quality for concreting, change in weather and temperature, congested reinforcement and complicated detailing, arrangement for providing correct cover to reinforcing bars, unskilled labour, inadequate facilities and equipment for maintaining good and consistent quality, inadequate supervision, awkward and inaccessible sections, aggressive environmental conditions.

The construction method and its sequence are often left to the site personnel and contractors.

Hence, the procedure for construction is followed as per the facilities available on site. There should not be any problems as long as the construction sequence and the method align with the design assumptions.

14.1.6 Cover to Bars

Providing proper cover to reinforcing bars in structural concrete elements is important to meet the design assumptions and the following requirements:

- Durability
- Fire resistance

Concrete cover must have a minimum thickness to develop a proper bond between concrete and steel by embedding sufficiently into the concrete so that the composite action takes place as per the assumptions in the design.

Hence, the cover in the structural elements will play an important role in both the serviceability condition and long-term performance of the structure.

Damages occurring during construction

Apart from other reasons, damages may occur during concreting due to Plastics shrinkage cracking, drying shrinkage cracking, plastic settlement cracking, thermal contraction cracking, cracking caused by poor workmanship – careless workmanship may cause cracks in immature concrete at the time of removal of formwork, tension cracks due to settlement of formwork.

14.1.7 Diagnostic Approach to Failures

It is important to plan a set of tests that will be suitable and may be adapted for the structures in distress, as mentioned. It would be prudent to examine the reasons and understand the causes of failures before undertaking any repair work. In this connection, some non-destructive testing approaches may be considered as diagnostic techniques for the distressed structure under consideration. There are various types of non-destructive tests available, as mentioned. Careful selection based on the requirements will be important.

14.1.8 Repairing Methods That May Be Considered

The methods that are often adopted for repair and rehabilitation work for reinforced concrete structures are [1,2]

Patch repair, repair of large portions/areas using formwork, repair of concrete columns using jacketing concrete, repair/strengthening of structural members by adopting fibre wrapping techniques, injection methods to seal cracks/voids, repair of large areas using guniting or shotcreting, repair of waterproofing on roofs and basements by using a suitable membrane or other methods, repair of different joints – the type of joint in the structure must be clearly understood before undertaking the repair work, use of protective coatings/treatments on the reinforcing bars and on the completed structures.

In many cases, the removal of damaged concrete will be required. This has to be carried out very carefully without damaging the remaining parts of the affected structural elements. Temporary propping will be required in many cases.

A case study using the fibre wrapping technique has been included in Chapter 7.

14.1.9 Repairing Materials

The basic materials for the repair have been mentioned already. However, some items are repeated below:

Cement, sand, aggregates, water, bonding agents, mineral admixtures, materials for injection methods, sealants, different chemical admixtures, protective treatments, etc.

The materials may be classified as

- Premixed cement concrete/mortars (modified with non-polymeric admixtures/additives)
- Polymers/latex modified cement additives for mortars/concrete/cement slurry
- Corrosion inhibitors, removal of rust
- Epoxy resin products

Manufacturer's technical literature must be studied carefully before recommending the repairing methods and materials.

Source:

- Ganguly [1]
- Ganguly [2]

14.2 REVIEW ON REHABILITATION WORK FOR CONCRETE

14.2.1 Review of Concrete Repair

The repair work on the concrete structure may be necessary to achieve the following:

- Restore structural integrity
- Control/arrest deterioration
- Restore original form/surface
- Hide or seal cracks and other blemishes
- Retain the original appearance

14.2.2 Deterioration of Concrete

To ensure the serviceability of a concrete structure, it is important to satisfy the following conditions:

- The mechanical properties of concrete must not reduce over time to an unacceptable level.
- There must not be appreciable loss of concrete section or framework, due to attack on concrete or spalling caused by corrosion.
- There should not be significant loss of the reinforcing steel area in the concrete section.
- The performance/serviceability of the structure must not be adversely affected due to cracking.

14.2.3 Tests on Damaged Concrete

The items mentioned below are relevant:

Strength testing, cover metre tests, chemical testing, permeability testing, etc.

Some comments are listed below following the cases studied on a few structures associated with the rehabilitation work:

- Carbonated concrete surface
- Cracked concrete cover
- Inadequate, and in some cases practically no, cover. However, some rendering work was done on the concrete surface
- Corroded reinforcing bars
- Voided and cracked concrete surface
- Signs of water seepage
- Cracks/spalling of concrete at the column–beam junctions
- Indications of poor concrete mix used initially

Deterioration generally occurs over time. A distressed structure will provide warnings regarding its serviceability condition in the form of cracking, increased deflection, spalling of concrete, etc. These warnings must be taken seriously, and proper checks/tests should be arranged in a timely manner.

It is important to consider a proper protective treatment for finished concrete structures from the outset in order to control deterioration in an unfriendly environment. Furthermore, this will increase the lifespan of the structure and reduce the cost of repair work.

14.2.4 Repair Methods

The following items should be noted:

- Removal of damaged/deteriorated concrete
- Treatment of exposed reinforcement
- Treatment for the cracks
  - Protection of the reinforcing bars
  - Prevention of water penetration
  - Prevention of staining
  - Sealing of the cracks
- Repair /reinstatement of concrete sections
- **Surface Treatment** – Different types of materials are available for treatment of concrete surfaces.

Source:

- O'Brien [3]

14.3 STEEL STRUCTURES

Local repairs to the damaged or corroded parts of the steel structures may be carried out by welding or bolting the required steel plates or pieces to the affected member. Adequate propping of the structure may be necessary prior to execution. No further comments will be made regarding the repair of steel structures.

14.4 REPAIR OF MASONRY STRUCTURES

Comments on the stitching technique will be made concerning masonry load-bearing members. The stitching method was employed to control the cracks, as shown below (Figures 14.1–14.5):

14.5 SEALING OF OLD MASONRY WALLS AGAINST DAMPNESS

14.5.1 Case Study

This case study deals with the sealing of dampness in masonry walls of a residential building.

It is essentially a single-storey building about 60–70 years old. The walls were constructed using lime–surki mortar. The walls developed dampness due to the seepage of water from below. A silicone grouting technique was adopted for rectification.

FIGURE 14.1 Stitching work in progress.

FIGURE 14.2 Stitching done.

FIGURE 14.3 Drilling holes.

FIGURE 14.4 Cracks in the wall.

FIGURE 14.5 Stitching done.

14.5.2 Procedure Followed for Rectification

The injection method was followed for rectification, as per the details given below:

Visual inspection of all walls was conducted to assess the extent of damage (i.e., dampness).

It was decided to inject the selected material into all the walls along their lengths.

For thicker walls (i.e., those greater than 250mm), injection would be carried out from both inside and outside.

Steps

- Holes were drilled inside the wall at 300 mm c/c along the bottom and just above the skirting level for the injection work.
- For thicker walls, injection was carried out from both inside and outside the wall, as mentioned.
- 12 mm GI nozzles were placed and fixed properly after drilling to prevent leakage from the outside surface of the nozzles. PVC pipes were connected to the nozzles at one end, and the other end was supported and tied properly at a height to provide sufficient head for the injection material to flow under low pressure.
- Injection liquid material was poured into the PVC pipe, which was kept hanging for a period of 24 hours.
- Continuous checks were made to ensure that the flow of liquid was occurring during this period. Generally, after 24 hours GI nozzles were removed and the holes were sealed with cement–sand mortar.

14.5.3 Injection Material

The injection material is a liquid silicone material from Wacker SMK 1311, which was diluted with water in a proportion of 1:10 (i.e., silicone-to-water ratio), as per the recommendations.

A sufficient quantity of injection material must be arranged to ensure that there will not be any shortage during the operation.

14.5.4 Photographs

A few photographs illustrating the procedure are presented below (Figures 14.6–14.16):

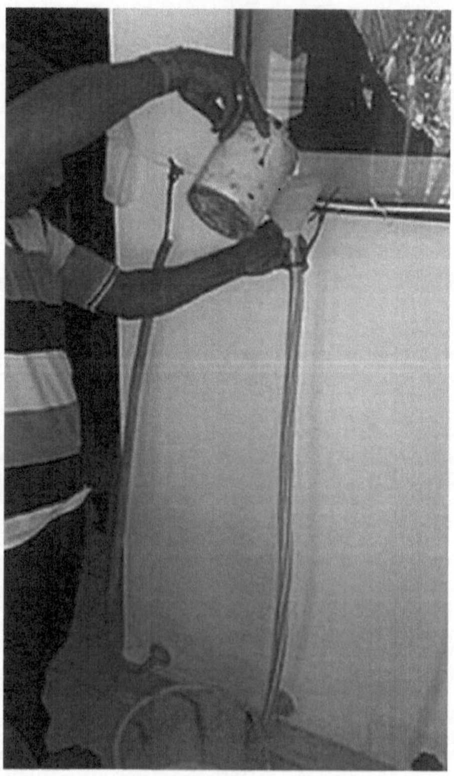

FIGURE 14.6 Pouring the liquid.

FIGURE 14.7 Treatment/sealing.

FIGURE 14.8 Treatment/sealing.

FIGURE 14.9 Treatment/sealing.

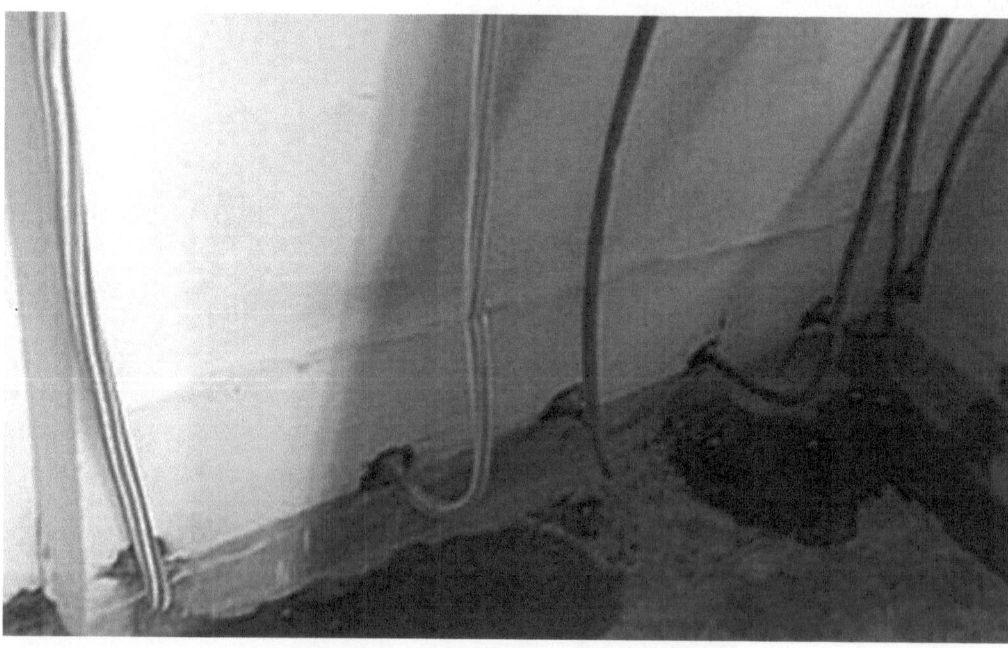

FIGURE 14.10 Treatment/sealing.

FIGURE 14.11 Treatment/sealing.

FIGURE 14.12 Treatment from outside.

FIGURE 14.13 Sealing the holes.

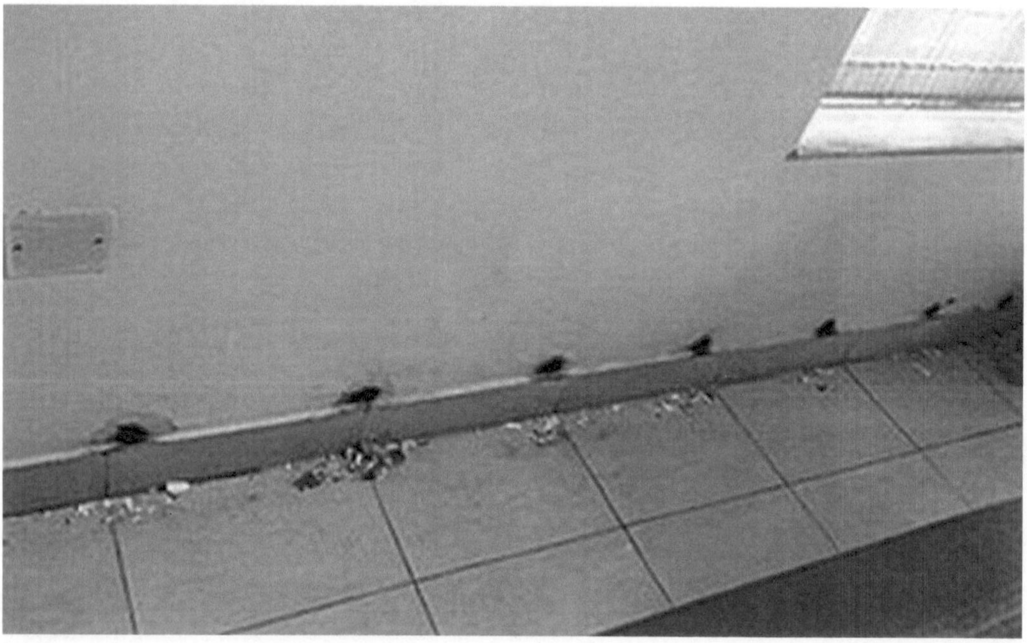

FIGURE 14.14 Sealing the holes.

FIGURE 14.15 Treatment from outside.

FIGURE 14.16 Treatment from outside.

REFERENCES

1. Ganguly K K, *Reinforced Concrete Design, Repair/Restoration, Strengthening of R C Structures and Concrete Technology*, Medtech, 2020.
2. Ganguly K K, A Brief Review of Different Applications with Reinforced Concrete, in *Industrial Projects, with Present and Future Trend, International Symposium on Innovative World of Concrete (ICI-IWC-93), proceedings*, Expressions in Concrete, pp. 1–12.
3. O'Brien T P, Concrete Deterioration and Repair, in *Proceedings of the Institution of Civil Engineers, Part* 1, Design and Construction, August 1980, Vol. 68(3), pp. 399–408.

15 Construction Photographs from Projects

15.1 PHOTOGRAPHS

A selection of photographs is presented in this chapter to demonstrate the various applications of concrete, steel, and alternative building materials used in different projects spanning an extended period. During the author's services with different employers, these projects were executed. They are mostly associated with power projects, fertiliser projects, material-handling projects, commercial, educational, housing, research units, sports complexes, and other projects.

The author is grateful to the clients (whose names are not mentioned) and DCPL for providing these opportunities. Unfortunately, the photographs were not adequately preserved for publication. However, some photographs could only be retrieved.

The photographs are connected with projects where the author was actively involved with the team related to design, construction supervision or periodic supervision (as applicable), co-ordination, and project management. Most of the photographs are related to projects handled by Development Consultants Private Limited (DCPL) during the author's long service with the organisation. Some photographs have been included from projects executed abroad.

A few photographs may have been repeated from other chapters for completeness.

15.1.1 Concrete Structures

FIGURE 15.1 Concrete circular tank.

DOI: 10.1201/9781003319108-15

FIGURE 15.2 Industrial structure.

FIGURE 15.3 Cable trench.

FIGURE 15.4 Buildings.

FIGURE 15.5 Building.

FIGURE 15.6 Folded plates.

FIGURE 15.7 Building.

FIGURE 15.8 Inside of a building.

FIGURE 15.9 Entrance to the complex.

FIGURE 15.10 Building.

FIGURE 15.11 Building under construction.

FIGURE 15.12 Building.

FIGURE 15.13 Pumping concrete.

FIGURE 15.14 Building.

FIGURE 15.15 Inside the building.

FIGURE 15.16 Inside the building.

FIGURE 15.17 Building.

FIGURE 15.18 Industrial structure.

FIGURE 15.19 Building.

FIGURE 15.20 Building.

FIGURE 15.21 Building.

FIGURE 15.22 Building.

FIGURE 15.23 Building.

FIGURE 15.24 Building.

FIGURE 15.25 Building.

FIGURE 15.26 Building.

FIGURE 15.27 Pumping concrete.

FIGURE 15.28 Industrial structure.

FIGURE 15.29 Buildings.

FIGURE 15.30 Inside the building.

FIGURE 15.31 Building.

FIGURE 15.32 Buildings.

FIGURE 15.33 Buildings.

FIGURE 15.34 Building.

FIGURE 15.35 Folded plate, inside.

FIGURE 15.36 Industrial building.

FIGURE 15.37 Concrete duct.

FIGURE 15.38 Water tower.

FIGURE 15.39 Workshop/storage building.

FIGURE 15.40 Storage structure.

FIGURE 15.41 Industrial unit.

FIGURE 15.42 Storage structure, inside.

FIGURE 15.43 Storage structure.

FIGURE 15.44 A group of buildings.

FIGURE 15.45 Building under construction.

FIGURE 15.46 Concrete construction.

FIGURE 15.47 Concrete construction.

FIGURE 15.48 Ribbed folded plates.

FIGURE 15.49 Building structure.

FIGURE 15.50 Elevated structures, in concrete.

FIGURE 15.51 Storage structure under construction.

FIGURE 15.52 Pipe trench.

FIGURE 15.53 Secondary building.

15.1.2 Steel Structures

FIGURE 15.54 Steel storage tank.

FIGURE 15.55 Steel structure under construction.

FIGURE 15.56 Steel structure.

FIGURE 15.57 Elevated steel structure under construction.

FIGURE 15.58 Space structure.

FIGURE 15.59 Space structure.

FIGURE 15.60 Steel structure.

FIGURE 15.61 Steel structure.

FIGURE 15.62 Steel structure.

FIGURE 15.63 Sport complex.

FIGURE 15.64 Lighting tower.

FIGURE 15.65 Sport complex.

15.1.3 Concrete and Steel Combinations

FIGURE 15.66 Industrial structure.

FIGURE 15.67 Industrial structure.

FIGURE 15.68 Industrial structure.

FIGURE 15.69 Industrial structure.

FIGURE 15.70 Buildings.

FIGURE 15.71 Building, treatment at the roof.

15.1.4 Stone and Rammed Earth

FIGURE 15.72 Short buildings.

15.1.5 Masonry and Burnt Brick Applications

FIGURE 15.73 Short building.

FIGURE 15.74 Short building.

FIGURE 15.75 Masonry walls.

FIGURE 15.76 Short building.

FIGURE 15.77 Burnt bricks and concrete frames.

FIGURE 15.78 Masonry construction.

15.1.6 Ferro-Cement

FIGURE 15.79 Door.

FIGURE 15.80 Tank.

FIGURE 15.81 Ferro-cement units.

FIGURE 15.82 Ferro cement units.

15.1.7 CEB and Ferro-Cement

FIGURE 15.83 Ferro-cement and CEB.

FIGURE 15.84 Ferro-cement and CEB.

FIGURE 15.85 Checking of CEB.

FIGURE 15.86 Ferro-cement and CEB.

FIGURE 15.87 Ferro cement and CEB combination.

FIGURE 15.88 CEB and ferro-cement roofing.

FIGURE 15.89 Lifting of a ferro-cement unit.

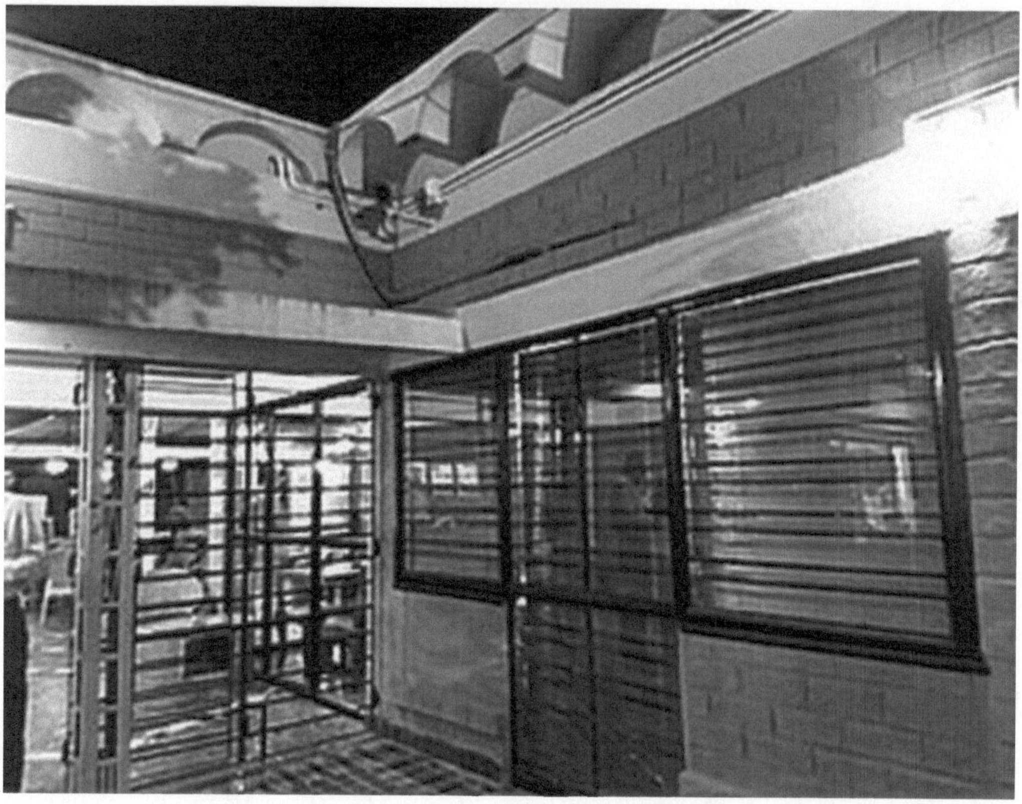

FIGURE 15.90 Ferro-cement and CEB.

FIGURE 15.91 Ferro-cement and CEB combination.

FIGURE 15.92 CEB corbelled wall.

FIGURE 15.93 Ferro-cement and CEB.

FIGURE 15.94 Ferro-cement and CEB.

FIGURE 15.95 Ferro-cement and CEB combination.

FIGURE 15.96 CEB applications.

FIGURE 15.97 Ferro-cement and CEB, inside.

FIGURE 15.98 Ferro-cement and CEB units.

REFERENCE

Construction photographs from different projects.

16 Summary and Comments from Review on Alternative Building Materials

16.1 INTRODUCTION

An attempt has been made to review different types of alternative materials that may be suitable for building construction. A study on common building materials, along with earth engineering and construction, has been carried out. Different types of alternative building materials have been considered to a limited extent. The focus is placed on selected alternative materials, which are covered in different chapters.

Efforts have been made to explore, study, and understand earth engineering/architecture. The emphasis is on increasing the use of earth engineering and construction techniques such as rammed earth, compressed earth blocks (CEBs), and other methods.

The construction materials that have already been tried, including stone construction, masonry construction, timber, steel, ceramic, concrete, reinforced concrete, ferro-cement, and fibre-reinforced concrete, have also been covered with photographs. Ferro-cement can be very efficient in projects, and its use may increase in the future. The advantages of using this material in projects have been outlined separately in Chapter 6. Apart from crack control and other advantages, fibres may be used effectively, especially for slabs, allowing for a reduction in steel reinforcement.

Natural and ancient materials, along with those used in India, are referenced. A review of clay and earth materials has also been conducted. Applications of some earth structures over a long period of time have been referenced. It is essential to understand traditional materials when using alternative materials alongside them.

Two traditional and common building materials are included to illustrate different requirements for building materials and to facilitate comparisons.

Maintenance of earth structures, along with the repair of earth buildings, has been highlighted. It is mentioned that earth design techniques and materials may be used alongside traditional building materials in many areas of building projects. This will be very relevant in reducing the consumption of traditional materials. The cost of earth construction will be important for many people.

It may not be possible for earth material to perform and compete with concrete and steel in modern development in many cases. The limitations of earth as a building material must be understood, and improvements should be attempted wherever possible.

Further discussion on earth engineering methods for structures has been added, along with alternative materials like cob, wattle-and-daub, and other applications of earth. Requirements for sustainable and green building design have been reviewed. Alternative and green building construction materials like straw bale, hempcrete, grasscrete, bamboo, wood, mycelium, timbercrete, and ferrock have also been mentioned. Low-cost building materials such as earth, coir, straw, jute, and fibre composite materials are also considered. Low-cost, energy-efficient building materials would be ideal for building construction. Low-cost materials suitable for India are also referenced. Seismic strengthening of bearing wall buildings and a summary of comments on earth technology are presented.

A few important items related to alternative materials and earth technology/construction will be mentioned.

DOI: 10.1201/9781003319108-16

16.2 CEBS

Review and comments have been made on one important use of earth, i.e., CEBs, along with their applications. The proper type of soil is very relevant for the production of CEB. Different types of soil, along with engineering properties, are mentioned. Soil identification, common tests, characteristics of earth, and values for design guidance are included. Stabilisation techniques for soil are highlighted and will be very useful. Types and choices of equipment for the production of CEB are discussed. CEB can be used effectively in rural areas. Applications of CEB from projects in West Bengal, India, with photographs, have been included.

Different types of bricks related to masonry reconstruction, such as adobe, common burnt clay bricks, burnt clay fly ash bricks, and fly ash–lime bricks, are referred to. Advantages of fly ash and clay bricks and machines for production are discussed. Comparisons of clay bricks and fly ash bricks have been made. A review of mortar, lime, and special mortar is included. Comments on improving earthquake resistance in design and construction, as per IS code for low-strength masonry buildings, are included. Repair and strengthening of masonry buildings have been covered. Applications of masonry construction, with photographs from projects, have been presented.

16.3 RAMMED EARTH STRUCTURES

The rammed earth method of design and construction, along with its history, manufacturing process, and raw materials, has been discussed. Guidance from IS code on rammed earth construction has been mentioned. Comments on the traditional rammed earth method covering different items have been made. Advantages and disadvantages have been highlighted. Rammed earth construction can also be very pleasant in appearance and may be used selectively for development work in urban areas. Repair methods for rammed earth structures have been covered, with a review of cost, material, structural design, and quality control.

A case study for the repair of cracks in the structure, along with the application from a project, has been included.

16.4 FURTHER ON RAMMED EARTH STRUCTURE – COST-EFFECTIVE, ECO-FRIENDLY ALTERNATIVE BUILDING MATERIAL

The critical and relevant items for rammed earth structures are

- Natural and local materials that may be available for use at the construction site.
- The soil and other associated items for construction will be available at a low cost.
- Rammed earth walls will be cheaper than stone wall construction.
- It will be a monolithic construction and will not need mortar joints, which may help resist earthquake effects.
- Rammed earth structures will be sustainable.
- The critical item will be the selection of the correct type of soil.
- The suitable mixture of materials for construction may be finalised by trial and error.
- The workmanship will be important.
- It has low resistance to water. Different additives may be used to stabilise the earth and increase its geotechnical properties.

16.4.1 Review of Advantages

- The construction process will need less water, which may be very important for dry climates.
- Additives may be required to improve the properties.

- It has good insulation properties when built with high thermal mass, especially for hot climates.
- No harmful emissions.
- Suitable for noise reduction.
- Earth will not burn; hence, rammed earth walls may be considered fireproof.
- It will be load-bearing, reducing the need for structural supports.
- A 400-mm-thick rammed wall may be used as load-bearing for construction up to four storeys high.
- Termites and other pests will not affect rammed earth walls.
- Rammed structures will be beneficial for human health.

16.4.2 Disadvantages

- Some carpentry knowledge/skill will be required for formwork.
- It will be a labour-intensive activity to construct.
- Ramming work will be relatively time-consuming.

16.4.3 Methodology

The activity starts with a site visit, excavation, screening, and various testing, etc.

16.4.4 Case Study, Field Visit

16.4.4.1 Advantages

- From the inspection of a building, it was observed that the temperature difference between the outside and inside would be about 2°C–3°C less without the use of HVAC.
- Rammed earth structures will be biodegradable.
- Good for human living.
- The house is ecologically sustainable and eco-friendly.

16.4.4.2 Further Improvement

- PCC may be required in the toilet and bathroom.
- Signs of leakage were observed in the toilet and bathroom walls.

16.4.4.3 Limitations

- No black cotton soil would be suitable. Laterite soil and red mud will be preferable.
- Shape constraints.

For further information on rammed earth, Chapter 5 may be referred.

Source:

- Totla, Sadwilkar, More, Kallada, Deshmukh, and puranik [1]

16.5 FERRO-CEMENT

Discussion on ferro-cement design and construction has been done, including historical development and potential. Construction materials for ferro-cement with a reinforcing system have been mentioned. Comments on the use of cement, aggregates, natural and alternative natural sand, admixtures, and fibres with mix proportions are covered. Mechanical properties covering various items for ferro-cement are also included. Advantages and disadvantages have been discussed. A comparison of ferro-cement and RCC construction has been provided.

Canal lining with ferro-cement has been completed; IS code guidance has been referred to, along with experimental work on ferro-cement lining. Repair and maintenance of ferro-cement have been covered. Case studies, including the protection of the embankment of a water body project, have been included. Some important overseas applications have been mentioned. The combination of ferro-cement and CEB can provide attractive structures. The photographs included will demonstrate this.

16.5.1 Some Additional Comments

These are presented below in items 16.6–16.10.

16.6 FIBRES FOR BRICK MANUFACTURE, SUSTAINABLE CONSTRUCTION

By adding natural and waste materials in the making of bricks, it may be possible to improve the strength and durability of the finished products.

This item has been reviewed for adobe bricks and presented in Chapter 4.

Source:

- Salih, Osofero, and Imbabi [2]

16.7 FIBRES FOR REDUCED TIME AND COST

The items that may be mentioned for using fibres are

- Cost-effective; the costs of construction and maintenance may be reduced.
- Construction time may be reduced.
- Better long-term serviceability.
- Efficient against temperature and shrinkage issues.
- Environmentally friendly.
- Recycled steel fibres may be considered.
- Their use will benefit the health and safety of construction personnel.

The uses of fibres, especially for earthen construction, are discussed in Sections 7.3–7.5.

Source:

- Mahoney [3]

16.8 REPAIR AND PROTECTION OF STRUCTURES, COASTAL ENVIRONMENTS

A few items in this connection are

16.8.1 Durability Issues

Reinforced concrete structures in coastal environments are subjected to premature deterioration due to corrosion of reinforcing bars.

The repair of these structures in a durable context is very important. Brief comments on the repair covering a few items have been given below:

16.8.2 Concrete Repairs

Fibres and pozzolanic materials may play an important role. By using them, the repairing mortar will have reduced porosity and permeability.

16.8.3 Silica Fume

Silica fume has been found to be very suitable for producing a cohesive mix with excellent washout resistance.

16.8.4 Protective Coatings

Spray or brush-applied 2 mm thick cementitious coatings may show satisfactory results in marine conditions. The coatings also exhibited good washout resistance. Tests indicate that a 2 mm coating under a hydrostatic head of 100 m will be impermeable to water and may provide total protection/barrier to chloride ions.

Source:

- Lloyd and James [4]

16.9 CONCRETE COATING SYSTEMS FOR CORROSION PREVENTION IN REINFORCED CONCRETE STRUCTURES

It should be noted that one of the main problems of failure in reinforced concrete structures may be due to corrosion of reinforcing bars, as mentioned before. The concrete structures need to be repaired and protected. Guidance may be available from BS EN 1504-9.

16.9.1 Further Comments

Corrosion of reinforcing bars is a primary concern for RCC structures related to durability.

The use of proper and suitable concrete coatings may be considered a simple and efficient method to protect concrete structures against environmental effects and aggressive attacks. Different coatings may be available for application. However, careful study and selection against environmental issues and specific requirements must be properly studied/considered. It appears from the study that zinc-rich paint may also be considered a suitable coating for concrete structures.

Source:

- Goyal, Pouya, Ganjian, and Tyrer [5]

16.10 FIBRE-REINFORCED CONCRETE

Fibre-reinforced concrete has been included with a broad classification of the composites. Fibre-reinforced materials are considered based on fibre potential. Information on different fibres has been covered.

The load–deflection pattern has been presented. Types of fibres including steel, stainless steel, glass fibres, synthetic/non-metallic fibres, polyester fibres, natural fibre materials, and polypropylene fibres are mentioned and discussed, offering mechanical and other properties along with limitations.

The advantages and disadvantages have also been covered. Fibre-reinforced concrete may be used for more projects. Its applications, particularly in floor slabs and ground floor applications in industrial projects, will significantly improve the performance of the concrete slab, as already stated. It can contribute to the improvement of durability in design. Special applications of translucent and bendable concrete are also referenced. There is a strong belief that alternative applications of fibres in concrete can replace traditional materials, especially for ground floor slabs, and can provide better design options and performance in the end.

Repair work with polymer composites is also included. Standard strengthening of structural members with carbon fibre-reinforced plates has been mentioned.

Design guidance for structural members with fibre composite materials has been provided. Near-surface mounted reinforcement for repair/retrofitting work has been included. Guidance from the TR 55 document has also been mentioned.

Different tests carried out on fibre-reinforced concrete have been referenced to some extent.

As a case study, the strengthening of a column in a residential building using carbon fibre wrapping polymer composite material techniques has been included, along with photographs.

16.11 FURTHER ON POTENTIAL OF GRC

16.11.1 Potential of Durability with GRC

16.11.1.1 Architectural Facades with GRC

GRC has been used for different applications and is becoming a suitable façade construction material.

The following items on GRC will be relevant:

16.11.1.1.1 Form/Shape

It may provide the freedom to create different curves, returns, and patterns, contributing to excellence, innovation, and aesthetically pleasing items for new buildings.

It can be thin and lightweight with different shapes, colours, structures, and surfaces. The strength at corners and edges can also be maintained. For special shapes and surfaces, a spray technique may be adopted.

16.11.1.1.2 Durable

Based on various tests, it may be mentioned that the material does not rot, erode, or dissolve.

Acid attacks on GRC may create problems. GRC facade may last over 100 years.

The performance of GRC facades has been found to be satisfactory under different temperature conditions.

16.11.1.1.3 Low Weight

The facade elements with a thickness of 12 mm will weigh about 25 kg/m^2. The supporting structural members required will be fewer due to the high strength and low weight of the material. The installation will be faster.

16.11.1.1.4 Non-combustible

GRC concrete can be considered a non-combustible building material as it does not burn or emit smoke when exposed to fire. These elements can be considered suitable materials for wind and rain.

16.11.1.1.5 Environment Friendly

The energy consumption for the production of GRC will be less. The main energy requirement will be for the production of cement, which will be used in the mix. Considering the low thickness, i.e., generally 12 mm, a small quantity of cement would be required for production.

Source:

- Brandt [6]

16.12 MORE ON GRC FAÇADE PANELS

The comments on a few items are

16.12.1 Construction

The panels may be prepared in factories and transported to the site for erection and sealing.

16.12.2 Properties

It may be designed as a rain screen arrangement. The requirements for waterproofing and thermal performance may be met by GRC. GRC may be manufactured as per grade 18P specifications.

16.12.3 Casting

Lightweight GRC may be produced as per the mentioned requirements.

A suitable colour may be chosen for the facade design. An acid-etched and textured finish may be provided.

Source:

- Casado [7]

16.12.4 Additional Comments

GRC is a very durable building material and may be available with different types of surface treatments, e.g., sand-blasted, ground and polished methods, or with painting, acid washing, plastering, or other coatings. This may satisfy the needs for waterproofing and anti-graffiti treatments. It is a strong material and may be used as thinner items for structural elements compared to normal concrete.

Source:

- Heindl [8]

GRC elements may be made attractive with excellent long-lasting finishes to satisfy current demands. The cost of GRC may be an important consideration, along with its stone-like elegant finish.

Source:

- Butler [9]

16.13 CASE STUDIES

Two case studies will be presented using GRC.

These are

16.13.1 Case Study No. - 1 Nyingma Institute of Higher Studies in Sikkim, India

The Nyingma Institute (Shedra), Martam, East Sikkim, India, was constructed using glass fibre-reinforced concrete (GRC) units.

Nyingma means ancient, and the Nyingma tradition may be considered the oldest of the four major schools in Tibetan Buddhism. The new Buddhist University would be the second largest institute in Sikkim, India.

The residential institute will accommodate 350 monks who will study and practice Higher Buddhist Studies and Philosophy. A 9-year course at the institute may be considered equivalent to an MA in Philosophy.

Use of GRC – The introduction of GRC in the design provided a pleasant look to the traditional Buddhist design and theme. The use of GRC was considered favourable for lightweight application, quality, environmental friendliness, easy handling, high strength-to-weight ratio, and no maintenance requirements. The inclusion of GRC may be regarded as the right choice considering the shortage of timber and the availability of traditional carvers. The Bhutan GRC team was involved in preparing the drawings, fixing details, etc.

Design – The application of GRC was carefully introduced to adhere to Buddhist norms and requirements. A total of 600 GRC cladding panels were used to achieve a high standard of decorative architectural treatments of different types.

The internal prayer hall featured decorative pillars, a decorative Golden Throne, and a decorative cornice profile at the ceiling perimeters.

The courtyard had decorative column capitals as per Buddhist tradition.

16.13.1.1 Production

A premix spray technique was used with a standard mix ratio. The installation of GRC was carried out by a trained in-house team and closely supervised/monitored by GRC experts. For fixing the units, HST stud anchors and anchors from Hilti International were used.

16.13.1.2 Special Comments

The incorporation of Buddhist norms and traditions in the design, including the historical features, posed challenges and was the most difficult part of the project. The project was completed in 2017 and inaugurated in May 2018.

Source:

- Lama [10]

16.13.2 Case Study No. - 2 Marmara Faculty of Technology, Theology Mosque, Istanbul, Turkey

Modern materials have been considered for the mosque's architecture, such as glass, steel, and fibre-reinforced concrete.

The project has a built area of 30,000 m², which includes an urban square, mosque, and cultural centre beneath the mosque and the square. The mosque accommodates 4,500 worshippers.

The dome structure was designed with steel for the Faculty of Theology Mosque.

The height of the mosque's dome is 34.7 m, while the height of the minaret is 57.8 m. The central dome has a diameter of 35 m and rests on steel supports that uniformly increase towards the connection details for the special geometrical structures. The supporting columns are well connected to the walls of the dome with steel beams to represent a transition from tradition.

Fibro-light sunshades were introduced using GRC material between the GRC column claddings. The sunshades were specially designed with openings measuring 7.3×7.6 m and were connected to the main steel construction.

About 8,250 m^2 of GRC materials were used in the mosque, both outdoors and indoors, covering the minaret and the shadirvan. Approximately 1,100 m^2 of GRC material were consumed for the interior part of the mosque and the mihrab.

The mosque was opened in October 2015.

Source:

- Kohen and Marasli [11]

16.14 FURTHER ON NEW BUILDING MATERIALS

Comments on New Building Materials are given below:
These may include

- Hardwood cross-laminated timber
- Self-healing concrete
- Bioplastic
- Artificial spider silk
- Homeostatic facades
- Aerographite
- 3D-printed graphene

Source:

- Materials that are set to Change the Building Industry [12]

16.14.1 Further on New Materials

The comments may be

- **Mass Timber –** The use of laminated timber, cross-laminated, and glue-laminated timber may be mentioned.
- **Air Cleaning Bricks –** Clean air is pulled into the building either mechanically or passively.
- **Self-Healing Materials –** This may refer to self-healing cement for concrete, which is becoming popular.
- **Passive Cooling Ceramics –** The use of clay composite and hydro-gel for cooling the building.
- **Strand Rods –** This refers to thermoplastic carbon fibre composite, known as strand rods. The composite is covered with inorganic and synthetic fibres and finished with thermoplastic resin.
- **Trash –** The use of recycled materials like scrap metal, cardboard, and plastic bottles in buildings.

Source:

- Richardson [13]

16.15 REVIEW AND SUMMARY OF NEW MATERIALS THAT MIGHT CHANGE THE BUILDING INDUSTRY/CONSTRUCTION

These may be mentioned as

- **Self-Healing Concrete –** With water-activated bacteria to seal cracks.
- **Aerographite –** Will be stronger when compressed.
- **3D Graphene –** Lightweight materials that may be suitable for tall buildings.
- **Laminated Timber –** May be suitable for tall buildings and will reduce carbon emissions.
- **Transparent Aluminium –** May be used for windows/domes and space vehicles.
- **Modular Bamboo –** May be suitable for earthquake zones.
- **Translucent Wood –** May be used for window glass replacements and solar panel cells. It may be coloured stripped wood, providing good insulation, strength, and biodegradability.

- **Light-Generating Concrete –** Minuscule glass balls embedded in concrete to reflect light. May be used for signage, artistic buildings, and marking dangerous areas.
- **Microbial Cellulose –** A mixture of bacteria, yeast, and microorganisms to create layered structures. Potential uses include building facades and signage.
- **Aluminium Foam –** It is produced by injecting air into molten metal. The foam is recyclable. Its uses include durable and decorative building cladding.
- **Spider Silk –** A very durable material with resonance properties. Its applications may be for acoustic building tiles and laboratories.
- **Wool Bricks –** These are produced by fusing wool and seaweed polymer. They are stronger than earth bricks and suitable for buildings in cold climates.
- **Nano-crystal –** The use of crystalline nano-particles allows natural light to pass through while blocking heat. Its applications include dynamic windows for efficient lighting and heat control.
- **Hydroceramics –** This is a membrane made of ceramic, fabric, and hydrogels. Its uses can include cooling building temperatures and reducing electricity costs.
- **Biochar –** Made from waste materials, this has high insulation quality. It may be produced by burning tree debris in pyrolysis kilns. Its applications may include sustainable and self-reliant buildings
- **Pollution-Absorbing Bricks –** These are double-layered insulated bricks capable of filtering 30% of fine and 100% of coarse pollutants. Their use may be for filtered air and self-sustaining ventilation in green buildings.
- **Invisible Solar Cells –** These allow non-visible wavelengths of light to reach solar cells located at the edges. Their use will be to generate energy from building components, windows, and doors.
- **Bioreactors –** These are algae-infused building panels that can generate energy. The material may be used for thermal control in sustainable and self-reliant buildings.

Source:

- New Building Materials that will Revolutionize Construction [14]

16.16 ENGINEERING BUILDINGS IN THE BUILT ENVIRONMENT

It is very important to carefully assess the current situation developing on our planet before deciding on the design of buildings in the built environment [15].

The following points may be relevant:

The continuous supply of irreplaceable construction materials and the use of cheap, non-renewable energy persists. This condition needs to be checked and controlled.

Population growth is very high; it is now over 6 billion and likely to reach 9 billion by 2050.

Some researchers estimate this figure may be even higher. The rate of increase is about 80 million extra persons per year worldwide. This is a serious problem we face.

Carbon dioxide loading is an indicator of pollutant potential. As per current estimates, it is about 5.5 tonnes of CO2 per person per day in the USA, 2.7 tonnes in Europe, and 0.8 tonne in China.

Building activities must take into account the geographical imbalance of wealth and population.

It is reported that 70% of forests have already been destroyed, and at the current rate, 100% destruction could occur by the year 2027 [15].

Global warming is already taking place, and according to some estimates, the temperature rise in the next 100 years may be around 5.8°C [15].

It has also been suggested that after 2020 [16]

- Winters will be reduced in length
- Wind speeds will be variable and may be stronger.
- Summers will have hotter spells with periods of drought.
- Tidal heights will increase.

We cannot pass our planet in this condition to our children and grandchildren. We must save the planet for future generations [15,16].

The challenges before the engineers/architects are [7,15]

Better whole life performance, using less primary materials, using less non-renewable energy, wasting less, causing less emission of greenhouse gases and disturbance to the natural environment, cutting down the harmful effects of construction activity, reducing the burden on the planet, and working out solutions that will fulfil social, environmental, and economic criteria.

It will be important for the engineering and management team to work together on all issues covering conception, planning, modelling and calibration, operation, construction, and use.

Source:

- Dickson [15]
- Stansfield [16]
- Ganguly [17]

16.17 DEVELOPING ARCHITECTURAL AWARENESS IN STRUCTURAL ENGINEERS – A CREATIVE RELATIONSHIP

In non-industrial projects like university complexes, hostels/resorts, hospitals, housing projects, and commercial complexes, the architects normally play a major role in the development of the layout, building form, aesthetics, etc. The cost of architectural finishes along with services may also be very high in such projects. The challenges are greater here for the structural designer, and there will be pressure to reduce the dimensions of the structural elements for visual and other requirements while producing economical designs for different structural schemes. The services of structural engineers are more demanding and more appreciated in this case compared to industrial projects. Engineers and architects are dependent on each other to design and construct innovative modern buildings [17].

Structural engineers must learn to work with architects in harmony for creative and innovative design. Each profession has different skills and carries different responsibilities. Structural engineers will have to learn to work in multidisciplinary teams with different professionals and understand how others think and work. Architectural creativity must be supported by engineering skills to produce innovative designs and work. Clients should be convinced that the innovation will be worthwhile and cost-effective. Engineering arguments should not hinder architectural solutions. Mutual respect and appreciation between architects and engineers will be required, and there will always be an engineering solution to good architecture.

The best buildings can be designed and constructed from a symbiotic relationship [19] between architects and engineers. Structural engineers have to shoulder a greater responsibility in both design and construction to produce a safe and durable structure in the end.

Source:

- Chrisp, Wilson, and Cairns [18]
- Stansfield [19]
- Ganguly [17]

16.18 REVIEW ON OTHER ASSOCIATED ITEMS

The review of some other items have been done like

16.18.1 Concrete as a Material

Different issues of concrete design including resilient design have been mentioned. The performance of concrete under fire has been mentioned. Hybrid construction with façade engineering is discussed briefly. Greenhouse gas emissions due to concrete and development of different special concrete have been included.

16.18.2 Steel as a Material

Designs/conceptual designs with steel have been included. Guidance from the IS code has been referenced. Site inspection of steel work and items like fire, fatigue, and connections, have been discussed. Drawing/sketches for structural arrangements have been included.

16.18.3 Use of Computer

The use of computers for engineering design calculations, detailing, etc. has been covered. Training and education of engineers and communication are addressed. Reliability and achievement of quality have been included.

16.18.4 Sustainability/Resilience

The discussion on sustainability/resilience has been included. Concrete structures in hazard protection are mentioned, including risks and structural safety. The effects of silica dust and the presence of Co_2 in the air are reviewed.

16.18.5 Construction Supervision, Site Safety, Accidents and Communication

Some basic comments have been made on construction materials, site supervision, site safety, accidents and other issues are reviewed. The comments have been made on communication which is so important among different disciplines in a project.

16.18.6 Rehabilitation, Masonry Wall/Members

Concrete structures with common problems developing during use have been reviewed with rehabilitation work. The repair of masonry structures, including examples of drilling/stitching of members from a project, has been included. Sealing of dampness in masonry walls has also been covered.

16.18.7 Construction Photographs

The construction photographs from different executed projects have been presented to demonstrate the performance of versatile building materials. Different alternative materials have been used for structures/buildings in specific applications.

16.19 STRUCTURAL SAFETY

This item has been discussed in Chapter 12.

Source:

- Ganguly [17]

16.20 SAFETY AND ETHICS - THE ENGINEER'S RESPONSIBILITY

The topic covers a vast area. Only the salient and relevant items will be mentioned briefly.

Ethical behaviour is an essential part of human society.

Working together with individual members can improve the quality of life and fulfilment in society. An atmosphere of trust is essential. Acceptance of responsibility for the consequences of one's actions and its influence on the quality of life of others is important. Our engineering activities may be associated with a socio-economic context—communication, accommodation, education, health and safety, infrastructure, food production, energy supply, etc.

Members of a profession should have

- Specialised knowledge and skill.
- Knowledge and capacity to affect society.
- Undergo an extensive period of training.
- Have a monopoly or near monopoly of a particular skill.
- Members of a professional body will be responsible for regulating standards, protecting rights of practice, and providing necessary training.

16.20.1 Professional Practitioners

They are expected to have

- **Integrity** – Honesty and openness with all concerned.
- **Impartiality** – Free from bias and unbalanced interest.
- **Independence** – To be free from secondary interest.
- **Competence** – Thorough knowledge of the work they do.
- **Responsibility** – To acknowledge and accept personal commitment.
- **Discretion** – Care with communication.

Professional Institutions may provide

- Development of moral awareness, skills, and responsibility.
- Regulation of the practice of individual professionals.
- Assistance in the development of moral responsibility
- An important role in communicating with the public
- Contribution as a learned society to the development of science and technology of engineering.
- Setting standards for admission to institutions.
- Providing continuous professional training.

Source:

- Armstrong [20]

16.21 WRAPPING COMPOSITE MATERIALS FOR STRENGTHENING/RETROFITTING STRUCTURES

Fibre-reinforced polymers (FRPs) may be used for strengthening/retrofitting structures.

The advantages of the FRP system over traditional methods are

- Lightweight
- Relatively easy to install

- Non-corrosive
- Shorter execution time

Further comments have been made on this item in Section 7.3.

Source:

- Ganguly [17]

16.22 RESEARCH AND DEVELOPMENT

16.22.1 Research/Study for Enhanced Performance

Further studies/research should include understanding fundamental information on the constituent materials of the product. The main emphasis should be to improve the performance of the building and FRP materials to produce efficient, durable, cost-effective, and environmentally friendly items. The improvements should be included in the codes after proper studies and verifications for updating.

16.22.2 Comments on FRP Material

16.22.2.1 Guidance for Use

16.22.2.1.1 Codes and Standards

For design guidance and to develop confidence, the following items will be required for engineers, applicators, and owners

- Detailed specifications of FRP material for identification, including mechanical and physical properties.
- Consideration of lower modulus and ductility compared to conventional materials.
- Standard test methods.
- Performance requirements to eliminate uncertainties.
- Methods of design and use related to applications for safety and confidence.
- Long-term experience of the material should be known.
- The material should be applicable for new structures as reinforcement and for repair/retrofitting of existing structures.
- Comments on cost advantages related to life cycle cost will be important.
- Emphasis on inspection, maintenance, and some basic routine tests.

It seems that FRP strengthening methods will continue to be the preferred technique for many repair and rehabilitation projects. With different applications, the material will become better known and confidence will develop. It is likely that the material may be selected for more applications, covering buildings, bridges, and other projects, including historical and important structures in the future. Applications may be particularly relevant in seismic areas.

Education and training of engineers, applicators, and owners will be important in relevant areas of FRP technology and practice.

Source:

- Bakis, Bank, Brown, Cosenza, Davalos, Lesco, Machida, Rizkalla, and Triantafillou [21]

16.22.3 Innovative Thinking

The social value model of the UK Government may be referenced. This is focused on the following themes:

- Helping local communities manage and recover from the impact of COVID-19.
- Addressing economic inequalities.
- Tackling climate change.
- Promoting equal opportunity
- Supporting well-being

Social enterprises can compete with procurement considerations such as price, quality, and risk. Contractors must ensure that subcontractors understand their social procurement policies.

Source:

- Pilgrim [22]

16.23 REVIEW OF COMMENTS

16.23.1 General Comments

- Various items and comments have been presented in different chapters of the publication. An attempt has been made to summarise the main comments in this chapter for easy understanding.
- The emphasis is on the use of alternative building materials, with particular reference to using earth wherever possible for projects. Knowledge of alternative building materials will be very important for architects and engineers.
- Some alternative materials have been identified for this purpose.
- Importance has been given to studying the technique and the materials used in the past. It would be beneficial to see more earth engineering and construction.
- Suggestions on the use of fibre-reinforced concrete and the repair of structures with fibre polymer composites have been made after studying different applications.
- The efficient use of ferro-cement has been highlighted with photographs from the projects.
- Some ancient structures have already demonstrated very good performance with earth construction.
- A few types of earth construction have been mentioned after study.
- The ancient materials that are still in use are also referenced.
- The discussion on CEBs has been conducted. Local materials near the site may be used for CEBs.
- Local people can be effectively used after some preliminary training for construction with earth.
- The constituent materials for production will also be available for earthwork [23].
- Clients may need to be convinced of the use of such alternative materials in some cases. We have to try hard to achieve this.
- Earth as a supporting material may become competitive in replacing traditional materials in some places/applications [23].
- With proper inspection and maintenance, an extended life may be achieved for earthen structures.
- The design with earth materials may also be integrated with other materials to work compositely to meet different design requirements.

16.23.2 Summary of Comments on Alternative Building Materials and Earth Technology and Applications

- The emphasis is on using alternative building materials for design and construction instead of traditional materials to the extent possible. The use of earth should be examined for applications in projects.
- Earth has been used as an alternative building material for a long time. The review has been conducted to study different methods of earth construction that may be suitable. Different combinations may be explored to meet specific requirements.
- Building with earth may be considered ecological and economical [24]. However, the low perception of earth construction poses a barrier to its acceptance as a building material for low-income group owners [24]. The poor performance of traditional earth houses in terms of water and weather resistance, termite infestation, frequent maintenance, and government policies has negatively affected the reputation of earth.
- CEB is comparable to conventional building materials, and its production is economical and sustainable. CEB blocks have already been used in some projects. Earth is still a material less preferred than concrete and fired bricks for constructing houses in rural and urban areas.
- Earth construction might be a practical choice for many rural areas that are disconnected from the building material supply chain network [24]. It must meet the desired specifications required by conventional materials in terms of durability, aesthetics, and economy.
- The economic, self-help construction potential, and recyclability aspects of traditional earth construction will be most valuable. A combination of the two construction types may be considered to produce structures that are economically, socially, and ecologically sustainable [24].
- Modern methods of stabilising earth materials can reduce the use of cement and hydraulic lime in favour of bio-based alternatives. Assisted self-help construction can also significantly reduce the cost of a building project.
- The demonstration of a successful earthen housing project will have a large impact on the people and the government.
- Earthen design and construction must satisfy contemporary requirements.
- The successful design and implementation of methods in large projects like schools, shopping centres, museums can change the perception of earth as a building material and increase confidence in its use for construction [24].
- Traditional earth construction may be considered the main barrier to accepting earth as a building material.
- This must be overcome with the developments and improvements that have taken place over the years. Protective measures may need to be taken, such as the proper use of stabilisers, fibres, and protective treatments to address weaknesses. These must be successfully demonstrated through applications.
- The new generation of architects and planners should receive proper support for innovative design using earth construction.
- For some people, earth buildings will be the only viable solution, especially for low-income groups and in developing countries.
- The use of ferro-cement and fibre-reinforced concrete will be useful and relevant.
- These are mentioned in different sections.
- Following different applications and advantages, it seems that the FRP strengthening technique will continue to be the preferred method for many repair and rehabilitation projects.
- For alternative building materials, earth, ferro-cement, fibre-reinforced materials, and their combinations look encouraging for future projects.

16.23.3 Future of Earth Engineering

- Earth engineering and architecture have been used by many countries for a long time, as stated. There are different techniques for meeting specific requirements. One must assess what will be suitable for a particular situation, considering advantages and limitations. The design and construction with earth have stood the test of time and proven to be reliable and practical in most cases [23]. The advantages and disadvantages of different techniques must be well understood in advance through studies. We need to see more earth engineering and construction carried out by architects and engineers.
- There may be different approaches for applications, such as constructing the exterior wall with cob and the interior walls (non-load-bearing) with wattle-and-daub for a project. Other alternatives and combinations may exist [23].
- We must learn and advance earth engineering without rejecting everything we are familiar with [23]. One should carefully consider the advantages of earth buildings, such as material cost, control of interior temperature, flexibility, strength, and aesthetics, which are especially important under present conditions, particularly for developing countries.
- We need to retain what is necessary in the present situation and improve the performance of earth engineering through research and applications without blindly copying past methods [23].
- In some situations, present alternatives may not be acceptable to local authorities and clients due to mismatches with existing structures in terms of design and aesthetics. The cost and availability of construction materials will be a very important factor in all cases.
- The major challenges of development for earth engineering may include proper engineering design methods, poor seismic resistance of vernacular construction, uncertified products, lack of education, and a weak regulatory system [25].
- The main focus should be on modern earth-based construction [25].
- Modern earth construction methods are expected to be suitable for low-cost housing in India [24].
- It can be noted that a combination of rammed earth/compressed earth blocks along with ferro-cement can produce sustainable and elegant building construction. The photographs from projects will demonstrate this.
- Recent developments will include wider acceptance of compressed earth blocks and rammed earth constructions [25].
- The future use of earth design/construction relies on entrepreneurs, designers, and research workers who can produce materials with appropriate properties and designs at minimum cost. A robust supply chain, manufacturing infrastructure, and training will provide superior quality of construction. Research in earth engineering and construction will rest on the pillars of desirability, durability, and affordability for wider acceptance [24].
- Let us hope to see more applications of alternative materials for projects, especially in earth design and construction for buildings in the future.

Source:

- Build your home with earth [23]
- Construction and Building Materials, Earth material for low-cost housing in rural India - Science Direct, Elsevier [24]
- Modern earth building materials & Technologies, Science Direct.com by Elsevier [25]

16.24 PHOTOGRAPHS

The photographs featuring alternative materials have already been included in this publication. A few more are presented below (Figures 16.1–16.7):

FIGURE 16.1 CEB and ferro-cement combination.

FIGURE 16.2 CEB and ferro-cement combination.

FIGURE 16.3 CEB and ferro-cement combination.

FIGURE 16.4 CEB and ferro-cement construction.

FIGURE 16.5 CEB construction.

FIGURE 16.6 Ferro-cement boundary wall.

FIGURE 16.7 CEB dome.

REFERENCES

1. Totla P, Sadwilkar M, More S, Kallada B, Deshmukh B and Puranik A, Sustainable rammed earth structure: a structurally integral, cost-effective and eco-friendly alternative to conventional construction material. *International Journal of Innovative Technology and Exploring Engineering (IJITEE)*, September 2019, Vol. 8.
2. Salih M M, Osofero A I and Imbabi M S, Critical review of recent development in fibre reinforced adobe bricks for sustainable construction, *Frontiers of Structural and Civil Engineering,* 2020, Vol. 14(4), pp. 839–854.
3. Mahoney M, *7 Key Benefits of Fibre-Reinforced Concrete*, For Construction Pros, September 2021, pp. 1–34.
4. Lloyd C and James G, Repair and protection of structures in coastal environments, *Concrete*, November 2018, pp. 14–16.
5. Goyal A, Pouya H S, Ganjian E and Tyrer M, Concrete coating systems for corrosion prevention in reinforced concrete structures, *Concrete*, November 2018, Vol. 52, pp. 22–24.
6. Brandt N R, Durable architectural facades with GRC, *Concrete*, March 2019, pp. 14–15.
7. Casado J M S, King's Cross T1, London, Concrete, March 2019, pp. 16–18.
8. Heindl S, Crossrail-Bond Street Station, London, Concrete, March 2019, pp. 21–22.
9. Butler J, Luxury living with GRC, *Concrete*, March 2019, pp. 23–24.
10. Lama K, Nyingma Institute of Higher Studies, Martam, East Sikkim, India, Concrete, August 2018, pp. 8–10.
11. Kohen B and Marasli M, Marmara Faculty of Theology Mosque, Istanbul, Turkey, Concrete, August 2018, pp. 12–14.
12. 7 Materials That Are Set to Change the Building Industry, Inter Focus, pp. 1–7.
13. Richardson A, *6 New Materials That Are Changing Commercial Construction*, Aquicore, pp. 1–7.
14. 18 New Building Materials that will Revolutionize Construction, BIG RENTZ, December 2018, pp. 1–7. https://www.bigrentz.com/blog/the-future-of-building-materials?srsltid=AfmBOorCk1Hk1rJC7MJKIe4s1BEoaCF6MvD2Njdt0en9HcF4tGm6kzmW Accessed Oct 09, 2024.
15. Dickson M, Engineering buildings for a small planet: towards construction without depletion, *The Structural Engineer*, 5 February 2002, Vol. 80(3), pp. 35–43. [Online]. https://www.istructe.org/journal/volumes/volume-80-(published-in-2002)/issue-3/engineering-buildings-for-a-small-planet-towards-c/. Accessed: Oct 08, 2024.

16. Stansfield K, Global warming: issues for engineers, *The Structural Engineer*, 17 July 2001, Vol. 79(14), pp. 13–14.
17. Ganguly K K, Applications of structural engineering in projects and other associated issues - a brief review, *The Indian Concrete Journal*, January 2008, Vol. 82(1), pp. 41–56.
18. Chrisp T M, Wilson L and Cairns J, Developing architectural awareness in structural engineers, *The Structural Engineer*, 1 April 2003, Vol. 81(7), pp. 27–32.
19. Stansfield K, Engineers and architects: a creative relationship, *The Structural Engineer*, 2 January 2002, pp. 13–14.
20. Armstrong J H, Safety and ethics: the engineer's responsibilities, *The Structural Engineers*, 2 January 2002, Vol. 80(1), pp. 28–31.
21. Bakis C E, Bank L C, Brown V L, Cosenza E, Davalos J F, Lesco J J, Machida A, Rizkalla S H and Triantafillou T C, Fiber-reinforced polymer composites for construction, American Society of Civil Engineers, Pennsylvania State University, US, 2003, pp. 4–78.
22. Pilgrim B, Think social, *New Civil Engineer*, January 2022, pp. 51.
23. "Earth Building Techniques – Build Your Home with Earth," *This Cob House*. https://www.thiscob-house.com/earth-building-techniques-build-home-earth/. Accessed Sep 30, 2024.
24. Earth Material for Low-Cost Housing in Rural India. In *Construction and Building Materials*. Science Direct, Elsevier, June 2020, pp. 28–29.
25. Hall M. R, Lindsay R, and Krayenhoff M, *Modern earth buildings: Materials, engineering, constructions, and applications*. Elsevier, 2012.

Index